Quasicrystals

Quasicrystals
Fundamentals and Applications

Enrique Maciá-Barber

CRC Press
Taylor & Francis Group
Boca Raton London New York

CRC Press is an imprint of the
Taylor & Francis Group, an **informa** business

First edition published 2021
by CRC Press
6000 Broken Sound Parkway NW, Suite 300, Boca Raton, FL 33487-2742

and by CRC Press
2 Park Square, Milton Park, Abingdon, Oxon, OX14 4RN

© 2021 Taylor & Francis Group, LLC

CRC Press is an imprint of Taylor & Francis Group, LLC

Reasonable efforts have been made to publish reliable data and information, but the author and publisher cannot assume responsibility for the validity of all materials or the consequences of their use. The authors and publishers have attempted to trace the copyright holders of all material reproduced in this publication and apologize to copyright holders if permission to publish in this form has not been obtained. If any copyright material has not been acknowledged please write and let us know so we may rectify in any future reprint.

Except as permitted under U.S. Copyright Law, no part of this book may be reprinted, reproduced, transmitted, or utilized in any form by any electronic, mechanical, or other means, now known or hereafter invented, including photocopying, microfilming, and recording, or in any information storage or retrieval system, without written permission from the publishers.

For permission to photocopy or use material electronically from this work, access www.copyright.com or contact the Copyright Clearance Center, Inc. (CCC), 222 Rosewood Drive, Danvers, MA 01923, 978-750-8400. For works that are not available on CCC please contact mpkbookspermissions@tandf.co.uk

Trademark notice: Product or corporate names may be trademarks or registered trademarks and are used only for identification and explanation without intent to infringe.

ISBN: 978-0-8153-8180-8 (hbk)
ISBN: 978-0-367-67893-7 (pbk)
ISBN: 978-1-351-20915-1 (ebk)

Typeset in Computer Modern font
by KnowledgeWorks Global Ltd.

Contents

Preface

During the early 1980s, a few people started to realize that certain arrays of atoms *lacking* periodic translation symmetry were surprisingly able to give rise to sharp diffraction patterns, fully constellated with a lot of discrete Bragg reflections, just as well-known periodic arrays of atoms generally do. In addition, their diffraction patterns revealed a *new crystallographic symmetry*, namely, scaling invariance, which endows them with an impressive esthetical appeal. This symmetry had not been previously considered in classical crystallography, and it is directly related to the emergence of self-similar, hierarchical patterns embodying atomic cluster aggregates. Therefore, from a structural viewpoint these compounds can be regarded as self-similar arrays of atoms, where the translation symmetry, characteristic of periodic crystals, is replaced by a scale invariance one. In this sense, they could be pictured as fractal solids representatives, with scale factors given by *irrational* numbers. Consequently, this novel condensed matter phase can be properly regarded as describing long-range ordered solids whose diffraction patterns exhibit symmetries that are incompatible with translational symmetry.

Upon closer scrutiny it was understood that these solids represent a natural extension of the periodic crystal notion when the location of atoms throughout the space is described in terms of *quasiperiodic functions* instead of the usual periodic ones. Accordingly, they were referred to as quasiperiodic crystals – *quasicrystals*, for short. In retrospective, one may consider the choice of this abbreviation as somewhat unfortunate. In fact, the very semantics of the term *quasicrystal*, albeit originally derived from the mathematically rigorous quasiperiodic function notion, suggested that quasicrystals (QCs) were occupying an intermediate position between well ordered arrangements of matter and glassy state ones, thereby denoting some sort of defective, flawed materials which failed to properly fulfil the basic tenets of well behaved condensed matter representatives. Quite on the contrary, it is now well established that QCs show up long-range orderings of atoms rendering essentially discrete diffraction patterns, which can be indexed in terms of a hyperspace generalization of classical crystallography, suitably including inflation symmetry operations. Therefore, the discovery of QCs ultimately led to the convenience of revamping the very notion of crystal in condensed matter.

Quasicrystals also exhibit unusual physical properties, closely related to the fractal nature of their energy and frequency spectra. Physically this feature means that some specific fragments of the spectra appear once and again at different scales. Thus, rather than trying to fit the electronic structure of QCs within the conceptual schemes introduced in order to describe classical periodic solids (namely, metals, semimetals, semiconductors, or insulators) we should expand these categories to properly deal with their highly-fragmented energy and frequency spectra. Indeed, it has been confirmed that transport properties of thermodynamically stable QCs of high structural quality resemble a more semiconductor-like than metallic character. Therefore, QCs provide an intriguing example of ordered solids made of typical metallic atoms which do not exhibit most of the physical properties usually signaling the presence of metallic bonding. Accordingly, by the light of the knowledge gained during the years elapsed, it is now clear that QCs are representatives of a novel phase of matter exhibiting a large number of remarkable specific properties.

This book aims at providing a detailed and updated introduction to the field of QCs in an interdisciplinary way, focusing on their fundamental aspects but also on their present and future applications. The role of quasiperiodic order in science and technology is also examined by highlighting the new design capabilities provided by this novel ordering of matter. This book is specifically devoted to promoting the very notion of quasiperiodic order in condensed matter, and to spur its physical implications and technological capabilities in materials science. It, therefore, explores the main topics related to intermetallic, photonic and phononic quasicrystals, as well as soft-matter quasicrystals, by the light of their intrinsic physical and structural properties. It also explores exciting applications of quasiperiodically ordered systems in new technological devices, including multilayered quasiperiodic systems, along with 2D and 3D designs. To this end, I will thoroughly discuss both experimental data and the most promising theoretical approaches aimed to explain them, properly comparing the main distinctive features of classical crystals and QCs from an analogical perspective.

Several topics on the role of aperiodic order in different domains of physical sciences and technology will be covered in this book. The first chapter addresses some basic notions and presents the most characteristic features of different kinds of aperiodic systems in a descriptive way. In Chapter 1, we introduce different orderings of matter and describe the progressive transition from periodic to aperiodic thinking in physical sciences. In Chapter 2, the very notion of aperiodic crystal is introduced, fully describing its historical roots as well as the paramount discovery of quasicrystalline alloys and their beautiful novel symmetries. The study of the unusual physical properties of quasicrystalline alloys is then presented in more detail in Chapter 3, paying a special attention to their intriguing electronic structure and the possible nature of chemical bonding in hierarchically arranged cluster based solids. In Chapter 4, we introduce the basic structural properties of man-made materials consisting of aperiodic sequences of layers such as Fibonacci semiconductor-based superlattices or Cantor-like dielectric multilayers. The main mathematical features of the substitution sequences defining their growth rule are also reviewed along with the possible signatures of quasiperiodicity in their physical properties. In Chapter 5, we focus on the possible applications of QCs, and in Chapter 6, we outlook new frontiers in the QCs science in the years to come.

The book contains 50 proposed exercises (highlighted in boldface through the text) accompanied by their detailed solutions, motivating and illustrating the different concepts and notions to provide readers with further learning opportunities. I have prepared the exercises mainly from results published and discussed in regular research papers during the last decade, in order to provide a glimpse into the main current trends in the field. Although the exercises and their solutions are given at the end of each chapter for convenience, it must be understood that they are an integral part of the presentation. Accordingly, it is highly recommended to the reader that he/she try to solve the exercises in the sequence they appear in the text, then check his/her obtained result with those provided at the end of the chapter, and only then to resume the reading of the main text. In this way, the readers (who are intended to be both graduate students as well as senior scientists approaching this rapidly growing topic from other research fields) will be able to extract the maximum benefit from the materials contained in this book in the shortest time. All the references are listed in the bibliography section at the end of the book. I have tried to avoid a heavily referenced main text by concentrating most references in the places where they are most convenient to properly credit results published in the literature, namely, in the figures and tables captions, in the footnotes, and in the exercises and their solutions.

This book is dedicated to the memory of Professors Esther Belin-Ferré, Patricia A. Thiel and An Pang Tsai, with my heartfelt gratitude for the continued interest in my research activities, valuable advice and motivating support they provided me during the time I had the fortune to share with them.

I am gratefully indebted to Professors Jose Luis Aragon, Marc de Boissieu, Arunava Chakrabarti, Jean Marie Dubois, Kaoru Kimura, Edmundo Lazo, Uichiro Mizutani, Gerardo G. Naumis, Juan M. Pérez-Mato, Manuel Torres, and Victor R. Velasco for their interest in my research work during the last 25 years. Their illuminating advice has significantly contributed to guide my scientific work in the science of QCs. It is a pleasure to thank Professors Claire Berger, Luca Bindi, Luca Dal Negro, Janez Dolinsek, Luis Elcoro, Sergey V. Gaponenko, Federico García-Moliner, Katzumoto Iguchi, Didier Mayou, Tsunehiro Takeuchi, and Chumin Wang for inspiring conversations, and to Professors Eudenilson L. Albuquerque, Emilio Artacho, Roberto Escudero, Javier García-Barriocanal, Gábor Gébay, Roland Ketzmerick, Luigi Moretti, Ruwen Peng, Michael E. Pollard, and José Reyes-Gasga for sharing with me very useful materials.

The author is deeply grateful to the following persons belonging to CRC Press, Taylor and Francis Group, namely, Ms. Francesca McGowan (Editor, Physics) for her kind invitation to prepare this book in 2017, as well as Ms. Rebecca Davis and Dr. Kirsten Barr for their continued help in dealing with all the related editorial matters. Last, but not least, I warmly thank Ms. Victoria Hernández for her invaluable support, unfailing encouragement and her continued care to the detail.

Enrique Maciá Barber
Madrid, October 2020

CHAPTER 1

Intermetallic Quasicrystals

1.1 WHAT IS A QUASICRYSTAL?

"A quasicrystal is the natural extension of the notion of a crystal to structures with quasiperiodic, rather than periodic, translational order". (Dov Levine and Paul Joseph Steinhardt, 1984) [486]

1.1.1 The Al$_{86}$Mn$_{14}$ alloy electron diffraction picture

"It took the right instrument, an electron microscope, because while X-ray diffraction is great for studying many aspects of crystals, it cannot show you rotational symmetry" (Daniel Shechtman, 70th Birthday Celebration Symposium, 27 July 2011, Iowa State University, Ames).

"As I was studying rapidly solidified aluminium alloy which contained 25% manganese by transmission electron microscopy, something very strange and unexpected happened. There were 10 bright spots in the selected area diffraction pattern, equally spaced from the center and from one another. I counted them and repeat the count in the other direction and said myself: 'There is no such animal'. In Hebrew: 'Ein Chaya Kazo'". (Daniel Shechtman commenting on his 8 April 1982 discovery [321], see Fig. 1.1)

To properly understand Daniel Shechtman's astonishment, we must recall that 20th century crystallography started in 1912 with the introduction of X-ray diffraction methods by Max von Laue (1879–1950), William Henry Bragg (1862–1942), and William Lawrence Bragg (1890–1971), who assumed that atoms were arranged in ordered crystals according to regular *periodic* arrays on a lattice. During the seventy years elapsed until Shechtman's discovery almost all reported crystalline structures were perfectly compatible with that assumption, and accordingly the main attribute of crystalline solid state was progressively identified with the presence of a periodic distribution of matter density, $\rho(\mathbf{r})$, inside the considered samples. Ideally assuming that atoms can be treated as material points, such a distribution can be envisioned as a lattice of points in Euclidean three-dimensional space, so that the vectors describing the lattice points have the general form $\mathbf{r} = n_1\mathbf{e}_1 + n_2\mathbf{e}_2 + n_3\mathbf{e}_3$, where $\{\mathbf{e}_i\}$ is a suitable vector basis and the coordinates are integer numbers (i.e., $n_i \in \mathbb{Z}$). Within this framework, lattice periodicity is given by the invariance condition $\rho(\mathbf{r} + \mathbf{R}_0) = \rho(\mathbf{r})$, where $\mathbf{R}_0 = m_1\mathbf{e}_1 + m_2\mathbf{e}_2 + m_3\mathbf{e}_3$ ($m_i \in \mathbb{Z}$) belongs to the lattice-related vectorial space. Physically, lattice periodicity guarantees that the local atomic environment we have at a given volume determined by the vector basis $\{\mathbf{e}_i\}$ inside the crystal exactly repeats itself throughout the space.

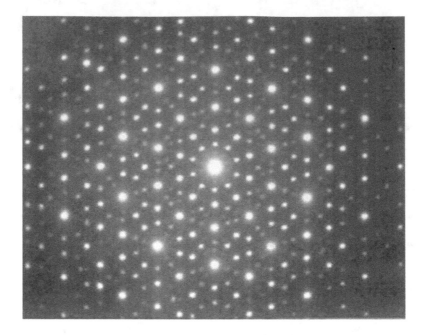

Figure 1.1 Electron diffraction pattern corresponding to the $Al_{86}Mn_{14}$ alloy quasicrystal discovered by Shechtman. A 10-fold symmetry axis around the origin can be clearly appreciated. (Reprinted figure with permission from Shechtman D, Blech I, Gratias D, and Cahn J W, Physical Review Letters 53, 1951, 1984. Copyright (1984) by the American Physical Society).

Periodic arrangements in Euclidean space involve rotations along with translations. We say that a given structure possesses a m-fold rotational symmetry if it is left unchanged when rotated by an angle $2\pi/m$, with respect to a certain axis, fixed a given point. The integer $m \in \mathbb{N}$ is called the order of the rotational symmetry (or of its related symmetry axis). Therefore, the lattice remains the same under rotations relating lattice points to each other, which are represented by a set of matrices of the form

$$\mathbf{R}_\alpha = \begin{pmatrix} n_{11} & n_{12} & n_{13} \\ n_{21} & n_{22} & n_{23} \\ n_{31} & n_{32} & n_{33} \end{pmatrix},$$

whose entries are integer numbers. On the other hand, on a properly chosen basis, pure rotations can be expressed in terms of the orthogonal matrices[1]

$$\mathbf{M}_X = \begin{pmatrix} 1 & 0 & 0 \\ 0 & c & -s \\ 0 & s & c \end{pmatrix}, \ \mathbf{M}_Y = \begin{pmatrix} c & 0 & s \\ 0 & 1 & 0 \\ -s & 0 & c \end{pmatrix}, \ \mathbf{M}_Z = \begin{pmatrix} c & -s & 0 \\ s & c & 0 \\ 0 & 0 & 1 \end{pmatrix}, \quad (1.1)$$

with $c \equiv \cos\varphi$ and $s \equiv \sin\varphi$, describing counterclock rotations by an angle φ around the X, Y, and Z Cartesian axes, respectively. Accordingly, $\mathbf{R}_\alpha = \mathbf{B}\mathbf{M}_\alpha\mathbf{B}^{-1}$, where $\mathbf{B}$ denotes the basis matrix. Now, since the trace of a matrix is invariant under a basis transformation, the periodicity condition $\rho(\mathbf{R}_\alpha\mathbf{r}) = \rho(\mathbf{r})$ relating lattice points, implies that $\text{tr}(\mathbf{R}_\alpha) = \text{tr}(\mathbf{M}_\alpha)$,

---

[1]Note that these matrices are unimodular (i.e., $\det \mathbf{M}_\alpha = 1$) as well.

TABLE 1.1 Allowed
symmetry axes in 3D
periodic crystals

n	φ	AXIS
-1	π	2-fold
0	$2\pi/3$	3-fold
1	$\pi/2$	4-fold
2	$\pi/3$	6-fold
3	0	Identity

so that we obtain the so-called crystallographical restriction [381, 840],

$$1 + 2\cos\varphi = n_{11} + n_{22} + n_{33} \equiv n \in \mathbb{Z}, \tag{1.2}$$

which can be rewritten in the convenient form $\cos\varphi = (n-1)/2$. The bond $|\cos\varphi| \leq 1$ then implies $n = \{-1, 0, 1, 2, 3\}$. In this way, we obtain the allowed rotation angles listed in Table 1.1. Therefore, a very small number of rotations are compatible with the periodicity condition, namely, only 2-, 3-, 4-, and 6-fold symmetry axes are allowed for 3D periodic lattices. Not surprisingly, Eq. (1.2) along with its corresponding version for two-dimensional (2D) periodic lattices (**Exercise 1.1**), have played a significant role in the foundations of classical crystallography. Accordingly, solid state physics textbooks properly emphasized that all rotational symmetries other than those listed in Table 1.1 were forbidden in crystals. On the basis of these allowed rotational symmetry elements, the so-called 14 Bravais lattices naturally followed, along with the related 230 space groups, thus providing the basic tool for crystal classification and the *International Tables of Crystallography* was the ultimate classification catalog for crystals. In this way, for many decades, it was firmly believed that 5-fold rotational symmetry could not exist in well ordered condensed phases of matter, so that the pentagon symmetry, which is widely found in the world of the living, was excluded from the mineral kingdom,[2] . . . until Shechtman took his now celebrated electron diffraction picture shown in Fig. 1.1.[3]

Quite remarkably, the electron diffraction study of the $Al_{86}Mn_{14}$ alloy performed by Shechtman not only revealed the presence of this unexpected symmetry axis. In fact, when the specimen was tilted and viewed from other directions, the characteristic fingerprints of 2- and 3-fold symmetry axes were also observed. The complete account of symmetry rotation axes yield fifteen 2-fold axes, ten 3-fold axes, and six 5-fold axes, which together with the corresponding mirror planes, amount to the 120 symmetry elements characteristic of the *icosahedral* group, as it is shown in Fig. 1.2.

The existence of a novel solid phase of matter, which diffracts electrons like a single crystal does but has icosahedral point group symmetry, which is inconsistent with periodic lattice translations, was announced to the scientific community by Daniel Shechtman, Ilan Blech, Denis Gratias, and John W. Cahn (1928–2016) in a paper entitled *"Metallic phase with long-range orientational order and no translational symmetry"*, published in the Physical Review Letters on 12 November 1984 [801]. The remarkable sharpness of the diffraction patterns shown in Figs. 1.1 and 1.2 clearly indicates a high coherency of the electrons spatial

---

[2]Some observations of minerals probably exhibiting quinary axes of symmetry were earlier reported by Rome d'Isle, but by all indications these early observations were subsequently ignored [700].

[3]Because diffraction patterns are always inversion symmetric, it is convenient to distinguish between *diffraction* symmetry axes (10-fold in this case) and *real space* rotational symmetry axes (5-fold in this case). See Secs. 6.1.3.2 and 6.3.3.

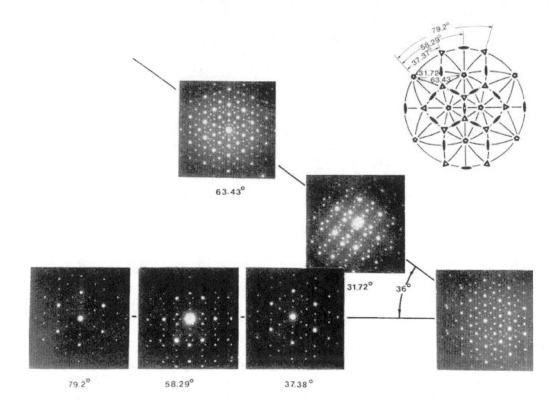

Figure 1.2 Selected-area electron diffraction patterns taken from a single grain of the $Al_{86}Mn_{14}$ alloy phase discovered by Shechtman. Rotations match those indicated in the stereographic projection of the symmetry elements of the icosahedral group m35 shown in the upper right corner. (Reprinted figure with permission from Shechtman D, Blech I, Gratias D, and Cahn J W, Physical Review Letters 53, 1951, 1984, Copyright (1984) by the American Physical Society).

interference, comparable to the one usually encountered in classical periodic crystals. Thus, this alloy exhibited well-defined long-range order, but one which is explicitly incompatible with the periodic translational one. What may be the nature of this ordering of matter which had remained so elusive to crystallographers until then?

Six weeks after Shechtman and co-workers paper publication, the nature of this long-range order was explained by Paul J. Steinhardt and Dov Levine in a paper entitled *"Quasicrystals: a new class of ordered structures"*, published in the Physical Review Letters on 24 December 1984 [486]. In this paper the notion of quasiperiodic crystals was introduced (see the quotation opening this Chapter) by invoking the mathematical notion of quasiperiodic functions (which will be described in Sec. 1.3.2), thereby widening the concept of a periodic distribution of atoms through the space to that of a quasiperiodic (QP) one.

This extension was performed on the basis of a detailed comparison between a computed diffraction pattern for a suitable QP atomic model and the electron diffraction pattern obtained by Shechtman for the $Al_{86}Mn_{14}$ alloy (Fig. 1.3). By inspecting both figures, it is apparent that the actual atomic arrangements in the alloy must be closely related to those adopted by the lattice points in the model. Indeed, the position of each electron diffraction reflection matches with the position of a peak in the calculated pattern, and a hierarchy of

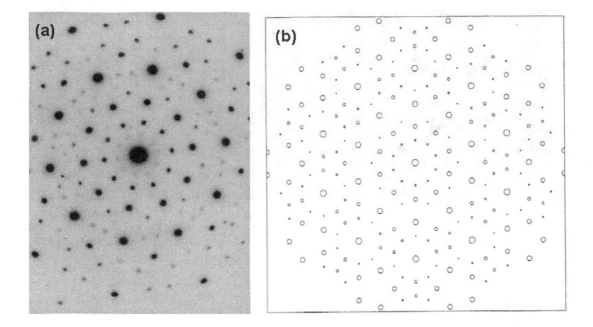

Figure 1.3 (a) Electron diffraction pattern corresponding to the $Al_{86}Mn_{14}$ QC discovered by Shechtman. (Reprinted figure with permission from Shechtman D, Blech I, Gratias D, and Cahn J W, Physical Review Letters 53, 1951, 1984. Copyright (1984) by the American Physical Society). (b) Computed diffraction pattern in a plane normal to a 5-fold axis for an ideal icosahedral QC, displaying only peaks above some given intensity. The circles are centered at the location of Bragg peaks and have a radius proportional to their intensity. (Reprinted figure with permission from Levine D and Steinhardt P J, Physical Review Letters 53, 2477, 1984. Copyright (1984) by the American Physical Society).

intensities characteristic of quasiperiodicity is observed at several orders [486]. In this way, the discovered novel phase was envisioned as a natural extension of the classical crystal notion, now embodying QP arrangements of matter as well. Accordingly, the shorthand *quasicrystal* (QC), standing for *quasiperiodic crystal*, was coined.[4]

Paul J. Steinhardt and Dov Levine were awarded the Oliver Buckley Condensed Matter Prize from the American Physical Society, in 2010 for *"pioneering contributions to the theory of quasicrystals, including the prediction of their diffraction pattern"*. Daniel Shechtman was awarded the Nobel Prize in Chemistry, in 2011 for the *"discovery of quasicrystals"*, almost thirty years after this discovery took place.

---

[4]The abstract of a paper submitted by David Ruelle on 13 January 1981 reads *"We discuss the possibility that, besides periodic and quasiperiodic crystals, there exist turbulent crystals as thermodynamic equilibrium states at non-zero temperature"* [760]. These quasiperiodic crystals exhibit *"long-range order with two independent modulations"* [760], thereby belonging to the family of incommensurate structures first described by Pim de Wolff and co-workers in Ref. [172]. Certainly, it is quite remarkable that D. Ruelle considers *"an ordinary crystal (periodic or quasiperiodic)"* as early as in 1981, hence anticipating the unifiying view given in the revamped definition of crystal proposed by the ICrU in 1992 [357] (see Sec. 1.1.5).

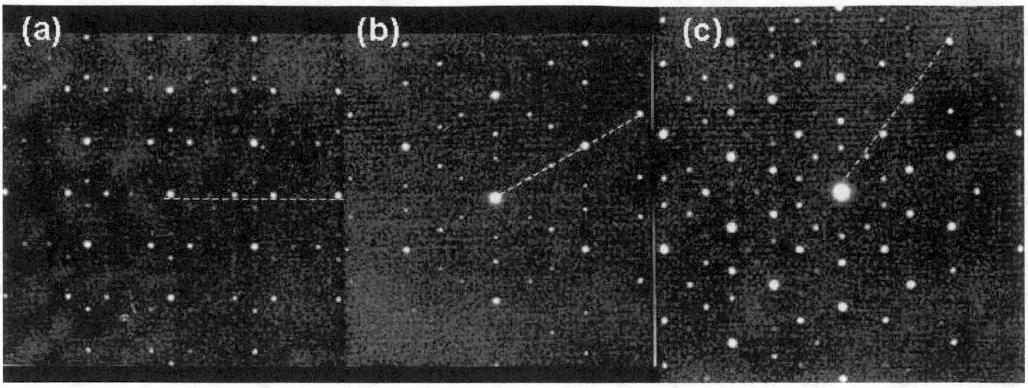

Figure 1.4 Electron diffraction patterns taken from an i-$Al_{86}Mn_{14}$ QC grain showing the 2-fold (a), 3-fold (b) and 5-fold (c) rotation axes of the icosahedral $m\bar{3}\bar{5}$ point group. (Reprinted figure with permission from Elser V, Physical Review B 32, 4892, 1985. Copyright (1985) by the American Physical Society).

1.1.2 A novel crystallographic symmetry

> "The diffraction pattern of icosahedral quasicrystals revealed a new crystallographic symmetry: scaling invariance". (Aloysio Janner, 2006) [374]

In order to gain a deeper understanding on the nature of the spatial order underlying atomic arrangements in QCs, let us consider the electron diffraction patterns shown in Figs. 1.4 and 1.5. By inspecting the distribution of reflection peaks in these diffraction patterns one can observe the presence of:

- The full set of icosahedral point group symmetry elements, namely 2-, 3-, and 5-fold axes, along with the corresponding mirror planes, characteristic of the so-called *icosahedral* QCs (iQCs).

- The fingerprints of scale invariance symmetry, which can be determined from the study of two main features, namely, (1) a series of linear arrays of diffraction spots along the radial directions from the center peak (dashed lines in Fig. 1.4 and solid radial lines in Fig. 1.5), and (2) the presence of nested polygonal Bragg reflection patterns, featuring the so-called *Pythagorean pentagram* design (Fig. 1.5).

In order to disclose this characteristic inflation symmetry, let us start by closely inspecting Fig. 1.5. To this end, we measure the distances of successive main diffraction spots along a radial direction outwards from the center, d_n, and calculate the ratios d_n/d_1. The d_n obtained values are listed in the second column of Table 1.2, and they clearly follow a non-periodic series, which obeys the recursion rule $d_{n+1} = d_n + d_{n-1}$ $(n > 1)$. The d_n/d_1 ratios listed in the third column of Table 1.2 are rational numbers whose numerators are given by consecutive *Fibonacci* numbers. These celebrated numbers are defined by the sequence $F_n = \{1, 1, 2, 3, 5, 8, 13, 21, ...\}$, with $F_0 = F_1 = 1$, where each number in the sequence is just the sum of the preceding two. Quite remarkably, we can appreciate a close relationship between the corresponding d_n/d_1 values and successive powers (rounded to the second decimal place) of the so-called golden mean $\tau = (\sqrt{5} + 1)/2 = 1.6180339887...$, as it is

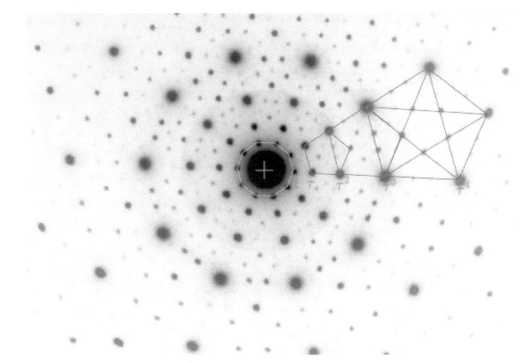

Figure 1.5 Close up view of the electron diffraction pattern shown in Fig. 1.1. In addition to a 10-fold symmetry axis around the origin, a conspicuous series of nested pentagons can be clearly appreciated. Two of them are highlighted at the upper right corner, displaying the so-called Pythagorean pentagram design, obtained by connecting all the opposite vertices by straight lines. The resulting nested pentagons exhibit a size ratio given by τ^{-2}. A sequence of diffraction peaks spaced according to a geometric series based on the golden mean (see Table 1.2) is indicated along the meridian radial path. (Adapted from D. Gratias, M. Quiquandon, Discovery of quasicrystals: The early days, C. R. Physique 20, 803-816 (2019). doi: 10.1016/j.crhy.2019.05.009. CC NY-NC-ND license 4.0)[307].

seen by comparing the data listed in the fourth and fifth columns of Table 1.2, respectively. This result indicates that the successive interplanar distances in the considered QC atomic distribution increase by the golden ratio factor.

The golden mean has been largely known from ancient times, since it frequently appears in pentagon and decagon based polygons and polyhedra. This celebrated ratio can be derived in several ways. In general, a segment is said to be divided in the golden mean if the ratio of the whole segment to the larger part is equal to the ratio of the larger to the smaller one. If we take the smaller segment as unit and label the larger part as the unknown x, this geometrical definition can be expressed as $(x+1)/x = x/1$, which leads to the algebraic equation $x^2 = x+1$, whose positive solution is the irrational number $x_+ = (\sqrt{5}+1)/2 = \tau$. Therefore, we get the following expressions relating the golden mean, its square and its reciprocal, namely, $\tau^2 = \tau + 1$, and $\tau^{-1} = \tau - 1$. By successively multiplying the basic

TABLE 1.2 The main Bragg reflections along the meridian radial direction in the diffraction pattern shown in Fig. 1.5 can be labeled according to a power series related to the golden mean.

n	d_n (mm)	d_n/d_1	d_n/d_1	τ^{n-1}	$|\Delta|$	(F_n, F_{n-1})
1	7.5	1	1.00	1.00	–	–
2	12.5	5/3	1.67	1.62	0.05	$(1,0)$
3	20.0	8/3	2.67	2.62	0.05	$(1,1)$
4	32.5	13/3	4.33	4.24	0.09	$(2,1)$
5	52.5	21/3	7.00	6.85	0.15	$(3,2)$

relation $\tau^2 = \tau + 1$ by τ we get,

$$\tau^3 = \tau^2 + \tau = 2\tau + 1,$$
$$\tau^4 = 2\tau^2 + \tau = 3\tau + 2,$$
$$\tau^5 = 3\tau^2 + 2\tau = 5\tau + 3,$$
$$\vdots$$
$$\tau^n = \tau F_{n-1} + F_{n-2}, \quad (n \geq 2). \tag{1.3}$$

Thus, any power of τ can be expressed as a linear combination of two successive Fibonacci numbers in the base $\{\tau, 1\}$. In this way, making use of Eq. (1.3), one can properly label the different diffraction reflections along the radial directions of the diffraction pattern shown in Fig. 1.5 in terms of a pair of Fibonacci numbers, as it is shown in the last column of Table 1.2.

Another useful mathematical relationship between the Fibonacci series and the golden mean is given by the asymptotic limit,[5]

$$\lim_{n \to \infty} \frac{F_n}{F_{n-1}} = \tau, \tag{1.4}$$

so that the ratio of two successive larger and larger Fibonacci numbers comes closer and closer to the golden mean. If one considers the ratio between next-neighbors in the Fibonacci series one gets,

$$\lim_{n \to \infty} \frac{F_{n+1}}{F_{n-1}} = \lim_{n \to \infty} \frac{F_n + F_{n-1}}{F_{n-1}} = 1 + \lim_{n \to \infty} \frac{F_n}{F_{n-1}} = 1 + \tau = \tau^2. \tag{1.5}$$

By proceeding in a similar way, one can readily see that higher powers of the golden mean are related to asymptotic limits involving ratios of Fibonacci numbers further and further apart along the series. This property is illustrated in the third column of Table 1.2.

At this point, it is important to highlight that the presence of scale invariance symmetry in iQCs can also be observed by studying the diffraction patterns taken along 2- and 3-fold symmetry axes, as it can be readily checked by measuring the distances between successive diffraction peaks along the dashed lines in Fig. 1.4a and Fig. 1.4b, respectively. In doing so, one obtains the same scale factor τ that we obtain when considering the diffraction pattern along the 5-fold axis shown in Fig. 1.4c. This result indicates that the scale invariance symmetry is an isotropic property of iQCs. Therefore, in iQCs the translation symmetry, characteristic of periodic crystals, is *replaced* by an inflation/deflation symmetry

---

[5]Discovered in 1611 by Johannes Kepler (1571–1630).

in 3D, which is measured in terms of a characteristic scale factor given by the golden mean. The same scale factor is obtained in the case of decagonal QCs (**Exercise 1.2**), whereas for octagonal and dodecagonal QCs one obtains the scale factors given by the irrational numbers $1 + \sqrt{2}/2$ and $2 + \sqrt{3}$, respectively (**Exercise 1.3**). We must note, however, that irrational numbers are beyond the reach of experimental measurement by definition. In fact, the analysis of the electron diffraction patterns performed in the above referred exercises, where all the obtained data can be expressed in terms of rational ratios (see the third column in Tables 1.2, 1.5, 1.6, and 1.7), clearly illustrates this point. Therefore, in order to justify the true irrational nature of the involved scale factors one must rely on some kind of mathematical construction.

The inflation symmetry scale factor can also be derived by constructing the Pythagorean pentagram depicted in Fig. 1.5, taking into account that the ratio between the diagonal and the side of a given pentagon is τ (**Exercise 1.4a**). To this end, one connects all the vertices of a regular pentagon by diagonals. At their intersecting points these diagonals form a smaller pentagon, and the diagonals of this pentagon will form a new pentagram enclosing a yet smaller pentagon. This progression can be continued *ad infinitum*, creating smaller and smaller pentagons and pentagrams in an endless succession exhibiting a self-similar nesting typical of fractal geometry (**Exercise 1.4b**). This self-similar property leads to the appearance of a *hierarchy* of Bragg reflections intensities rendering highly dense diffraction patterns. Indeed, in the ideal case of a perfect QC structure, characterized by the presence of exact irrational scale factors, the distribution of peaks will densely fill the reciprocal space. In actual samples, however, most of the weakest intensity peaks will become extremely dim, making it possible to distinguish individual Bragg spots experimentally. In this way, attending to the relative intensities of the reflections in the diffraction pattern one can readily appreciate the existence of a set of spots which dominate the diffraction pattern, and can be taken to construct a quasi-Brillouin zone. Nevertheless, due to the incommensurate nature of the scale factor we cannot use these peaks to define an average structure compatible with any sort of translation symmetry.[6]

At this point, it is worth mentioning that mineralogists and materials scientists have discovered materials containing icosahedral building blocks, either isolated or linked, well before the discovery of QCs, but these materials exhibited symmetries compatible with periodic order at a large enough scale, as it occurs in fullerite molecular crystal [921], or in organic crystals composed of icosahedral virus particles studied to date (see Sec. 6.1.4.5). In fact, evidence of icosahedral symmetry in condensed matter was early reported in the study of the geometry of virus capsids from X-ray diffraction measurements of several virus crystals. Arguments supporting the possible presence of a high symmetry degree in virus capsids were put forward by Francis H. C. Crick (1916–2004) and James D. Watson, who in 1956, suggested that most of them could be built up from a relatively small number of protein subunits arranged in a symmetrical way, and noticed that of all the possible types of symmetry arrangements only the cubic point groups were likely to lead to an isometric structure [144]. The first experimental evidence for icosahedral symmetry in a virus came from the X-ray diffraction studies of Donald L. D. Caspar on tomato bushy stunt virus [112], soon followed by that of Aaron Klug (1926–2018) and Rosalind E. Franklin (1920–1958) on turnip yellow mosaic virus [428, 429]. These investigations confirmed that the X-ray diffraction patterns from certain virus crystals showed intensity distributions characteristic of a *periodic* arrangement of *icosahedral particles* throughout the space.[7]

---

[6] At variance with the incommensurately modulated phases case. See, for instance, [381].

[7] For this reason, these virus particles are generally known as icosahedral viruses. Aaron Klug was awarded

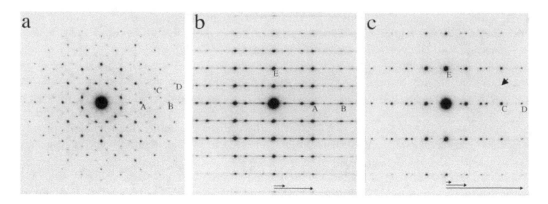

Figure 1.6 Electron diffraction pattern corresponding to a dd-$Mn_{72.0}Si_{17.5}Cr_{5.5}Ni_{5.0}$ QC taken along the 12-fold axis (a), and along two different kinds of perpendicular 2-fold axes (b) and (c). Arrows at the bottom of (b) and (c) pictures illustrate the characteristic $2+\sqrt{3}$ scaling factor of this axial QC (**Exercise 1.3**). (Taken from Dodecagonal quasicrystal in Mn-based quaternary alloys containing Cr, Ni and Si, Iwami S and Ishimasa T, Philosophical Magazine Letters, 85, 229–236, 2005. Reprinted by permission of Taylor & Francis Ltd., http://www.tandfonline.com).

Therefore, neither the presence of icosahedral symmetry alone (previously observed in virus assemblies), nor the presence of a scale invariance symmetry (also present in fractal structures) is specific of iQCs. It is the *simultaneous* presence of both types of symmetries in the same piece of matter which makes the difference. Accordingly, in order to properly understand the origin of specific physical properties of QCs, one must properly take into account the characteristic scaling invariance and point group symmetries altogether, introducing a geometrical picture able to blend inflation–deflation relations with rotational and mirror symmetry elements [374].

1.1.3 The sequel: axial quasicrystals

Shortly after the publication of the two seminal papers by D. Shechtman, I. Blech, D. Gratias and J. W. Cahn, and by D. Levine and P. J. Steinhardt, respectively, a number of works reporting the existence of new QCs characterized by the presence of different types of non-crystallographic symmetry axes were published in quick succession. Thus, a new ordered state of matter exhibiting a 12-fold symmetry axis, referred to as dodecagonal QC (ddQC) was found in small particles of a Ni-Cr alloy by T. Ishimasa, H. U. Nissen, and Y. Fukano on 25 July 1985 [367]. An example of the corresponding electron diffraction patterns is shown in Fig. 1.6 for a recently synthesized ddQC in the Mn-Si-Cr-Ni quaternary alloy phase.

Two months later, L. Bendersky reported on another non-crystallographic phase, characterized by the presence of a 10-fold symmetry axis, in rapidly solidified Al-Mn alloys with higher manganese content (18–22 at.%) than those originally studied by Shechtman (10–20 at.%), which was referred to as decagonal QC (dQC) [50, 269]. An example of the corresponding electron diffraction pattern along a 10-fold axis is shown in Fig. 1.7 for a sample

---

the Nobel Prize in Chemistry, in 1982, for his "development of crystallographic electron microscopy and his structural elucidation of biologically important nucleic acid-protein complexes".

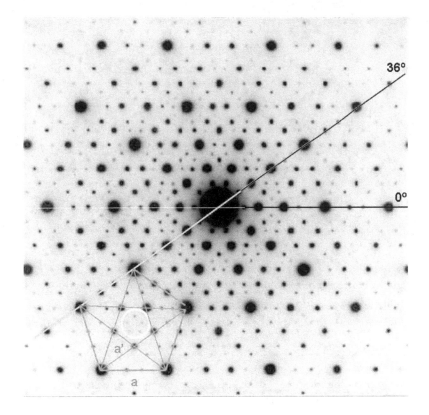

Figure 1.7 Electron diffraction pattern corresponding to a d-$Al_{70}Cu_{15}Co_{15}$ QC (http://www.tohoku.ac.jp/en/research/research_highlights/images/rh07_01.jpg). A Pythagorean pentagram formed by a large pentagon of side a and a smaller pentagon of side a' is highlighted at the left lower corner. A second generation smaller pentagon is encircled in the pentagram's central region.

belonging to the ternary Al-Cu-Co alloy.[8] Finally, in 1987, N. Wang, H. Chen, and K. H. Kuo reported the existence of QCs with 8-fold rotational axis in rapidly solidified V-Ni-Si and Cr-Ni-Si alloys, which were referred to as representatives of the octagonal QC class (oQC) [950].

Octagonal, decagonal, and dodecagonal QCs share a common structural design: they consist of transversal planes where the atoms are quasiperiodically arranged displaying 8-, 10-, or 12-fold symmetry axes, respectively, and these planes pile up along the non-crystallographic axis leading to octagonal and dodecagonal platelets or decagonal prisms growth morphologies, as that shown in Fig. 1.8a. Accordingly, the representatives of these three QC classes can be generically referred to as *axial* QCs.

In these materials, two kinds of long-range orders, namely, periodic and QP, coexist in the same sample, at variance with iQCs which display quasiperiodic order (QPO) all

---

[8]It is interesting to note that the observation of a diffraction pattern with a pseudo-pentagonal symmetry had earlier been reported in an investigation by transmission electron microscopy (the same technique used by Shechtman) of rapidly solidified AlPd alloys in 1978. Unfortunately, not enough attention was paid to this system, which was later recognized as belonging to the decagonal phase [210].

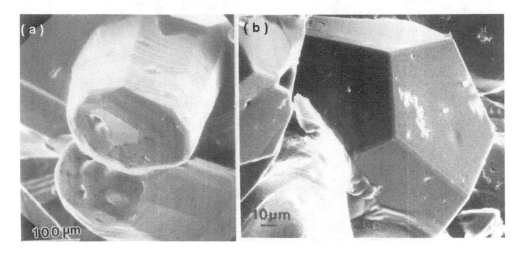

Figure 1.8 Scanning electron images showing the growth morphology of two representative thermodynamically stable QCs. (a) A decagonal prism habit corresponding to the d-AlNiCo QCs. (b) A regular pentagonal dodecahedral habit corresponding to the i-AlCuFe QCs. (Courtesy of An Pang Tsai).

over the space, exhibiting dodecahedral (Fig. 1.8b) or triacontahedral (Fig. 1.9) growth morphologies. This interesting feature suggested that other kinds of *mixed order* structures may also exist. In fact, a one-dimensional QC derived from decagonal phase representatives was found in rapidly solidified AlNiSi, AlCuMn, and AlCuCo alloys. In this class of QCs, in addition to the periodic translation along the 10-fold axis present in axial QCs, the translation along one of the 2-fold axes normal to the 10-fold one becomes periodic as well [329].

In Fig. 1.9, we present a classification of QCs attending to the number of spatial dimensions in which the QPO occurs. For any given solid its 3D spatial dimension is expressed as the sum of its periodic dimension (labelled by the corresponding row number) plus its QP dimension (labelled by the corresponding column number). In this way, the coordinates of a given solid in the chart indicate the relative importance of QP versus periodic order in its structure. For example, the iQC shown at the left upper corner exhibits QPO along the three dimensions of space, so that it can be regarded as an example of isotropic QC. On the contrary, dQCs located at the middle of the chart exhibit QPO in the planes perpendicular to the decagonal axis, but are periodically arranged along this axis, so that they must be regarded as a representative example of anisotropic QCs. At the right lower corner we have another interesting instance of a hybrid-order system: a Fibonacci heterostructure. In this system a series of periodic crystal layers are stacked following a QP 1D array. As we will see in Sec. 4.2.1, this kind of structures have been artificially grown and they properly illustrate the technological potential related to the very notion of hybrid-order QP designs in materials science.

1.1.4 Thermodynamically stable quasicrystals grow large

> *"At the end of the 18th century, it was realized that the shape of natural crystals and the angles between exposed facets reflected their internal symmetry.*

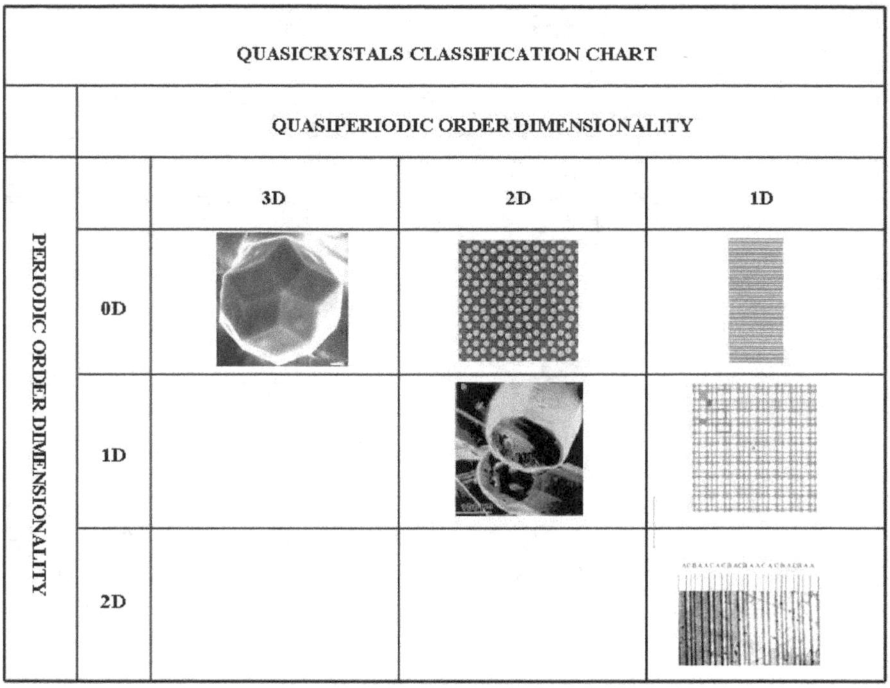

Figure 1.9 Classification of different types of QCs attending to their QP dimensionality. From left to right and from top to bottom we see a triacontahedral iQC (https://blog.stephenwolfram.com/data/uploads/2018/12/quasicrystals-rhombic-triacontahedra-2.png), a photonic oQC, a dQC, a Fibonacci code bar[246], a Fibonacci-periodic hybrid square, and a dielectric Fibonacci multilayer (Adapted from Maciá E, The role of aperiodic order in science and technology, Rep. Prog. Phys. 69, 397–441, 2006; doi:10.1088/0034-4885/69/2/R03 © IOP Publishing. Reproduced with permission. All rights reserved).

> *Similarly, faceted samples have been observed in quasicrystalline materials grown in the laboratory".* (Julian Ledieu and Vincent Fournée, 2014) [483].

First generation (1984–1986) QCs were obtained by means of far from equilibrium techniques and exhibited small sizes, typically a few micrometers, which is too small to precisely investigate their intrinsic physical properties. In fact, they were not even thermodynamically stable and transformed into periodic crystals after heating at moderate temperatures (Fig. 1.10). In addition, their X-ray diffraction lines were rather broad, resembling those observed in highly faulted periodic crystals. Thus, during the early years of the field, it was naturally assumed that QCs were located at a somewhat intermediate position between periodic crystals and amorphous materials, so that they should be inherently unstable. This assumption proved, however, to be wrong.

In 1955, H. K. Hardy and J. M. Silcock classified the intermetallic compounds in equilibrium with the aluminum solid solution in the Al-Li-Cu system by using a Debye-Sherrer power diffraction method. At $500°$C, they found several compounds corresponding to hexagonal and cubic structures, in addition to a phase they called T2, with composition close to Al_6Li_3Cu, which exhibited a X-ray diffraction spectrum escaping any crystalline indexation

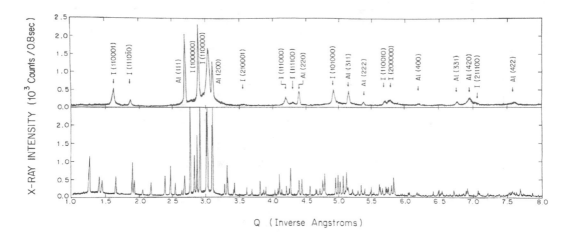

Figure 1.10 (Top panel) High-resolution X-ray diffraction pattern of quenched i-$Al_{86}Mn_{14}$ quasicrystalline powder. Some contaminating fcc Al phase is present. Icosahedral phase peaks are labeled using generalized Miller indices in 6D. (Bottom panel) Diffraction pattern of annealed i-$Al_{86}Mn_{14}$ powder by heating to 410°C for 75 minutes in an argon atmosphere. All peaks can be indexed as belonging to the Al_6Mn orthorhombic phase. (Reprinted figure with permission from Bancel P A, Heiney P A, Stephens P W, Goldman A I, and Horn P M, Physical Review Letters 54, 2422, 1985. Copyright (1985) by the American Physical Society).

[316, 320]. This oddity remained dormant for more than three decades, until B. Dubost, J. M. Lang, M. Tanaka, P. Sainfort, and M. Audier proved that this phase actually belongs to the iQC class. In a paper entitled *Large AlCuLi single quasicrystals with triacontahedral solidification morphology*, published in Nature on 6 November 1986, they announced the discovery of the first thermodynamically stable iQC, thus being able to grow large grains up to the millimeter size by conventional solidification techniques in close to equilibrium conditions [213]. In this way, large quasicrystalline grains, exhibiting the growth morphology of a multi-faceted rhombic triacontahedron (Fig. 1.9 left top panel), were demonstrated in i-$Al_{56}Li_{33}Cu_{11}$ QCs by several groups. Such a triacontahedral morphology was a novelty in crystallography and mineralogy.

Shortly after, two new stable iQC representatives, now exhibiting pentagonal dodecahedral solidification morphologies, were reported by A. P. Tsai, A. Inoue, and T. Masumoto in the i-$Al_{65}Cu_{20}Fe_{15}$ QC (September 1987) [904], and by W. Ohashi and F. Spaepen in i-$Zn_{43}Mg_{37}Ga_{20}$ QC (December 1987) [660], respectively.[9]

The structural quality of a sample can be estimated from the X-ray diffraction spectra in terms of the so-called correlation length, given by the relation 2π/FWHM, where FWHM is the diffraction peak full width at half maximum. According to this criterion, first generation QCs did not possess a very good structural quality, showing correlation lengths in the approximate range 50–100 nm. During the period 1987–2000, a remarkable number of thermodynamically stable iQCs of high structural quality were discovered by A. P. Tsai and co-workers in the AlCu(Fe,Ru,Os), AlPd(Mn,Re), Zn-Mg-(Y,Gd,Tb,Dy,Ho,Er),

---

[9]For the sake of historical curiosity both manuscripts were received on 14 July 1987 in their respective journals.

TABLE 1.3   Main representatives of thermodynamically stable iQCs
[840, 502, 68, 145, 297, 462, 874]. RE stands for rare-earth atoms. The star marked sample
corresponds to the mineral QC icosahedrite (see Sec. 1.2).

Al-based	Zn-based	Ti-based	Cd-based	Ag/Au-based
$Al_{70.5}Pd_{21}Mn_{8.5}$	$Zn_{88}Sc_{12}$	$Ti_{45}Zr_{38}Ni_{17}$	$Cd_{88.3}RE_{11.7}$	$Ag_{62.7}Al_{27}Ca_{14.3}$
$Al_{70}Pd_{20}Re_{10}$	$Zn_{84}Ti_8Mg_8$	$Ti_{40}Hf_{40}Ni_{20}$	$Cd_{85}Yb_{15}$	$Au_{60.0}Sn_{26.7}Yb_{13.3}$
$Al_{65}Cu_{25}Fe_{15}$	$Zn_{80}Sc_{15}Mg_5$		$Cd_{85}Ca_{15}$	$Au_{46}Al_{38}Tm_{17}$
$Al_{65}Cu_{25}Ru_{15}$	$Zn_{77}Sc_{15}Mg_5$		$Cd_{88.3}RE_{11.7}$	$Ag_{44.2}In_{41.7}Ca_{14.1}$
$Al_{65}Cu_{25}Os_{15}$	$Zn_{75}Sc_{16}Fe_7$		$Cd_{66.8}Mg_{21.4}RE_{11.8}$	$Ag_{42}In_{42}Yb_{16}$
$*Al_{63}Cu_{24}Fe_{13}$	$Zn_{71.5}Sc_{16.2}Cu_{12.3}$		$Cd_{65}Mg_{20}Ca_{15}$	$Au_{42}In_{42}Ca_{16}$
$Al_{56}Li_{33}Cu_{11}$	$Zn_{60}Mg_{30}RE_{10}$		$Cd_{54.6}Mg_{33.0}RE_{12.4}$	
$Al_{54}Pd_{30}Sc_{16}$	$Zn_{43}Mg_{37}Ga_{20}$			

TABLE 1.4   Main representatives of thermodynamically stable dQCs attending
to the number of layer periods along the 10-fold axis [490, 502, 661, 840]. RE
stands for rare-earth atoms. The star marked sample corresponds to the mineral
QC decagonite (see Sec.1.2).

2	4	6	8
$*Al_{71}Ni_{24}Fe_5$	$Al_{70}Ni_{15}Fe_{15}$	$Al_{70.5}Mn_{16.5}Pd_{13}$	$Al_{75}Pd_{15}Os_{10}$
$Al_{70}Ni_{15}Fe_{15}$	$Al_{70}Ni_{20}Rh_{10}$	$Al_{65}Rh_{20}Cu_{15}$	$Al_{73}Ir_{14.5}Os_{12.5}$
$Al_{70}Ni_{20}Rh_{10}$	$Al_{65}Cu_{20}Co_{15}$	$Co_{47}Ga_{43}Cu_{10}$	$Al_{70}Ni_{20}Ru_{10}$
$Al_{65}Rh_{20}Cu_{15}$	$Al_{65}Cu_{20}Ir_{15}$		
$Mn_{47.4}In_{37.8}Ni_{14.8}$	$Al_{65}Rh_{20}Cu_{15}$		
$In_{44.6}Ni_{27.2}Mn_{23.2}$	$Zn_{68}Mg_{40}RE_2$		

and Cd(Yb,Ca) alloy systems [904, 906, 908]. A number of thermodynamically stable dQCs
was also discovered by this team in the AlCo(Cu,Ni) alloy system [905]. In 2005, An Pang
Tsai (1958–2019) received the Jean Marie Dubois Award *"in recognition of fundamentally
important discoveries of new quasicrystalline phases"*. On the other hand, new thermodynamically stable iQCs were reported by K. F. Kelton, Y. J. Kim, and R. M. Stroud in the
Ti-Zr-Ni alloy system in 1997 [409], and by John D. Corbett (1926–2013) and Q. Lin in
the Zn-Sc-Cu and (Au/Ag)-In-(Ca/Yb) systems in 2003 and 2007, respectively. The first
stable dodecagonal phase, dd-$Ta_{62}Te_{38}$, was discovered in 1998 by M. Conrad, F. Krumeich,
and B. Harbrecht. Currently, about two hundred QC representatives belonging to a broad
variety of alloy systems have been reported to be thermodynamically stable up to their respective melting points and to exhibit diffraction patterns of extraordinary quality, some of
them comparable to those observed in the best monocrystalline samples ever grown. For the
sake of information some representative compounds belonging to the iQC and dQC classes
are listed in Tables 1.3 and 1.4, respectively. With a few exceptions most stable dQCs are
Al-based and they are classified attending to the number of atomic layers (i.e., 2, 4, 6, 8)
in the period along the 10-fold axis. The growth of large size, high structural quality stable
QC samples allowed for the study of the physical properties of these compounds, which we
will discuss in detail in Chapter 3.

1.1.5   Revamping the very crystal notion

The discovery of QCs was a terrible shock for the crystallographical, solid state, and
condensed matter communities. Indeed, quasicrystalline alloys were unexpectedly found,
once the theoretical framework aimed to understand the nature of crystal kingdom from a

microscopic viewpoint was regarded to be essentially completed on rigorous mathematical grounds, within a conceptual scheme which had proved very successful in accounting for the structure of every form of matter studied during seven decades. And such a rigorous and successful theory explicitly forbade the growth morphologies and atomic arrangements exhibited by icosahedral, decagonal, octagonal, and dodecagonal QCs!

This apparent paradox naturally spurred a hectic period of intellectual storm, and the solution fortunately arrived relatively soon. It turned out to be quite simple in retrospective: our previous theoretical understanding of ordered condensed matter was entirely based on the very notion of periodic arrangements of atoms in space. But what about aperiodic ones? We had simply missed them in constructing classical crystallography.

> "Because no exceptions were found in almost 200 years, it became accepted that all crystals were regular arrays on a lattice, and became a law or paradigm that arose from experience, rather than from fundamental principles". (John Werner Cahn, 2014) [104]

Consequently, all the conceptual troubles stemming from the discovery of this new form of matter were rooted on an epistemological shortcoming rather than pointing to any sort of pathological state of matter. The message from nature was that matter can display well-ordered arrangements well beyond the rules imposed by strict periodicity, thereby lifting the restrictions regarding the rotational symmetries that can leave the atomic order invariant. In this way, the puzzle originally risen by the crystallographical restriction theorem was eventually clarified, allowing for new progresses in our understanding of the atomic arrangements in solids by the light of the notion of *aperiodic crystal*, as it was heralded by E. Schrödinger (1887–1961):

> "In physics we have dealt hitherto only with periodic crystals. To a humble physicist's mind, these are very interesting and complicated objects; they constitute one of the most fascinating and complex material structures by which inanimate nature puzzle his wits. Yet, compared with the aperiodic crystal, they are rather plain and dull". (Erwin Schrödinger, 1944) [789]

Accordingly, the International Crystallographic Union approved in April 1991, the establishment of a *Commission on Aperiodic Crystals* with the membership of J. M. Pérez-Mato (Chairman), G. Chapuis, M. Farkas-Jahnke, M. L. Senechal, and W. Steurer. According to the terms of reference introduced by this Commission:

> In the following by 'crystal' we mean any solid having an essentially discrete diffraction diagram, and by 'aperiodic crystal' we mean any crystal in which three-dimensional lattice periodicity can be considered to be absent [357].

Thus, QCs belong to the novel aperiodic crystals category,[10] whereas usual periodic crystals are now known as *classical* crystals. The revamped crystal definition reflects our current understanding that periodicity at the atomic scale is a sufficient but not necessary condition for crystallinity. Instead, the presence of a long-range atomic order *able to diffract* must be regarded as the essential attribute of crystalline matter rather than mere periodicity. Consequently, within the crystalline family we can now distinguish between periodic crystals, which display periodic arrangements of atoms, and aperiodic crystals, lacking such

---

[10]Along with incommensurate modulated crystals and incommensurate composites [381].

a periodicity, which is replaced by other kinds of symmetries, such as scale invariance (inflation symmetry) or long-range repetitiveness. Thus, the existence of a mathematically well defined long-range atomic order should be properly regarded as the *generic* attribute of ordered solid state matter.

In this context, it is convenient to highlight that, although both random structures and aperiodically ordered lattices lack strict translational symmetry, the *absence of periodicity* characteristic of aperiodic systems (e.g., QCs, incommensurate phases, or fractal structures) is not the same in nature as the *complete lack* of *long-range order* characteristic of amorphous matter. In fact, as we mentioned above, in aperiodic systems translational symmetry is replaced by other symmetry properties which amorphous matter does not possess.

The importance of the scale invariance property in defining the very QC notion is illustrated by the fact that one can consider the existence of QCs exhibiting rotational symmetries which are *allowed* by the classical restriction theorem (namely, 2-, 3-, 4-, and 6-fold). In fact, albeit the most striking symmetry feature of QCs at their discovery was the presence of 5-, 8-, 10-, or 12-fold rotation axes in their diffraction patterns, it is natural to question as to whether this is an essential condition for a solid to be properly considered as a QC. It is currently agreed that the presence of non-crystallographic axes is not a necessary condition for a solid to be regarded as a QC, and that the key feature to this end is just to exhibit scale invariance symmetry [381]. Indeed, several examples of QCs exhibiting 2-, 3-, and 4-fold symmetry axes *along with* scale invariance symmetry characterized by *irrational* scale factors have been reported, being referred to as "cubic QCs" [229]. These findings support the view that QC definition *should not include* the requirement that they must display a classically *forbidden* axis of symmetry, as it is stated in the *Online Dictionary of Crystallography* of the International Union of Crystallography, were one reads:

> *Often, quasicrystals have crystallographically 'forbidden' symmetries (...). However, the presence of such a forbidden symmetry is not required for a quasicrystal.*[11]

An illustrative example of 2D QC without forbidden symmetries is the so-called square Fibonacci tiling, with tetragonal point group symmetry (4 mm) [494]. This square can be generated by superposition of two Fibonacci line grids, which are orthogonal to each other. In this way we obtain an infinite set of lines whose inter-line spacing follows the Fibonacci sequence of short (unity) and long (τ) distances. The resulting two-dimensional QP structure is shown in Fig. 1.11a, and it consists of three different tiles: (a) a large square of dimensions $\tau \times \tau$, (b) a small square of dimensions 1×1, and (c) a rectangle of dimensions $1 \times \tau$ (see **Exercise 2.7**). A similar construct can be extended to obtain a cubic Fibonacci tiling in 3D (see Fig. 4.9) [579]. An experimental realization of the Fibonacci square grid structure was obtained by selectively depositing C_{60} fullerene molecules atop Mn atoms on the 2-fold surface of an i-AlPdMn QC, as it is shown in Figs. 1.11b–c [136].

1.1.6 The reciprocal space take-over

> *"It was a misfortune of intellectual history that the space-group classification scheme of crystallography was developed in real space to categorize periodic structures, rather than in Fourier space to categorize diffraction patterns consisting of sharp Bragg peaks"*. (N. David Mermin, 1992) [600]

---

[11]https://reference.iucr.org/dictionary/Quasicrystal. Last accessed on 10 January 2020. Notwithstanding this, it is fair to say that some authors consider that the term QC should denote crystals with diffraction patterns showing non-crystallographic symmetry axes in order to distinguish them from other representatives belonging to the aperiodic crystals class [840].

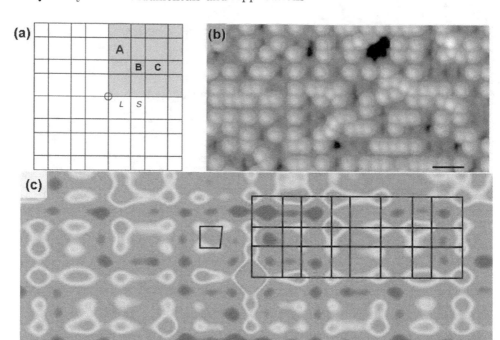

Figure 1.11   (a) The Fibonacci square grid with the constituent A, B, and C tiles. The center of rotation of 4-fold symmetry is marked by a circle. (b) Fullerene molecules deposited atop Mn atoms on the 2-fold surface of an i-AlPdMn single grain QC imaged by scanning tunneling microscopy. (c) Autocorrelation function of the C_{60} overlayer shown in (b). A distorted square is marked along with the Fibonacci square grid with $L = 2.04$ nm and $S = 1.26$ nm ($L/S = 34/21 \simeq 1.619$). (Adapted from Coates S, Smerdon J A, McGrath R, Sharma H R, A molecular overlayer with the Fibonacci square grid structure, Nature Commun. 9 3435 (2018), doi:10.1038/s41467-018-05950-7. CC NY-NC-ND license 4.0).

According to its very definition the diffraction pattern of an aperiodic crystal must mainly consist of sharp Bragg reflections, like those of periodic classical crystals, albeit lacking 3D periodicity. Thus, a fundamental aspect of the revamped definition of crystal is that the conceptual focus is shifted from direct space, where the atoms are physically located, to reciprocal space, where their underlying order is unveiled by means of diffraction phenomena.[12] In fact, diffraction patterns are closely related to the Fourier transform of the lattice atomic distribution $\rho(\mathbf{r})$, given by the expression

$$F(\mathbf{k}) = \int_V \rho(\mathbf{r}) e^{i\mathbf{k}\cdot\mathbf{r}} \, d\mathbf{r}, \tag{1.6}$$

---

[12]It is likely that the transition from X-ray crystal structure analysis to high resolution electron microscopy as a paradigmatic method for the investigation of crystal structures, had played a significant role in promoting this conceptual shift [568].

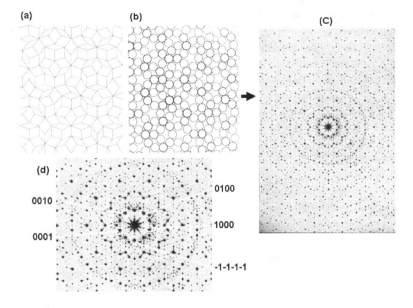

Figure 1.12   (a) A portion of the 2D Penrose tiling made of acute and obtuse rhombi. (b) Arrangement of circles centered at each vertex of a Penrose tiling (c) Optical diffraction pattern of this arrangement made by G. Harburn. The annular objects show in the circular strong and weak modulations of the pattern, which itself exhibits local 10-fold symmetry. (d) Magnification of the central region of the pattern shown in (c) with the indexing of the main spots in terms of four integers. (Reprinted from Physica A: Statistical Mechanics and its Applications, 114, Mackay A L, Crystallography and the Penrose pattern, 609–613, Copyright (1982), with permission from Elsevier).

where $\mathbf{k}$ denotes reciprocal space vectors and the volume integral extends over all the space. Thus, Fourier transform can be envisioned as a linear transformation relating two different mathematical domains: that corresponding to the space vector $\mathbf{r}$ where the atoms reside, and that corresponding to the $\mathbf{k}$ vectors related reciprocal space. The key question regarding the very nature of possible orderings of matter (either periodic or aperiodic ones) can then be formulated as follows: given a certain arrangement of atoms in terms of a distribution function $\rho(\mathbf{r})$, which are the necessary and sufficient conditions in order for its Fourier transform to exhibit an essentially discrete spectrum?

A significant step forward in order to solve this fundamental issue was given by Alan L. Mackay, who in 1982 reported that a distribution of circles arranged according to the 2D QP Penrose pattern was able to diffract (Fig. 1.12).

> "(...) Alan (Mackay) presented his optical diffraction pattern of Penrose tiling vertices, heralding the once-unthinkable, now ever-expanding research field of aperiodic order. A few months later, Dan Shechtman discovered AlMn crystals that had a strikingly Penrose-like diffraction pattern, and crystallography changed for ever". (Marjorie Senechal, 2017)[796]

The diffraction pattern shown in Fig. 1.12 displays local 10-fold axis, and thus represents a structure outside the formalism of classical crystallography, which was designated as a *quasi-lattice* by Mackay [566]. The quasilattice notion was introduced as an array of points in *reciprocal space*, where each point has integer indices with respect to a base of vectors and where the base dimension is larger than the dimensionality of the physical space [565]. For instance, in order to index the diffraction pattern obtained by Mackay one needs four integer indices. The four basis vectors are mapped on positions with integer indices under a $2\pi/5$ rotation (note the pentagonal arrangement of the peaks highlighted in Fig. 1.12d), as $\{\cos(2\pi n/5), \sin(2\pi n/5), \cos(4\pi n/5), \sin(4\pi n/5)\}$, $n = 1, \ldots, 4$. This gives a four-dimensional matrix, which on an orthogonal basis combines a two-dimensional rotation over $2\pi/5$ with another two-dimensional rotation over $4\pi/5$ [382]. Note that in the term quasilattice the prefix "quasi" is used to design a perfectly ordered structure in higher dimensions, not a defective or flawed one. Thus, although Mackay was completely aware of the perfectly ordered nature of the structure described in terms of the quasilattice notion, he did not use other possible terms, such as "super-lattice" or "hyper-lattice" to this end.[13] In 2010, Alan L. Mackay was awarded the Oliver E. Buckley Condensed Matter Prize from the American Physical Society for his *"pioneering contributions to the theory of quasicrystals, including the prediction of their diffraction pattern"*.

In 1987, it was noted by Rokhsar, Mermin, and Wright that if a QC diffraction pattern contains reflections corresponding to two wave vectors, say $\mathbf{k}_1$ and $\mathbf{k}_2$, one in general will find reflections corresponding to the sum, $\mathbf{k}_1 + \mathbf{k}_2$, and difference $\mathbf{k}_1 - \mathbf{k}_2$ as well, unless such reflections are forbidden by symmetry constraints. QCs, like periodic crystals, can therefore be indexed by a set of wave vectors that is closed under addition (sums and differences). For a periodic crystal this vector set reduces to the so-called reciprocal lattice, the set of wave vectors of plane waves with the periodicity of the direct lattice in physical space. For a QC this set still forms a lattice (in the sense that it is closed under addition), but unlike a periodic reciprocal lattice a quasicrystalline reciprocal lattice has no minimum distance between points, and the corresponding set of plane waves has no common real-space period. Although quasicrystalline wave vectors form a dense set in $\mathbf{k}$ space, a quasicrystalline diffraction pattern admits a countable indexing, unlike the diffraction pattern of an amorphous material [751].

It is thus reasonable to address the crystallographic classification of QCs from the reciprocal space viewpoint by cataloging the allowable lattices attending to the point group symmetry of the considered QC. These lattices are generated by *six* or *five* vectors for icosahedral and decagonal lattices, respectively, which are derived from the corresponding diffraction patterns. The icosahedral lattice can be defined in terms of a base of dimension 3 and rank 6 given by six vectors in the reciprocal space $\mathbf{a}_1^*$, $i = 1, \ldots, 6$, pointing from the center to the non-aligned vertices of an icosahedron (Fig. 1.13). The integral linear combinations of the $\mathbf{a}_i^*$ vectors define the points of the icosahedral lattice

$$\Lambda^* = \left\{ \sum_{i=1}^{6} n_i \mathbf{a}_i^* \mid n_i \in \mathbb{Z} \right\}, \tag{1.7}$$

---

[13] As a matter of fact, the term "super-lattice" had been previously introduced by Leo Esaki and R. Tsu in 1970, in order to describe an artificial periodic structure consisting of alternate layers of two different semiconducting materials, with layer thickness of the order of nanometers [235]. They called this synthetic structure a *superlattice* because periodic order appears at two different scale lengths: at the atomic level, we have the usual crystalline order determined by the periodic arrangement of atoms in each layer (lattice structure), whereas at longer scales, we have the periodic order determined by the sequential deposition of the different layers (superlattice structure).

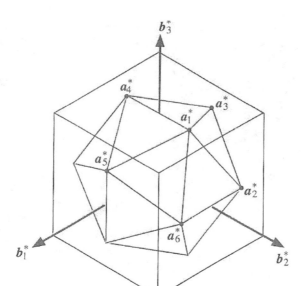

Figure 1.13   Perspective view of the two alternative reciprocal bases generally used to study icosahedral lattices in reciprocal space: the cubic and the icosahedral settings, represented by the bases $\mathbf{b}_i^*$ and $\mathbf{a}_i^*$, respectively. The Cartesian coordinate system is centered at the icosahedron center and oriented along three mutually orthogonal 2-fold axes of the icosahedron.

where the basis vectors $\mathbf{a}_i^*$ are given by

$$
\begin{pmatrix} \mathbf{a}_1^* \\ \mathbf{a}_2^* \\ \mathbf{a}_3^* \\ \mathbf{a}_4^* \\ \mathbf{a}_5^* \\ \mathbf{a}_6^* \end{pmatrix} = \begin{pmatrix} 0 & 1 & \tau \\ -1 & \tau & 0 \\ -\tau & 0 & 1 \\ 0 & -1 & \tau \\ \tau & 0 & 1 \\ 1 & \tau & 0 \end{pmatrix} \begin{pmatrix} \mathbf{b}_1^* \\ \mathbf{b}_2^* \\ \mathbf{b}_3^* \end{pmatrix},
\tag{1.8}
$$

in the $\mathbb{R}^3$ orthonormal basis $\mathbf{b}_i^*$, $i = 1, \ldots, 3$ [840]. Accordingly, the diffraction reflections can be labeled by six generalized Miller indices and the diffraction pattern is spanned by six linearly independent reciprocal lattice vectors. The space group classification of QCs belonging to the icosahedral point groups Y and Y_h within the reciprocal space framework was addressed by David Mermin and co-workers in 1988, who showed that taking into account the scaling properties of QCs is crucial for a correct enumeration of their related space groups [751].

1.2   MINERAL QUASICRYSTALS

> *"Had we found a crystal? Many definitions of crystals are in use, some have changed over the centuries. Our solid (...) can be grown to form 'a convex solid enclosed by symmetrically arranged plane surfaces, intersecting at definite and characteristic angles'. According to the latter of these older definitions, quasicrystals are crystals".* (John W. Cahn, 2001) [103]

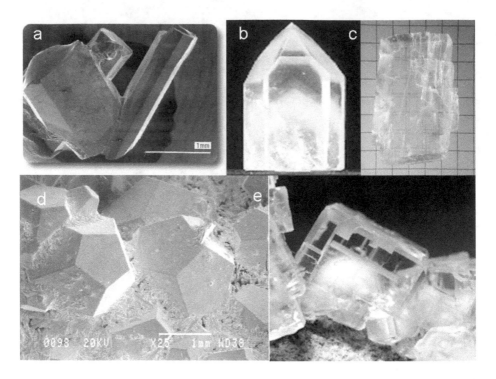

Figure 1.14   Periodic and QP crystals gallery. (a) d-AlCuCo QC exhibiting a decaprismatic growth habit, (b) Quartz (SiO_2) crystal exhibiting a hexagonal prismatic habit, (c) Calcite ($CaCO_3$) crystal exhibiting a hexagonal growth morphology, (d) i-AlCuRu QC exhibiting a dodecahedral growth habit, and (e) Fluorite (CaF_2) crystal exhibiting a cubic growth morphology (Maciá-Barber E, *Aperiodic Structures in Condensed Matter: Fundamentals and Applications* (CRC Press, Boca Raton, FL, 2009). With permission).

The hallmark of *mineral* crystals are regular polyhedral shapes and flat polygonal surfaces. According to these morphological criteria thermodynamically stable QCs, exhibiting pentagonal dodecahedral, triacontahedral or decaprismatic growth habits, should be properly regarded as typical mineral crystals by all standards. This point is illustrated in Fig. 1.14, were we compare different representatives of both periodic and QP crystal classes. As we can see, all samples nicely meet the phenomenological criteria for a piece of matter to be considered a full-fledged "crystal". Accordingly, had quasicrystalline alloys been found to spontaneously occur in nature from ancient times, then they would most probably have been regarded as another instance of mineral crystals representatives, for they exhibit all the basic features most of them show up [548]. However, QCs were not actually discovered from the study of a newly found mineral, but from the analysis of synthetic alloys produced in the laboratory. Notwithstanding this, if QCs are energetically stable and robust as periodic crystals, then it is conceivable that they formed under natural conditions, just like usual periodic crystalline minerals do. So, would it be possible that QCs were formed through natural geological processes long before they were discovered in the laboratory?

Following a decade-long search using power diffraction data included in the powder diffraction file published by the International Center for Diffraction Data (which includes over 80,000 patterns for 9,000 minerals in addition to synthetic inorganic and organic

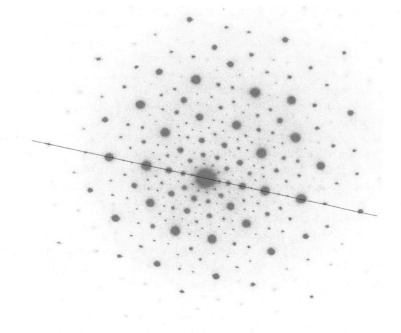

Figure 1.15 Electron diffraction pattern corresponding to the mineral i-$Al_{63}Cu_{24}Fe_{13}$ QC found in the Khatyrka meteorite[76]. (Courtesy of Luca Bindi).

phases), the first mineral with an icosahedral quasicrystalline structure was reported by L. Bindi, P. J. Steinhardt, N. Yao, and P. J. Lu in a paper entitled *"Natural Quasicrystals"*, published in *Science* on 5 June 2009. It was found in a rock within a meteorite collected from Khatyrka region of the Koryak mountains in the Chukotka oblast in the northeastern part of the Kamchatka peninsula (Russia), and deposited in the Mineralogical Collection of the Museo di Storia Naturale, Università di Firenze, Italy (catalogue number 46407/G) [67, 68]. In this sample several homogeneous grains, with a stoichiometry close to $Al_{63}Cu_{24}Fe_{13}$, were examined by transmission electron microscopy. The obtained diffraction patterns exhibited sharp Bragg reflections arranged in a QP self-similar pattern with 5-, 3-, and 2-fold symmetry axes: the unmistakable signature of iQCs (Fig. 1.15). This first mineral QC was officially accepted by the Commission on New Minerals, Nomenclature and Classification of the International Mineralogical Association (IMA) on 1 October 2010, and named *icosahedrite* (IMA No. 2010-42).

Interestingly, diffraction patterns demonstrated a high degree of structural perfection in icosahedrite, as compared with that observed in similar composition QCs produced in the laboratory (**Exercise 1.5**). The fact that icosahedrite shows no visible phason strain (see Sec. 1.4.5) means that either the mineral sample was originally formed without this sort of defects, or subsequent thermal processing was sufficient for such a strain to relax away [73]. Yet, icosahedrite grains were not formed under pristine conditions but rather intergrown in a complex aggregation with other metallic phases, such as khatyrkite ($CuAl_2$) and cupalite (CuAl), along with metallic aluminium, which requires highly reducing conditions to form, hence indicating that the formation of icosahedrite occurred under quite unusual environments [69, 835].

In 2015, a new mineral QC with decagonal symmetry and stoichiometry close to $Al_{71}Ni_{24}Fe_5$ (hence containing less (more) iron (nickel) than the d-$Al_{70}Ni_{15}Fe_{15}$ QC obtained in the laboratory) was also identified in ~60 μm sized grains belonging to the Khatyrka meteorite, and named *decagonite* by the IMA (No. 2015-017) [70, 71]. One year later, the occurrence of a third iQC mineral, with composition $Al_{62}Cu_{31}Fe_7$, was discovered in Khatyrka meteorite grains also containing icosahedrite [72]. Its stoichiometry, with a more Cu-rich/Fe-poor content than that observed in synthetic AlCuFe iQCs, located this sample outside the measured equilibrium stability range at standard pressure in the Al-Cu-Fe QC region phase diagram. About two years later, the first known mineral periodic crystalline approximant to decagonite was found in fragments from Khatyrka meteorite. This approximant, with chemical formula $Al_{59}Ni_{34}Fe_7$, does not correspond to any previously recognized synthetic or natural phase. This mineral was approved by the IMA (No. 2018–038) and officially named *proxidecagonite* [74].[14] Therefore, the definition of mineral, that previously included periodic crystals, incommensurate structures, and amorphous phases, should henceforth include QCs as well, expanding the catalog of structures formed by nature and raising an interesting challenge to explain how they formed naturally [67].

Although mineral QCs naturally formed *on earth* have not been reported to date, it is quite conceivable that some of these minerals may exist on our planet. In fact, synthesis experiments and crystallochemical analyses suggest that a number of minerals could be transformed into QCs under extreme pressure and temperature conditions usually occurring in geological processes [28, 83, 279]. Hence, searching for such mineral QCs of terrestrial origin is relevant to the investigation of the formation conditions of aperiodic crystals in nature and would constitute a new contribution coming from the mineral kingdom to the advancement of crystallography. This appealing issue has recently been addressed by several authors, who have discussed current investigations aimed at the search for new possible quasicrystalline minerals in nature [73, 700].

What is the significance of the discovery of naturally formed QCs? In geology this discovery has opened a new chapter in the study of mineralogy, altering the conventional classification of mineral forms. In condensed-matter physics, the discovery has pushed back the age of the oldest QCs by orders of magnitude and has had an important impact on our view of how difficult it is for QCs to form, allowing to study QCs stability over annealing times and physical conditions not accessible in the laboratory [73]. Indeed, icosahedrite and decagonite samples found in the Khatyrka CV3 carbonaceous chondrite meteorite are suspected to have formed from materials originally present in a parent minor body (maybe the asteroid 89 Julia) during the early solar system stages (~4.564 Gy ago), in which the first AlCuFe alloys formed and subsequently experienced a more recent (i.e., about 600 million years) impact-induced shock event able to create a heterogeneous distribution of high pressure and temperature, followed by rapid cooling conditions leading to the formation of both i-$Al_{62}Cu_{31}Fe_7$ grains and the high-pressure mineral phases found in the meteorite [597]. These findings certainly settle the fundamental issue about QCs as a form of matter that is effectively stable over geologic time scales.

---

[14]It has an orthorhombic structure (space group Pnma) with parameters $a = 29.013$, $b = 8.156$, and $c = 12.401$ Å.

1.3   HIERARCHY AND QUASIPERIODICITY

> "*Hierarchy could offer an alternative to lattice repetition, in providing an assembly of atoms with an infinite number of almost identical, or quasi-equivalent, sites*". (Alan Mackay, 1982) [566]

In the previous sections, we have learnt that both synthetic and mineral QCs are characterized by the presence of novel symmetries which periodic crystals lack, namely, scale invariance (which replaces translation symmetry) and long-range QPO, which distinguishes QCs from those simpler fractal structures exhibiting a rational, instead of an irrational, scale factor.[15] In this Section, we will study these new ordering principles, based on hierarchical designs, in more detail.

1.3.1   Scale invariance symmetry

Let S_λ denote the scale transformation operation defined as $S_\lambda f(\mathbf{r}) = f(\lambda \mathbf{r})$, so that a scaling transformation with scaling factor $\lambda \in \mathbb{R}^+$ transforms by λ-multiplication all the components of vector $\mathbf{r} \in \mathbb{R}^N$. Then, a function $f(\mathbf{r})$ is said to be scale invariant when $S_\lambda f(\mathbf{r}) = \mu f(\mathbf{r}), \forall \lambda \in \mathbb{R}^+$, where $\mu \in \mathbb{R}^+$. For instance, monomial functions in $\mathbb{R}$, $f(x) = x^n$, satisfy the scale invariance property

$$S_\lambda f(x) \equiv f(\lambda x) = (\lambda x)^n = \lambda^n f(x),$$

where $\mu \equiv \lambda^n$, which is characteristic of the so-called homogeneous functions of degree n. A representative example of scale invariant functions in 2D are provided by the Archimedean, $r = a\varphi$, Fermat (also known as parabolic), $r = a\sqrt{\varphi}$, or logarithmic, $r = ae^{b\varphi}$, spiral curves in polar coordinates, which can be respectively expressed as

$$\varphi_1(r) = \frac{r}{a}, \quad \varphi_2(r) = \left(\frac{r}{a}\right)^2, \quad \varphi_3(r) = b^{-1}\ln\left(\frac{r}{a}\right),$$

where $a > 0$ has the dimension of a length and $b > 0$ is a dimensionless parameter. By applying the scale transformation we have

$$S_\lambda\varphi_1(r) = \frac{\lambda r}{a} = \lambda\varphi_1(r), \quad S_\lambda\varphi_2(r) = \left(\frac{\lambda r}{a}\right)^2 = \lambda^2\varphi_2(r),$$

$$S_\lambda\varphi_3(r) = b^{-1}\ln\left(\frac{\lambda r}{a}\right) = b^{-1}\ln\lambda + b^{-1}\ln\left(\frac{r}{a}\right) = \varphi_0 + \varphi_3(r).$$

Thus, the original $\varphi_1(r)$ and $\varphi_2(r)$ spiral curves are simply scaled by the factors $\mu_1 = \lambda$ and $\mu_2 = \lambda^2$, respectively, whereas the transformed logarithmic spiral describes a *rotated* version of the original curve by an angle $\varphi_0 = b^{-1}\ln\lambda$ with respect to the adopted reference frame.[16] All these examples illustrate *global* scale invariant systems, where the scaling factor can take on any $\mu \in \mathbb{R}^+$ value.

It may happen that a system is scale invariant only for an enumerable set of scaling factors $\{\lambda_n\}$, $n \in \mathbb{N}$. Thus, discrete scale invariance is a sub-symmetry of continuous scale invariance. In this case, one usually talks of *self-similarity*, characterized by the invariance of the structure under the dilatations set $\{\lambda_n\}$. For instance, in the iQC diffraction patterns

---

[15]One can also define Cantor-like structures exhibiting irrational scale factors.

[16]Indeed, by applying the rotation $\varphi' = \varphi_0 + \varphi$ to $r = ae^{b\varphi}$, we get $r = e^{b\varphi_0}ae^{b\varphi} = e^{b\varphi_0}r \equiv \lambda r \Rightarrow \lambda = e^{b\varphi_0}$, so that the scale factor is determined by the rotation above.

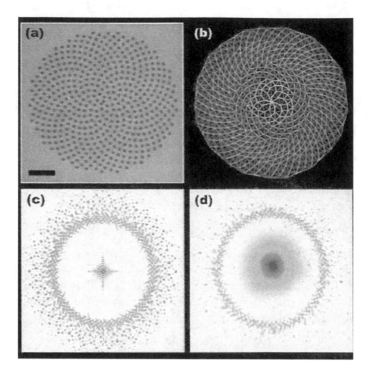

Figure 1.16 (a) Scanning electron microscope image of a sunflower fabricated by electron beam lithography containing 500 holes of radius $r = 0.539$ μm. The pattern, highlighting the so-called parastichies spiral curves, has a strong modal nearest-neighbor pitch $a = 2.2$ μm derived from a Delaunay triangulation of the points set shown in (b). The number of parastichies in each rotational sense are consecutive numbers in the Fibonacci series. The calculated and experimental diffraction patterns are shown in (c) and (d), respectively [702]. (Courtesy of Michael E. Pollard).

shown in Figs. 1.1–1.7 we can see conspicuous arrays of bright reflections along the radial directions, whose relative distances from the origin obey the geometric series $\lambda_n = \lambda_0^n$, with a ratio given by the golden mean for icosahedral and decagonal QCs, or $\lambda_0 = 2 + \sqrt{3}$ for ddQCs. In this case, the discrete nature of the self-similar symmetry operation stems from the fact that the denser diffracting atomic planes in the QC solid follow a discrete succession of layers.

On the other hand, one can also find discrete lattices exhibiting a global scale invariance symmetry, rather than a discrete one. An illustrative example is given by the so-called spiral lattices, which provide an interesting example of perfectly ordered systems where both translational and orientational symmetries are absent. These lattices are based on the application of a simple mathematical algorithm. We start from a *generating* spiral *curve* in polar coordinates, such as the previously considered Archimedean ($r = a\varphi$), parabolic ($r = a\sqrt{\varphi}$), or logarithmic curves ($r = ae^\varphi$). The spiral *lattice* is then obtained by restricting the possible values of r and φ according to a quantization condition of the form [101]

$$r_n = an^\nu, \qquad \varphi_n = \phi_d n, \quad n = 0, 1, 2..., \tag{1.9}$$

where $a > 0$ and ν are real numbers, and ϕ_d is also a real number measuring the angle between adjacent radius vectors r_n and r_{n+1}, which is referred to as the divergence angle.

Many spiral lattices generated in this way exhibit arrangements analogous to those observed in many botanical structures [746]. For instance, the so-called Vogel's spirals, which describe the arrangement of florets in the sunflower head, are obtained when $\nu = 1/2$ in Eq. (1.9) and are characterized by self-similar inflation or deflation operations, like those observed in QCs (**Exercise 1.6**) [544]. In Fig. 1.16a, we show a 2D parabolic spiral lattice derived from the equation

$$x_n = a\sqrt{n}\cos(n\phi_d), \quad y_n = a\sqrt{n}\sin(n\phi_d), \tag{1.10}$$

where $a \in \mathbb{R}^+$ is a scaling factor. Adopting irrational values for ϕ_d we obtain lattices entirely lacking rotational symmetry. Accordingly, their diffraction patterns do not show sharp reflection spots, but rather they feature broad and diffuse patchy rings, whose spectral features are determined by the particular spiral lattice geometry [155, 973]. For the sake of illustration, the calculated and experimental diffraction patterns of a Vogel spiral lattice with a divergence angle $\phi_d = 2\pi/\tau^2 \simeq 137.508°$ (referred to as the golden angle), are shown in Fig. 1.16c and 1.16d, respectively. A close inspection to these diffraction patterns reveal a finer structure in the concentric ring. Using analytical Fourier-Hankel decomposition it was found that the Fourier transform of a Vogel spiral density given by

$$\rho(r, \varphi) = \sum_{n=1}^{N} \frac{1}{r}\delta(r - a\sqrt{n})\delta(\varphi - n\phi_d), \tag{1.11}$$

can be written as [153],

$$f_m(k_r) = \frac{1}{2\pi}\sum_{n=1}^{N} J_m(k_r a\sqrt{n})e^{inm\phi_d}, \tag{1.12}$$

where $k_r = 2\pi q_r$ is the radial wave vector associated to the radial spatial frequency q_r and J_m is the m-th order cylindrical Bessel function of the first kind. The key result is that the exponential factor in Eq. (1.12) will contribute with azimuthal peaks when the product $m\phi_d$ is an integer. Since ϕ_d is an irrational number this will never occur exactly, but the situation can be reasonably fulfilled by the rational approximants of ϕ_d. Therefore, for spirals generated using an arbitrary irrational number ϕ_d azimuthal Bragg reflections of order m will appear in the Fourier spectrum due to the denominators of these rational approximants [153].

Let us conclude with a higher dimensional example by considering the six vectors $\{\mathbf{a}_i\}$, $i = 1,\ldots,6$, pointing from the center to the non-aligned vertices of an icosahedron. The integral linear combinations of $\{\mathbf{a}_i\}$ define the points of an icosahedral lattice $\Lambda = \left\{\sum_{n=1}^{6} n_i \mathbf{a}_i \mid n_i \in \mathbb{Z}\right\}$ in physical space. In the orthonormal basis $\{\mathbf{e}_1, \mathbf{e}_2, \mathbf{e}_3\}$ of $\mathbb{R}^3$ the vectors $\{\mathbf{a}_i\}$ can be chosen as

$$\begin{pmatrix} \mathbf{a}_1 \\ \mathbf{a}_2 \\ \mathbf{a}_3 \\ \mathbf{a}_4 \\ \mathbf{a}_5 \\ \mathbf{a}_6 \end{pmatrix} = \begin{pmatrix} 1 & 0 & \tau \\ \tau & 1 & 0 \\ 0 & \tau & 1 \\ -1 & 0 & \tau \\ 0 & -\tau & 1 \\ \tau & -1 & 0 \end{pmatrix} \begin{pmatrix} \mathbf{e}_1 \\ \mathbf{e}_2 \\ \mathbf{e}_3 \end{pmatrix} \rightarrow \tilde{\mathbf{a}} = \tilde{\mathbf{B}}\tilde{\mathbf{e}}. \tag{1.13}$$

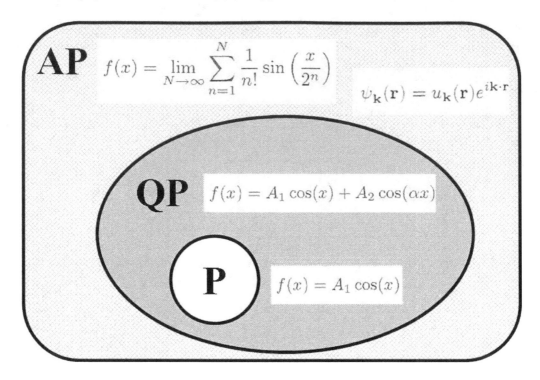

Figure 1.17 Graphical representation of the hierarchically nested relationship among almost periodic (AP), quasiperiodic (QP) and periodic (P) functions. Illustrative examples are given.

In the icosahedral basis $\{\mathbf{a}_i\}$ the scaling transformation $\mathbf{S}_\tau$ is then represented by the 6D matrix (**Exercise 1.7**) [374]

$$\mathbf{S}_\tau = \frac{1}{2}\begin{pmatrix} 1 & 1 & 1 & 1 & 1 & 1 \\ 1 & 1 & 1 & -1 & -1 & 1 \\ 1 & 1 & 1 & 1 & -1 & -1 \\ 1 & -1 & 1 & 1 & 1 & -1 \\ 1 & -1 & -1 & 1 & 1 & 1 \\ 1 & 1 & -1 & -1 & 1 & 1 \end{pmatrix}. \tag{1.14}$$

1.3.2 Almost periodic and quasiperiodic functions

"Periodic solids give discrete diffraction, but we did not know then that certain kinds of aperiodic objects can also give discrete diffraction; these objects conform to a mathematical concept called almost- or quasi-periodicity". (John W. Cahn, 2001)[103]

"The theory of almost periodic functions has from its beginning attracted the strong interest of many mathematicians because of its inner beauty, its diverse connections with various branches of mathematics, and its many applications to celestial mechanics, physics, and engineering". (M. A. Shubin, 1978) [803]

In the previous sections, we have seen that QCs are pretty well ordered structures. In this section, we will learn that QCs are actually endowed with a *higher rank* of structural order than classical periodic crystals have [544]. This feature can be readily grasped by considering the hierarchical relationship between almost periodic (AP), QP and periodic functions shown in Fig. 1.17. In fact, from a mathematical viewpoint periodic functions are just a special case of QP functions which are, in turn, a special case of AP functions.[17] This nested relationship naturally suggests that it is more insightful to think of different *hierarchies of order* in the arrangements of matter [555], rather than maintaining the dichotomous vision which regarded periodic lattices as the paradigm of ordered matter, whereas random atomic distributions represented that of amorphous matter. Indeed, the importance of adopting a hierarchical ordering principle to properly describe well ordered structures beyond classical crystals was earlier heralded by John D. Bernal (1901–1983) [54], and this approach has been properly formulated, within the framework of the so-called generalized crystallography, by Alan L. Mackay in his long-term search for a wide-scoped science of atomic structures [111, 322, 564, 567, 569, 571].

Almost periodic functions were introduced as a generalization of periodic ones, and Bohr's original methods for establishing the fundamental results of the theory were always based on reducing the issue to a problem of purely periodic functions [57, 81, 82]. For the sake of clarity, let us start by first recalling some mathematical notions. A continuous bounded function $f(\mathbf{x})$ on $\mathbb{R}^n$ is called almost periodic if, for any arbitrary small number $\varepsilon > 0$, there exists a so-called almost-period $\mathbf{P} \in \mathbb{R}^n$ such that the shifted function differs less than ε from the unshifted one, namely, $|f(\mathbf{x} + \mathbf{P}) - f(\mathbf{x})| < \varepsilon, \forall \mathbf{x} \in \mathbb{R}^n$ [381]. In general, the smaller the value of ε, the larger becomes the required translation $\mathbf{P}$, although they are relatively dense in $\mathbb{R}^n$. By this we mean that, for each ε, there are two real values R_1 and R_2 ($R_2 > R_1$) such that every sphere of radius R_2 contains at least one almost-period $\mathbf{P}$ satisfying the above condition, and in every sphere of radius R_1 around any translation $\mathbf{P}$ satisfying the condition, there is no other translation but $\mathbf{P}$. Hence, when considering AP functions we must deal with many suitable almost periods rather than just one, as occurs for periodic functions. Physically, this means that in the study of aperiodic structures one must consider more than one relevant physical scale. The concept of AP function can straightforwardly be extended to $\mathbb{C}$-valued functions, the so-called Bloch functions being a very important representative in solid state physics (see **Exercise 2.14**).

Almost periodic functions can be uniformly approximated by Fourier series containing a countable infinity of pairwise incommensurate reciprocal periods [381]. More precisely, for a given almost periodic function $f(\mathbf{x})$ there exists a countable subset Λ_f of $\mathbb{R}^n$ such that

$$f(\mathbf{x}) = \sum_{\lambda \in \Lambda_f}^{\infty} A_\lambda e^{i\lambda \cdot \mathbf{x}}, \tag{1.15}$$

that is, the series converges uniformly towards $f(\mathbf{x})$, where A_λ can be real or complex. The elements $\{\lambda\}$ of the subset Λ_f are referred to as the set of reciprocal periods of $f(\mathbf{x})$. Two illustrative examples of AP function in $\mathbb{R}$ and $\mathbb{C}$ are given in Fig. 1.17. From an empirical viewpoint we have so far not been concerned with almost periodicity, since in any experiment the dimension of reciprocal space is less than or equal to the number of observed reflections, and thus is finite [103].

When the set of reciprocal periods required can be generated from a finite set, instead of a countable infinite one, the resulting function is referred to as a QP one, and Eq. (1.15) takes the form

$$f(\mathbf{x}) = \sum_{\nu=1}^{m} A_v e^{i\lambda_v \cdot \mathbf{x}}. \tag{1.16}$$

---

[17]The theory of almost periodic functions was developed in the 1920s by Harald Bohr (1887–1951), brother of the well-known physicist Niels Bohr (1885–1962) [82].

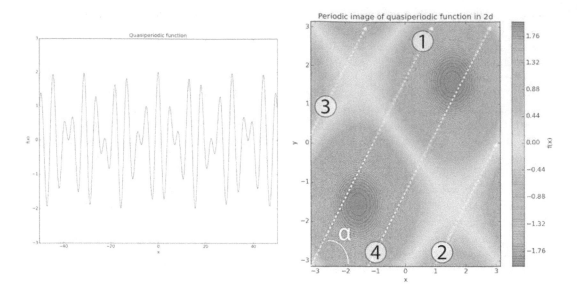

Figure 1.18 On the left we plot the 1D QP function $f(x) = \sin x + \cos(\sqrt{2}x)$. On the right the periodic image of $f(x,y) = \sin x + \cos y$ is ploted in the 2D XY plane. Moving along the lines labeled $1 \to 2 \to 3 \to 4$ one gets the QP profile shown on the left panel. The slope of these parallel lines are equal to $\sqrt{2}$, which measures the ratio between the two incommensurate periods of the QP function. (Adapted from Blinov I V, Periodic almost-Schrödinger equation for quasicrystals. Sci. Rep. 5, 11492; doi: 10.1038/srep11492 (2015). Work licensed under a Creative Commons Attribution 4.0 International License).

The simplest one-dimensional example of a QP function can be written as (see Fig. 1.17).

$$f(x) = A_1 \cos(x) + A_2 \cos(\alpha x), \tag{1.17}$$

where α is an irrational number and A_1 and A_2 are real numbers. It is interesting to note that this QP function can be obtained as a one-dimensional *projection* of a related *periodic* function in *two* dimensions

$$f(x, y) = A_1 \cos x + A_2 \cos y, \tag{1.18}$$

through the restriction $y = \alpha x$. An illustrative example is shown in Fig. 1.18 for the QP function $f(x) = \sin(x) + \cos(\sqrt{2}x)$ (**Exercise 1.8**). This property is at the basis of the so-called cut-and-project method, which is widely used in the study of QCs. In fact, since any QP function can be thought of as deriving from a periodic function in a space of higher dimension, most of the basic notions of classical crystallography can be properly extended to the study of QCs by considering suitable lattices in appropriate hyperspaces [381, 840]. as we will describe in Sec. 1.4.1. According to their very definition, QCs have an atomic arrangement which can be described by a sum of two or more periodic functions whose periods are mutually incommensurate. This fundamental property is illustrated in **Exercise 1.9** for the case of a 2D QC obtained by placing dielectric pillars at certain positions given by the maxima of a multi-beam interference holographic pattern. This structure is an example of a man-made photonic QC, which will be thoroughly described in Chapter 4.

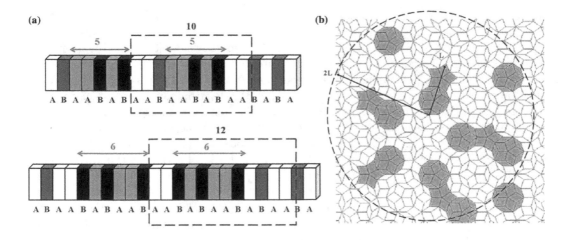

Figure 1.19 (a) llustration of Conway's theorem in a Fibonacci multilayer: given an arbitrary sequence of layers (highlighted in the picture) one will always find a replica of it at a distance smaller than twice its length (dashed boxes). (b) Penrose tiling illustrating the local isomorphism property prescribed by Conway's theorem in two dimensions. (Adapted from Maciá E, Exploiting aperiodic designs in nanophotonic devices, Rep. Prog. Phys. **75**, 036502 (2012); © IOP Publishing. Reproduced with permission. All rights reserved).

1.3.3 Local isomorphism

> *"Conway had proved much of what could be proved from the tiling's scaling, or substitutions, properties, including the fact that every patch of tiles, of every size, appears 'everywhere' in every Penrose tiling"*. (Marjorie Senechal, 2008) [795]

Due to the presence of unavoidable imperfections along with the limited resolution of instruments, the diffraction patterns of actual QC samples, albeit highly populated, will always display a finite number of reflections. As an illustrative example we can consider the electron diffraction pattern shown in Fig. 1.7 for a dQC of high structural quality. Thus, by attending to the relative intensities of the reflections constellating the diffraction pattern, one can readily appreciate the existence of a set of spots which dominate the diffraction pattern, which could be taken to construct a quasi-Brillouin zone.[18] Notwithstanding this, although the number of diffraction peaks one can appreciate in any actual experiment is finite in number, one should keep in mind that the diffraction pattern of an *ideal* QP distribution of atoms would exhibit an infinite number of diffraction reflections, since densely filled diffraction patterns can be regarded as a fingerprint of long-range QPO in the system.

The physical reason for the presence of so many diffraction reflections arises from the fact that QPO guarantees the existence of suitable interference conditions at multiple scales. In the mathematical literature this important feature is referred to as *local isomorphism* and its significant role in interference processes can be described in terms of a theorem put forward by John H. Conway (1937–2020), stating that distances between identical local arrangements of atoms scale with their own characteristic size [275]. That is, any finite-size region

---

[18]However, at variance with the incommensurately modulated phases case, we cannot use these reflections to define an average structure compatible with translation symmetry.

reappears again and again in a non-periodic fashion, but always having slightly different surroundings. Thus, the notion of translation symmetry, typical of periodic arrangements, should be replaced by that of local isomorphism or repetitiveness [32], which expresses the occurrence of any bounded region of the structure infinitely often across the whole volume. In the particular cases of Fibonacci chain and Penrose tiling, Conway's theorem states that given any local pattern having a certain characteristic length, say ℓ, at least one identical pattern can be found within a distance of 2ℓ, as it is illustrated in Fig. 1.19.

As we will see in Sec. 2.2 the existence of critical wave functions with a power law decay in QCs is closely related to the combined effects of local isomorphism and long range scale invariance. Thus, the absence of periodicity favors the localization of the wave function in a local pattern. However, according to Conway's theorem this local pattern must have duplicates that extend throughout the whole structure. Then, quantum resonances between such localized states, at identical local configurations, favor a tunneling effect which gives rise to a hopping mechanism. As a result, the wave functions become more extended in nature, ultimately resulting in electronic and phonon states exhibiting a self-similar structure, in the sense that certain portions of the wave function amplitudes extending over identical configurations only differ by a scale factor [37].

1.4 CRYSTALLOGRAPHY IN THE HYPERSPACE

"The aim of the present paper is to stress that quasicrystals are three-dimensional drawings by nature of higher-dimensional objects, which evoke in our minds the presence of still more symmetries and harmonies". (Aloysio Janner, 1991)[373]

"All types of aperiodic crystals (excluding almost periodic structures) can be described entirely by the same n-dimensional approach. Crystallography of quasicrystals, therefore, corresponds mainly to crystallography extended to n dimensions". (Walter Steurer and Torsten Haibach, 1999)[837]

1.4.1 The cut-and-project method

The so-called superspace formalism envisions aperiodic crystals as *projections* of higher-dimensional *periodic* structures onto $3D$ physical space. In this way, the notion of periodic order is recovered at the cost of working in higher dimensions. When discussing projection methods from superspace, one must distinguish if the considered high dimensional space describes the reciprocal lattice, directly probed by the diffraction experiments, or the direct lattice where the diffracting atoms are located. Since crystallographers are mainly interested in solving an indexation problem, they naturally address the reciprocal lattice viewpoint in the first place, and subsequently try to understand the obtained indexed pattern in terms of a suitable distribution of diffracting centers in physical space [59, 600]. Thus, it was first shown by Pim M. de Wolff (1911–1998) in 1974 that the diffraction patterns of incommensurate modulated structures in $3D$ can be described as projections of appropriate $4D$ reciprocal lattices onto physical space [172], in such a way that the direct space incommensurate modulated structure results from cutting $4D$ periodic hypercrystals with physical space. Following this approach a global construction of the Penrose tile, interpreted as the projection of a $5D$ lattice structure into a $2D$ subspace, was subsequently given by Nicolaas Govert de Bruijn (1918–2012) in 1981 [167].

In the case of iQCs the superspace periodic lattice is defined in $6D$ and it is decomposed into two $3D$ subspaces: the so-called *physical* (or parallel) space, $E_\parallel$, and the *perpendicular* (or internal) space, $E_\perp$. The QP atomic arrangement in physical space is obtained by an irrational cut through the $6D$ periodic hyperlattice properly decorated with $3D$ objects, lying in $E_\perp$, called occupation domains (or atomic surfaces), which describe the local atomic environment in the physical space. These domains in higher dimensional space are described in terms of *finite* lines (in $4D$), surfaces (in $5D$) or volumes (in $6D$), respectively, which are centered in the hyperlattice points. The main goal of the projection method in the general case is to construct a QP lattice in a physical subspace $E_\parallel$ (of dimension $D < N$) of a Euclidean space $\mathbb{R}^N$ by projecting the vertices of a hypercubic lattice L onto $E_\parallel$ (such that $E_\parallel \oplus E_\perp = \mathbb{R}^N$), so that $E_\perp$ has dimension $N - D$. The canonical projection method considers a hypercube in $\mathbb{R}^N$, centered at the origin, and considers a set $W \subset \mathbb{R}^N$ which defines the so-called *projection window*. This window is introduced in order to account for a condition of minimum distance between atoms in physical space. The quasilattice Q in the 3D physical space $E_\parallel$ is achieved by projecting onto $E_\parallel$ all the points of the hypercubic lattice L inside the projection window, that is $Q \equiv \pi(L \cap W)$ where π denotes a suitable projection operator. A convenient choice for the projection window is when it spans exactly one unit cell of the hyperlattice L and $E_\parallel$ coincides with the lower boundary of the window. The lattice points in $\mathbb{R}^N$ which lie inside the volume $W \times E_\parallel$ are then projected onto $E_\parallel$.[19]

In this way, iQCs can be regarded as projections of decorated hypercubic lattices in $\mathbb{R}^6$ (equipped with the canonical base $\{\mathbf{e}_1, \mathbf{e}_2, \mathbf{e}_3, \mathbf{e}_4, \mathbf{e}_5, \mathbf{e}_6\}$) into the physical $\mathbb{R}^3$ space.[20] In the $6D$ Euclidean space there are three Bravais lattices compatible with the icosahedral point group symmetry $2\bar{3}5/m$, namely, simple (P), face-centered (F), and body-centered (I) cubic lattices, respectively [751]. AlLiCu, ZnMgGa, CdYb, ZnScMg, or AgInYb iQCs are of the P type, whereas iQCs belonging to the systems AlCuTM, AlPdTM, ZnMgRE, or CdMgRE are of F type (both types having point group symmetry $m\bar{3}\bar{5}$). QCs belonging to the I type (with $2\bar{3}\bar{5}$ point group symmetry) have not yet been observed. A primitive, P-type lattice in $\mathbb{R}^6$ is given by

$$L_P = \left\{ \sum_{i=1}^{6} n_i \mathbf{e}_i \mid n_i \in \mathbb{Z} \right\}, \tag{1.19}$$

so that all vectors with integral coefficients appear. A face-centered cubic lattice in $\mathbb{R}^6$ is defined as

$$L_F = \left\{ \sum_{i=1}^{6} n_i \mathbf{e}_i \mid \sum_{i=1}^{6} n_i = 0 \ (\mathrm{mod}2) \right\}, \tag{1.20}$$

so that the sum of the coefficients of a given vector is even. Finally, a body-centered cubic lattice in $\mathbb{R}^6$ is defined as the lattice dual to L_F, and is given explicitly by

$$L_I = \left\{ \sum_{i=1}^{6} n_i \mathbf{e}_i \mid \sum_{i=1}^{6} n_i = 0 \ \text{or} \ \sum_{i=1}^{6} n_i = 1 \ (\mathrm{mod}2) \right\}, \tag{1.21}$$

so that all coefficients of a given vector have the same parity, that is, either all even or all odd [16]. The three icosahedral lattices are known to exhibit the following relationships of scale invariance: $\tau^3 L_P = L_P$, $\tau L_F = L_F$, and $\tau L_I = L_I$, where τ is the golden mean. In this way, the primitive and face-centered iQCs can be experimentally distinguished by considering their twofold electron diffraction patterns and measuring the relative distances

---

[19]Note that cut-and-project tilings generally do not possess an inflation symmetry, and inflation tilings generally do not allow an embedding into a higher dimensional lattice with a bounded window [795].

[20]A $6D$ hypercube has $2^6 = 64$ corners, 192 edges, 240 $2D$ faces, 160 $3D$ faces (cells), 60 $4D$ faces, and 12 $5D$ faces.

of the reflections along the fivefold direction: the distance between these reflections shows an inflation by τ^3 for the P-type iQCs and by τ for the F-type ones [909]. The classification of QCs is then based on the point and space groups in higher dimensions. To this end, the restriction that the projection of the nD ($n > 3$) point symmetry group onto $3D$ physical space has to be isomorphous to the point group of the $3D$ structure decreases the number of relevant symmetry groups drastically. For instance, out of 7104 point groups in a $6D$ hyperspace there are only 2 groups for QCs with icosahedral diffraction symmetry [840].

As we previously mentioned, in the framework of the hyperspace crystallography the $6D$ unit cell is decorated by $3D$ objects known as atomic surfaces. Thus, binary and ternary iQCs are usually described in terms of several atomic surfaces associated with the various atomic species and their geometric environments. In doing so, one generally requires an infinite number of continuous parameters to be characterized, thus ruining the hope of describing *ab initio* a QC structure with a finite amount of data. To this end, additional constraints have to be introduced to reduce the number of parameters to a finite set, which should be inferred from plausible physical condition. The first constraint that can be introduced is structural simplicity: like in periodic crystals the structure should show finite local complexity in the sense that each atom has a finite number of different local environments to any finite size. This requirement implies the atomic surfaces are $N - D$ volumes piecewise parallel to $E_\perp$. For instance, in the $6D$ description the atomic surface is a rhombic triacontahedron [115]. The second requirement is based on the idea that QCs can be viewed as possible ground states for some hypothetical finite-range interaction Hamiltonians (see Chapter 2). This leads to the existence of certain matching rules leading to enforce the projection strip to be oriented in a unique way in the embedding hyperspace. The third constraint assumes that the atomic jumps necessary to locally restore the ideal QPO in actual QCs with phasonic strain should be of low energy, so that these jumps occur on small or very small distance. All together these constraints suggest constructing atomic surfaces for the icosahedral phase as polyhedra resulting from intersections, unions and/or rescaling by any number of the form $(n + m\tau)/d$, $n, m \in \mathbb{Z}$, $d \in \mathbb{N}^+$ of the eight polyhedra defined by the elementary tessellation under the point group $m35$ of the mirror planes perpendicular to the 2-fold axes [306].

In order to illustrate the essential features of the cut-and-project method let us consider the square lattice $\mathbb{Z}^2$ of points having integer Cartesian coordinates (n_x, n_y) playing the role of the hyperlattice L with a unit cell parameter a_L. Rotated by an angle φ with respect to the square lattice coordinate system XY, we introduce another orthogonal coordinate system with axes labeled $E_\parallel$ and $E_\perp$, as it is shown in Fig. 1.20. The coordinates of the square lattice points orthogonally projected onto the physical space line $E_\parallel$ can be obtained from the expression[21]

$$\mathbf{x}_\parallel = \begin{pmatrix} 1 & 0 \\ 0 & 0 \end{pmatrix} \begin{pmatrix} \cos\varphi & -\sin\varphi \\ \sin\varphi & \cos\varphi \end{pmatrix} \begin{pmatrix} n_x a_L \\ n_y a_L \end{pmatrix}, \tag{1.22}$$

leading to a series of points along the space line $E_\parallel$ given by the relationship $x_\parallel(n_x, n_y) = (n_x \cos\varphi - n_y \sin\varphi)a_L$, which explicitly depends on the Cartesian coordinates of the point in the original square lattice XY. In general, if the slope of the physical space line with respect to the X axis, $\tan\varphi$, is a rational number one obtains a periodic distribution of projected lattice points along the physical space. On the contrary, if the slope is an irrational number one gets an aperiodic sequence of points along the $E_\parallel$ line, determining an aperiodic series of line segments of different length between successive points. Only in the case of $\tan\varphi$ being a quadratic irrational number will this sequence of segments be a QP one [231, 232, 486].

---

[21]We note that the projection and rotation matrices in Eq. (1.22) do not commute in general.

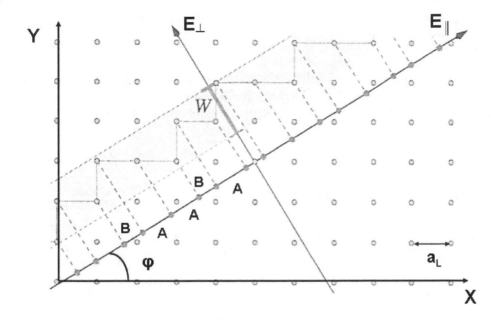

Figure 1.20   Construction of the Fibonacci chain through the cut-and-project method. The slope of the $E_\parallel$ axis equals the reciprocal of the golden mean ($\tan\varphi = \tau^{-1}$). In this diagram the projection window W defines the so-called acceptance domain, since only those lattice points inside the strip of width W are projected onto $E_\parallel$ (dashed lines).

The window strip condition imposes certain restrictions on the allowed coordinate values (n_x', n_y') for the nearest neighbors of a given (n_x, n_y) lattice point. For instance, in the particular case $\tan\varphi = \tau^{-1}$, the distance between two consecutive points along the $E_\parallel$ line can take on two different values only, namely, $|n_x' - n_x| = \{1,0\}$ and $|n_y' - n_y| = \{0,1\}$, so that we obtain a series of long ($A = |\cos\varphi|a_L = \frac{\tau a_L}{\sqrt{1+\tau^2}}$) and short ($B = |\sin\varphi|a_L = \frac{a_L}{\sqrt{1+\tau^2}}$) segments arranged according to the Fibonacci sequence (see Fig. 1.20), which can be obtained recursively from the substitution rule $A \to AB$, $B \to A$, starting from letter A. In this way, we realize that the long-range order of the QP distribution of projected points in the $1D$ physical space can be traced back to the underlying periodicity of the hyperlattice in $2D$. To this end, it is convenient to make use of the matrix relationship

$$\begin{pmatrix} x_\parallel \\ x_\perp \end{pmatrix} = \frac{a_L}{\sqrt{1+\tau^2}} \begin{pmatrix} \tau & -1 \\ 1 & \tau \end{pmatrix} \begin{pmatrix} n_x \\ n_y \end{pmatrix}$$

From a physical viewpoint the convenience of introducing the projection window is motivated by the observed intensity distribution of reflections in the diffraction patterns. In fact, a close inspection of the electron diffraction pattern shown in Fig. 1.3, for instance, reveals that the intensity of the diffracted beams does not systematically decrease as a function of the distance to the centre, but in an alternating way. This property can be accounted for in terms of the projection framework if one assumes that the Bragg reflections intensities decay with the magnitude of the *perpendicular* distance to the parallel space $E_\parallel$ measured along the long-dashed lines in the $E_\perp$ direction sketched in Fig. 1.20. In this way, experimental resolution is represented by the pair of short-dashed lines parallel to the $E_\parallel$

space (also shown in Fig. 1.20), so that only the lattice points comprised between these short-dashed lines give rise to observable reflection peaks. Moving the dashed lines outward from each other, thereby widening the acceptance domain window, has the effect of increasing the resolution of the spectrometer, thus densely filling in the regions between peaks on the $E_\parallel$ axis with weaker peaks. We note that two projections yielding very close points in $E_\parallel$ can arise only at the expense of having their corresponding square lattice points widely separated in the orthogonal $E_\perp$ direction. Thus, the main qualitative features observed in the QC diffraction patterns can be satisfactorily reproduced. As we see, the lattice points coordination number included in the projection window plays a significant role in obtaining the projected series of node points in $E_\parallel$. This dependence should be taken into account when considering the role of chemical bonding effects in the physical properties of QCs.

1.4.2 The Radon transform method

"One possible approach is to regard a quasi-crystalline solid as the Radon transform (...) The connection to Radon transform, a powerful tool of analysis, should help further studies in the properties of quasi-crystals". (R K P Zia and W J Dallas, 1985)[1004]

A QP sequence of lattice points can be alternatively obtained making use of a Radon transform instead of using a cut-and-project method. The Radon transform is named after the Austrian mathematician Johan August Radon (1887–1956) who proved that a function on $\mathbb{R}^3$ can be determined explicitly by means of its integrals over planes in $\mathbb{R}^3$. This result was subsequently generalized for a broad class of functions in $\mathbb{R}^N$ ($N > 3$), as follows: The Radon transform of a real function $f(\mathbf{x})$, where $\mathbf{x} = (x_1, \ldots, x_N)$ is the vector position of an arbitrary point in $\mathbb{R}^N$, is defined by the integral [243, 887],

$$\mathcal{R}f(\mathbf{x}) \equiv \int_{-\infty}^{+\infty} f(\mathbf{x})\delta(p - \xi \cdot \mathbf{x})d\mathbf{x}, \qquad (1.23)$$

where $\xi = (\xi_1, \ldots, \xi_N)$ is a unitary vector (i.e., $|\xi| = 1$) describing the orientation of a hypersurface $\mathbb{R}^D \subset \mathbb{R}^N$ ($D < N$), p measures the perpendicular distance from the origin of the $\mathbb{R}^N$ hyperlattice to the considered hypersurface, and $\delta(\ldots)$ stands for the δ-Dirac distribution. More explicitly, we have

$$\mathcal{R}f(x_1, \ldots, x_N) \equiv \int_{-\infty}^{+\infty} \cdots \int_{-\infty}^{+\infty} f(x_1, \ldots, x_N)\delta\left(p - \sum_{i=1}^{N} \xi_i x_i\right) dx_1 \ldots dx_N. \qquad (1.24)$$

Since the integration is performed over all the Cartesian coordinates the resulting Radon transform will depend on the parameters $\{p, \xi_1, \ldots, \xi_N\}$, so that we can formally express $\mathcal{R}f(x_1, \ldots, x_N) \equiv \breve{f}(p, \xi_1, \ldots \xi_N)$. From its very definition it follows that the Radon transformation is a linear one, satisfying the relation $\mathcal{R}(af + bg) = a\mathcal{R}f + b\mathcal{R}g$, for any pair of real functions f and g, where $a, b \in \mathbb{R}$. In the simplest 2D case the Radon transform of a real function $f(x, y)$ can be defined as [243, 887],

$$\mathcal{R}f \equiv \int_{-\infty}^{+\infty} f(\xi \cos\varphi - \eta \sin\varphi, \xi \sin\varphi + \eta \cos\varphi)\, d\eta, \qquad (1.25)$$

where $\{\xi, \eta\} \in \mathbb{R}$ denote the coordinates of points (the so-called Radon space) referred to a Cartesian system which is rotated an angle $\varphi \in [0, 2\pi]$ with respect to the original XY one, so that

$$\begin{pmatrix} x \\ y \end{pmatrix} = \begin{pmatrix} \cos\varphi & \sin\varphi \\ -\sin\varphi & \cos\varphi \end{pmatrix} \begin{pmatrix} \xi \\ \eta \end{pmatrix}, \qquad (1.26)$$

and comparing with Fig. 1.20 we see that the axis ξ (alternatively η) plays a role analogous to $E_\parallel$ (alternatively $E_\perp$) in the cut-and project method, respectively.

In order to grasp the geometrical meaning of Radon transform, as well as its relation to the cut-and-project method introduced in Sec. 1.4.1, it is convenient to consider a square lattice of points given by the Cartesian product of Dirac delta functions in the $\mathbb{Z}^2$ plane

$$L(x,y) = \sum_{n_x,n_y}^{\infty} \delta(x - n_x a_L)\delta(y - n_y a_L), \tag{1.27}$$

where a_L is the lattice parameter. In order to make use of Eq. (1.25) we first express Eq. (1.27) above in terms of Radon coordinates by means of the transformation given by Eq. (1.26), to get

$$\mathcal{R}L = \sum_{n_x,n_y}^{\infty} \int_{-\infty}^{+\infty} \delta(\eta \sin\varphi - (n_x a_L - \xi \cos\varphi))\delta(\eta \cos\varphi - (n_y a_L + \xi \sin\varphi))\, d\eta. \tag{1.28}$$

Taking into account the scaling property $\delta(\lambda z) = |\lambda|^{-1}\delta(z)$, $\lambda \in \mathbb{R} - \{0\}$, along with the integral

$$\int_{-\infty}^{+\infty} \delta(z-a)\delta(z-b)dz = \delta(a-b) = \delta(b-a), \tag{1.29}$$

we can express (1.28) in the form

$$\mathcal{R}L = \sum_{l=1}^{\infty} \delta(\xi - \xi_l), \tag{1.30}$$

where $\xi_l \equiv (n_x \cos\varphi - n_y \sin\varphi)a_L$, and l is a suitable integer which depends on n_x and n_y (**Exercise 1.10**). Therefore, the obtained Radon transform gives a sum of δ-Dirac functions, which can be interpreted as a set of points $\{\xi_l\}$ arranged along the ξ-axis in the Radon space. In the particular cases $\varphi = 0$ (i.e., $\xi_l = n_x a_L$) and $\varphi = \pi/2$ (i.e., $\xi_l = -n_y a_L$) we simply obtain a periodic sequence of lattice points along the axis X or Y, respectively. In general, if the slope of the ξ-axis line with respect to the Cartesian X axis is rational one has $\tan\varphi = p_1/p_2$, where $p_1, p_2 \in \mathbb{Z}$ are coprime integers. So, we can express $\sin\varphi = \alpha p_1$, $\cos\varphi = \alpha p_2$, with $\alpha \in \mathbb{R}$, so that $\xi_l \equiv \alpha(n_x p_2 + n_y p_1) \equiv \alpha l$, with $l \in \mathbb{Z}$. On the contrary, if the slope is irrational then one obtains an aperiodic sequence of points along the ξ-axis line. For instance if we choose $\tan\varphi = \tau^{-1}$ we get $\sin\varphi = \pm 1/\sqrt{1 + \tau^2}$ and $\cos\varphi = \pm\tau/\sqrt{1 + \tau^2}$, so that

$$\xi_{n_x,n_y} = \pm \frac{\tau n_x - n_y}{\sqrt{1 + \tau^2}} a_L. \tag{1.31}$$

Since the Radon transform extends by definition over all the hyperplane, one obtains an infinitely numerable delta Dirac comb, as given by Eqs. (1.30)–(1.31), which must be subsequently restricted to a countable set in order to describe actual (i.e., finite resolution) diffraction patterns.

In summary, within the Radon transform approach, a QC lattice can be obtained from a suitable periodic hyperlattice

$$L(x_1, \ldots, x_N) = \sum_{\nu_i}^{\infty} \prod_{i=1}^{N} \delta(x_i - \nu_i a_L),$$

in the hyperspace $\mathbb{R}^N$ by applying the following steps: (1) a rotation by an angle φ so that $\tan\varphi \in \mathcal{I}$, (2) a Radon transform of the rotated lattice to get $\mathcal{R}L$, and (3) a suitable restriction of the resulting infinite countable set of lattice points in order to compare with finite resolution diffraction patterns.

To conclude it is worth mentioning that the nD Fourier transform of a real function $f(\mathbf{x})$ is equivalent to the Radon transform of $f(\mathbf{x})$, for example, $\mathcal{R}f(x_1,\ldots,x_N) = \check{f}(p,\xi)$, followed by a $1D$ Fourier transform on the variable p. This general result is known as the *central-slice theorem*. This designation follows from the observation that the $1D$ Fourier transform of a projection of $f(\mathbf{x})$ for a fixed angle is a slice of the nD Fourier transform of $f(\mathbf{x})$ for the same fixed angle [788].

1.4.3 Statistical methods

The higher dimensional method has gained a growing acceptance since its early introduction in 1985 and currently provides the basic mathematical framework for structure description of QCs. However, even in the first years of research in the field, it was noticed that the choice of appropriate atomic surfaces, the adjustment of distance in the perpendicular space, and the orientation and size of the projection strip in the hyperlattice are free parameters that have an important impact on the resulting lattice structure in the physical space. Indeed, in order to accurately determine the position of the atoms, the nD approach must be supported with additional information regarding the local arrangement of atoms and the specific nature of the long-range order. The former is directly related to cluster aggregates analysis, whereas the latter is associated with tiling models [266].

An alternative method of QC structural investigation, referred to as the average unit cell statistical method, allows for constructing a distribution function of atomic positions in physical space. In many aspects, this approach is equivalent to the nD one, as a mathematical correspondence can be found between them. However, the statistical method gives a valuable tool for disorder analysis in the statistical language of distribution functions and standard deviations. In addition, combined with the tiling and cluster description, it gives insights into the spatial atomic positions during the structure refinement procedure [846, 960].

1.4.4 Approximant crystals

In most quasicrystalline forming alloy systems the true QC is accompanied by a number of compositionally related periodic crystals, having huge unit cell sizes, which are called *approximant* phases, because these crystals not only have very similar compositions, but also atomic structures closely resembling that of the true QC, from which they can nevertheless be distinguished. Approximants are important for understanding the structure of the related QCs since the structures of approximant crystals can be determined with high accuracy using conventional techniques, which is very helpful in obtaining a first-approach QC structure model. Within the hyperspace formalism these periodic structures can be derived by projection from a parent hyperlattice onto the $3D$ physical space, as it occurs in the case of QCs. However, whereas for QCs the projection involves irrational numbers, in the case of approximant crystals the projection is expressed in terms of *rational approximants* to the former irrational quantities. Accordingly, approximant crystals can be classified by a rational number, which is a ratio of two consecutive terms F_n/F_{n-1} in the Fibonacci sequence, in the case of icosahedral and decagonal QCs. The approximant crystal lattice constant is related to the iQC quasilattice constant, a_{QC}, by the relationship [840],

(Exercise 1.11)

$$a_{F_n/F_{n-1}} = \sqrt{\frac{2}{2+\tau}}(F_{n-1} + \tau F_n)a_{QC} = \sqrt{\frac{2}{2+\tau}}\tau^{n+1}a_{QC}. \qquad (1.32)$$

Up to now, the cubic 2/1 approximants are the highest order cases reported with structural details, the majority of cubic crystal approximants structural reports being for the order 1/1. In some systems, however, only the approximants are thermodynamically stable and the QCs need to be produced by rapid cooling [369]. Studies using variable temperature and/or pressure to induce transformations between QCs and their approximants can shed very valuable light onto their structural relationship [690], although not all transformations of QCs lead to some of their expected approximant phases. For example, studies of an AlCuFe iQC have shown that increasing the temperature reversibly transforms the iQC to crystalline phases, none of which corresponds to an approximant structure [715]. Therefore, identifying the main features of periodic structures that give rise to QCs or vice versa remains as an appealing open question. In this regard, the recent observation of an *in-situ* transformation of an i-CaAuAl QC to its cubic 2/1 crystal approximant phase, allowing for the crystallographic solution of the approximant structure as an overall packing of interpenetrating and edge-sharing icosahedral atomic clusters, represents an important step forward to this end [690].

Approximant phases should not be confused with giant-unit-cell intermetallics exhibiting complex structures that contain some hundred up to several thousand atoms in the unit cell [551]. A representative example is provided by the $Mg_{32}(Al,Zn)_{49}$ compound, first described by Linus Pauling (1901–1994) and co-workers in 1952, with 162 atoms in the unit cell. The structure of the $Mg_{32}(Al,Zn)_{49}$ crystal is based on a body-centered lattice composed of the so-called Bergman clusters (see Fig. 1.22a in Section 1.5.1). Other representative examples are orthorhombic ξ'-$Al_{74}Pd_{22}Mn_4$ (258 atoms in the unit cell) [89, 187], λ-Al_4Mn (586 atoms in the unit cell) [451], cubic β-Al_3Mg_2 (1168 atoms in the unit cell) [769], and the heavy-fermion compound $YbCu_{4.5}$, comprising as many as 7448 atoms in the unit cell [116]. These giant unit cells, with lattice parameters of several nanometers, strikingly contrast with the unit cells of elementary metals and simple intermetallics, which in general comprise from single up to a few tens atoms only. In addition, all these giant unit cells have a substructure based on polyhedral atom arrangements or clusters that partially overlap or are linked by bridging elements. Accordingly, they exhibit translational periodicity on the scale of many interatomic distances, whereas on the few nm scale, the atoms are arranged in clusters, where icosahedrally-coordinated environments play a prominent role [551, 190].

1.4.5 Phasonic freedom degrees

"So a symmetry in a higher dimension can sometimes explain a rather baffling transformation in a lower dimension". (Ian Stewart, 2007) [844]

A lot of information is encoded in the mathematical toolbox of higher dimension crystallography. A half of this information is used to describe the three dimensional appearance of QCs in the physical space through the projection from hyperspace. What about the information stored in the perpendicular space? This extra information is related to a new concept specific to higher dimensional crystallography, namely, the notion of phasonic freedom degrees. The term phason was coined in 1971 by A. W. Overhauser (1925–2011) in the context of lattice excitations of an incommensurate charge density wave system, where he showed that, because the free energy of the system is invariant under a phase shift of the

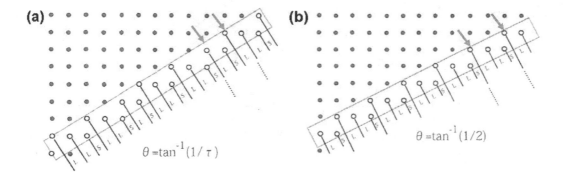

Figure 1.21 The cut-and-project method is used to illustrate the notion of phasonic defect. In (a) the Fibonacci chain composed of long (L) and short (S) segments is obtained by projecting the atoms contained within the boxed window strip where the slope of the paralel space $E_{\parallel}$ axis is given by $\tan\theta = \tau^{-1}$. In (b) the slope of the window strip is given by $\tan\theta = 1/2$, and the obtained sequence is periodic with a unit cell LLS. The arrows highlight the underlying 2D square lattice atoms that enter or abandon the corresponding projection windows as a consequence of the slope change, thereby producing the emergence of a phason flip within the region indicated by the dotted lines. (Adapted from Tsai A P, Icosahedral clusters, icosahedral order and stability of quasicrystals - a view of metallurgy, Sci. Tech. Adv. Mater. 9 013008 (30pp) 2008. Courtesy of An Pang Tsai).

modulation, there are new lattice excitations (termed phasons) whose dispersion relation involves only a purely imaginary contribution of the form $\omega = -i\tilde{D}\mathbf{q}^2$, where $\tilde{D}$ is the phason diffusion constant and $\mathbf{q}$ is the wavevector. Therefore, phasons behave diffusively (i.e., they are not propagative modes) and obey an acoustic-like dispersion relation.

Soon after the discovery of QCs several groups pointed out the importance of phason modes in these compounds as well. In fact, the term phason has been used with at least three different related meanings in the study of QCs. A *phason jump* refers to collective atomic rearrangements throughout the structure, a *phason strain* describes a specific type of distortion present in QCs, and *phason modes* describe elementary excitations similar to those previously considered in the study of incommensurate aperiodic systems [261, 380]. The signature of phason strain can be appreciated in diffraction patterns as: (1) a systematic shift in the Bragg reflections positions from the ideal ones by an amount that increases as the peak intensity decreases (phason strain can be observed as a slight deviation of the dimmer peaks from straight lines), and (2) an anisotropic broadening of the reflection peaks (compare Figs. 1.5 and 1.15).

Within the cut-and-project framework phason strain can be visualized as a shear transformation acting on the hyperlattice. To this end, let us consider the constructions depicted in Fig. 1.21 for the particular case of a 2D lattice, where we consider two acceptance domains of the same width, but with different slopes. In Fig. 1.21a the slope is $\tan\theta = \tau^{-1}$ and the projection leads to a Fibonacci chain, characterized by the presence of long (L) and short (S) segments arranged according to the Fibonacci sequence. In Fig. 1.21b the slope is a rational approximant of the golden mean ($\tan\theta = 1/2$) and we obtain a periodic lattice, whose unit cell is given by the sequence LLS. As we see, the projection window slope change results in an effective translation of the square lattice along the perpendicular space $E_{\perp}$ direction, leading to a structure which is not identical, but which is energetically indistinguishable from the original. As a result, some lattice points move outside the original acceptance window, while some others move into the new one (some of them are indicated by arrows in Fig. 1.21). In this way, a number of local rearrangements in the original projected sequence take place. For example, some local strings of the form LSL in the Fibonacci sequence turn to the form LLS in the approximant, as indicated by the dotted lines. This sort of rearrangements are called phason jumps (or *phason flips*) and result in correlated atomic jumps involving characteristic distances of the order of 1 nm. A phason flip in the case of 2D Penrose tiling can be understood as a rearrangement of tiles such that groups of two thick and one thin rhombuses swap their positions at a certain orientation in physical space. Thus, in a phason jump a few atoms can leave an atomic surface and join a neighboring one leading to atomic motions not involving vacancies. These atomic rearrangements are believed to play an important role in growth processes (see Sec. 1.5.4) and in the mechanical properties of QCs (see Sec. 3.9). Therefore, while a distortion of the lattice along the $E_\parallel$ space corresponds to a phonon propagation, a distortion of the lattice along the $E_\perp$ space defines a kind of diffusive motion which is specific to QCs. For instance, in the case of i-AlPdMn QCs, it has been reported that phason modes decay exponentially with time, with a characteristic time proportional to λ^2, where λ is the phason wavelength.

Note that square lattice points near the edges of the acceptance window fall very easily outside it, while points close to its center are very robust against phason effects, and are responsible of the system stability [641]. In this sense, one may think of introducing phason flips by changing the geometry of the acceptance window itself. For example, by defining it in terms of a sinusoidal function instead of a straight line, we can introduce a periodic modulation in the overall phason distribution. Thus, different classes of models can be derived by the projection method, namely, the perfect QC, a QC with linear phason strain, or QCs with modulated or even random strains.

An interesting physical consequence of the phason flip notion is that one may describe a continuous transformation of a 1D lattice from QP to periodic order by simply changing the slope of the $E_\parallel$ line. In this way, one realizes that the very notion of lattice defect can be addressed from a completely different viewpoint. In fact, while one usually considers the presence of defects as destroying periodic order, in this case one starts with a quasiperiodically ordered structure and, by increasing the number of phasonic defects, one actually improves the periodic order in the system. Quite interestingly, the progressive transition from QP to periodic order could be assessed, at least in principle, by performing a systematic experimental study of the acoustic (or optical) response of suitable QP multilayers as a number of planar defects (mimicking phasonic defects) is progressively introduced in their way towards the periodic order limit [30].

1.5 QUASICRYSTALS AS ATOMIC CLUSTERS AGGREGATES

"Several intermetallic compounds with complicated crystal structures have the common feature that they may be derived from a cubic body-centered lattice if one replaces the points of this lattice by 'clusters' of metal atoms. Such a cluster is built up from different nested polyhedra with a common center which may be occupied or not". (Erwin Hellner and Elke Koch, 1981) [331]

1.5.1 Nested polyhedra building blocks

The idea of explaining certain orderings in nature by means of a hierarchical geometrical construction based on the systematic use of Platonic solids can be probably traced back to Johannes Kepler (1571–1630) who, in his book *Mysterium Cosmographicum* (1596), tried to account for the relative size of the orbits of the six planets known at the time by successively inscribing an octahedron, an icosahedron, a dodecahedron, a tetrahedron, and an hexahe-

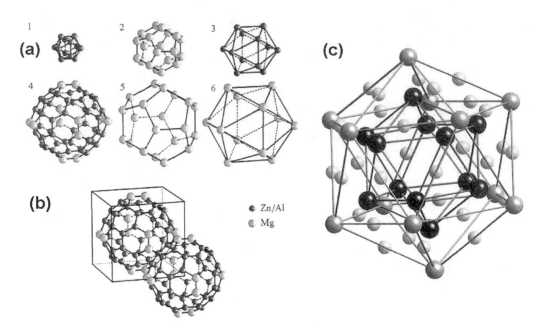

Figure 1.22 (a) A Bergman cluster contains 137 atoms arranged within six nested shells with icosahedral symmetry: an icosahedron (1), a pentagonal dodecahedron (2), an icosahedron (3), a truncated icosahedron (4), a dodecahedron (5) and an icosahedron (6), respectively. (b) Body-centered packing of the Bergman clusters, sharing a hexagonal face of the fourth shell produce the structure of the $(Al,Zn)_{49}Mn_{32}$ Bergman phase. (Courtesy of Prof. Janez Dolinšek). (c) A Mackay cluster contains 54 atoms arranged within three successive shells: 12 atoms at the vertices of an inner icosahedron (in black), 12 atoms at the vertices of a larger icosahedron (in dark gray), and 30 atoms on the two-fold axes of an outer icosidodecahedron (light gray). The diameter of this atomic cluster is about 0.96 nm. (From Maciá-Barber E, *Aperiodic Structures in Condensed Matter: Fundamentals and Applications* (CRC Press, Boca Raton, FL, 2009). With permission).

dron of appropriate sizes in a series of concentric spheres. This basic idea has reappeared in several scientific domains since then, in particular in crystallography and condensed matter physics, generally involving building blocks with a high point group symmetry [570]. For instance, cubic crystals based on icosahedral structural units were reported for $MoAl_{12}$, WAl_{12}, or $MnAl_{12}$ alloys, whose structure is given by a body-centered cubic lattice where a regular icosahedron of twelve aluminium atoms cluster around the smaller transition metal atom, which occupies the central position at each lattice point [3]. In 1953, Linus Pauling (1901–1994) and co-workers successfully described a complex phase corresponding to the ternary $(Al,Zn)_{49}Mn_{32}$ alloy by using a series of nested polyhedra as basic building blocks (Fig. 1.22a–b). The structure contains 162 atoms in its unit cell and it is based on a body-centered lattice [51]. Following a different line of thought John D. Bernal (1901–1971) was extremely keen on hierarchy as a principle of building atomic structures and generalizing crystallography [321], and this guiding principle inspired Alan L. Mackay to seek hierarchic structures as an alternative to lattice repetition. In so doing, he devised an atomic cluster based on an arrangement of three successive shells with icosahedral symmetry (Fig. 1.22c) [513].

A next step towards a fully nested polyhedra based structure is provided by the β-boron phase, which can be conveniently described in terms of a B_{84} unit that consists of a central B_{12} icosahedron with each vertex linked by a single boron atom to the pentagonal faces of an outer B_{60} shell having exactly the same buckyball structure as C_{60} [563]. In this material, we find an incipient hierarchical design which is further exploited in the case of the YB_{66} structure, described in terms of *supericosahedra* consisting of 13 B_{12} icosahedra. The resulting $B_{156} = B_{12}(B_{12})_{12}$ units are composed of a central B_{12} icosahedron with twelve B_{12} icosahedra bonded to each of the twelve vertices of the central subunit cluster [687]. More recently, the crystal structure of B_6O oxide has been described as a hierarchical periodical packing of icosahedral boron clusters with oxygen occupying the holes between them [353]. Commenting on the structure of the B_6O compound, A. L. Mackay elaborated on two important concepts, namely, (1) clusters of clusters are an alternative to strictly periodic crystalline arrangements which could give rise to a novel type of condensed matter phases, and (2) iQCs can probably be described as icosahedral clusters, themselves clustered icosahedrally in successive hierarchical levels, allowing for appropriate overlappings in order to preserve a finite density value for the resulting solid under the inflation growing scheme [570].

1.5.2 Cluster based classification of intermetallic quasicrystals

Most structural models of QCs and their approximants are based on one or more characteristic structural units adopting well defined polyhedral shapes determined by regular arrangements of atoms in nested shells, generally adopting point group icosahedral symmetries (dodecahedron, icosahedron, icosidodecahedron, and triacontahedron). Accordingly, iQCs can be classified into three classes according to different atomic clusters as building blocks, namely: (1) the AlMnSi class, whose prototype is given by the α-$Mn_{12}(Al,Si)_{57}$ phase, (2) the MgAlZn class, whose prototype is given by the $Mg_{32}(Al,Zn)_{49}$ Bergman phase, and (3) the CdYb class, whose prototype is given by the Cd_6Yb phase. The nested polyhedra geometries for the three classes are shown in Fig. 1.23. The structures of the AlMnSi and MgAlZn classes, derived from iQCs approximants, are characterized by a bcc packing of Mackay (Fig. 1.22c) and Bergman (Fig. 1.22a) clusters, respectively. The Tsai-type cluster contains randomly oriented tetrahedral clusters as the innermost shell and an icosidodecahedral fourth shell, along with a fifth outermost shell consisting of a 92-atoms bearing rhombic triacontahedron (Fig. 1.25a) [975]. The central tetrahedron is believed to play a crucial role in the structural stability of this class of QCs, as it relaxes locally a frustration arising from the icosahedral symmetry. Indeed, studies of Zn_6Sc, isostructural to the Cd_6Yb approximant to i-$YbCd_{5.7}$, showed that the Zn tetrahedron reorients at room temperature and its motion freezes along the [100] direction below a critical temperature of about 160 K [368].

Thermodynamically stable iQCs found in the AlCu(Fe,Ru,Os), AlPd(Mn,Re,Tc), and AlPd(Ru,Os) systems were grouped in the α-AlMnSi class, along with their related 1/0, 1/1, and 2/1 approximants [840]. In this case, pseudo-Mackay icosahedra (PMI) clusters (containing 51 atoms instead of the 54 atoms included in the standard Mackay cluster shown in Fig. 1.22c) were experimentally identified in i-AlPdMn and i-AlCuFe QCs structures, as deduced from X-ray and neutron diffraction data [90]. Each PMI is made of three centrosymmetrical atomic shells: a cubic core of nine atoms, an intermediate icosahedron of 12 atoms, and an external icosidodecahedron of 30 atoms (Fig. 1.24a). The last two shells

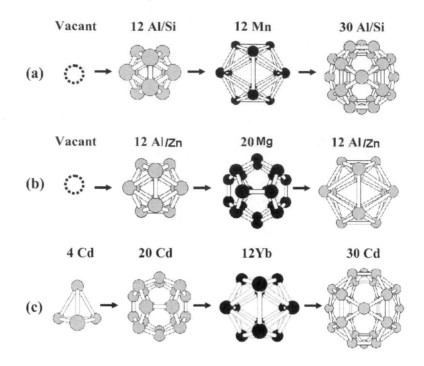

Figure 1.23 Concentric structures of three types of icosahedral clusters derived from 1/1 approximants of iQCs. (a) The AlMnSi class is based on the Mackay icosahedral cluster: the center is vacant, the first shell is an Al/Si icosahedron, the second shell is a Mn icosahedron, and the third shell is an AlSi icosidodecahedron. (b) The ZnMgAl class is based on the Bergman cluster: the center is vacant, the first shell is an Al/Zn icosahedron, the second shell is a Mg dodecahedron, and the third shell is a larger Al/Zn icosahedron. (c) The CdYb class is based on the Tsai cluster class: the center is a Cd tetrahedron, the second shell is a Cd dodecahedron, the third shell is a Yb icosahedron, and the fourth shell is a Cd icosidodecahedron. (Adapted from Tsai A P, Icosahedral clusters, icosahedral order and stability of quasicrystals - a view of metallurgy, Sci. Tech. Adv. Mater. 9 013008 (30pp) 2008. Courtesy of An Pang Tsai).

have practically equal radii and constitute altogether the surface of the PMI whose diameter is very close to 0.96 nm. Two families of chemically different PMI can be distinguished attending to the chemical decoration of the outer shell [375, 377, 379]. The calculated atomic density of an individual PMI is 64 at/nm^3, which is close to the measured density of the bulk materials within the experimental accuracy. Starting from an individual PMI the entire self-similar atomic structure can be grown following an inflation process via successive substitutions of atoms by PMIs with proper rescaling by a scale factor τ^3 (Fig. 1.24b). Note that, in order to preserve the density and stoichiometry of the solid under the inflation growing scheme some overlapping is required.

Stable iQCs in the AlLiCu, ZnMgZr, ZnMgRE, ZnMg(Al,Ga), and MgAlPd systems were grouped in the MgAlZn Bergman-type cluster class. Finally, stable iQCs found in the binary Cd(Yb,Ca) and the ternary ZnScMg, (Ag,Au)In(Yb,Ca), CaAu(In,Ga), and CdMgRE systems have been grouped in the CdYb Tsai-type cluster class, which constitutes the largest of the three classes of iQCs by number [840]. In the core of each Tsai cluster

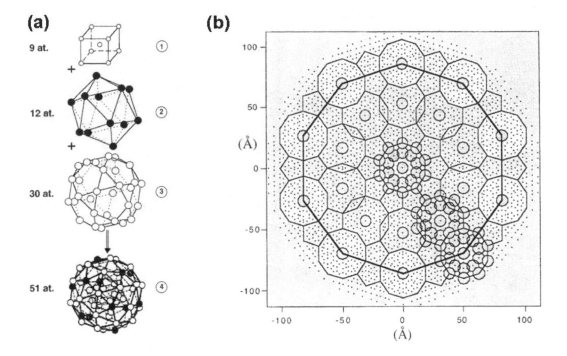

(a)

9 at. ①

+

12 at. ②

+

30 at. ③

51 at. ④

(b)

Figure 1.24 (a) Structure of a pseudo-Mackay cluster. Similar clusters can be identified in the bulk of AlPdMn and AlCuFe iQCs. (b) Arrangement of pseudo-Mackay clusters showing the hierarchical, self-similar arrangement of overlapping clusters. For simplicity, only the equatorial ring of atoms is presented for each cluster, the smallest ring representing the PMI cluster. Three successive levels of the hierarchy are drawn [376]. (Reprinted figure with permission from Janot C, Physical Review B 53, 181, 1996. Copyright (1996) by the American Physical Society).

there is a tetrahedron created by four positionally disordered Cd atoms (Fig. 1.23c), which breaks the icosahedral symmetry. The entire self-similar atomic structure can be grown, following an inflation process via successive substitutions of atoms by icosidodecahedral clusters with proper rescaling, as it is illustrated in Fig. 1.25, derived from detailed X-ray diffraction analysis [859].

Regarding the minor family of ddQCs, which includes a few representatives in the Ni-V(-Si) [367], Bi-Mn [980], and Ta-Te systems [139], we can consider the case of tantalum tellurides, whose basic structural element is a $M_{151}Te_{74}$ cluster (where M stands for Ta- or V-atoms). Such clusters are grouped around the vertices of tilings composed of squares and/or triangles forming a heavily corrugated, approximately 1 nm thick lamella [454].

1.5.3 Beyond purely geometric designs?

> *"We would prefer, however, to restrict the use of the term 'metallic cluster' to compounds containing groups of three or more metal atoms, arranged in a polygonal or polyhedral array and bonded to each other by metal-to-metal bonds".*
> (Erwin Hellner and Elke Koch, 1981)[331]

(a) **(b)**

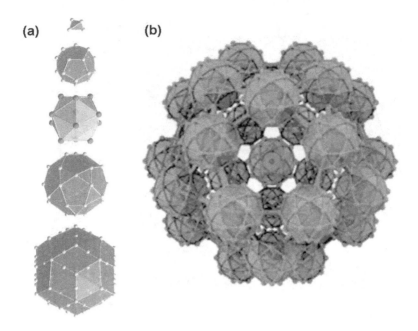

Figure 1.25 (a) Successive shells of the Tsai-type cluster. (b) A three-dimensional perspective of the τ^3 inflated cluster forming the basic icosidodecahedron structural motif of i-CdYb phase. Its radius is about 12 nm. (Adapted from Tsai A P 2013, *Chem. Soc. Rev.* **42** 5352 with permission of the Royal Society of Chemistry).

Certainly, one may consider the systematic use of nested polyhedral clusters in QCs structure determination as a matter of mere convenience, though the role of clusters as possible structural units is supported by the fact that such clusters can be universally identified in bulk structures of most QCs and their related approximants. Then, a major question in the field is whether clusters are physically significant, chemically stable entities, or simply convenient geometrical designs [882]. So, is it reasonable to take a step further and to consider QCs as molecular solids composed of actual nested atomic clusters arranged in a self-similar hierarchical way?

Some evidences favoring the existence of clusters as stable physicochemical entities come from direct imaging techniques, such as secondary electron imaging [1014], X-ray photoelectron diffraction [646], or STM [219]. These studies support the picture of QCs as cluster aggregates, namely, a 3D QP lattice properly decorated by atomic clusters which have the same point symmetry as the whole QC [1]. In the same vein, experimental electron density maps of crystalline approximants have shown that the covalent bonding character is larger for intra- than for inter-cluster linkages, thereby supporting the idea of clusters as chemical entities [422]. In fact, several physical and chemical properties, which will be described in Chapter 3, strongly suggest that local atomic order, on the scale of a few nanometers, plays an essential role in the emergence of certain peculiar electronic properties of these materials, probably due to the formation of a number of covalent bonds among different atoms grouped in clusters [318, 847]. Anyway, if atomic clusters really describe the actual architecture of QCs in physical space, it is reasonable to expect that these clusters should be obtained from a projection involving suitable arrangements of lattice points in the hyperspace. This issue was addressed by Gratias and co-workers [817].

When thinking of clusters as physical entities, rather than convenient geometrical designs, one should properly address the following issues, which are the focus of intensive research nowadays:

- What is the number and structure of the different atomic clusters which are compatible with the chemistry of QCs belonging to a given alloy system?

- What is the nature of the chemical bonding among the atoms belonging to a given cluster, as well as among different clusters themselves? For example, it may occur that a given cluster may act as a chemically stable structure when isolated, but it progressively loses its identity when assembled to form a solid, due to strong interactions with close neighbors [491]. Then, along with the stability of clusters we should also consider those aspects related to their reactivity.

- What is the more appropriate packing rule (including possible overlappings) among clusters at different hierarchical stages?

- What mechanical behavior should be expected for a solid mainly consisting of clusters?

In this regard, cleavage and annealing experiments have been interpreted as proof for the existence of clusters in QCs with a high mechanical stability [220], though it seems that most mechanical properties reported for QCs (with the exception of a brittle–ductile transition at elevated temperatures) can be properly accounted for in terms of usual processes (including the additional phasonic freedom degree). Probably the crucial proof supporting the view of QCs as cluster aggregates will come from a bottom up synthesis route, yielding the growth of a full-fledged bulk QC starting from its constituent cluster(s) building block(s) from the scratch [292, 912].

1.5.4   Intermetallic quasicrystals growth and stability

> *"but how on earth would Nature do it?... the* spontaneous *growth of large regions of such quasi-crystalline five-fold symmetric substances had seemed to me virtually insurmountable".* (Roger Penrose, 1989)[684]

> *"If stable quasicrystals are possible, Nature should have formed them without any human intervention, and they should have survived for eons, just like crystals".* (Luca Bindi and Paul J. Steinhardt, 2018) [75]

According to the classical crystal growth theory, established about a hundred years ago, a crystal is developed by nucleation and the repeated attachment of building blocks, which can be ions, atoms, or molecules. The polyhedral morphology characteristic of periodic crystals at a macroscopic scale can then be explained by different growth rates along different crystal orientations. By the light of this knowledge several question naturally arise when considering QCs growth conditions, namely:

- Do intermetallic QCs grow from the melt just the same way as periodic crystals do?

- Are the growth processes dependent on the QC symmetry class (i.e., icosahedral, decagonal, dodecagonal)?

- How do these processes differ from those leading to related approximant periodic crystals?

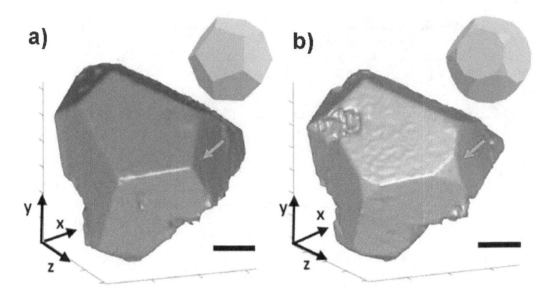

Figure 1.26 3D renderings of an i-Al$_{71}$Pd$_{19}$Mn$_{10}$ QC growth at (a) an early stage (~420 s after nucleation), and (b) when interfacial velocities are near-zero (~720 s after nucleation). The growth shape corresponds to a pentagonal dodecahedron at the first stage, and the equilibrium shape is a truncated dodecahedron (both ideal polyhedra are shown as upper insets). Scale bars are 100 μm. (Adapted from Scripta Materalia 146, Senabulya N, Xiao X, Han I, Shahani A J, On the kinetic and equilibrium shapes of icosahedral Al$_{71}$Pd$_{19}$Mn$_{10}$ quasicrystals, 218–221, Copyright (2018), with permission from Elsevier).

- What are the local driving forces ultimately leading to the formation of intermetallic QCs?

- Which factors determine whether a structure grows periodically or quasiperiodically?

- What stabilizes a QP structure over energetically competing periodic ones?

- What guides the growth of a QP structure?

Earlier empirical observations indicated that quasicrystalline phases can be easily obtained by rapid solidification, such as melt spinning, with cooling rates greater than 10^6 Ks^{-1}. This raises the question of how long-range QPO is achieved in such a short timescale, albeit it is also true that most of the QCs obtained by rapid solidification are metastable and transform into periodic crystalline phases upon annealing. The easy formation of metastable iQCs has been explained by their preferred nucleation from undercooled melts due to the icosahedral short-range order *already* present in liquid alloys. This does not explain, however, how long-range QPO is obtained. In particular, which parameters favor the growth of QCs against the high-order approximants that do not differ locally from QCs? Indeed, metastable iQCs may be regarded as intermediate phases, and by incorporating structural information of related approximants and QC compounds, one may gain insight as to why nature prefers the quasicrystalline structure in certain cases [909].

By all indications thermodynamically stable intermetallic QCs form via nucleation and growth, starting from a microscopic nucleus, which spontaneously arises from the liquid phase, and spreads outward, converting the system from liquid to solid. On the basis of *in situ* synchrotron X-ray tomography experiments of i-AlPdMn QCs from the liquid phase, under relatively slow solidification conditions, revealed that iQC nucleates at 900°C and grows into the liquid with five-fold symmetry as time proceeds. The growth process occurs for ~600 s until the composition of the liquid phase reaches that of the liquidus. During this regime a facetted pentagonal dodecahedron is observed. Afterwards, during the transition from growth regime to equilibrium state, the pointed vertices located at the intersection of three pentagonal planes in the dodecahedron are gradually truncated leading to a $\{10, 3\}$ Archimedean polyhedron (see arrows in Fig. 1.26). At this stage, a rough interface displaying porosity is observed as well [791].

The solid–liquid interface velocities tend to decay over time due to the depletion of solute in the liquid phase as the QC grows, varying from as high as 1 μms^{-1} following nucleation to near zero at the later stages of solidification [792]. On the other hand, the facets velocities are highly anisotropic: the facets that have their plane vector approximately anti-parallel to the direction of gravity grow much faster than those facets misaligned with the gravity direction. Most facet areas increase during continuous cooling, but a few facets decrease in area to make way for the 3-fold facet of the truncated icosahedron that forms during the last growth stage. On the atomic scale, the clusters located at the vertices of the pentagonal dodecahedron are loosely bound to those in the bulk. Hence, the clusters are in a thermodynamically unstable configuration and dissolve readily, leaving a 3-fold facet behind. The reported anisotropic velocities suggest that iQC growth is governed by the kinetics of bulk transport (leading to a higher rate of incorporation of the heavy-atom clusters for facets that are oriented anti-parallel to gravity) rather than by interfacial attachment processes [791]. Consequently, growth of the iQC phase from a liquid phase is largely governed by gravity-induced convection. Once the supersaturation in the liquid has been almost entirely depleted, the pentagonal dodecahedron habit growth evolves into a truncated dodecahedron near equilibrium configuration, indicating that the pentagonal dodecahedron is not the lowest energy structure in this iQC [791].

Similar *in situ* synchrotron X-ray tomography measurements, in which both growth and dissolution from the liquid phase of a d-AlCoNi QC and a related approximant phase were observed as a function of time, indicated that growth of both solids proceed via the attachment of large clusters of atoms from the liquid, rather than by attachment of individual atoms [319]. This result strongly supports the importance of clusters (as independent physical entities) in stabilizing intermetallic QCs [85]. According to the obtained results the solid–liquid interfacial normal velocity scales linearly with the undercooling rate, $v = \beta_m \Delta T$, where the kinetic coefficient β_m was found to be approximately 6 to 9 (2 to 8) orders of magnitude smaller than those of simple metals (intermetallic) periodic crystals, respectively. Thus, β_m decreases as the complexity of the building unit increases, as expected. Quite remarkably, however, significant differences in the nucleation and growth rates between dQC and approximant phases were found. Thus, the kinetic coefficient for the approximant phase, $\beta_m = 4.5 \times 10^{-7}$ cms^{-1}K^{-1}, is about five times smaller than that of the QC in the aperiodic $\{00001\}$ plane. This result was interpreted as indicating that dQC possess a structural flexibility that is not present in the approximant, since it stems from the QC higher (phasonic) degrees of freedom [2, 319, 786]. Indeed, grain boundaries can migrate in dQC through an "error-and-repair" type process, wherein phason strain is first introduced and then relaxed to generate ideal quasicrystalline order. In fact, misfit cluster attachments increase the phason elastic energy, which in turn is reduced through tile flips, as it was directly observed in dQCs via *in situ* High resolution transmission electron microscopy (HRTEM) [632]. In contrast, the same clusters cannot attach to the periodic approximant crystal in a flawed way, requiring instead extensive cluster spatial rearrangements to preserve the translational symmetry of the underlying periodic lattice. Consequently, despite having similar building blocks the two phases can have very different growth kinetic signatures [319].

Among the driving forces for the growth of intermetallic QCs which have been identified to date we can mention: (1) a precise chemical composition providing a specific valence electron concentration, (2) a number of energetically favorable structural subunits (clusters) with composition differing from the overall QC composition and well defined overlapping rules among them, and (3) fluctuations of the local composition and, consequently, of the chemical potential, with respect to the global average, which can lead to QP growth if the stoichiometry allows for it [843]. Notwithstanding this, so far it is not clearly known what specific qualities (if any) a system must possess in order to form QP instead of periodic crystals.

In the same vein, up to now we have observed binary, ternary, quaternary, and higher multinary QCs to exist. Are QCs made of only one chemical element possible? The QP dimensionality (see Fig. 1.9) seems to play a role in the thermodynamic stability of aperiodic structures. Thus, most of the iQCs found to date are thermodynamically stable, whereas the majority of dQCs are metastable. All octagonal and, with one exception, all dodecagonal phases are metastable as well [840]. The geometrical constraints for the formation of iCQs seem to be much stronger than those of dQCs. Nevertheless, dQCs must not be seen as a mere stacking of 2D QP layers, since it has been observed that the chemical bonding is similar in both parallel and perpendicular direction to the QP atomic layers [191]. Indeed, the decaprismatic crystals do not only show prisms faces but also facets inclined to the 10-fold axis. Consequently, the periodic growth along the 10-fold axis somehow forces the QP growth in the planes perpendicular to it.

A long-lasting discussion in the field concerns as to whether QCs are energy or entropy stabilized. In other words, whether structural QPO can be a ground state of condensed matter at absolute zero temperature, or has to be stabilized by entropic contributions coming from phonons, phasons or other structural disorder possible sources. In the case of an entropy stabilization mechanism QCs would be high temperature phases, only stable above a specific threshold temperature. Generally, it is still not clear whether entropy plays a decisive role for the stabilization of QCs. During the growth processes intrinsic disorder is generally created. Most of the metastable iQCs obtained by rapid solidification are QCs with linear phason strain, and those obtained by the crystallization of glass states are QCs with random strain. In the case of finite real QCs there will be no infinite but a large number of configurationally different but energetically almost equal structures contributing to the entropy of the system. On the other hand, the fact that intermetallic QCs can be regarded as Hume-Rothery phases (see Sec. 3.2.1), strongly supports the hypothesis of these QCs being energy-stabilized. However, it does not exclude the possibility that in addition to the Brillouin zone – Fermi surface electronic stabilization and the hybridization between d and sp states in TM bearing QCs, some entropic contribution may be necessary as well (see Sec. 3.2.1) [842]. High-order approximant crystals should have similar Gibbs free energies, $G = H - TS$, as the related QCs. Nonetheless, there is no known case of a QC transforming to an approximant crystal with exactly the same chemical composition according to a reversible transition QC $\leftrightarrow$ AC, either as a function of temperature or with increasing pressure [843].

Growth model calculations are very important in order to gain some insight into the factors controlling the formation and stability of QCs. Most commonly used are Monte

Carlo or molecular dynamics simulations employing specific pair potentials, though whether periodic or QPO results in a low enough free energy for a given mass density and chemical composition of the melt largely depends on the cluster packing optimization rather than on pair potentials directly [841]. Due to multibody interactions and the existence of itinerant electrons, the pair potentials of atoms in intermetallic phases strongly depend on the kind and arrangement of neighboring atoms. Thus, Al-TM based QCs, where d-orbitals play a crucial role can be described in terms of Friedel oscillations, whereas simpler Lennard-Jones potentials of the form

$$U(r) = U_0 \left[\left(\frac{r_0}{r} \right)^{12} - \left(\frac{r_0}{r} \right)^6 \right], \tag{1.33}$$

where r_0 measures the location of the potential minimum, are appropriate in order to model Mg-Zn based QCs with s- and p-electrons dominating metallic bonding [758]. Local disorder sources, such as phason flips, can be studied quite well in terms of these classical simulation methods, which can deal with system sizes up to millions of atoms. In turn, quantum mechanical first principles calculations can shed light onto the nature of local chemical bonding of energetically favorable, optimal cluster building blocks. Nevertheless, these calculations can only consider model systems with a size below a thousand atoms, which corresponds to just a few clusters per unit cell. Consequently, the role of long-range QPO can not be properly modeled and the information obtained from these calculations refers to low-order approximant model structures rather than to QP ones[22] [841].

1.6 EXERCISES WITH SOLUTIONS

1.1 *Determine the allowed rotation axes in two-dimensional periodic lattices.*

In two dimensions the orthogonal matrix describing rotations by an angle φ reads

$$\mathbf{M} = \begin{pmatrix} \cos\varphi & -\sin\varphi \\ \sin\varphi & \cos\varphi \end{pmatrix}, \tag{1.34}$$

so that the crystallographical restriction in 2D can be analogously written as

$$\mathrm{tr}\mathbf{M} = 2\cos\varphi \equiv n \in \mathbb{Z}, \tag{1.35}$$

which can be rewritten in the simple form $\cos\varphi = n/2$. The condition $|\cos\varphi| \leq 1$ implies $n = \{-2, -1, 0, 1, 2\}$ in this case. Quite remarkably, however, by plugging these integer numbers in the expression $\varphi = \cos^{-1}(n/2)$, we obtain exactly the same solutions listed in Table 1.1. Therefore, the crystallographical restriction leads to the *same* set of allowed rotation axes for 2D and 3D periodic lattices [32]. This is a general feature of successive $2n$D and $(2n+1)$D couples of Euclidean hyperspaces (see [840] for more details).

1.2 *Following the procedure described in Section 1.1.2 study the relative distances to the center of the main diffraction peaks along a radial direction for the (a) decagonal, and (b) icosahedral QCs shown in Figs. 1.7 and 1.27, respectively. Compare the obtained results with those reported in Table 1.2.*

(a) Firstly, we measure the distances of successive main diffraction spots along the radial direction outwards from the center, d_n, and calculate the ratios d_n/d_1. The obtained values

---

[22]This shortcoming may be overcome by implementing a suitable renormalization scheme able to exploit the underlying self-similar structure of full-fledged QCs.

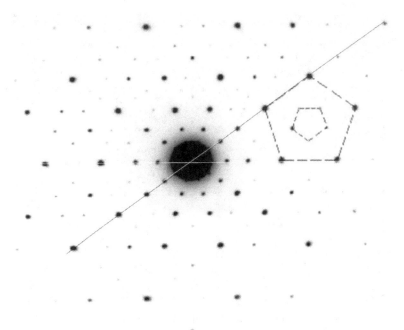

Figure 1.27 Electron diffraction pattern corresponding to an i-AlCuFe QC. A 10 fold symmetry axis around the origin can be clearly appreciated along with a series of nested pentagons, two of which are highlighted at the upper right corner (Courtesy of J. Reyes-Gasga).

for the d-$Al_{70}Cu_{15}Co_{15}$ QC diffraction pattern displayed in Fig. 1.7 are listed in the second column of Table 1.5 and they clearly follow a non-periodic series, obeying the Fibonacci-like recursion rule $d_{n+1} = d_n + d_{n-1}$. Also, we can appreciate a close correspondence between the listed d_n/d_1 values and successive powers ($n > 1$) of the golden mean $\tau = (\sqrt{5} + 1)/2 \simeq 1.62$ (rounded to the second decimal place), as it is seen in the fourth and fifth columns of Table 1.5. In the third column we list the rational approximants to the decimal expressions given in the fourth column. As we see, all of them share the same denominator, while the numerators are given by the Fibonacci series $\{5, 8, 13, 21\}$, in agreement with the asymptotic limit prescribed by Eq. (1.5), and their natural extensions, as described in Sec. 1.1.2. Finally, taking into account the relation between τ^n powers and Fibonacci numbers given by Eq. (1.3), the corresponding diffraction peaks can be labeled in terms of Fibonacci couples, as it is indicated in the last column of Table 1.5.

TABLE 1.5 The main reflections in the diffraction pattern shown in Fig. 1.7 along a radial direction can be labeled according to a power series related to the golden mean.

n	d_n (mm)	d_n/d_1	d_n/d_1	τ^{n-1}	$|\Delta|$	(F_n, F_{n-1})
1	6.0	1	1.00	1.00	–	–
2	10.0	5/3	1.66	1.62	0.04	$(1,0)$
3	16.0	8/3	2.66	2.62	0.04	$(1,1)$
4	26.0	13/3	4.33	4.24	0.09	$(2,1)$
5	42.0	21/3	7.00	6.85	0.15	$(3,2)$

(b) Proceeding in a completely analogous way with the diffraction pattern displayed in Fig. 1.27 we obtain the data listed in Table 1.6 for an i-AlCuFe QC.

TABLE 1.6   The main reflection spots along a radial direction in the diffraction pattern shown in Fig. 1.27 can be labeled according to a power series related to the golden mean.

n	d_n (mm)	d_n/d_1	d_n/d_1	τ^{n-1}	$|\Delta|$	(F_n, F_{n-1})
1	9.25	1	1.00	1.00	–	–
2	14.50	58/37	1.57	1.62	0.05	$(1,0)$
3	23.75	95/37	2.57	2.62	0.05	$(1,1)$
4	38.75	155/37	4.19	4.24	0.05	$(2,1)$
5	63.50	254/37	6.86	6.85	0.01	$(3,2)$

By comparing the results presented in Tables 1.2, 1.5, and 1.6, we realize that the main features displayed by the electron diffraction patterns of decagonal and icosahedral QCs, measured along the 5- and 10-fold axes, respectively, are very similar. In particular, the scale factor is given by the golden mean in both cases (within the experimental accuracy). Nevertheless, by inspecting the data listed in the sixth column of Tables 1.2, 1.5, and 1.6, we note that the i-AlCuFe sample displays a higher structural quality than the i-Al$_{86}$Mn$_{14}$ sample, measured in terms of the smaller $|\Delta|$ values.

1.3 *Determine the characteristic inflation scale factor of a dodecagonal QC making use of the Bragg reflections arranged along the 15° radial lines in the diffraction pattern shown in Fig. 1.28.*

The main peaks along the 15° radial lines are labelled F, C, D outwards from the center. The F and D peaks average distances, measured along each of the twelve radial lines, read $d_F = 11.7 \pm 0.2$ mm, and $d_D = 44.1 \pm 0.3$ mm, respectively. Thus, we get the ratio

$$\frac{d_D}{d_F} = \frac{44.1}{11.7} = \frac{49}{13},$$

that is, $\frac{d_D}{d_F} \simeq 3.77 \pm 0.09$, which is reasonably close to the exact value $2+\sqrt{3} = 3.732\,050\,81 \simeq 3.73$, within the experimental accuracy.

1.4 *(a) Making use of the relation $\tau = 2\cos\frac{\pi}{5}$ deduce the ratio between the diagonal and the side of a regular pentagon (Fig. 1.29). Compare this exact result with the experimental value obtained from the diffraction pattern shown in Fig. 1.7b. (b) Determine the scale factor between the large and small pentagons in the Pythagorean pentagram highlighted in Fig. 1.7.*

(a) Making use of the cosine theorem we have $d^2 = 2a^2 - 2a^2\cos\varphi = 2a^2(1-\cos\varphi) = 4a^2\sin^2\frac{\varphi}{2} = 4a^2\sin^2\left(\frac{3\pi}{10}\right)$, where we have used the value $\varphi = \frac{3\pi}{5}$ for the angles of a regular pentagon. Thus, we get the ratio

$$\frac{d}{a} = 2\sin\left(\frac{3\pi}{10}\right) = 2\sin\left(\frac{\pi}{2} - \frac{\pi}{5}\right) = 2\cos\frac{\pi}{5} = \tau.$$

On the other hand, measuring the length of the five sides and diagonals of the large pentagon highlighted in Fig. 1.7 we obtain the average values $a = 16.2 \pm 0.2$ mm and $d = 26.1 \pm 0.2$ mm, respectively, which yields the ratio $d/a = 1.61 \pm 0.03$, a value which agrees well with that of the golden mean within the experimental accuracy.

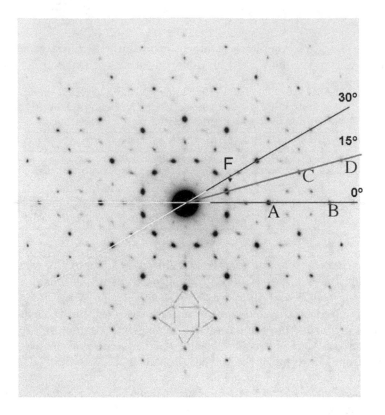

Figure 1.28 Electron diffraction pattern corresponding to a MnCrNiSi dodecagonal quasicrystal taken along the 12-fold axis. Characteristic motives (one of them highlighted at the bottom) based on squares and equilateral triangles sharing a common edge can be readily appreciated. (Adapted from *Dodecagonal quasicrystal in Mn-based quaternary alloys containing Cr, Ni and Si*, by Iwami S, and Ishimasa T, Philosophical Magazine Letters, 85, 229–236, 2005. Reprinted by permission of Taylor & Francis Ltd, http://www.tandfonline.com).

(b) Measuring the length of the five sides of the first generation small pentagon drawn at the left bottom corner of Fig. 1.7 we obtain the average value $a' = 6.4 \pm 0.1$ mm, so that we obtain the scale factor $a/a' = 2.53 \pm 0.07 \simeq \tau^2$, within the experimental accuracy (see Table 1.5). For the sake of illustration, we have also measured the average side length of the second generation smaller pentagon encircled in Fig. 1.7, obtaining $a'' \simeq 2.5 \pm 0.2$ mm, which leads to the second order scale factor $a/a'' = 6.5 \pm 0.6$, which falls somewhat short from the expected value $\tau^4 \approx 6.85$ (see Table 1.5).

1.5 *Following the procedure described in Section 1.1.2 study the relative distances to the center along a radial direction of the main diffraction peaks of the icosahedrite natural QC shown in Fig. 1.15. Compare the obtained results with those reported in Tables 1.2 and 1.6.*

Proceeding in a way completely analogous to that described in the **Exercise 1.2a** above, we obtain the data listed in Table 1.7. By comparing the results listed in Tables 1.2, 1.7, and 1.6, we conclude that the natural QC sample (the mineral icosahedrite) displays a

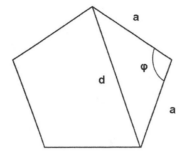

Figure 1.29   Geometrical elements of a regular pentagon.

TABLE 1.7   The main reflection spots along a radial direction in the diffraction pattern shown in Fig. 1.15 can be labeled according to a power series related to the golden mean.

n	d_n (mm)	d_n/d_1	d_n/d_1	τ^{n-1}	$|\Delta|$	(F_n, F_{n-1})
1	4.25	1	1.00	1.00	–	–
2	6.75	27/17	1.59	1.62	0.03	$(1,0)$
3	11.00	44/17	2.59	2.62	0.03	$(1,1)$
4	18.00	72/17	4.23	4.24	0.01	$(2,1)$
5	29.00	116/17	6.82	6.85	0.03	$(3,2)$
6	46.75	187/17	11.00	11.09	0.09	$(5,3)$

significantly higher degree of structural quality than the i-AlMn QC found by Schechtman or typical i-AlCuFe QCs produced in the laboratory samples, as measured in terms of: (1) its smaller $|\Delta|$ values, and (2) the fact that we can observe Bragg reflections corresponding to the τ^5 terms in its electron diffraction pattern. In addition, the more intense Bragg peaks display no discernible evidence of phason strain (note the remarkably isotropic shape of the peaks, as compared with those shown in Fig. 1.27, for instance) [67].

1.6 *Determine the global or discrete nature of the scale transformation symmetry acting on the spiral lattice given by the relationships $r_n = a\sqrt{n}$ and $\varphi_n = n\alpha$, where a and α are positive real numbers.*

By combining both expressions we get $r_n = a\sqrt{\frac{\varphi_n}{\alpha}} \rightarrow \varphi_n = \frac{\alpha}{a^2}r_n^2$. The action of the scale transformation operator is given by

$$S_\lambda \varphi_n(r_n) = \frac{\alpha}{a^2}(\lambda r_n)^2 = \lambda^2 \varphi_n(r_n),$$

so that we obtain for the discrete spiral lattice the same scale factor $\mu = \lambda^2$ we obtained for the related (Fermat) spiral curve. In this way, we have a discrete lattice which exhibits a global scale invariance symmetry.

1.7 *Making use off Eqs. (1.13) and (1.14) prove the scale invariance relationship $S_\lambda \tilde{a} = \tau \tilde{a}$, involving the icosahedral lattice vectors $\{a_i\}$ in 6D.*

$$
\mathbf{S}_\tau \tilde{\mathbf{a}} = \frac{1}{2}
\begin{pmatrix}
1 & 1 & 1 & 1 & 1 & 1 \\
1 & 1 & 1 & -1 & -1 & 1 \\
1 & 1 & 1 & 1 & -1 & -1 \\
1 & -1 & 1 & 1 & 1 & -1 \\
1 & -1 & -1 & 1 & 1 & 1 \\
1 & 1 & -1 & -1 & 1 & 1
\end{pmatrix}
\begin{pmatrix}
a_1 \\ a_2 \\ a_3 \\ a_4 \\ a_5 \\ a_6
\end{pmatrix}
$$

$$
= \frac{1}{2}
\begin{pmatrix}
1 & 1 & 1 & 1 & 1 & 1 \\
1 & 1 & 1 & -1 & -1 & 1 \\
1 & 1 & 1 & 1 & -1 & -1 \\
1 & -1 & 1 & 1 & 1 & -1 \\
1 & -1 & -1 & 1 & 1 & 1 \\
1 & 1 & -1 & -1 & 1 & 1
\end{pmatrix}
\begin{pmatrix}
1 & 0 & \tau \\
\tau & 1 & 0 \\
0 & \tau & 1 \\
-1 & 0 & \tau \\
0 & -\tau & 1 \\
\tau & -1 & 0
\end{pmatrix}
\begin{pmatrix}
\mathbf{e}_1 \\ \mathbf{e}_2 \\ \mathbf{e}_3
\end{pmatrix}
$$

$$
=
\begin{pmatrix}
\tau & 0 & \tau+1 \\
\tau+1 & \tau & 0 \\
0 & \tau+1 & \tau \\
-\tau & 0 & \tau+1 \\
0 & -\tau-1 & \tau \\
\tau+1 & -\tau & 0
\end{pmatrix}
\begin{pmatrix}
\mathbf{e}_1 \\ \mathbf{e}_2 \\ \mathbf{e}_3
\end{pmatrix}
= \tau \tilde{\mathbf{B}} \tilde{\mathbf{e}} = \tau \tilde{\mathbf{a}}.
$$

where we have used the relation $1 + \tau = \tau^2$.

1.8 *Show that the function* $f(x) = \sin(x) + \cos(\sqrt{2}x)$ *satisfies the condition* $|f(x + P_n) - f(x)| < \varepsilon$, $\forall\ x \in \mathbb{R}$, *for a countable set of almost periods* P_n, *and estimate their values.*

Let $P_n \equiv nP$, where $n \in \mathbb{Z}$, and $P \in \mathbb{R}$, such that $\sin P_n = 0$. Then,

$$
\begin{aligned}
|f(x + P_n) - f(x)| &= \left| \sin(x + P_n) + \cos(\sqrt{2}x + \sqrt{2}P_n) - \sin(x) - \cos(\sqrt{2}x) \right| \\
&= \left| \cos(\sqrt{2}x + \sqrt{2}P_n) - \cos(\sqrt{2}x) \right| \\
&= \left| -2\sin\left(\frac{\sqrt{2}}{2}P_n\right) \sin\left(\frac{\sqrt{2}}{2}(2x + P_n)\right) \right| \leq 2 \left| \sin\left(\frac{\sqrt{2}}{2}nP\right) \right|.
\end{aligned}
$$

Because $\sqrt{2}$ is irrational, the sequence $\sqrt{2}nP/2$ (mod 1) is dense in the interval $[0, 1)$. It is thus clear that for any $\varepsilon > 0$, there are infinitely many integers $n \in \mathbb{Z}$, so that the relationship $|f(x + P_n) - f(x)| < \varepsilon$ holds for all $x \in \mathbb{R}$ (qed).

In order to estimate the possible values for the almost periods P_n we note that $|\sin(x)| \leq |x|$ on $\mathbb{R}$, so that we can write $\varepsilon = 2\left|\sin\left(\frac{\sqrt{2}}{2}P_n\right)\right| \leq \sqrt{2}P_n$. Now, $\sqrt{2}$ can be approximated by the members of the series

$$
\left\{ 1, \frac{3}{2}, \frac{7}{5}, \frac{17}{12}, \frac{41}{29}, \frac{99}{70}, \frac{239}{169}, \cdots \right\},
$$

which are of the form $\frac{a_n}{b_n}$, with $a_n = a_{n-1} + 2b_{n-1}$ and $b_n = a_{n-1} + b_{n-1}$, for $n \geq 1$, with $a_0 = b_0 = 1$. Thus, for a given n value we have $P_n \simeq \frac{b_n}{a_n}\varepsilon$ [32].

1.9 *A quasiperiodic tiling of the plane is obtained by placing dielectric pillars at the maxima of an octagonal multi-beam interference holographic pattern given by*

$$I(\mathbf{r}) = \left| \sum_{m=1}^{N=8} A_m e^{i(\mathbf{k}_m \cdot \mathbf{r} + \varphi_m)} \right|^2,$$

where $\mathbf{r} = (x,y)$ *is the position vector in the plane,* $A_m \equiv 1 \ \forall m$,

$$\mathbf{k}_m = \frac{2\pi}{\lambda} n \sin\theta \left(\sin\left(\frac{2\pi m}{N}\right), \cos\left(\frac{2\pi m}{N}\right) \right),$$

defines a wave vector set pointing from the vertices of an octagon to the center, forming an angle θ *with respect to the normal to the plane, in a medium of refractive index* n, *and we adopt the parameters* $\varphi_1 = \varphi_5 = 0$, $\varphi_3 = \varphi_7 = \pi$, *and* $\varphi_2 = \varphi_4 = \varphi_6 = \varphi_8 = \pi/2$. *[1011] Determine the intensity distribution pattern and prove that it is described in terms of a quasiperiodic function in the XY plane.*

Making use of the adopted parameters we have

$$\sum_{m=1}^{N=8} A_m e^{i(\mathbf{k}_m \cdot \mathbf{r} + \varphi_m)}$$

$$= e^{i\mathbf{k}_1 \cdot \mathbf{r}} + ie^{i\mathbf{k}_2 \cdot \mathbf{r}} - e^{i\mathbf{k}_3 \cdot \mathbf{r}} + ie^{i\mathbf{k}_4 \cdot \mathbf{r}} + e^{i\mathbf{k}_5 \cdot \mathbf{r}} + ie^{i\mathbf{k}_6 \cdot \mathbf{r}} - e^{i\mathbf{k}_7 \cdot \mathbf{r}} + ie^{i\mathbf{k}_8 \cdot \mathbf{r}}$$

$$= e^{ik'(x+y)} + ie^{ik_0 x} - e^{ik'(x-y)} + ie^{-ik_0 y} + e^{-ik'(x+y)} + ie^{-ik_0 x} - e^{-ik'(x-y)} + ie^{ik_0 y}$$

$$= -4\sin(k'x)\sin(k'y) + 2i\left[\cos(k_0 x) + \cos(k_0 y)\right] \equiv -4A(x,y) + 2iB(x,y),$$

where $k' \equiv \frac{\sqrt{2}}{2}k_0$, and $k_0 \equiv \frac{2\pi}{\lambda} n \sin\theta$. Therefore,

$$\begin{aligned} I(x,y) &= \left[4A(x,y) - 2iB(x,y)\right]\left[4A(x,y) + 2iB(x,y)\right] \\ &= 16\sin^2(k'x)\sin^2(k'y) + 4\left[\cos(k_0 x) + \cos(k_0 y)\right]^2. \end{aligned} \tag{1.36}$$

Thus, the holographic pattern can be regarded as resulting from the superposition of two contributions, namely $I_1(x,y) = (4\sin(k'x)\sin(k'y))^2$ and $I_2(x,y) = 4\left[\cos(k_0 x) + \cos(k_0 y)\right]^2$, each one characterized by a different wave number, and both exhibiting a characteristic 4-fold symmetry. For the sake of illustration, in Fig. 1.30a–b we plot the $I_1(x,y)$ and $I_2(x,y)$ contributions for the case $k_0 \equiv 1$, respectively. Now, since the wavenumbers k' and k_0 are incommensurate to each other, the field given by Eq. (3.42) exhibits an overall QP pattern characterized by the presence of both 4- and 8-fold local arrangements of relative maxima, as it is illustrated in Fig. 1.30c.

1.10 *Obtain Eq. (1.30).*

In order to properly express Eq. (1.28) in the form given by Eq. (1.29), we make use of the scaling property $\delta(\lambda z) = |\lambda|^{-1}\delta(z)$, $\lambda \in \mathbb{R} - \{0\}$, to get $\delta(\eta \sin\varphi - (n_x a_L - \xi \cos\varphi)) = \csc\varphi \, \delta(\eta - (n_x a_L \csc\varphi - \xi \cot\varphi)) \equiv \csc\varphi \, \delta(\eta - a)$, and $\delta(\eta \cos\varphi - (n_y a_L + \xi \sin\varphi)) = \sec\varphi \, \delta(\eta - (n_y a_L \sec\varphi + \xi \tan\varphi)) \equiv \sec\varphi \, \delta(\eta - b)$. Making use of these Dirac delta expressions in Eq. (1.28) we have

$$\mathcal{RL} = \sec\varphi \csc\varphi \sum_{n_x,n_y}^{\infty} \int_{-\infty}^{+\infty} \delta(\eta - a)\delta(\eta - b) \, d\eta = \sec\varphi \csc\varphi \sum_{n_x,n_y}^{\infty} \delta(b - a), \tag{1.37}$$

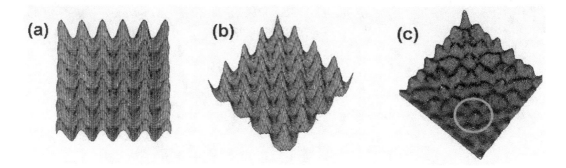

Figure 1.30   Intensity field pattern corresponding to the contributions (a) $I_1(x,y)$, and (b) $I_2(x,y)$, for the values $k_0 \equiv 1$ and $k' = \sqrt{2}/2$. The considered frames cover the range $-12 \leq x \leq 12$ and $-12 \leq y \leq 12$. (c) Overall intensity field pattern $I(x,y) = I_1(x,y) + I_2(x,y)$. Encircled in the right bottom we highlight a local octagonal ring of relative maxima. The considered frame covers the range $-16 \leq x \leq 16$ and $-16 \leq y \leq 16$.

where we have used Eq. (1.29). Now,

$$b - a = \xi(\cot \varphi + \tan \varphi) + n_y a_L \sec \varphi - n_x a_L \csc \varphi = \frac{\xi - (n_x \cos \varphi - n_y \sin \varphi) a_L}{\sin \varphi \cos \varphi} \quad (1.38)$$

and plugging Eq. (1.38) into Eq. (1.37), we finally obtain

$$\mathcal{R}L = \sum_{l=1}^{\infty} \delta(\xi - \xi_l), \quad (1.39)$$

where $\xi_l \equiv (n_x \cos \varphi - n_y \sin \varphi) a_L$.

1.11 *In the Zn-Yb-Au alloy system the presence of an i-AlYbAu QC ($a_{QC} = 7.383$ Å) and a 2/1 approximant crystal has been reported [369]. Making use of Eq. (1.32) determine the size of the approximant unit cell.*

By plugging the 6D lattice parameter of the iQC, a_{QC}, into Eq. (1.32), we get

$$a_{2/1} = \sqrt{\frac{2}{2+\tau}\tau^3} \times 7.383 = 23.252\,(7) \text{ Å}.$$

The experimentally measured cell parameter is $a_{2/1} = 23.271(2)$ Å[369], which reasonably agrees with the theoretically expected value.

CHAPTER 2

Hamiltonian Quasicrystals

2.1 FROM HYPERLATTICES TO QUASIPERIODIC SOLIDS

In Chapter 1, we grabbed the limelight onto purely structural aspects of intermetallic QCs, emphasizing the role of their constituent atoms as point diffracting centers. In so doing, we focused on the mass distribution function $\rho(\mathbf{r})$, and its related atomic arrangement was either described in terms of QP point lattices in physical space, or periodic hyperlattices in the crystallographical superspace. In this Chapter intermetallic QCs are no longer regarded as mere collections of points, but as bulk solids composed of atoms chemically bounded to each other, thereby adopting the viewpoint of solid state physics.

In order to illustrate the main theme, let us consider the aperiodic system shown in Fig. 2.1. During the 1940s DNA was regarded as a trivially periodic tetranucleotide macromolecule, say $(ACTG)_n$,[1] unable to store the amount of information required for the governance of cell function. At variance with this view Schrödinger suggested that a gene consists of a long sequence of a few repeating elements exhibiting a well defined order *without* the recourse of periodic repetition, and illustrated the vast combinatorial possibilities of such a structure (See Sec. 1.1.5). In this way, the notion of a one-dimensional (1D) *aperiodic solid* was originally introduced by considering DNA as a 1D aperiodic chain, with four different nucleotides properly arranged to store the required genetic information [789]. Thus, along with a periodic atomic arrangement in the sugar-phosphate DNA backbone (Fig. 2.1b), one also has an aperiodic distribution of informative chemical order at the *molecular scale*, as determined by the base pairs sequence along the double-helix axis (Fig. 2.1a). The chemical order of the base pairs sequence itself is beyond the reach of diffraction techniques, but it can be properly characterized via *ab-initio* quantum chemistry calculations, which nicely highlight the emergence of molecular orbitals when going from the atomic to the molecular scale. Therefore, from a structural point of view DNA fiber crystals could be classified as a periodic crystal (with helical symmetry!) if one attends to the sugar-phosphate scaffold exclusively, but they could be regarded as aperiodic solid representatives if we focus on the nucleobase system *electronic structure* instead. Accordingly, one may think of DNA as a sort of hybrid order system exhibiting both aperiodic (nucleobase subsystem) and periodic (sugar-phosphate backbone double-helix subsystem) order features [547].

In this way, the DNA case conveniently illustrates our main point: in the conceptual route leading from a QP lattice to a QP solid we must endow lattice points with both physical (mass, size, charge, and spin) and chemical (s, p, d, or f nature of electronic states; ionic,

---

[1]Here the letters A, C, T, and G respectively stand for Adenine, Cytosine, Thymine, and Guanine nucleotides.

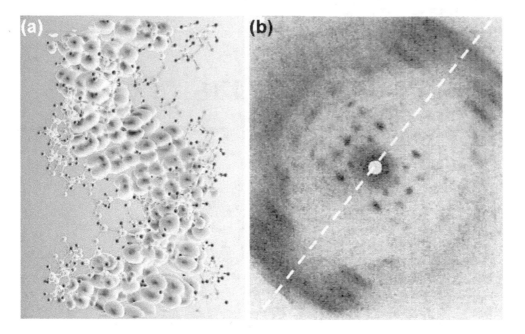

Figure 2.1 Two different views of DNA aperiodic solid: (a) molecular scale view illustrated by surfaces of constant charge density for the states corresponding to the lowest unoccupied band and highest occupied bands of a short DNA molecule in the A form in dry conditions [23]. These molecular orbitals are related to the guanine and cytosine nucleobases located in the core of the double-helix structure, and around them we can see the atoms belonging to the outer sugar-phosphate backbone (Courtesy of Emilio Artacho). (b) Atomic scale view illustrated by an X-ray diffraction pattern from A-DNA fibers consisting of a regular distribution of Bragg reflections describing the regular arrangement of DNA molecules in the fiber crystal. The dashed line indicates the fiber's meridiane axis (With permission from Zimmermann S B and Pheiffer B H, Helical parameters of DNA do not change when DNA fibers are wetted: X-ray diffraction study, Proc. Natl. Acad. Sci. *USA* 76, 2703–2907, 1979).

covalent or metallic nature of bonds) attributes, and properly incorporate these ingredients in a Hamiltonian model relating the underlying geometrical order of the QP lattice point to the resulting physical and chemical properties of the related QP solid. To this end, in the present Chapter, we will introduce suitable model Hamiltonians, which allow us to describe: (1) the interactions among the different sorts of atoms composing metallic QCs (chemical bonding), (2) their relative motions (phonons, phasons, and atomic diffusion effects), and (3) the dynamics of electrons and other elementary excitations propagating throughout them. In this way, we address a key question in the theory of aperiodic crystals, regarding the relationship between their atomic spatial order, determined by a given aperiodic distribution of atoms *and* bonds all over space, and the physical properties stemming from the resulting physicochemical network. In fact, since the vibrational and electronic features of a solid depend on both the spatial distribution of its constituent atoms and their bonding properties one may reasonably expect that aperiodic crystals belonging to different structural classes

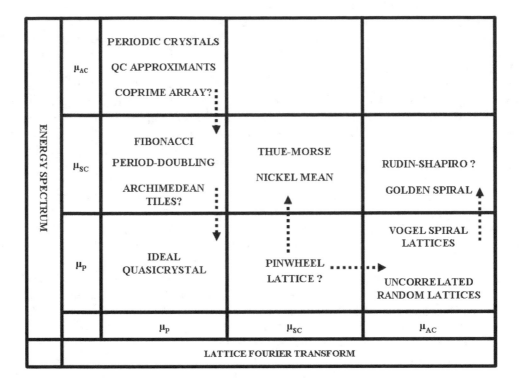

ENERGY SPECTRUM	μ_{AC}	PERIODIC CRYSTALS QC APPROXIMANTS COPRIME ARRAY?		
	μ_{SC}	FIBONACCI PERIOD-DOUBLING ARCHIMEDEAN TILES?	THUE-MORSE NICKEL MEAN	RUDIN-SHAPIRO ? GOLDEN SPIRAL
	μ_P	IDEAL QUASICRYSTAL	PINWHEEL LATTICE ?	VOGEL SPIRAL LATTICES UNCORRELATED RANDOM LATTICES
		μ_P	μ_{SC}	μ_{AC}
	LATTICE FOURIER TRANSFORM			

Figure 2.2 Classification of aperiodic systems attending to the spectral measures of their lattice Fourier transform (abscissas) and their Hamiltonian spectrum energy (ordinates). (Updated from Ref. [559] with permission. © 2017 by WILEY-VCH Verlag GmbH & Co. KGaA, Weinheim).

would exhibit some characteristic distinctive features in their electron and phonon spectra structures as well.

2.2 SPECTRAL CLASSIFICATION CHARTS

A mathematically convenient way to account for both the structure and the energy (or frequency) related properties of solids is provided by the use of *measures*, which are linear maps that associate a number to an appropriate function.[2] From the view point of condensed matter physics there are two important measures one can consider when studying the properties of solid materials. For one thing, we have a measure related to the atomic density distribution Fourier transform, which discloses the main features of X-ray, electron or neutron diffraction patterns resulting from the spatial structure of the solid. On the other hand, we have a measure related to the energy (frequency) spectra of the system, respectively describing its electronic structure, or the frequency distribution of the atomic vibrations in the case of the phonon spectrum.

In order to gain a deeper insight on the relationship between the structural order present in an aperiodic solid (as determined by its diffraction pattern) and its related transport

---

[2]A well known and widely used measure is the Lebesgue measure $\mu(f) \equiv \int f(x)dx$, which is commonly used in integration of functions $f(x)$ on the real numbers $\mathbb{R}$ [311].

properties, stemming from its energy and frequency spectra main features and the nature of its eigenstates, we will exploit the Lebesgue's decomposition theorem, which states that any measure μ can be uniquely decomposed in terms of three kinds of spectral components (and mixtures of them), namely: pure point (μ_P), absolutely continuous (μ_{AC}), and singularly continuous (μ_{SC}) spectra, in the form $\mu = \mu_P \cup \mu_{AC} \cup \mu_{SC}$ [32, 311]. To this end, we introduce the spectral charts depicted in Figs. 2.2 and 2.3, where we provide a graphical classification scheme of aperiodic systems based on the Lebesgue's spectral components of their diffraction spectra (in abscissas) and their energy spectra (in ordinates), respectively. In this way, the old-fashioned classification scheme of solids based on the periodic-amorphous dichotomy is replaced by a much richer one, including nine different entries [80, 152, 154, 156, 544, 555].

Let us inspect Fig. 2.2. In the upper left corner we have the periodic crystals exhibiting pure point Fourier spectra (discrete Bragg diffraction peaks) and an absolutely continuous energy spectrum (Bloch wave functions in allowed bands). In the same box we place the periodic approximants to QCs, I have also tentatively included the pure-point diffracting co-prime array (whose nodes are given by prime number coordinates in the 2D plane) [154] on the basis that the existence of short-range correlation given by the so-called twin-primes (i.e., 11–13, or 101–103) guarantees the existence of minibands containing extended states in the energy spectrum[3] [762, 770]. Fourier spectra exhibiting typical singular continuous features have been described from the study of 2D point arrays generated by using the distribution of the prime numbers on complex quadratic fields and quaternions rings [951]. In the lower right corner we have amorphous matter representatives, described in terms of uncorrelated random lattices, exhibiting an absolutely continuous Fourier spectrum (diffuse spectra) and a pure point energy spectrum (exponentially localized wave functions) [122, 984]. In the right bottom corner I have tentatively located Vogel spiral lattices[4], which provide an interesting instance of perfectly ordered systems where both translational and orientational symmetries are discarded by construction (see Sec. 1.4.1). Accordingly, their Fourier transform does not show sharp reflections, but a continuous background, along with circularly symmetric concentric scattering rings containing patchy features (see Fig. 1.16), thereby displaying a lot of spatial frequency components [154, 156], the radii of the rings indicating the dominant spatial frequencies of the structure [902]. Detailed numerical calculations of the spatial structure of a number of optical modes in different spiral lattices with a varying degree of structural order have disclosed the existence of modes with different localization degrees, ranging from modes confined within rings of different radii to modes located near the boundary of the spiral lattice, hence allowing for some light leakage through the border [109, 134, 155]. Light localization in Vogel spirals is driven by collective electromagnetic coupling effects that involve multiple length scales, leading to a preferential spatial localization of radiation in the radial direction [798]. This rich localization behavior suggests these optical modes may be related to a singular continuous frequency spectrum, rather than to an absolutely continuous one. Thus, in the absence of rigorous mathematical results on the nature of these spectra, I have tentatively located the golden-angle based spiral lattice in the central row of the spectral chart. In the left lower corner we include the ideal quasicrystals (i.e., those exhibiting a perfect scale invariance symmetry), since theoretical arguments suggest that in a perfect self-similar structure the presence of coherent resonance effects among the electronic states may efficiently induce their localization, leading to a perfectly insulating solid phase [928, 930].

---

[3]To the best of my knowledge, the possible presence of a singular continuous component in the energy spectrum of the 2D coprime array has not been rigorously excluded, which is indicated by a dotted arrow in the spectral chart.

[4]Please, note that in previous versions of this spectral chart I tentatively placed spiral lattices at the upper right corner instead.

TABLE 2.1   Substitution sequences most widely considered in the study of aperiodic crystals based on a binary alphabet, where n and m are positive integers.

SEQUENCE	SUBSTITUTION RULE
Fibonacci	$g(A) = AB \quad g(B) = A$
Silver mean	$g(A) = AAB \quad g(B) = A$
Bronze mean	$g(A) = AAAB \quad g(B) = A$
Precious means ($n > 3$)	$g(A) = A^n B \quad g(B) = A$
Octonacci	$g(A) = B \quad g(B) = BAB$
Fibonacci-class	$g(A) = B^{n-1}AB \quad g(B) = B^{n-1}A$
Period-doubling	$g(A) = AB \quad g(B) = AA$
Copper mean	$g(A) = ABB \quad g(B) = A$
Nickel mean	$g(A) = ABBB \quad g(B) = A$
Metallic means ($m > 3$)	$g(A) = AB^m \quad g(B) = A$
Mixed means	$g(A) = A^n B^m \quad g(B) = A$
Thue–Morse	$g(A) = AB \quad g(B) = BA$
Kolakoski	$g(A) = A^n B^n \quad g(A) = A^m B^m$
Periodic	$g(A) = AB \quad g(B) = AB$

TABLE 2.2   Substitution sequences considered in the study of aperiodic crystals based on ternary and quaternary alphabets.

SEQUENCE	SET $\mathcal{A}$	SUBSTITUTION RULE
Ternary Fibonacci	$\{A, B, C\}$	$g(A) = AC \quad g(B) = A$ $g(C) = B$
Tribonacci	$\{A, B, C\}$	$g(A) = AB \quad g(B) = AC$ $g(C) = A$
Kolakoski	$\{A, B, C\}$	$g(A) = ABC \quad g(B) = AB$ $g(C) = B$
Rudin-Shapiro	$\{A, B, C, D\}$	$g(A) = AB \quad g(B) = AC$ $g(C) = DB \quad g(D) = DC$
Paper folding	$\{A, B, C, D\}$	$g(A) = AB \quad g(B) = CB$ $g(C) = AD \quad g(D) = CD$

In the central row we include several 1D structures based on substitution sequences which have been extensively studied in the literature during the last three decades. A substitution sequence is formally defined by the action of a substitution rule on an alphabet $\mathcal{A} = \{A, B, C, ...\}$, consisting of a certain number of letters. The corresponding aperiodic sequence is then obtained by iterating the substitution rule starting from a given letter of the set $\mathcal{A}$ in order to obtain an aperiodic string of letters, as it is shown in Tables 2.1 and 2.2 for binary, ternary, and quaternary alphabets.[5]

By inspecting the central row of Fig. 2.2, one realizes that although Fibonacci, Thue-Morse, nickel mean, and Rudin-Shapiro based systems share the same kind of energy spectrum (a purely singular continuous one), they have different lattice Fourier transforms, so that these aperiodic crystals must be properly classified in separate categories.

---

[5] We recall that most metallic QCs found to date are ternary alloys containing three different chemical elements. The few examples of binary metallic QCs representatives are provided by $iCd_{85}Yb_{15}$, $iCd_{85}Ca_{15}$, $Cd_{88}RE_{12}$, $Zn_{88}Sc_{12}$, and $ddTa_{1.6}Te$.

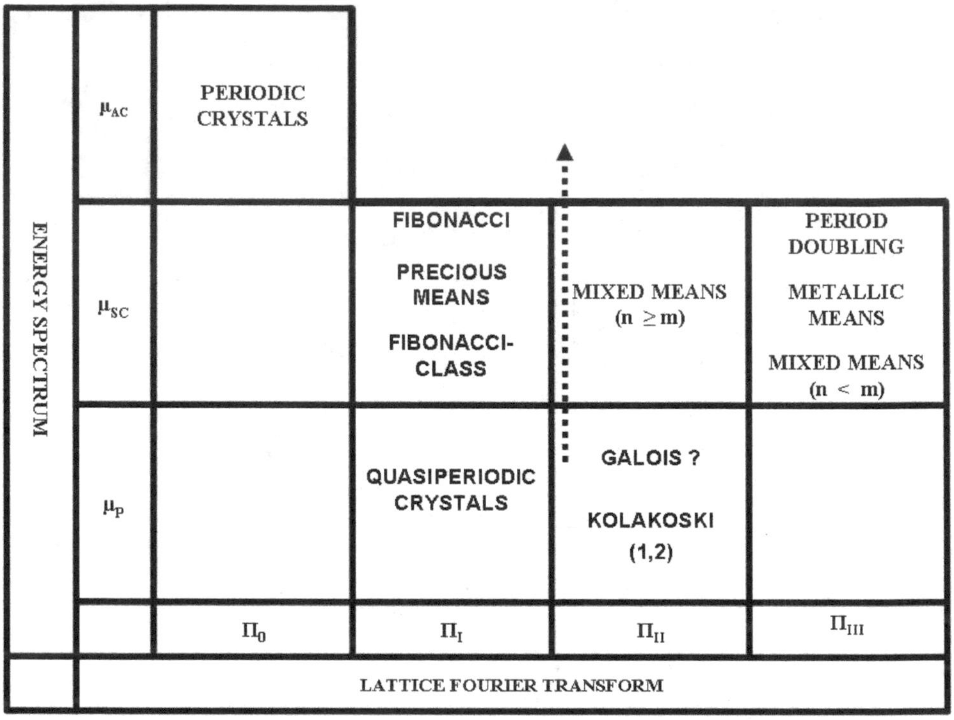

ENERGY SPECTRUM		Π_0	Π_I	Π_{II}	Π_{III}
	μ_{AC}	PERIODIC CRYSTALS			
	μ_{SC}		FIBONACCI PRECIOUS MEANS FIBONACCI-CLASS	MIXED MEANS (n ≥ m)	PERIOD DOUBLING METALLIC MEANS MIXED MEANS (n < m)
	μ_P		QUASIPERIODIC CRYSTALS	GALOIS ? KOLAKOSKI (1,2)	

LATTICE FOURIER TRANSFORM

Figure 2.3   Zoom out of the pure point lattice Fourier spectrum column in Fig. 2.2 showing the splitting of the pure point Fourier spectral measure μ_P in four separate classes labeled Π_0, Π_I, Π_{II}, and Π_{III} (more details in the text). The tentative assignation of Galois sequence is based on the localization properties reported in Ref. [942] from numerical studies of on-site and transfer models (see Sec. 2.3.1). More recent studies, however, disclose the presence of extended states depending on the adopted model parameters [479, 480]. (Updated from Ref. [559] with permission. © 2017 by WILEY-VCH Verlag GmbH & Co. KGaA, Weinheim).

Finally, it has been rigorously shown that there are no discrete components in the diffraction spectrum of the so-called pinwheel lattice, which is a tiling generated by successive decomposition of a right triangle into five smaller, congruent copies of itself, and has the property that their tiles appear in infinitely many orientations [718]. However, it is not currently (2020) known if its Fourier spectrum is absolutely continuous or singular continuous. Accordingly, I have tentatively located the pinwheel lattice in the central box of the bottom row in the spectral chart, also assuming that its states are mainly localized in nature.

By attending to finer details in their related diffraction patterns, pure point lattice Fourier measures can be further split into four separate groups, namely, the so-called periodic (Π_0), QP (Π_I), limit-QP (Π_{II}), and limit-periodic (Π_{III}) pure point classes, respectively [233, 439, 440, 518, 650]. The diffraction spectra of Π_0 and Π_I classes representatives both consist in Bragg reflections supported by a *finite* Fourier module whose rank either equals the physical space dimension (Π_0 class) or is larger than it (Π_I class). In the limit-QP and limit-periodic classes Π_{II} and Π_{III} one also finds a diffraction spectrum consisting of a dense distribution of Bragg reflections. However, this distribution is supported by a Fourier module with a countably *infinity* of generators over the integers (i.e., the reciprocal space has infinite dimensions) [270, 518]. The point-group symmetry of the diffraction spectrum of the limit-periodic structures belonging to the Π_{III} spectral class is compatible with periodicity (unlike QCs) and their overall atomic structure can be described in terms of a union of periodic substructures with ever increasing lattice constants, forming a sequence of successive sublattices [582, 825]. Analogously, the diffraction spectrum of the limit-QP class representatives can be generated by a superposition of spectra of an infinite number of QP patterns. Geometrically, a limit-QP structure can be regarded as a section of a limit-periodic lattice in a higher dimension space, just as QP structures can be obtained as sections of periodic lattices in high dimensions [270, 651].

In Fig. 2.3, we conveniently zoom out the left column of Fig. 2.2 spectral chart to provide some illustrative examples of several binary aperiodic crystals (BACs) exhibiting pure point spectra classified according to the above scheme. In order to properly assign to a given aperiodic structure based on a substitution sequence on a given alphabet $\mathcal{A} = \{A, B, C, \ldots\}$ the spectral class to which it belongs one usually relies on certain algebraic properties of its related substitution matrix:

$$\mathbf{S} = \begin{pmatrix} n_A[g(A)] & n_A[g(B)] & n_A[g(C)] & \cdots \\ n_B[g(A)] & n_B[g(B)] & n_B[g(C)] & \cdots \\ n_C[g(A)] & n_C[g(B)] & n_C[g(C)] & \cdots \\ \cdots & \cdots & \cdots & \cdots \end{pmatrix}, \tag{2.1}$$

where $n_i[g(j)]$ indicates the number of times a given letter i appears in the substitution sequence $g(j)$, irrespective of the order in which these letters occur. The dimension of the substitution matrix is then determined by the number of different letters included in the basic alphabet $\mathcal{A}$. In Tables 2.3 and 2.4, we list the substitution matrices for the sequences listed in Tables 2.1 and 2.2, respectively, along with pertinent algebraic information [32, 479, 480, 811]. By inspecting these Tables we realize that the substitution matrix does not give a complete description of a substitution rule, because different substitution rules can have the same substitution matrix (e.g., Thue-Morse and binary periodic sequences). Notwithstanding this, one can obtain relevant information about the diffraction properties of different aperiodic crystals by considering two algebraic magnitudes related to the corresponding substitution matrix, namely, its determinant and its characteristic polynomial, $P_\lambda = |\mathbf{S} - \lambda \mathbf{I}|$, from which we obtain the characteristic eigenvalues and eigenvectors. We note that different substitution sequences can have the same characteristic polynomial, albeit having different substitution matrices, as it occurs for the silver mean and Octonacci sequences, or period-doubling and copper mean sequences, for instance, or more generally with the representatives of precious means and Fibonacci-class, respectively.

The components of the eigenvector related to the substitution matrix eigenvalue with larger modulus (sometimes referred to as Frobenius eigenvalue), say λ_+, once normalized, read

$$v_A^+ = \frac{\lambda_+ - n_B[g(B)]}{\lambda_+ + n_B[g(A)] - n_B[g(B)]} \qquad v_B^+ = \frac{n_B[g(A)]}{\lambda_+ + n_B[g(A)] - n_B[g(B)]} \tag{2.2}$$

for binary sequences (analogous expressions are obtained for sequences containing three or more letters) and they respectively indicate the frequencies of letters A and B in the infinite sequence $N \to \infty$ limit. For instance, making use of the data included in Table 2.3, we readily obtain the well-known results $v^+ = (\tau^{-1}, \tau^{-2})$, $v^+ = (1/2, 1/2)$, and $v^+ = (2/3, 1/3)$, for the Fibonacci, Thue-Morse, and period-doubling sequences, respectively.

TABLE 2.3 Substitution matrix $\mathbf{S}$, its determinant value, characteristic polynomial P_λ, and its eigenvalues for the sequences listed in Table 2.1. The corresponding spectral class is included in the last column. Those sequences satisfying the Pisot property are highlighted (bold face) in the first column.

SEQUENCE	$\mathbf{S}$	det $\mathbf{S}$	P_λ	EIGENVALUES	CLASS
Fibonacci	$\begin{pmatrix} 1 & 1 \\ 1 & 0 \end{pmatrix}$	-1	$\lambda^2 - \lambda - 1$	$\lambda_\pm = \frac{1\pm\sqrt{5}}{2}$	Π_I
Silver mean	$\begin{pmatrix} 2 & 1 \\ 1 & 0 \end{pmatrix}$	-1	$\lambda^2 - 2\lambda - 1$	$\lambda_\pm = 1 \pm \sqrt{2}$	Π_I
Bronze mean	$\begin{pmatrix} 3 & 1 \\ 1 & 0 \end{pmatrix}$	-1	$\lambda^2 - 3\lambda - 1$	$\lambda_\pm = \frac{3\pm\sqrt{13}}{2}$	Π_I
Precious means	$\begin{pmatrix} n & 1 \\ 1 & 0 \end{pmatrix}$	-1	$\lambda^2 - n\lambda - 1$	$\lambda_\pm = \frac{n\pm\sqrt{n^2+4}}{2}$	Π_I
Octonacci	$\begin{pmatrix} 0 & 1 \\ 1 & 2 \end{pmatrix}$	-1	$\lambda^2 - 2\lambda - 1$	$\lambda_\pm = 1 \pm \sqrt{2}$	Π_I
Fibonacci-class	$\begin{pmatrix} 1 & 1 \\ n & n-1 \end{pmatrix}$	-1	$\lambda^2 - n\lambda - 1$	$\lambda_\pm = \frac{n\pm\sqrt{n^2+4}}{2}$	Π_I
Period-doubling	$\begin{pmatrix} 1 & 2 \\ 1 & 0 \end{pmatrix}$	-2	$\lambda^2 - \lambda - 2$	$\lambda_+ = 2 \quad \lambda_- = -1$	Π_{III}
Copper mean	$\begin{pmatrix} 1 & 1 \\ 2 & 0 \end{pmatrix}$	-2	$\lambda^2 - \lambda - 2$	$\lambda_+ = 2, \quad \lambda_- = -1$	Π_{III}
Nickel mean	$\begin{pmatrix} 1 & 1 \\ 3 & 0 \end{pmatrix}$	-3	$\lambda^2 - \lambda - 3$	$\lambda_\pm = \frac{1\pm\sqrt{13}}{2}$	Γ
Metallic means $(m = \ell(\ell+1))$	$\begin{pmatrix} 1 & 1 \\ m & 0 \end{pmatrix}$	$-m$	$\lambda^2 - \lambda - m$	$\lambda_+ = \ell+1, \quad \lambda_- = \ell$	Π_{III}
Mixed means $(n \geq m)$	$\begin{pmatrix} n & 1 \\ m & 0 \end{pmatrix}$	$-m$	$\lambda^2 - n\lambda - m$	$\lambda_\pm = \frac{n\pm\sqrt{n^2+4m}}{2}$	Π_{II}
Mixed means $(n < m)$	$\begin{pmatrix} n & 1 \\ m & 0 \end{pmatrix}$	$-m$	$\lambda^2 - n\lambda - m$	$\lambda_\pm = \frac{n\pm\sqrt{n^2+4m}}{2}$	Π_{III}
Thue-Morse	$\begin{pmatrix} 1 & 1 \\ 1 & 1 \end{pmatrix}$	0	$\lambda^2 - 2\lambda$	$\lambda_+ = 2, \quad \lambda_- = 0$	Γ
Periodic	$\begin{pmatrix} 1 & 1 \\ 1 & 1 \end{pmatrix}$	0	$\lambda^2 - 2\lambda$	$\lambda_+ = 2, \quad \lambda_- = 0$	Π_0
Galois	$\begin{pmatrix} 2 & 2 \\ 1 & 1 \end{pmatrix}$	0	$\lambda^2 - 3\lambda$	$\lambda_+ = 3, \quad \lambda_- = 0$	Π_{II}
Kolakoski	$\begin{pmatrix} n & m \\ n & m \end{pmatrix}$	0	$\lambda(\lambda - n - m)$	$\lambda_+ = n+m, \quad \lambda_- = 0$	Π_{II}

TABLE 2.4 Substitution matrix $\mathbf{S}$, its determinant value, and its eigenvalues for the sequences listed in Table 2.2. The corresponding spectral class is included in the last column. Those sequences satisfying the Pisot property are highlighted (bold face) in the first column.

SEQUENCE	$\mathbf{S}$	det$\mathbf{S}$	EIGENVALUES	CLASS
Plastic	$\begin{pmatrix} 0 & 0 & 1 \\ 1 & 0 & 1 \\ 0 & 1 & 0 \end{pmatrix}$	1	$\lambda_+ = 1.324\ldots, \ |\lambda_{2,3}| = \lambda_+^{-1/2}$	Π_I
Ternary Fibonacci	$\begin{pmatrix} 1 & 1 & 0 \\ 0 & 0 & 1 \\ 1 & 0 & 0 \end{pmatrix}$	1	$\lambda_+ \simeq 1.466\ldots, \ |\lambda_{2,3}| = \lambda_+^{-1/2}$	Π_I
Tribonacci	$\begin{pmatrix} 1 & 1 & 1 \\ 1 & 0 & 0 \\ 0 & 1 & 0 \end{pmatrix}$	1	$\lambda_+ \simeq 1.839\ldots, \ |\lambda_{2,3}| = \lambda_+^{-1/2}$	Π_I
Kolakoski (3,1)	$\begin{pmatrix} 1 & 1 & 0 \\ 1 & 1 & 1 \\ 1 & 0 & 0 \end{pmatrix}$	1	$\lambda_+ \simeq 2.205\ldots, \ |\lambda_{2,3}| = \lambda_+^{-1/2}$	Π_I
Kolakoski (4,2)	$\begin{pmatrix} 2 & 0 & 1 \\ 1 & 1 & 1 \\ 0 & 2 & 1 \end{pmatrix}$	0	$\lambda_\pm = 2 \pm \sqrt{7}, \ \lambda_3 = 0$	Π_{II}
Paper folding	$\begin{pmatrix} 1 & 0 & 1 & 0 \\ 1 & 1 & 0 & 0 \\ 0 & 1 & 0 & 1 \\ 0 & 0 & 1 & 1 \end{pmatrix}$	0	$\lambda_+ = 2, \ \lambda_2 = 1, \ \lambda_{3,4} = 0$	Π_{III}
Rudin-Shapiro	$\begin{pmatrix} 1 & 1 & 0 & 0 \\ 1 & 0 & 1 & 0 \\ 0 & 1 & 0 & 1 \\ 0 & 0 & 1 & 1 \end{pmatrix}$	0	$\lambda_+ = 2, \ \lambda_2 = 0, \ \lambda_{3,4} = \pm\sqrt{2}$	μ_{AC}

The knowledge of the determinant and the characteristic eigenvalues of a given $\mathbf{S}$ matrix allow us to properly classify the related BAC within the different pure point spectral classes [84, 96, 294]. To this end, it is convenient to introduce the so-called Pisot-Vijayaraghavan (PV) numbers, which are real algebraic numbers (i.e., a number which is obtained from the solution of an algebraic equation) greater than one, all of whose conjugate elements (the other solutions of the algebraic equation) lie inside the open unit disk (i.e., they have absolute value *strictly* less than unity). If at least one of the conjugate elements equals unity then it will be referred to as a Salem number. The smallest PV number (known as the *plastic number*) is related to the substitution sequence given by $g(A) = B$, $g(B) = C$, $g(C) = AB$ (see Table 2.4), with characteristic polynomial $P_\lambda = \lambda^3 - \lambda - 1$, so that $\lambda_+ = [(9 + \sqrt{69})^{1/3} + (9 - \sqrt{69})^{1/3}]/18^{1/3} = 1.3247\ldots$ [32].

The Bombieri-Taylor conjecture provides a sufficient condition for the absence of quasiperiodicity in a sequence generated from the successive application of a given substitution rule, and reads as follows: if the spectrum of the substitution matrix $\mathbf{S}$ contains a PV number, then the lattice can be QP; otherwise it is not [84, 294]. On the basis of this result one can introduce a useful classification scheme as follows: if the substitution matrix characteristic polynomial P_λ contains a PV number, *and* $|\det \mathbf{S}| = 1$, then the lattice belongs to the QP class Π_I. Alternatively, if the sequence satisfies the PV property but $|\det \mathbf{S}| \neq 1$, then the lattice belongs to the limit-QP Π_{II} class instead [439, 440]. Finally, if the sequence does not satisfy the Pisot property the lattice either belongs to the limit-periodic Π_{III} class or it must be located elsewhere in the Fig. 2.2 spectral chart. For instance, the nickel mean and the Thue-Morse sequences belong to the so-called Γ class, since they have pure singular continuous diffraction spectra instead of pure point ones. Indeed, it is important to highlight that the Pisot property provides a necessary condition for a pure point diffraction spectrum, but it is not sufficient. Accordingly, the Pisot property $\lambda_+ > 1$ and $|\lambda_j| < 1$ provides a first criterion which demarcates certain aperiodic crystals possessing Bragg reflections from the rest. The unimodular condition $|\det \mathbf{S}| = 1$ gives a second criterion to distinguish between QCs (Π_I class) and limit-QP structures (Π_{II} class). Thus, among the mixed mean sequences we can find representatives belonging to both the Π_{II} and Π_{III} spectral classes, depending on their n and m values. We note that the substitution matrices corresponding to the copper mean and the period-doubling sequences are mutually transposed, so that they have the same eigenvalues and determinant values, thereby belonging to the same spectral class Π_{III}. Analogously, albeit the silver mean and Octonacci substitution sequences lead to quite different letter strings they have the same characteristic polynomial, so that both belong to the same Π_I class (**Exercises 2.1 and 2.2**).

All substitution sequences listed in Tables 2.3 and 2.4 are primitive. A substitution rule is called primitive when all entries of the substitution matrix power $\mathbf{S}^N$ are strictly positive integers (i.e., $s_{ij} \neq 0$), for some $N \geq 1$. This means that there exists an integer n such that every letter, say A_i, in the alphabet is contained in each word $g^n(A_j)$, that is, for some order of the substitution the images of the letter share the same symbol at least at one position. The primitive nature of a given substitution guarantees that, (i) the word resulting from the successive application of the corresponding substitution sequence is self-similar in the $N \to \infty$ limit, and (ii) λ_+ is real, positive and larger than one [518]. Accordingly, the substitution rule acts as a dilatation operator with a scaling factor which can be either integer (as occurs in the case of the copper, period doubling, Thue-Morse, Rudin-Shapiro and paper folding sequences with $\lambda_+ = 2$) or irrational, as it occurs in all Π_I class QCs (**Exercises 2.3 and 2.4**). A very representative example of non-primitive substitution sequence is provided by fractal lattices, which otherwise are prototypical representatives of self-similar structures (**Exercise 2.5**). For instance, one may consider the triadic Cantor set, which is obtained through the repetition of a simple rule: divide any given segment into three equal parts, then eliminate the central one, and continue this process. Though this is a usual way of obtaining a Cantor set, it is by no means the only one. More general Cantor sets can be generated by iterating the operation consisting in the division of a segment in $s = 2r - 1$ equal parts ($r > 2$) and the removal of $r - 1$ of its pieces. The resulting structures are self-similar and have a fractal dimension $D = \ln r / \ln s$. Alternatively, a Cantor structure can be obtained by successively applying the substitution rule $A \to ABA \dots BA$ (containing r A's) and $B \to BBB \dots B$ (containing s B's). Thus, the triadic Cantor set is obtained from the inflation process $A \to ABA$ and $B \to BBB$. We note that from a structural viewpoint QP lattices and fractal lattices belong to two different classes of aperiodic systems. In fact, one can readily check that the substitution matrix corresponding to a general Cantor lattice

$$\mathbf{S} = \begin{pmatrix} r & 0 \\ s - r & s \end{pmatrix}, \qquad (2.3)$$

has the eigenvalues $\lambda_+ = r > 1$ and $\lambda_- = s > 1$, so that none of the related sequences satisfies the PV property. The lack of QPO, however, does not prevent diffraction at all, since the self-similar arrangement of layers is perfectly able to do the job.

The substitution matrix formalism is very useful in order to introduce new substitution sequences in terms of previously known ones, on the basis of purely algebraic procedures. For instance, the Π_I class sequence referred to as squared Fibonacci sequence is directly obtained as[6]

$$\mathbf{S}^2 \equiv \begin{pmatrix} 1 & 1 \\ 1 & 0 \end{pmatrix}^2 = \begin{pmatrix} 2 & 1 \\ 1 & 1 \end{pmatrix}. \tag{2.4}$$

The resulting matrix can then interpreted in terms of the substitution rule $g(A) = AAB$ and $g(B) = AB$.[7] This process can be readily extended by considering general powers of the Fibonacci substitution matrix to get (by induction)

$$\mathbf{S}^n = \begin{pmatrix} 1 & 1 \\ 1 & 0 \end{pmatrix}^n = \begin{pmatrix} F_n & F_{n-1} \\ F_{n-1} & F_{n-2} \end{pmatrix}, \tag{2.5}$$

where all the entries of the power matrix are Fibonacci numbers, which guarantees this is a primitive substitution. One can verify that the related substitution sequences belong the Π_{I} class (**Exercise 2.6**).

The recourse to substitution sequences is not restricted to 1D lattices at all, but it can be readily extended to 2D and 3D aperiodic structures as well. For instance, the 2D square Fibonacci tiling introduced in Sec. 1.1.5 obeys the substitution rule (see Fig. 1.11) $g(A) = ABCC$, $g(B) = A$, and $g(C) = AC$. The substitution matrix reads,

$$S = \begin{pmatrix} 1 & 1 & 1 \\ 1 & 0 & 0 \\ 2 & 0 & 1 \end{pmatrix},$$

with the characteristic polynomial $P_\lambda = \lambda^3 - 2\lambda^2 - 2\lambda + 1 = (\lambda + 1)(\lambda^2 - 3\lambda + 1) \equiv (\lambda + 1)\tilde{P}_\lambda$, and the eigenvalues of the irreducible component $\tilde{P}_\lambda$ are $\lambda_\pm = \tau^{\pm 2}$, which satisfy the PV property. The scale factors characterizing the self-similar symmetry of the entire square Fibonacci tile can be obtained in terms of these eigenvalues by simply taking the corresponding square roots, namely, $\lambda_+ = \tau$ and $\lambda_- = \tau^{-1}$. The frequencies for the large squares, the small squares, and the rectangular tiles are τ^{-2}, τ^{-4}, and $2\tau^{-3}$, respectively. Alternatively, a 2D Fibonacci planar tiling can be obtained by exploiting the 1D substitution matrix $\mathbf{S}_{1D} = \begin{pmatrix} 1 & 1 \\ 1 & 0 \end{pmatrix}$ to form the Fibonacci-like *block* matrix (**Exercise 2.7**)

$$\mathbf{S}_{2D} = \begin{pmatrix} \mathbf{S}_{1D} & \mathbf{S}_{1D} \\ \mathbf{S}_{1D} & \mathbf{0} \end{pmatrix} = \begin{pmatrix} 1 & 1 & 1 & 1 \\ 1 & 0 & 1 & 0 \\ 1 & 1 & 0 & 0 \\ 1 & 0 & 0 & 0 \end{pmatrix}, \tag{2.6}$$

where $\mathbf{0} \equiv \begin{pmatrix} 0 & 0 \\ 0 & 0 \end{pmatrix}$.

2.3   ONE-DIMENSIONAL MODEL HAMILTONIANS

Broadly speaking, an obvious motivation for considering 1D models in solid state physics is the complexity of the full-fledged 3D problem. In the particular case of QCs this general motivation is further strengthened by the replacement of translational symmetry by a scale invariance one, along with the presence of long-range QPO all over the structure. This results in an involved QP network of atoms and bonds throughout space, which renders the 3D problem not readily amenable to analytical treatment. Fortunately enough, most

---

[6]Note that this result shows that the Fibonacci sequence is primitive.

[7]Other possible choices include $g(A) = ABA$, $g(B) = AB$, or $g(A) = BAA$ and $g(B) = BA$, for instance.

characteristic features of QP systems, such as the fractal structure of their energy and frequency spectra, as well as the spatial distribution of their related eigenstates, can be explained in terms of resonant coupling effects, which rely on the local isomorphism property discussed in Sec. 1.3.2. Therefore, the physical mechanisms at work are not so dependent on the dimension of the system, but are mainly determined by the repetitiveness of the underlying structure [32]. Consequently, the recourse to 1D model Hamiltonians can be considered as a promising starting point in order to relate the properties of QCs to the kind of order present in their structure, for such models encompass, in the simplest possible manner, most of the novel symmetries of QCs.

Thus, several studies have demonstrated that the electronic structure and vibrational spectrum of most 1D QP systems can be understood in terms of resonance effects involving a relatively small number of atomic clusters of progressively increasing size. In earlier works this scenario was discussed in terms of real-space based renormalization group approaches describing the mathematically simpler, but chemically unrealistic, diagonal (different types of atoms connected by the *same* type of bond) or off-diagonal (the same type of atom but *different* types of bonds between them) models [653, 654, 775, 884, 885, 777, 943, 772].[8] Later on an increasing number of works have been devoted to the mathematically more complex general case, in which both diagonal and off-diagonal terms are present in the system [117, 118, 119, 121, 125, 282, 456, 527, 774, 779]. In fact, since the properties of chemical bonds linking two different atoms generally depend on their chemical nature, any realistic treatment must explicitly consider that the aperiodic sequence of atoms along the chain naturally induces an aperiodic sequence of bonds in the considered solid. Indeed, a growing number of both experimental measurements and numerical simulation results highlight the important role of chemical bonding in the emergence of some specific physical properties of QCs (see Chapter 3). Accordingly, we will focus on 1D *general* QC models explicitly considering the role of a QP arrangement of both atoms *and* chemical bonds in the resulting electronic structure and frequency spectrum, in terms of Hamiltonians of the form

$$H_e = \sum_{k=1}^{N} V_k |k\rangle \langle k| + \sum_{m,k}^{N-1} t_{m,k} |m\rangle \langle k|, \quad (2.7)$$

$$H_{ph} = \frac{1}{2} \sum_{k=1}^{N} \frac{p_k^2}{m_k} + \frac{1}{2} \sum_{k=1}^{N-1} K_{k,k+1} (\eta_k - \eta_{k+1})^2, \quad (2.8)$$

for the electron and phonon dynamics, respectively, where N is the number of atoms in the lattice, V_k accounts for the atomic potentials at the k-th lattice site, $|k\rangle$ is the Wannier state at site k, the transfer integrals $t_{m,k}$ measure the hopping amplitudes of the electron between atoms, η_k is the displacement of the k-th atom from its equilibrium position, m_k is its mass, $p_k = m_k \dot{\eta}_k$ is the linear momentum, and $K_{k,k\pm 1}$ denotes the strength of the harmonic coupling between neighbor atoms.

Within the electron independent approximation the electron dynamics in a system described by Eq. (2.7) can be written in terms of a tight-binding model given by the Schrödinger equation (in $\hbar = 2m = 1$ units)

$$(E - V_k)\phi_k - t_{k,k-1}\phi_{k-1} - t_{k,k+1}\phi_{k+1} = 0, \quad (2.9)$$

---

[8]A possible realization of this model is trans-polyacetylene polymer where identical CH_2 units are linked by alternating simple (C-C) and double (C=C) bonds. The energy spectrum of a model polymer whose bonds alternate according to the Fibonacci sequence is shown in Fig. 2.4a. Another instance is provided by polyynic carbyne atomic wires where simple (C-C) and triple (C≡C) bonds alternate along the chain [468].

where ϕ_k stands for the amplitude of the wave function at site k, E is the electron energy, and we have considered, as a first approximation, next-neighbors interactions only. In an analogous way, the dynamical equation for normal modes of the form $\eta_k = u_k e^{i\omega t}$, related to Eq. (2.8), reads

$$(K_{k,k-1} + K_{k,k+1} - m_k\omega^2)u_k - K_{k,k-1}u_{n-1} - K_{k,k+1}u_{k+1} = 0, \tag{2.10}$$

where ω is the vibration frequency and u_k the k-th mode amplitude.

The QPO of the underlying structure can be introduced in several ways in the Hamiltonians and dynamical equations above, namely: (1) via the atomic potentials $\{V_k\}$ or atomic masses $\{m_k\}$ distributions (on-site or diagonal models), (2) by spatial modulation of hopping amplitudes $\{t_{m,k}\}$ or elastic constants $\{K_{k,k+1}\}$ sequences (transfer or off-diagonal models), or (3) by simultaneously considering an atomic and chemical bonding network, generally correlating both $\{V_k, m_k\}$ and $\{t_{m,k}, K_{k,k+1}\}$ values respectively (mixed or general models).

2.3.1 Energy and frequency spectra in 1D quasicrystals

"A fundamental question is: given a Schrödinger wave equation with a quasicrystalline potential field, how is the quasicrystalline symmetry manifest in the eigenvalue spectrum and the eigenfunctions?". (Julian D. Maynard, 2001)[592]

The dynamical equations given by Eqs. (2.9) and (2.10) are formally analogous to other similar expressions that are usually obtained when considering the propagation of other kinds of elementary excitations throughout aperiodic lattices, all of them adopting the canonical form [8, 532, 548],

$$v_k\phi_k - t_{k,k-1}\phi_{k-1} - t_{k,k+1}\phi_{k+1} = 0, \tag{2.11}$$

along with an appropriate set of boundary conditions. In Eq. (2.11), v_k depends on the excitation energy E, or frequency ω, as well as on other characteristic physical magnitudes of the system, such as atomic masses m_k, elastic constants $K_{k,k\pm1}$, or electronic binding energies V_k, as it is summarized in Table 2.5. In the mathematical literature, Eq. (2.11) is called a substitution Jacobi operator, which is related to the tridiagonal matrix

$$\begin{pmatrix} v_1 & t_{1,2} & 0 & \cdots & & \xi^*t_0 \\ t_{2,1} & v_2 & t_{2,3} & \ddots & & \vdots \\ 0 & t_{3,2} & v_3 & \ddots & & 0 \\ \vdots & \ddots & \ddots & \ddots & & t_{N-1,N} \\ \xi t_{N+1} & \cdots & & 0 & t_{N,N-1} & v_N \end{pmatrix}, \tag{2.12}$$

where the upper-right and lower-left corner elements account for boundary conditions, ξ being a suitable phase factor, and $t_{k,k\pm1} = t_{k\pm1,k} \neq 1$ and $V_k \neq 0$ in general. In realistic QCs models, the aperiodic sequence of atoms $\{v_k\}$ along the chain naturally induces an aperiodic sequence of bonds $\{t_{k,k\pm1}\}$ in the considered system, so that both sequences are correlated.

After Eq. (2.11) the elementary excitations amplitudes can be recursively obtained from the matrix expression

$$\begin{pmatrix} \phi_{N+1} \\ \phi_N \end{pmatrix} = \prod_{k=N}^{1} \mathbf{T}_{k,k\pm1} \begin{pmatrix} \phi_1 \\ \phi_0 \end{pmatrix} \equiv \mathcal{M}_N(E) \begin{pmatrix} \phi_1 \\ \phi_0 \end{pmatrix}, \tag{2.13}$$

TABLE 2.5 Values adopted by the different coefficients appearing in the general Eq. (2.11) depending on the considered elementary excitation.

Parameters	Electrons	Phonons
ϕ_k	ψ_k	u_k
v_k	$E - V_k$	$K_{k,k-1} + K_{k,kn+1} - m_k\omega^2$
$t_{k,k\pm1}$	$t_{k,k\pm1}$	$K_{k,k\pm1}$

where

$$\mathbf{T}_{k,k\pm1} = \begin{pmatrix} \frac{v_k}{t_{k,k+1}} & -\frac{t_{k,k-1}}{t_{k,k+1}} \\ 1 & 0 \end{pmatrix}, \qquad (2.14)$$

are known as local transfer matrices, $\mathcal{M}_N(E)$ is the so-called global transfer matrix, and where ϕ_1 and ϕ_0 set the initial conditions. Since the arrangement of transfer integrals $\{t_{k,k\pm1}\}$ is synchronized with the atomic on-site energies sequence $\{V_k\}$, the underlying aperiodic order of the atoms and bonds in general QC models naturally determines: (1) the number of different *kinds* of local transfer matrices, which must be considered in the $\mathcal{M}_N$ product, and (2) the specific *order* of appearance of these local transfer matrices in $\mathcal{M}_N(E)$. This illustrates the suitability of the transfer matrix formalism in order to encode the QPO of atoms and bonds in a sequence of matrices describing the dynamics of elementary excitations propagating through the system. In this way, the transfer matrix formalism provides a simple mathematical tool allowing for a unified treatment of such diverse problems as electron and phonon dynamics in both periodic and aperiodic lattices, the optical properties of dielectric multilayers (which we will describe in Chapter 4), the propagation of acoustic waves in semiconductor heterostructures and metallic superlattices [245, 686, 986], or the charge transport through DNA chains [302, 548, 581].

From the knowledge of the $\mathcal{M}_N(E)$ matrix elements several magnitudes of physical interest, like the dispersion relation, the density of states, the transmission coefficient, or the localization length, can be readily evaluated. For instance, the dispersion relation of a periodic approximant to a given QP lattice can be obtained by imposing the cyclic boundary conditions $\phi_{N+1} = e^{iqNa_0}\phi_1$ and $\phi_0 = e^{-iqNa_0}\phi_N$ to Eq. (2.13), where q is the wave number and a_0 is the lattice constant. The motion equation then reads

$$\begin{pmatrix} e^{iqNa_0}\phi_1 \\ \phi_N \end{pmatrix} \equiv \begin{pmatrix} M_{11} & M_{12} \\ M_{21} & M_{22} \end{pmatrix} \begin{pmatrix} \phi_1 \\ e^{-iqNa_0}\phi_N \end{pmatrix}, \qquad (2.15)$$

where M_{ij} are the elements of the global transfer matrix $\mathcal{M}_N(E)$. Equation (2.15) leads to the system

$$e^{iqNa_0}\phi_1 = M_{11}\phi_1 + M_{12}e^{-iqNa_0}\phi_N \qquad (2.16)$$

$$\phi_N = M_{21}\phi_1 + M_{22}e^{-iqNa_0}\phi_N, \qquad (2.17)$$

so that, by solving for ϕ_N in Eq. (2.17) and plugging the obtained expression in Eq. (2.16) one gets

$$e^{iqNa_0} + e^{-iqNa_0}\det\mathcal{M}_N = \mathrm{tr}\mathcal{M}_N. \qquad (2.18)$$

Eq. (2.18) significantly simplifies when $\det\mathcal{M}_N = 1$,[9] in which case one obtains the dispersion relation

$$\cos(qNa_0) = \frac{1}{2}\mathrm{tr}\mathcal{M}_N(E). \qquad (2.19)$$

---

[9]This is generally the case when H is Hermitian.

By systematically increasing the N value in Eq. (2.19) one can disclose the characteristic features related to the emergence of long-range QPO in approximant crystals approaching the QP limit $N \rightarrow \infty$. In an analogous way, Eq. (2.19) can be used to study the structural complexity effects resulting from progressively increasing the number of different atoms in the approximant QC unit cell.[10]

The most relevant results regarding the energy and frequency spectra of different QP systems described in terms of Eq. (2.11) can be summarized as follows:

- The energy spectrum exhibits an infinity of gaps and the total bandwidth of the allowed states vanishes in the $N \rightarrow \infty$ limit. This has been proven rigorously for systems based on the Fibonacci [47, 363], Thue-Morse and period doubling sequences [47].

- The position and number of the gaps in the energy/frequency spectrum are related to the singularities of the spatial Fourier (diffraction) spectrum in terms of the so-called gap labeling theorem [49, 517]. In the particular case of the Fibonacci chain this theorem states that the integrated density of states (IDOS) takes the simple form $\mathcal{N}(k) = p + q\tau^{-1}$, where $p, q \in \mathbb{Z}$, so that N keeps normalized within $[0, 1]$, τ is the golden mean, and k is the wave number. This formula has been experimentally observed for a polariton gas confined in a Fibonacci cavity [876]. The integers p and q can be regarded as topological invariants, that is, the relation $\mathcal{N}(k) = p + q\tau^{-1}$ holds irrespective of the precise form of the Hamiltonian as long as the gaps remain open. These features have been recently revisited in the wake of the growing interest in topological properties of condensed matter systems [160, 161, 450, 643–645].

- Scaling properties of the energy spectrum can be described using the formalism of multifractal geometry [241, 434].

An illustrative example of the spectrum structure corresponding to two finite realizations of QP systems is shown in Fig. 2.4. By inspecting these spectra, we clearly appreciate the following prefractal signatures:

- The spectra exhibit a highly-fragmented structure generally composed of as many subbands as the number of atoms present in the chain.

- The main subbands (which concentrate a high number of states) are separated by relatively wide forbidden intervals.

- The degree of internal structure inside each subband depends on the total length of the chain, and the longer the chain, the finer the structure, which displays distinctive features of a self-similar distribution of levels.

Taken altogether, these features provide compelling evidence about the intrinsic fractal nature of the numerically obtained spectra, which will eventually show up with increasing mathematical accuracy in the thermodynamic limit $N \rightarrow \infty$. Now, as one approaches this limit it is legitimate to question whether a given energy value actually belongs to the energy spectrum. Indeed, this is not a trivial issue in the case of highly fragmented spectra supported by a Cantor set of zero Lebesgue measure and can be only guaranteed on the basis of *exact analytical* results. In fact, because of the presence of extremely narrow bands, special care is required in order to avoid studying states belonging to a gap and erroneously interpreting their features as those proper of system eigenstates.

---

[10]The interested reader is addressed to Ref [469]. for an illustrative systematic treatment of periodic 1D lattices, with a generic unit cell containing up to six atoms, within this framework.

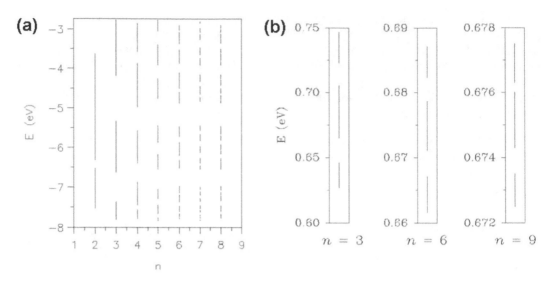

Figure 2.4 (a) Fragmentation pattern of the energy spectrum of a trans-polyacetylene QP chain. The number of allowed subbands increases as a function of the system size expressed in terms of the Fibonacci order n as $N = F_n$. (Reprinted from Physica B, 216, Maciá E and Domínguez-Adame F, Three-dimensional effects on the electronic structure of quasiperiodic systems, 53, Copyright (1995), with permission from Elsevier). (b) Self-similarity in the energy spectrum of a InAs/GaSb Fibonacci superlattice. The left panel shows the whole spectrum for an order $n = 3$ superlattice, whereas the central and right panels show a detail of the spectrum for superlattices of orders $n = 6$ and $n = 9$, respectively. (Reprinted from Domínguez-Adame F, Maciá E, Méndez B, Roy C L, and Khan A, Fibonacci superlattices of narrow-gap III-V semiconductors, Semiconductor Science Technology, 10, 797, 1995 © IOP Publishing. Reproduced with permission. All rights reserved).

In this regard, it should also be noted that irrational numbers cannot be explicitly included in a computing code as such, but only in terms of approximate truncated decimal expressions. Accordingly, one must very carefully check that the obtained results are not appreciably affected by the truncation. To this end, one should consider the systematic use of successive approximants of an irrational number in order to explore the possible influence of its irrational character (if any) in the physical model under study. In fact, one can implement numerically an empirical scaling analysis in which the QP system is approximated by a sequence of periodic systems with progressively larger unit cells of size q_n defined by the optimal rational approximants to the considered irrational number α, namely $\alpha_n = p_n/q_n$. In this way, by checking that finer discretization produce almost the same results one can be confident enough on the reliability of the obtained results [630].

2.3.2 General binary aperiodic crystals

In this Section, we will consider 1D binary aperiodic crystals (BACs) where two kinds of atoms, say A and B, are arranged aperiodically according to certain substitution sequences. After Eq. (2.14), any local transfer matrix is completely determined by a triplet formed by three consecutive atoms along the lattice. The role played by each atom of the triplet can be disclosed making use of the decomposition property [453, 467].

$$\mathbf{T}_{k,k\pm1} = \begin{pmatrix} t_{k,k+1}^{-1} & 0 \\ 0 & 1 \end{pmatrix} \begin{pmatrix} v_k & -1 \\ 1 & 0 \end{pmatrix} \begin{pmatrix} 1 & 0 \\ 0 & t_{k,k-1} \end{pmatrix} \equiv \mathbf{G}_R \mathbf{A}_k \mathbf{G}_L, \qquad (2.20)$$

where $\mathbf{A}_k$ describes the central atom contribution, and the matrices $\mathbf{G}_R$ ($\mathbf{G}_L$) describe its right (left) interaction with the neighboring atoms, respectively. We note that $\mathbf{A}_k$ belongs to the $SL(2,\mathbb{R})$ group, whereas the bond matrices $\mathbf{G}_R$ ($\mathbf{G}_L$) generally do not. By inspecting Eq. (2.20), we readily see that there exist eight possible local transfer matrices in a general BAC. Four of them are related to triplets containing a central A type atom, namely,

$$\mathbf{T}_{BAB} = \begin{pmatrix} v_A & -1 \\ 1 & 0 \end{pmatrix}, \quad \mathbf{T}_{AAB} = \begin{pmatrix} v_A & -\gamma_A \\ 1 & 0 \end{pmatrix},$$

$$\mathbf{T}_{BAA} = \begin{pmatrix} \frac{v_A}{\gamma_A} & -\frac{1}{\gamma_A} \\ 1 & 0 \end{pmatrix}, \quad \mathbf{T}_{AAA} = \begin{pmatrix} \frac{v_A}{\gamma_A} & -1 \\ 1 & 0 \end{pmatrix}, \qquad (2.21)$$

where $\gamma_k \equiv t_{k,k-1}/t_{k,k+1}$, and we have adopted the energy scale $t_{AB} \equiv 1$, so that $\gamma_A \equiv t_{AA}$ without loss of generality. Note that the local transfer matrices corresponding to the AAA and BAB triplets belong to the $SL(2,\mathbb{R})$ group. This is a natural consequence of the mirror symmetry of the bonds with respect to the center atom. The other four local transfer matrices are related to triplets containing a central B atom, and they can be straightforwardly obtained from Eq. (2.21) by performing the conjugation operation $A \leftrightarrow B$ to get

$$\mathbf{T}_{ABA} = \begin{pmatrix} v_B & -1 \\ 1 & 0 \end{pmatrix}, \quad \mathbf{T}_{BBA} = \begin{pmatrix} v_B & -\gamma_B \\ 1 & 0 \end{pmatrix},$$

$$\mathbf{T}_{ABB} = \begin{pmatrix} \frac{v_B}{\gamma_B} & -\frac{1}{\gamma_B} \\ 1 & 0 \end{pmatrix}, \quad \mathbf{T}_{BBB} = \begin{pmatrix} \frac{v_B}{\gamma_B} & -1 \\ 1 & 0 \end{pmatrix}, \qquad (2.22)$$

where $\gamma_B \equiv t_{BB}$.

Within this framework, a measure of the structural *complexity degree* of a given BAC can be estimated by counting the number of *different kinds* of local transfer matrices which are included in $\mathcal{M}_N$ when $N \to \infty$. For the sake of illustration in Fig. 2.5, we pictorially classify several BACs attending to the number of different local transfer matrices required to this end. As we see, BACs based on the precious means and the Fibonacci-class substitution sequences (both including the standard Fibonacci sequence as a particular case) require four different local transfer matrices each. In this regard their local topological complexity is comparable to that of *periodic* lattices with a relatively large unit cell (e.g., $AAAB$, see the bottom panel in Fig. 2.5).[11] The BACs based on the period doubling and the mixed means sequences with $n \geq m$, both requiring the knowledge of five different local transfer matrices, are examples of the next level of local complexity. The BACs based on the copper mean sequence (the first representative of the so-called metallic means family) and the mixed mean sequence with $n < m$, belong to the following complexity level, which requires the knowledge of six different local transfer matrices. Finally, BACs based on metallic mean substitution rules with $m \geq 3$ need seven different types of $\mathbf{T}_{k,k\pm1}$ matrices, thereby exhibiting a remarkable degree of topological complexity at the triplet atomic scale. We note that none of the BACs based on the substitution rules listed in Table 2.1 exhibits the maximum possible complexity degree.[12] The information provided by the classification

---

[11] We note that the $\mathcal{M}_N$ global matrices of periodic lattices with unit cell $AAAB$ and aperiodic lattices based on precious means ($n \geq 2$) contain exactly the same types of local transfer matrices.

[12] An example of periodic lattice fulfilling this condition has a unit cell containing ten atoms arranged as follows: $ABABBBAAAB$.

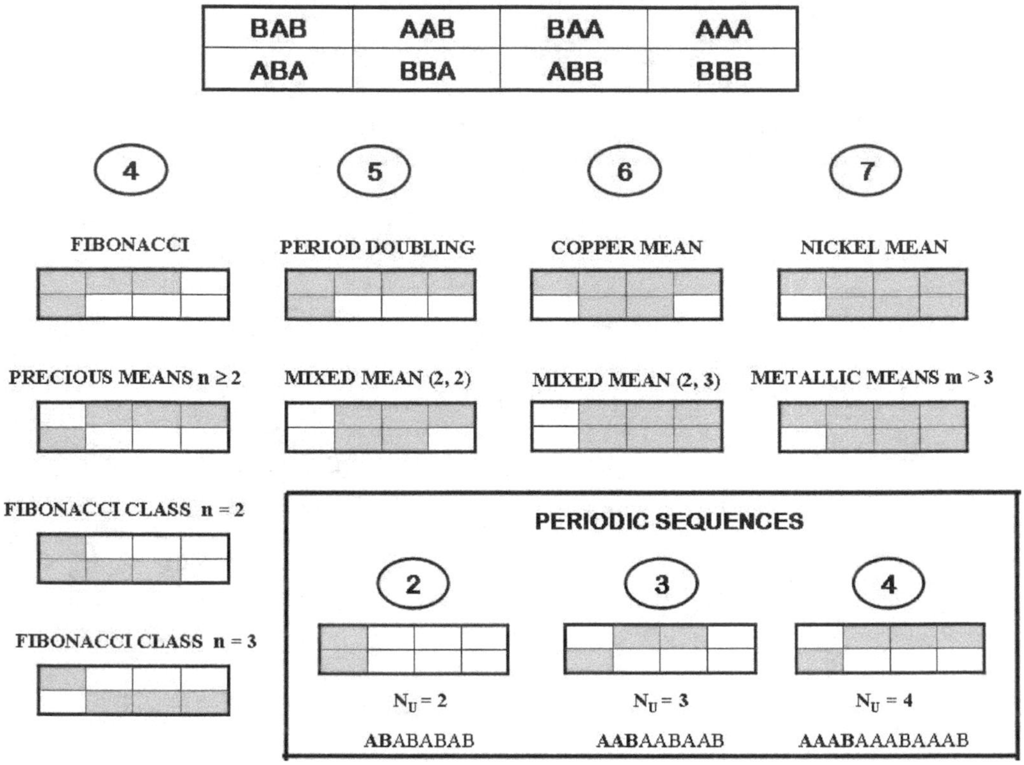

| BAB | AAB | BAA | AAA |
| ABA | BBA | ABB | BBB |

Figure 2.5 Classification of periodic (bottom right panel, where N_U indicates the unit cell size) and several aperiodic lattices considered in the literature according to the number (encircled) of different local matrices (shadowed boxes) required to fully describe their topological order at the triplet scale. The key for the local transfer matrices subscript labeling is given in the top panel. (Reprinted from Ref. [559] with permission. © 2017 by WILEY-VCH Verlag GmbH & Co. KGaA, Weinheim).

boxes in Fig. 2.5 can be complemented by including the frequency percentage of each local transfer matrix in the asymptotic QPO limit, as it was done in Ref. [470].

2.3.3 Fibonacci quasicrystal tridiagonal model

In this Section, we will consider 1D binary alloys where two kinds of atoms, say A and B, are arranged according to the Fibonacci sequence, so that the total number of atoms in the lattice is $N = F_j$, where $F_{j+1} = F_j + F_{j-1}$, with $F_0 = F_1 = 1$, is a Fibonacci number. The QP arrangement of atoms naturally induces a QP sequence of bonds in our system, whose strength is measured in terms of the transfer integrals $t_{AB} = t_{BA}$ and t_{AA}. The possible presence of two consecutive B atoms is not allowed in a Fibonacci sequence, resulting in the existence of two possible kinds of bonds only, namely, $A - A$ and $A - B$ $(B-A)$. Hereafter, we will refer to this system as a Fibonacci QC (FQC).[13] Accordingly, our

---

[13] We note that the term *Fibonacci quasicrystal* has also been used in the literature to refer to models, which are not described in terms of a tridiagonal Hamiltonian.

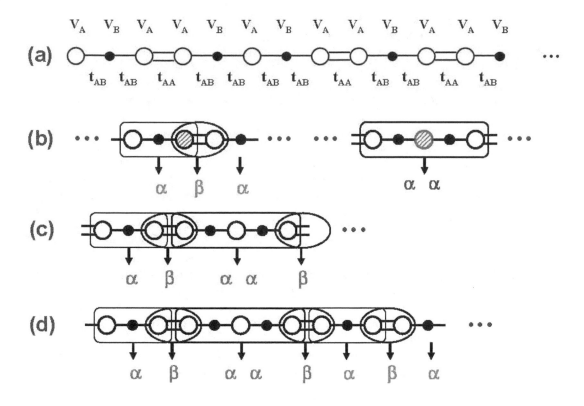

Figure 2.6 The structure of the original FQC chain shown in (a) is resolved in terms of overlapping AA and ABA clusters, respectively labeled β and α, as it is shown in (b), where shared neighboring atoms are dashed. The resulting sequence of atomic clusters α and β is arranged according to the Fibonacci sequence, as it is illustrated in (c) and (d). (Reprinted from Ref. [560] with permission. © 2017 by WILEY-VCH Verlag GmbH & Co. KGaA, Weinheim).

QC model is characterized by the existence of *both* atomic and chemical bonding QP distributions. The resulting sequences of atomic potentials $\{V_k\}$ and bonds $\{t_{k,k+1}\}$ is illustrated in Fig. 2.6a for a FQC containing $N = F_6 = 13$ atoms.

In order to disclose the nature of the correlation between both sequences, the lattice shown in Fig. 2.6a is resolved in terms of $A - A$ dimers and $A - B - A$ trimers, respectively labeled β and α in Fig. 2.6b, to obtain the sequence of α and β atomic *clusters* shown in Fig. 2.6c–d. In so doing, overlapping of neighboring $A - A$ and $A - B - A$ clusters is allowed by sharing their respective ending atoms, as it is highlighted in Fig. 2.6b.[14] In this way, we realize that the sequence of bonds in a FQC containing F_j atoms can be grouped in terms of F_{j-1} clusters, which are arranged themselves according to the Fibonacci sequence given by the substitution rule $g(\alpha) = \alpha\beta$ and $g(\beta) = \alpha$ starting with α.

Making use of Eqs. (2.13) and (2.14), it was proved that the spectrum of the general FQC described by the tridiagonal matrix given by Eq. (2.12) is a Cantor set of zero Lebesgue measure belonging to the purely singular continuous type (see Sec. 2.2) [434, 664]. thereby extending the results previously proved for diagonal and off-diagonal models [596].

---

[14]We note that these overlapping molecules play a role similar to that played by overlapping atomic clusters in the structural models proposed for the binary iQC Cd$_{85}$Yb$_{15}$, as discussed in [859].

Furthermore, the spectrum has a multifractal Hausdorff dimension, instead of a simple fractal single-valued one, which depends on the adopted v_k and $t_{k,k\pm1}$ values. These results indicate that the structure of the energy spectrum of a tridiagonal FQC model is significantly richer than those corresponding to the diagonal and off-diagonal simpler models.

2.4 HIERARCHICAL STRUCTURE OF ELECTRONIC SPECTRA

The hierarchical structure of energy spectrum of 1D BACs can be properly described in terms of an algebraic approach allowing for a unified and systematic mathematical derivation of the spectrum splitting patterns by means of commutators involving a number of basic matrices related to certain atomic clusters. In this way, we obtain closed analytical expressions for the energy spectrum structure in terms of the roots of energy dependent polynomials.

2.4.1 Self-similar blocking schemes

Our strategy will be to exploit the hierarchical structure of BACs based on substitution sequences in order to properly disclose their long-range correlations. To this end, we will zoom out the considered spatial scale, going from the nearest-neighbors scale related to the triplet level description all the way up to the entire lattice size. To achieve this goal, we will rewrite the global transfer matrix $\mathcal{M}_N(E)$ of *any* given BAC in terms of *just two cluster* matrices $\mathbf{R}_A^{(n,m)}$ and $\mathbf{R}_B^{(n,m)}$ in the form [527, 555, 952],

$$\mathcal{M}_N(E) = \prod_{k=N}^{1} \mathbf{T}_{k,k\pm1} \equiv \ldots \mathbf{R}_B^{(n,m)} \mathbf{R}_A^{(n,m)} \mathbf{R}_A^{(n,m)} \mathbf{R}_B^{(n,m)} \mathbf{R}_A^{(n,m)} \qquad (2.23)$$

according to certain *blocking schemes,* which are shown in Figs. 2.7–2.9 for BACs based on different substitution sequences. A close comparison of the lettering corresponding to the BACs depicted in Figs. 2.7–2.9 with the information provided in Table 2.6 clearly indicates that the topological order present in the original *atomic* lattice is preserved by the cluster matrices sequence appearing in Eq. (2.23).

Thus, albeit the number of $\mathbf{T}_{k,k\pm1}$ matrices we must consider at the triplet scale level can vary from four to eight in general BACs, the recourse to appropriate cluster matrices sets allows one to express the global $\mathcal{M}_N(E)$ matrix in a considerably simplified way, corresponding to an effective binary alloy, which preserves the original BAC atomic order. By comparing the last column of Table 2.1 with the third column of Table 2.6, it can be readily appreciated that the elemental cluster matrices obey exactly the same substitution rules as those followed by atoms themselves. In other words, the topological order of the sequence of atoms $\{V_k\}$ is naturally translated to the cluster matrices sequence, which obeys the same substitution rule. This characteristic feature of the cluster matrices is shown in Table 2.6 for the BACs based on the substitution sequences listed in Table 2.1. Accordingly, the matrices $\mathbf{R}_{A,B}^{(n,m)}$ will be referred to as *elemental cluster* matrices thereafter.

The total number of local transfer matrices in $\mathcal{M}_N(E)$ is naturally reduced due to the blocking process. In fact, by properly clustering the original chain in terms of successively longer block matrices, one is able to fully exploit the inflation/deflation symmetry characteristic of *any* BAC generated from a substitution rule in order to describe the electronic states at sites more and more farther apart. Thus, by inspecting Figs. 2.7–2.9, we can see that the number of cluster matrices decreases by two generation orders as compared to the number of atoms present in the original lattice. For instance, if $N = F_j$ is the number of atomic

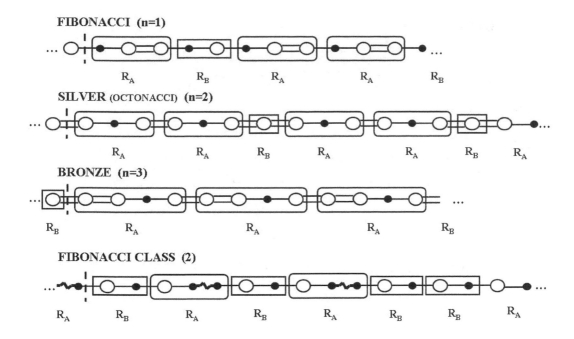

Figure 2.7 Elemental cluster matrices schemes for BACs based on substitution sequences belonging to the quasiperiodic Π_{I} class. White (black) circles denote A (B) type atoms and t_{AB}, t_{AA}, and t_{BB} transfer integrals are represented by single bonds, double bonds and jigjag bonds, respectively. In order to complete the blocking schemes the atoms at both ends are connected to each other by adopting cyclic boundary conditions. The vertical dashed line indicates the reading frame origin (from left to right). We note that the elemental cluster matrices appear in reversed order in the global matrix product given by Eq. (2.23). (Adapted from Ref. [559] with permission. © 2017 by WILEY-VCH Verlag GmbH & Co. KGaA, Weinheim).

lattice sites in the standard FQC, one can show by induction that the elemental cluster matrices sequence contains $n_A = F_{j-3}$ matrices $\mathbf{R}_A^{(1)}$ and $n_B = F_{j-4}$ matrices $\mathbf{R}_B^{(1)}$. Similar results can be readily obtained for the other considered BACs. Accordingly, the choice of appropriate elemental cluster matrices obeying certain self-similar schemes can be regarded as an effective *renormalization* of the original local transfer matrices sequence [527, 555]. In fact, from a physical point of view cluster matrices describe the electron dynamics through the basic atomic clusters in terms of which the entire BAC can be decomposed via an exact deflation process. For instance, in the case of a FQC, these atomic clusters are BA and BAA. In this regard, it is worth noticing that a decomposition scheme based on the mirror reversed AB and AAB atomic clusters has been considered in order to study the dynamics of classical particles in a time-driven Fibonacci lattice [959]. In terms of the blocking schemes shown in Figs. 2.7–2.9 one can usefully exploit the renormalization method within the framework of the transfer matrices approach in order to analytically obtain the energy spectrum structure in general substitution based BACs.

PERIOD DOUBLING

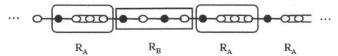

METALLIC MEANS

COPPER (m=2)

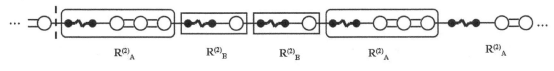

NICKEL (m=3)

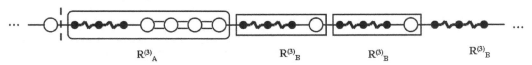

Figure 2.8 Elemental cluster matrices schemes for BACs based on the period doubling sequence and the two first representatives of the metallic mean sequences, all of them belonging to the limit-periodic Π_{III} class. The adopted notation is the same as in Fig. 2.7. (Reprinted from Ref. [559] with permission. © 2017 by WILEY-VCH Verlag GmbH & Co. KGaA, Weinheim).

2.4.2 Inflation symmetry commutators

Let us start by analyzing the role of the characteristic scale-invariance symmetry in the energy spectrum structure of FQCs. To this end, we will consider commutators containing progressively longer string products of elemental cluster matrices $\mathbf{R} \equiv \mathbf{R}_A^{(1)}$ and $\mathbf{r} \equiv \mathbf{R}_B^{(1)}$, as prescribed by the successive application of the Fibonacci substitution rule. The resulting Fibonacci matrix strings can be recursively obtained from the concatenation formula $\mathbf{F}_n = \mathbf{F}_{n-1}\mathbf{F}_{n-2}$, $n \geq 2$, starting with $\mathbf{F}_0 \equiv \mathbf{r}$, $\mathbf{F}_1 \equiv \mathbf{R}$. Therefore, the matrix strings express the characteristic inflation symmetry of FQCs in a natural way, and we will refer to them as *Fibonacci strings*. According to Table 2.6 the corresponding elemental cluster matrices can be written

$$
\mathbf{R} = \gamma_A^{-1}\begin{pmatrix} v_B(v_A^2 - \gamma_A^2) - v_A & \gamma_A^2 - v_A^2 \\ v_A v_B - 1 & -v_A \end{pmatrix} = \gamma_A^{-1}\begin{pmatrix} R_{11}(E) & \gamma_A^2 - (E-\epsilon)^2 \\ E^2 - \epsilon^2 - 1 & \epsilon - E \end{pmatrix},
$$

$$
\mathbf{r} = \begin{pmatrix} v_A v_B - 1 & -v_A \\ v_B & -1 \end{pmatrix} = \begin{pmatrix} E^2 - \epsilon^2 - 1 & \epsilon - E \\ \epsilon + E & -1 \end{pmatrix}, \tag{2.24}
$$

with $R_{11}(E) = (E + \epsilon)[(E - \epsilon)^2 - \gamma_A^2] + \epsilon - E$, where we have used Eqs. (2.21) and (2.22), and the origin of energies is defined in such a way that $V_A = \epsilon = -V_B$, so that $v_A = E - \epsilon$, and $v_B = E + \epsilon$, without loss of generality. Since $\det \mathbf{r} = \det \mathbf{R} = 1$, all Fibonacci strings belong to the $SL(2, \mathbb{R})$ group by construction.

The first member of the commutator series we are interested in reads

$$
g^1[\mathbf{R}, \mathbf{r}] \equiv [g^1(\mathbf{R}), g^1(\mathbf{r})] = [\mathbf{R}\mathbf{r}, \mathbf{R}] = -\mathbf{R}[\mathbf{R}, \mathbf{r}]. \tag{2.25}
$$

MIXED MEANS

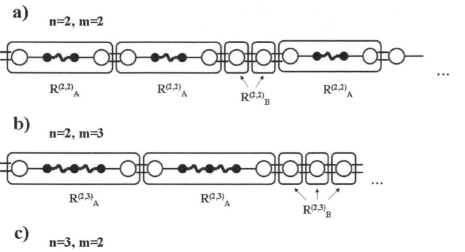

a) n=2, m=2

b) n=2, m=3

c) n=3, m=2

Figure 2.9 Elemental cluster matrices schemes for BACs based on the lowest order mixed means representatives, belonging to either the limit-quasiperiodic Π_{II} ($n \geq m$) or the limit-periodic Π_{III} ($n < m$) classes, respectively. The adopted notation is the same as in Fig. 2.7. (Reprinted from Ref. [559] with permission. © 2017 by WILEY-VCH Verlag GmbH & Co. KGaA, Weinheim).

To obtain the next members in a systematic way we will exploit the concatenation formula of Fibonacci strings, so that any commutator involving two arbitrary *successive* strings can be expressed in the form

$$g^n[\mathbf{R}, \mathbf{r}] \equiv [\mathbf{F}_{n+1}, \mathbf{F}_n] = \mathbf{F}_n \mathbf{F}_{n-1} \mathbf{F}_n - \mathbf{F}_n \mathbf{F}_n \mathbf{F}_{n-1} = -\mathbf{F}_n [\mathbf{F}_n, \mathbf{F}_{n-1}], \qquad (2.26)$$

where $n \geq 1$ indicates the inflation stage. The expression above can be recursively iterated to obtain

$$g^n[\mathbf{R}, \mathbf{r}] = (-1)^n \prod_{j=n}^{1} \mathbf{F}_j[\mathbf{R}, \mathbf{r}] \equiv (-1)^n \mathbf{\Phi}_n[\mathbf{R}, \mathbf{r}]. \qquad (2.27)$$

Therefore, commutators including progressively longer successive Fibonacci strings can be expressed in terms of a product of Fibonacci strings given by $\mathbf{\Phi}_n$, along with the *elemental commutator* $[\mathbf{R}, \mathbf{r}]$. After some algebra from Eq. (2.24), we get

$$[\mathbf{R}, \mathbf{r}] = \gamma_A^{-1} p_1(E) \begin{pmatrix} 1 & 0 \\ \epsilon + E & -1 \end{pmatrix} \equiv \gamma_A^{-1} p_1(E) \mathbf{F}, \qquad (2.28)$$

where

$$p_1(E) = \left(\gamma_A^2 - 1\right) E + \epsilon \left(\gamma_A^2 + 1\right), \qquad (2.29)$$

and the $\mathbf{F}$ matrix satisfies the relationships $\mathbf{F}^2 = \mathbf{I}$, where $\mathbf{I}$ is the 2×2 identity matrix, and

$$\mathbf{r}\mathbf{F}\mathbf{r} = \mathbf{F}, \quad \mathbf{R}\mathbf{F}\mathbf{R} = \mathbf{F}. \qquad (2.30)$$

TABLE 2.6 Structure of the elemental cluster matrices $R_A^{(n,m)}$ and $R_B^{(n,m)}$ in terms of local transfer matrices $T_{k,k\pm1}$, along with the related substitution rules defining the global transfer matrix $M_N(E)$ for several BACs. (Reprinted from Ref. [559] with permission. Copyright 2017 by WILEY-VCH Verlag GmbH Co. KGaA, Weinheim).

SEQUENCE	CLUSTER MATRICES	SUBSTITUTION RULE
Fibonacci $(n=1)$	$R_A^{(1)} = T_{AAB}T_{BAA}T_{ABA}$ $R_B^{(1)} = T_{BAB}T_{ABA}$	$g(R_A^{(1)}) = R_A^{(1)}R_B^{(1)}$ $g(R_B^{(1)}) = R_A^{(1)}$
Silver mean $(n=2)$	$R_A^{(2)} = T_{BAA}T_{ABA}T_{AAB}$ $R_B^{(2)} = T_{AAA}$	$g(R_A^{(2)}) = R_A^{(2)}R_A^{(2)}R_B^{(2)}$ $g(R_B^{(2)}) = R_A^{(2)}$
Bronze mean $(n=3)$	$R_A^{(3)} = T_{BAA}T_{ABA}T_{AAB}T_{AAA}$ $R_B^{(3)} = T_{AAA}$	$g(R_A^{(3)}) = R_A^{(3)}R_A^{(3)}R_A^{(3)}R_B^{(3)}$ $g(R_B^{(3)}) = R_A^{(3)}$
Precious means $(n>2)$	$R_A^{(n)} = R_A^{(n-1)}T_{AAA}$ $R_B^{(n)} = T_{AAA}$	$g(R_A^{(n)}) = \left[R_A^{(n)}\right]^n R_B^{(n)}$ $g(R_B^{(n)}) = R_A^{(n)}$
Fibonacci-class $(n=2)$	$R_A^{(2)} = T_{BBA}T_{ABB}T_{BAB}$ $R_B^{(2)} = T_{ABA}T_{BAB}$	$g(R_A^{(2)}) = R_B^{(2)}R_A^{(2)}R_B^{(2)}$ $g(R_A^{(2)}) = R_B^{(2)}R_A^{(2)}$
Fibonacci-class $(n>2)$	$R_A^{(n)} = T_{BBA}T_{BBB}^{n-2}T_{ABB}T_{BAB}$ $R_B^{(n)} = R_A^{(n-1)}$	$g(R_A^{(n)}) = \left[R_B^{(n)}\right]^{n-1} R_A^{(n)}R_B^{(n)}$ $g(R_B^{(n)}) = \left[R_B^{(n)}\right]^{n-1} R_A^{(n)}$
Period-doubling	$R_A = T_{ABA}T_{AAB}T_{AAA}T_{BAA}$ $R_B = T_{ABA}T_{BAB}T_{ABA}T_{BAB}$	$g(R_A) = R_A R_B$ $g(R_B) = R_A R_A$
Copper mean $(m=2)$	$R_A^{(2)} = T_{AAB}T_{AAA}T_{BAA}T_{BBA}T_{ABB}$ $R_B^{(2)} = T_{BAB}T_{BBA}T_{ABB}$	$g(R_A^{(2)}) = R_A^{(2)}R_B^{(2)}R_B^{(2)}$ $g(R_B^{(2)}) = R_A^{(2)}$
Nickel mean $(m=3)$	$R_A^{(3)} = T_{AAB}T_{AAA}^2T_{BAA}T_{BBA}T_{BBB}T_{ABB}$ $R_B^{(3)} = T_{BAB}T_{BBA}T_{BBB}T_{ABB}$	$g(R_A^{(3)}) = R_A^{(3)}R_B^{(3)}R_B^{(3)}R_B^{(3)}$ $g(R_B^{(3)}) = R_A^{(3)}$
Metallic means $(m>1)$	$R_A^{(m)} = T_{AAB}T_{AAA}^{m-1}T_{BAA}T_{BBA}T_{BBB}^{m-2}T_{ABB}$ $R_B^{(m)} = T_{BAB}T_{BBA}T_{BBB}^{m-2}T_{ABB}$	$g(R_A^{(m)}) = R_A^{(m)}\left[R_B^{(m)}\right]^m$ $g(R_B^{(m)}) = R_A^{(m)}$
Mixed mean $(n=m=2)$	$R_A^{(2,2)} = T_{AAB}T_{ABB}T_{BBA}T_{BAA}$ $R_B^{(2,2)} = T_{AAA}$	$g(R_A^{(2,2)}) = \left[R_A^{(2,2)}\right]^2\left[R_B^{(2,2)}\right]^2$ $g(R_B^{(2,2)}) = R_A^{(2,2)}$

By plugging Eq. (2.28) into Eq. (2.27), we have

$$g^n[\mathbf{R}, \mathbf{r}] = (-1)^n \gamma_A^{-1} p_1(E)\, \mathbf{\Phi}_n\mathbf{F}. \qquad (2.31)$$

Thus, in order to look for the possible existence of resonance energies satisfying the condition $g^n[\mathbf{R}, \mathbf{r}] = \mathbf{0}$ for any $n \geq 1$, one must consider the algebraic equation $p_1(E) = 0$, along with the matrix equation $\mathbf{\Phi}_n\mathbf{F} = \mathbf{0}$. Now, the necessary condition for the matrix equation $\mathbf{M}_1\mathbf{M}_2 = \mathbf{0}$ to be satisfied by two arbitrary square matrices ($\mathbf{M}_1 \neq \mathbf{0}$ and $\mathbf{M}_2 \neq \mathbf{0}$) is that one of the matrices appearing in the product has zero determinant. Since $\mathbf{F}_j \in SL(2, \mathbb{R})$ $\forall j$, this condition guarantees the string product $\mathbf{\Phi}_n \neq \mathbf{0}$, and we have $\det \mathbf{\Phi}_n = 1$ as well. On the other hand, $\mathbf{F} \neq \mathbf{0}$ and $\det \mathbf{F} = -1$ for any energy value. Thus, we finally conclude that $\mathbf{\Phi}_n\mathbf{F} \neq \mathbf{0}$ $\forall n \geq 1$. Consequently, the commutation condition $g^n[\mathbf{R}, \mathbf{r}] = \mathbf{0} \Rightarrow p_1(E) = 0$,[15] thereby leading to the *same* resonance energy for any arbitrary inflation stage n. Therefore,

---

[15]We note that, in the case $\gamma_A = 1$, which corresponds to the on-site diagonal model, this polynomial

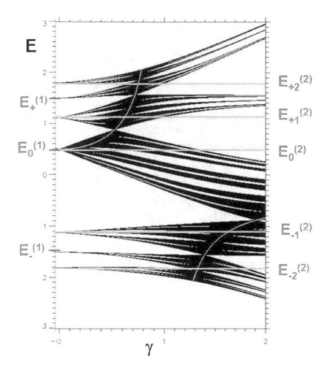

Figure 2.10 Numerically obtained phase diagram for a binary FQC with $N = 34$ atoms and on-site energy $\epsilon = 1/2$, over the range $0 \leq \gamma_A \leq 2$. The zeroth-order energy spectrum given by Eq. (2.32) is highlighted with a thin white line. The regions closer to this line are densely populated as a consequence of the high degeneration of the resonance energy E_*. (Reprinted from Ref. [560] with permission. © 2017 by WILEY-VCH Verlag GmbH & Co. KGaA, Weinheim).

for any realization of the general FQC (i.e., for any combination of ϵ and $\gamma_A \neq 1$ values) there *always* exists one energy satisfying the relation $p_1(E, \epsilon, \gamma_A) = 0$, namely [456, 527],

$$E_* = \epsilon \frac{1 + \gamma_A^2}{1 - \gamma_A^2}. \tag{2.32}$$

The parametric curve $E_*(\gamma_A, \epsilon)$ defines the zeroth-order energy spectrum structure of the considered FQC. For a given γ_A value, E_* scales linearly with ϵ, whereas for a given ϵ value E_* varies with γ_A in a significant non-linear way, and displays vertical asymptotes in the limits $\gamma_A = \pm 1$. An illustrative example of the energy spectra of general FQCs for different values of the model parameter γ_A is shown in Fig. 2.10. The $E_*(\epsilon, \gamma_A)$ curve is even with respect to γ_A (i.e., $E_*(-\gamma_A, \epsilon) = E_*(\gamma_A, \epsilon)$) and odd with respect to ϵ (i.e., $E_*(\gamma_A, -\epsilon) = -E_*(\gamma_A, \epsilon)$). It also exhibits the symmetry $E_*(\gamma_A^{-1}, \epsilon) = -E_*(\gamma_A, \epsilon)$, $\forall \gamma_A \neq 1$, when the bond strength values between $A - A$ and $A - B$ pairs are permuted.

---

reduces to the constant value $p_1 = 2\epsilon$, so that there are no resonance energies for this model, as it is well known [548].

Making use of the Cayley-Hamilton theorem,[16] the global transfer matrix can be explicitly evaluated for this resonance energy value in the closed form [527], (**Exercise 2.8**)

$$\mathcal{M}_N^{(1)}(E_*) \equiv \mathbf{R}(E_*)^{n_A} \mathbf{r}(E_*)^{n_B} = \begin{pmatrix} U_N & -\gamma_A U_{N-1} \\ \gamma_A^{-1} U_{N-1} & -U_{N-2} \end{pmatrix}, \qquad (2.37)$$

where $U_k(z)$, with $z \equiv \sqrt{E_*^2 - \epsilon^2}/2 = \epsilon \gamma_A (1 - \gamma_A^2)^{-1} \equiv \cos\varphi$, are Chebyshev polynomials of the second kind. Taking into account the relationship $U_k - U_{k-2} = 2T_k$ between Chebyshev polynomials of the first and second kinds,[17] from Eq. (2.37) we get $\mathrm{tr}[\mathcal{M}_N^{(1)}(E_*)] = 2\cos(N\varphi) \leq 2 \; \forall N$. Consequently, we can ensure that the bi-parametric set of energies $E_*(\epsilon, \gamma_A)$ belongs to the spectrum in the QP limit $N \to \infty$, albeit the energy spectrum becomes a Cantor set of zero Lebesgue measure in this case.[18] One can appreciate that the resonant states given by Eq. (2.32) are located across the densest region of the phase diagram. Indeed, since E_* is the only solution to the resonance conditions $g^n[\mathbf{R}, \mathbf{r}] = 0$, for any $n \geq 1$, the resonance energy value given by Eq. (2.32) becomes extremely degenerate as one considers commutators involving progressively longer Fibonacci strings, approaching the QP limit $N \to \infty$. Therefore, the global scale inflation symmetry, described in terms of the successive application of the characteristic Fibonacci substitution rule g, *is not* responsible for the progressive fragmentation of the energy spectrum in FQCs. To this end, we must consider long-range correlations among different Fibonacci strings, as we describe in Sec. 2.4.4.

2.4.3   Zeroth-order energy spectra of binary aperiodic crystals

The algebraic approach described in the previous section can be straightforwardly extended to consider other aperiodic systems based on substitution sequences, namely:

- *Precious means based QCs belonging to the* Π_I *class.*

Making use of the elemental cluster matrices definitions listed in Table 2.6 for the silver and bronze means based QC, we obtain $\left[\mathbf{R}_A^{(2)}, \mathbf{R}_B^{(2)}\right] = -\gamma_A^{-1} p_1(E)\sigma_x$, and $\left[\mathbf{R}_A^{(3)}, \mathbf{R}_B^{(3)}\right] = -\gamma_A^{-1} p_1(E)\,\sigma_x \mathbf{T}_{AAA}$, respectively, where σ_x is a Pauli matrix. As we see, both commutators have a closely related algebraic structure, namely, $\left[\mathbf{R}_A^{(3)}, \mathbf{R}_B^{(3)}\right] = \left[\mathbf{R}_A^{(2)}, \mathbf{R}_B^{(2)}\right] \mathbf{T}_{AAA}$. In

---

[16]The Cayley-Hamilton theorem states that any $n \times n$ square matrix $\mathbf{M}$ over the real or complex field is a root of its own characteristic polynomial $\det(\mathbf{M} - \lambda\mathbf{I}) = 0$ [272], so that one can write

$$\mathbf{M}^2 - 2z\mathbf{M} + \mathbf{I}\det\mathbf{M} = \mathbf{0}, \qquad (2.33)$$

where $z \equiv \frac{1}{2}\mathrm{tr}\mathbf{M}$. If $\mathbf{M}$ belongs to the $SL(n, \mathbb{R})$ group, one can readily use Eq. (2.33) in order to properly express any higher power of $\mathbf{M}$ as a linear combination of matrices $\mathbf{I}$ and $\mathbf{M}$. In the case of 2×2 unimodular matrices, one obtains by induction the expression

$$\mathbf{M}^N = U_{N-1}(z)\mathbf{M} - U_{N-2}(z)\mathbf{I}, \qquad (2.34)$$

where

$$U_N \equiv \frac{\sin(N+1)\varphi}{\sin\varphi}, \qquad (2.35)$$

with $\varphi \equiv \cos^{-1} z$, are Chebyshev polynomials of the second kind satisfying the recursion relation

$$U_{k+1} - 2zU_k + U_{k-1} = 0, \quad k \geq 1 \qquad (2.36)$$

with $U_0(z) = 1$ and $U_1(z) = 2z$.

[17]The Chebyshev polynomials of the first kind are defined as $T_k(z) = \cos(k\cos^{-1} z)$.

[18]We highlight that this result cannot be obtained by relying on numerical calculations only.

fact, this is a common feature shared by all of the commutators related to the precious means based QCs. To show this, we will exploit the relationships $\mathbf{R}_A^{(n)} = \mathbf{R}_A^{(n-1)}\mathbf{T}_{AAA}$ and $\mathbf{R}_B^{(n)} = \mathbf{T}_{AAA}, \forall n > 2$, listed in Table 2.6, in order to express the commutator corresponding to the general case in the form

$$\left[\mathbf{R}_A^{(n)}, \mathbf{R}_B^{(n)}\right] = \left[\mathbf{R}_A^{(n-1)}\mathbf{T}_{AAA}, \mathbf{T}_{AAA}\right] \equiv \mathbf{R}_A^{(n-1)}\mathbf{T}_{AAA}\mathbf{T}_{AAA} - \mathbf{T}_{AAA}\mathbf{R}_A^{(n-1)}\mathbf{T}_{AAA}$$
$$= \mathbf{R}_A^{(n-1)}\mathbf{R}_B^{(n-1)}\mathbf{T}_{AAA} - \mathbf{R}_B^{(n-1)}\mathbf{R}_A^{(n-1)}\mathbf{T}_{AAA} = \left[\mathbf{R}_A^{(n-1)}, \mathbf{R}_B^{(n-1)}\right]\mathbf{T}_{AAA}.$$
$$(2.38)$$

Then, iterating Eq. (2.38) backwards, we obtain

$$\left[\mathbf{R}_A^{(n)}, \mathbf{R}_B^{(n)}\right] = \left[\mathbf{R}_A^{(2)}, \mathbf{R}_B^{(2)}\right]\mathbf{T}_{AAA}^{n-2} = -\gamma_A^{-1}p_1(E)\sigma_x\mathbf{T}_{AAA}^{n-2}, \quad n \geq 2. \qquad (2.39)$$

Accordingly, the fundamental structure of commutators corresponding to QCs based on the precious means sequences can be expressed as a product involving the polynomial factor $p_1(E)$, the Pauli matrix σ_x, and a power of the local transfer matrix $\mathbf{T}_{AAA}$, which is naturally related to the existence of longer and longer strings of A type atoms in the corresponding QCs when n is progressively increased (see Fig. 2.7). Thus, the energy dependent polynomial $p_1(E, \epsilon, \gamma_A)$, originally appearing in the commutator of the standard FQC, is inherited by all the precious mean QCs commutators in a natural way.

Making use of Eq. (2.34), we can express Eq. (2.39), in the closed from

$$\left[\mathbf{R}_A^{(n)}, \mathbf{R}_B^{(n)}\right] = -\gamma_A^{-1}p_1(E)\begin{pmatrix} U_{n-3}(z) & -U_{n-4}(z) \\ U_{n-2}(z) & -U_{n-3}(z) \end{pmatrix}, \qquad (2.40)$$

where $z \equiv \text{tr}\mathbf{T}_{AAA}/2 = \gamma_A^{-1}(E - \epsilon)/2$. Now, the relationships given by Eqs. (2.35)–(2.36) guarantee that the three consecutive Chebyshev polynomials appearing in Eq. (2.40) can not equal zero simultaneously. Consequently, the commutation condition $\left[\mathbf{R}_A^{(n)}, \mathbf{R}_B^{(n)}\right] = \mathbf{0}$, reduces to the algebraic equation $p_1(E) = 0 \; \forall n$, leading to the *same* resonance energy E_* that was obtained in the study of the FQC. This is a very remarkable result, for it indicates that the zeroth-order structure of the electronic energy spectra of all precious means based QCs relies on a common resonance energy state at their very fundamental level. Therefore, the set of states highlighted by the thin white lines in the standard FQC phase diagram shown in Fig. 2.10, will also be present in *every* energy spectrum of *any* QC based on a precious mean sequence, as is illustrated in Fig. 2.11a.

It is natural to wonder as to whether this result also holds for the so-called Fibonacci-class lattices FC(n), obtained from the application of the substitution rule $g(A) = B^{n-1}AB$, $g(B) = B^{n-1}A$ (see Table 2.1), which trivially reduces to the standard FQC in the case $n = 1$,[19] and also belong to the QP Π_I class (see Table 2.3). Making use of the elemental cluster matrices definitions listed in Table 2.6 for the first representative FC(2), we obtain

$$\left[\mathbf{R}_A^{(2)}, \mathbf{R}_B^{(2)}\right]^{FC} = \gamma_B^{-1}\bar{p}_1(E, \epsilon, \gamma_B)\begin{pmatrix} 1 & 0 \\ E - \epsilon & -1 \end{pmatrix} \equiv \gamma_B^{-1}\bar{p}_1(E, \epsilon, \gamma_B)\tilde{\mathbf{F}}_1, \qquad (2.41)$$

where $\bar{p}_1(E, \epsilon, \gamma_B) \equiv -\epsilon(1 + \gamma_B^2) - E(1 - \gamma_B^2)$, and the superscript FC is introduced to distinguish this commutator from that corresponding to the silver mean lattice. By comparing

---

[19]Indeed, this substitution sequence was originally introduced in order to generalize the standard Fibonacci lattices [264], and we note that by defining $A' \equiv B^{n-1}A$ the proposed substitution rule reduces to the standard Fibonacci one for $n \neq 1$ as well.

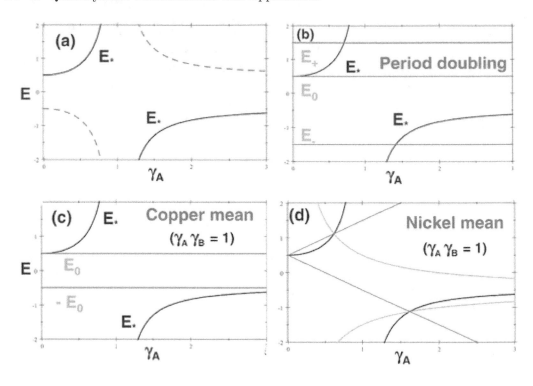

Figure 2.11 Phase diagrams showing the analytical zeroth-order spectra of (a) general precious mean (solid line) and Fibonacci-class FC(n) (dashed line) Π_I class QCs, the (b) period doubling ($E_0= 1/2$, $E_\pm= \pm3/2$), and (c) copper mean ($E_0= 1/2$) Π_{III} class representatives, and the (d) nickel mean based BAC for the particular case $\gamma_B=\gamma_A^{-1}$ (more details in the text). (Reprinted from Ref. [559] with permission. © 2017 by WILEY-VCH Verlag GmbH & Co. KGaA, Weinheim).

with Eq. (2.28), we realize that the FC(2) commutator can be obtained from the standard Fibonacci one upon the conjugation operation $\epsilon \to -\epsilon$ and $\gamma_A \to \gamma_B$, so that we have the formal relationship

$$\left[\mathbf{R}_A^{(2)}, \mathbf{R}_B^{(2)}\right]^{FC} = \mathcal{C} \cdot \left[\mathbf{R}_A^{(1)}, \mathbf{R}_B^{(1)}\right] \equiv \gamma_B^{-1}\bar{p}_1(E, \epsilon, \gamma_B)\, \mathcal{C} \cdot \mathbf{F}_1, \qquad (2.42)$$

with $\bar{p}_1(E,\epsilon,\gamma_B) = \mathcal{C} \cdot p_1(E,\epsilon,\gamma_A)$, where $\mathcal{C} \cdot$ denotes the conjugation operation. Thus, the resonance condition for any FC(n) QC realization (i.e., for any combination of ϵ and $\gamma_B \neq 1$ values) is given by the energy values satisfying the relation $\bar{p}_1(E,\epsilon,\gamma_B) = 0$, namely,

$$E_*^{FC} = -\epsilon\,\frac{1+\gamma_B^2}{1-\gamma_B^2} = \mathcal{C} \cdot E_*, \qquad (2.43)$$

where E_* is the standard Fibonacci resonance energy given by Eq. (2.32). Accordingly, the roots of the polynomial $p_1(E,\epsilon,\gamma_A)$ and its conjugate $\bar{p}_1(E,\epsilon,\gamma_B)$ play a very fundamental role in the energy spectrum structure of these families of Π_I class QCs, as it is illustrated in Fig. 2.11a.

• *Period doubling limit-periodic crystal (Π_{III} class).*

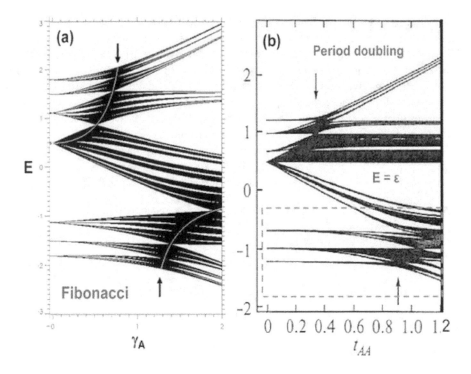

Figure 2.12 Comparison between the mumerically obtained phase diagrams of (a) a standard FQC with $N = 34$, $t_{AB} = 1$, and $\epsilon = 0.5$ over the bond strength range $0 \leq \gamma_A \leq 2$, and (b) a period doubling BAC with $N = 256$, $t_{AB} = 0.6$ and $\epsilon = 0.5$ over the same bond strength range (adapted from Ref. [446]). The vertical arrows in the period doubling diagram indicate the positions of the resonance energy curve $E_*(\epsilon, \gamma_A)$ ends given by the $p_1 = 0$ condition, which are highlighted with a thin white line in the Fibonacci phase diagram. Note that the energy units in (b) must be scaled up by a 5/3 factor in order to compare with the spectral features shown in (a). (Reprinted from Ref. [559] with permission. © 2017 by WILEY-VCH Verlag GmbH & Co. KGaA, Weinheim).

Making use of the elemental cluster matrices listed in Table 2.6 for the period doubling based BAC, we obtain

$$[\mathbf{R}_A, \mathbf{R}_B] = \gamma_A^{-2}\,(E - \epsilon)\,p_1(E)p_2(E) \begin{pmatrix} -1 & E + \epsilon \\ 0 & 1 \end{pmatrix}, \qquad (2.44)$$

where $p_2(E) \equiv E^2 - \epsilon^2 - 2$. As we see, the commutator exhibits the characteristic algebraic structure consisting of an energy dependent polynomial factor multiplying a null trace matrix but, in this case, the degree of the polynomial is four, instead of the first degree polynomials we obtained in the study of both precious means and Fibonacci-class based QCs. Accordingly, the zeroth-order energy spectrum of the period doubling BAC can be regarded as a superposition of the spectrum corresponding to the standard FQC, given by the $p_1(E, \epsilon, \gamma_A)$ polynomial root, plus the additional resonance energies $E_0 = \epsilon$ and $E_\pm = \pm\sqrt{\epsilon^2 + 2}$, as it is shown in Fig. 2.11b. The energies E_0 and $E_\pm$ can be traced back to the molecular levels of the ABA clusters present in the chain, and they do not depend on the bond strength γ_A. Thus, the higher order of the polynomial factor appearing in the period doubling based BAC commutator, naturally leads to an energy spectrum exhibiting a *richer* zeroth-order structure than that previously obtained for representatives of the Π_{I} spectral class. This feature is further illustrated in Fig. 2.12, where we see that FQC and period doubling limit-periodic crystal phase diagrams exhibit a close resemblance over the broad energy window below about $E \simeq -0.5$ (boxed frame in Fig. 2.12b), but when we consider higher energies, approaching the resonance energy $E_0 = \epsilon = 0.5$, both spectra significantly differ. In particular, one can readily appreciate a relatively broad, nearly straight horizontal band related to this resonance energy in the period doubling spectrum, which is completely absent in the Fibonacci one. Significant differences between both spectra can also be appreciated close to the ABA cluster resonance energy value $E_+ = 0.9$, where one can observe the presence of a similar straight band in the period doubling spectrum.

- *Copper (Π_{III} class) and Nickel (Γ class) means based aperiodic crystals.*

Making use of the elemental cluster matrix definitions listed in Table 2.6 for the copper mean, we obtain

$$\left[\mathbf{R}_A^{(2)}, \mathbf{R}_B^{(2)}\right]^{\mathrm{C}} = (\gamma_A \gamma_B)^{-2} \Lambda_{AB}^{\mathrm{C}}(E)(E - \epsilon) \begin{pmatrix} -v_B & 1 \\ \gamma_B^2 - v_B^2 & v_B \end{pmatrix}, \tag{2.45}$$

where

$$\Lambda_{AB}^{\mathrm{C}}(E, \epsilon, \gamma_A, \gamma_B) \equiv \left(1 - \gamma_A^2\right) E^2 - 2\epsilon\gamma_A^2 E - (1 + \gamma_A^2)\epsilon^2 + \gamma_A^2\gamma_B^2 - 1, \tag{2.46}$$

is a second degree polynomial in energy. As we see, the resulting degree of the commutator polynomial factor for the copper mean BAC is three, one order lower than that obtained for the period doubling BAC polynomial factor, but two orders higher than those corresponding to both precious means and Fibonacci-class QCs related commutators. From Eq. (2.45), we also see that the resonance energy $E_0 = \epsilon$, which we previously found in the period doubling BAC energy spectrum, is also present in the copper mean one. By inspecting Eq. (2.46), we note that it depends on both $A - A$ and $B - B$ bond strength parameters, at variance with the previously considered BACs, although γ_B only appears in the independent term. The $\Lambda_{AB}^{\mathrm{C}}$ polynomial roots read

$$E_{\pm}^{\mathrm{C}} = \frac{\gamma_A^2\epsilon \pm \sqrt{\epsilon^2 + \left(1 - \gamma_A^2\right)\left(1 - \gamma_A^2\gamma_B^2\right)}}{1 - \gamma_A^2}, \tag{2.47}$$

so that, by equating Eqs. (2.32) and (2.47), we readily check that the resonance energy E_* only belongs to the zeroth-order energy spectrum of the copper BAC in the particular case $\gamma_B = \gamma_A^{-1}$. In this case, the polynomial $\Lambda_{AB}^{\mathrm{C}}$ can be factorized in the form $\Lambda_{AB}^{\mathrm{C}} = \left(1 - \gamma_A^2\right)(E + \epsilon)(E - E_*)$. Therefore, the resonance energy E_* present in the energy spectra of both the precious means and Fibonacci-class QCs, is also present in the copper mean limit-periodic crystal spectrum when the bonds strengths satisfy the relationship $t_{AA}t_{BB} = 1$. The zeroth-order energy spectrum corresponding to this particular case is shown in Fig. 2.11c. For other choices of the $B - B$ bond strength value, we obtain more featured phase diagrams, such as those shown in Fig. 2.13a–b for the sake of illustration. We note the presence of two curves which are located close to those corresponding to the E_* resonance energy, exhibiting a similar qualitative behavior.

Let us now consider the nickel mean based BAC. Making use of the corresponding elemental cluster matrix definitions listed in Table 2.6, we obtain

$$\left[\mathbf{R}_A^{(3)}, \mathbf{R}_B^{(3)}\right]^{\mathrm{N}} = \gamma_A^{-3}\gamma_B^{-4}\Lambda_{AB}^{\mathrm{N}}(E)\left[\left(\gamma_A^2 - (E - \epsilon)^2\right)\right] \begin{pmatrix} v_B^2 - \gamma_B^2 & -v_B \\ v_B \left(v_B^2 - 2\gamma_B^2\right) & \gamma_B^2 - v_B^2 \end{pmatrix}, \tag{2.48}$$

where

$$\Lambda_{AB}^{N}(E, \epsilon, \gamma_A, \gamma_B) \equiv \left(1 - \gamma_A^2\right) E^3 + \left(1 - 3\gamma_A^2\right) \epsilon E^2 - rE - \epsilon q, \tag{2.49}$$

with $r \equiv \epsilon^2 \left(1 + 3\gamma_A^2\right) + \gamma_B^2 \left(1 - 2\gamma_A^2\right) + 1$, and $q \equiv \epsilon^2 \left(1 + \gamma_A^2\right) - \gamma_B^2 \left(1 + 2\gamma_A^2\right) + 1$, is a third degree polynomial in energy. As we see, the degree of the commutator polynomial factor is one order higher than that obtained for the period doubling BAC, two orders higher than that obtained for the copper mean BAC, and four orders higher than those corresponding to both precious means and Fibonacci-class QCs commutator polynomial factors, respectively. The polynomial Λ_{AB}^{N} can also be factorized in the particular case $\gamma_B = \gamma_A^{-1}$ to adopt the simpler form $\Lambda_{AB}^{N} = \left(1 - \gamma_A^2\right) (E - E_*) \left(v_B^2 - \gamma_A^{-2}\right)$. Accordingly, the resonance energy E_* also belongs to the nickel mean limit-periodic crystal zeroth-order energy spectrum in this case. In fact, by imposing the condition $\Lambda_{AB}^{N}(E_*) \equiv 0$, it can be shown that E_* belongs to the spectrum of the nickel mean based BAC iff $\gamma_A = \pm\gamma_B^{-1}$. In Fig. 2.11d, we show the phase diagram for the case $\gamma_B = \gamma_A^{-1}$, which is characterized by the presence of six main spectral features related to the resonance energies $E = E_*$ (in black), $E_{\pm}^{\alpha} = -\epsilon \pm \gamma_A^{-1}$ (in dark gray), and $E_{\pm}^{\beta} = \epsilon \pm \gamma_A$ (in light gray). We see that three of these curves intersect each other at two points whose coordinates read $(\tau^{\mp 1}, \pm(\tau - \frac{1}{2}))$, where τ is the golden mean.

- *Mixed mean limit-QP crystal (Π_{II} class).*

Making use of the elemental cluster matrix definitions listed in Table 2.6, we obtain $\left[\mathbf{R}_A^{(2,2)}, \mathbf{R}_B^{(2,2)}\right] = (\gamma_A\gamma_B)^{-1}\Lambda_{AB}^{X}(E, \epsilon, \gamma_A, \gamma_B)\sigma_x$, where

$$\Lambda_{AB}^{X} \equiv v_A^2(v_B^2 - 1)(\gamma_A^{-1} - 1) - v_B^2 + v_A v_B \gamma_A \gamma_B (2 - \gamma_A^{-1}) - \gamma_A^2 \gamma_B^2 + 1, \tag{2.50}$$

is a fourth degree polynomial in energy; the same degree we obtained for the period doubling BAC commutator. We note that Eq. (2.50) is invariant upon the simultaneous transformation $v_A \to -v_A$ and $v_B \to -v_B$, related to the energy sign change $E \to -E$ and $\epsilon \to -\epsilon$. Consequently, the considered mixed mean energy spectrum will exhibit reflection symmetry with respect to the abscissas axis. By imposing the condition $\Lambda_{AB}^{X}(E_*) \equiv 0$, it can be shown that E_* belongs to the spectrum *if* $\gamma_B = \pm\gamma_A^{-1}$ and $\epsilon = \pm\frac{1-\gamma_A^2}{2\gamma_A}\sqrt{2 + \gamma_A^2 + \gamma_A^{-1}}$ *only*. For the sake of illustration, in Fig. 2.13c we plot the phase diagram for this limit-QP crystal for the particular case $\gamma_A = \epsilon = 1/2$.

2.4.4 Anatomy of the energy spectrum in Fibonacci quasicrystals

Once we have studied the zeroth-order energy spectrum of several BAC representatives belonging to the Π_I Π_{II}, Π_{III}, and Γ spectral classes, we will now consider the origin of the energy levels splitting and band fragmentation which occurs as soon as long-range QPO effects are progressively taken into account. To this end, we must consider string matrices involving progressively longer atomic clusters at different scales. In so doing, we will focus on the FQC case for the sake of illustration of the general procedure. In order to proceed in a systematic way, we will begin by considering blocking schemes involving the elemental unimodular matrices $\mathbf{R}$ and $\mathbf{r}$, along with their related doublets $\mathbf{Rr}$, $\mathbf{rR}$, and $\mathbf{RR}$ (note that $\mathbf{rr}$ pairs are forbidden by construction in FQCs). Afterwards, we will study ensembles consisting of increasingly longer blocks. As the considered scale length linearly increases with the block size itself, this approach allows us to study the emergence of long-range QPO fingerprints in the resulting energy spectrum as we approach the asymptotic $N \to \infty$ and $n \to \infty$ limits in a natural way.

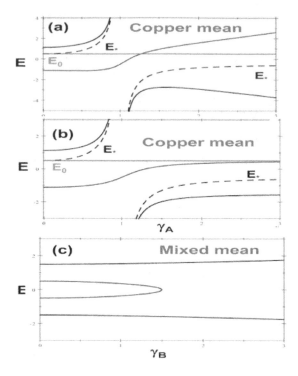

Figure 2.13 Phase diagrams showing the analytical zeroth-order spectra of the copper mean limit periodic crystals for (a) $\gamma_B = 1$, and (b) $\gamma_B = \gamma_A$. For the sake of comparison the curves for the resonance energy E_* (which does not belong to the spectrum) are also plotted (dashed lines). In (c) we plot the analytical phase diagram for the mixed mean $n = m = 2$ based limit-quasiperiodic crystal with $\gamma_A = \epsilon = 1/2$. (Reprinted from Ref. [559] with permission. © 2017 by WILEY-VCH Verlag GmbH & Co. KGaA, Weinheim).

2.4.4.1 First-order fragmentation pattern

Let us consider the FQC renormalized global transfer matrix

$$\mathcal{M}_N = \ldots \mathbf{RrRRrRrRRrRRrRrRRrRrR}.$$

There exist six blocking schemes in which this matrix string can be grouped in terms of single and double elemental matrices strings. Two of them, namely, $\ldots(\mathbf{Rr})\mathbf{R}(\mathbf{Rr})(\mathbf{Rr})\mathbf{R}(\mathbf{Rr})\mathbf{R}(\mathbf{Rr})$, and $\ldots \mathbf{R}(\mathbf{rR})\mathbf{R}(\mathbf{rR})(\mathbf{rR})\mathbf{R}(\mathbf{rR})\mathbf{R}(\mathbf{rR})$, are respectively related to the commutators $[\mathbf{Rr}, \mathbf{R}] = -\mathbf{R}[\mathbf{R}, \mathbf{r}]$ and $[\mathbf{rR}, \mathbf{R}] = -[\mathbf{R}, \mathbf{r}]\mathbf{R}$, which only contribute to the resonance energy E_* belonging to the zeroth-order energy spectrum, as prescribed by Eq. (2.28). The remaining blocking schemes include $\mathbf{RR}$ pair blocks, and the related commutators are respectively given by $[\mathbf{RR}, \mathbf{r}]$, $[\mathbf{RR}, \mathbf{Rr}]$, and $[\mathbf{RR}, \mathbf{rR}]$, along with $[\mathbf{RR}, \mathbf{R}]$, which trivially vanishes. The non-trivial commutators above can be easily evaluated by using Eq. (2.33) to express the $\mathbf{RR}$ product in the power form

$$\mathbf{R}^2 = (\mathrm{tr}\mathbf{R})\mathbf{R} - \mathbf{I} = \gamma_A^{-1} p_3(E)\mathbf{R} - \mathbf{I}, \tag{2.51}$$

where

$$p_3(E) = (E + \epsilon)[(E - \epsilon)^2 - \gamma_A^2] + 2(\epsilon - E). \qquad (2.52)$$

Accordingly, the commutators we are interested in read

$$[\mathbf{RR}, \mathbf{r}] = \gamma_A^{-2} p_1 p_3 \mathbf{F}, \quad [\mathbf{RR}, \mathbf{Rr}] = \gamma_A^{-2} p_1 p_3 \mathbf{RF}, \quad [\mathbf{RR}, \mathbf{rR}] = \gamma_A^{-2} p_1 p_3 \mathbf{FR}, \qquad (2.53)$$

where we have used Eqs. (2.51) and (2.28).[20] Now, we know that $\mathbf{F} \neq \mathbf{0}$, and since $\det \mathbf{F} \neq 0$ and $\det \mathbf{R} \neq 0$, then $\mathbf{RF} \neq \mathbf{0}$ and $\mathbf{FR} \neq \mathbf{0}$ $\forall E$, so that the resonance condition for the commutators (2.53) requires that $p_1(E) = 0$ and/or $p_3(E) = 0$. Therefore, in addition to the resonance energy E_* given by Eq. (2.32), we have three *new* resonance energy values given by the roots of $p_3(E)$, which define the first-order hierarchy of the energy spectrum fragmentation pattern. We note that the simultaneous presence of the polynomials p_1 and p_3 in the factorized commutators in Eq. (2.53) naturally accounts for the overall *nested* structure of the resulting spectrum.

The total number of bands present in the energy spectrum depends on the adopted γ_A value. In the weak coupling limit $\gamma_A \to 0$ we have $\lim_{\gamma_A \to 0} E_*(\epsilon, \gamma_A) = \epsilon \equiv E_0^{(0)}$ (where the superscript stands for the fragmentation order of the spectrum), and $p_3(E, \epsilon, 0) = (E - \epsilon)(E^2 - \epsilon^2 - 2)$, whose roots $E_0^{(1)} = \epsilon$, $E_\pm^{(1)} = \pm\sqrt{\epsilon^2 + 2}$, coincide with the energy levels of the $A - B - A$ molecule, so that the resulting energy spectrum only displays three main bands. As the $A - A$ atomic coupling increases the doubly degenerated level $E_0^{(1)} = \epsilon = E_0^{(0)}$ splits and the energy spectrum consists of four bands instead, except for those γ_A values satisfying the condition $E_*(\epsilon, \gamma_A) = E_\pm^{(1)}(\epsilon, \gamma_A)$, leading to double degenerate states in the upper and lower bands, thereby rendering a trifurcated spectrum as well. The corresponding crossing points (encircled in Fig. 2.14) can be analytically obtained from the condition $p_3(E_*, \epsilon, \gamma_A) = 0$, to get $\gamma_A^\pm = \sqrt{b \pm \sqrt{b^2 - 1}}$, with $b \equiv 1 + 2\epsilon^2/3$. We note that in the $\gamma_A \to 0$ limit the FQC overall energy spectrum (i.e., including both zeroth-order and first order contributions) closely resembles the zeroth-order spectrum of the period doubling BAC depicted in Fig. 2.11b. On the other hand, we note that in the very strong $A - A$ bonding regime $\lim_{\gamma \to \infty} E_*(\epsilon, \gamma_A) = \lim_{\gamma \to \infty} E_0^{(1)}(\epsilon, \gamma_A) = -\epsilon$, which also leads to a nearly trifurcation pattern of the energy spectrum. For the sake of illustration the first order energy spectrum, determined by the parametric curves $p_1(E, \epsilon, \gamma_A) = 0$ and $p_3(E, \epsilon, \gamma_A) = 0$, is plotted in Fig. 2.14 for a FQC composed of atoms with on-site energies $V_A = 1/2 = -V_B$, within the $A - A$ bond strength range $0 \leq \gamma_A \leq 2$.

2.4.4.2 Second-order fragmentation pattern

In order to describe the next hierarchy level in the energy spectrum structure we will consider blocking schemes including suitable triplets of elemental clusters matrices $\mathbf{r}$ and $\mathbf{R}$. To this end, we recall that the blocking scheme consisting of the Fibonacci strings $\mathbf{RrR} = \mathbf{F}_3$ and $\mathbf{R} = \mathbf{F}_1$ do not contribute to the fragmentation of the energy spectrum (see Sec. 2.4.2). Thus, we have identified three possible blocking schemes, namely: (1) $\mathbf{RrR}$ triplets along with single $\mathbf{r}$ matrices, (2) $\mathbf{rRr}$ triplets along with single $\mathbf{R}$ matrices, and (3) $\mathbf{rRr}$ triplets along with $\mathbf{RR}$ pairs. The corresponding commutators read (**Exercise 2.9**),

$$[\mathbf{RrR}, \mathbf{r}] = \mathrm{tr}(\mathbf{Rr})[\mathbf{R}, \mathbf{r}] = \gamma_A^{-2} p_1 p_5 \mathbf{F}, \quad [\mathbf{rRr}, \mathbf{R}] = -\mathrm{tr}(\mathbf{rR})[\mathbf{R}, \mathbf{r}] = -\gamma_A^{-2} p_1 p_5 \mathbf{F}, \qquad (2.54)$$

$$[\mathbf{rRr}, \mathbf{RR}] = [\mathbf{rRr}, \gamma_A^{-1} p_3 \mathbf{R} - \mathbf{I}] = \gamma_A^{-1} p_3 [\mathbf{rRr}, \mathbf{R}] = -\gamma_A^{-3} p_1 p_3 p_5 \mathbf{F}, \qquad (2.55)$$

---

[20]We recall the commutator properties $[\mathbf{A} - \mathbf{B}, \mathbf{C}] = [\mathbf{A}, \mathbf{C}] - [\mathbf{B}, \mathbf{C}]$ and $[\mathbf{A}, \mathbf{B} - \mathbf{C}] = [\mathbf{A}, \mathbf{B}] - [\mathbf{A}, \mathbf{C}]$.

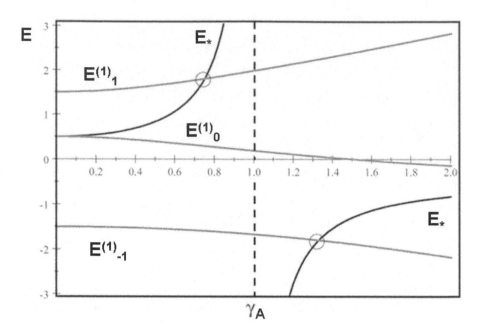

Figure 2.14 Analytical phase diagram for a FQC with $\epsilon = 1/2$, over the $A-A$ bond strength range $0 \leq \gamma_A \leq 2$, showing the zeroth order (in black) and the first order (in dark gray) energy spectrum fragmentation pattern contributions. In the weak coupling limit $\gamma_A \to 0$ the spectrum consists of three bands located at the ABA molecule energy levels $E_0^{(1)} = 1/2$ and $E_{\pm}^{(1)} = \pm 3/2$. As soon as the $A - A$ atomic coupling becomes non-negligible the spectrum consists of four bands due to the splitting of the degenerate $E_0^{(1)} = E_*$ level, except for those γ_A values satisfying the band crossing condition $p_3(E_*, \epsilon, \gamma_A) = 0$ (encircled points). The crossing points exact coordinates in the phase diagram are $(\frac{\sqrt{3}}{6}(\sqrt{13} - 1), \frac{\sqrt{13}}{2})$ and $(\frac{\sqrt{3}}{6}(\sqrt{13} + 1), -\frac{\sqrt{13}}{2})$. We note that the ordinate values of the encircled crossing points coincide with the energy values of the second-order fragmentation energies $E_{\pm 2}^{(2)}$ in the weak coupling limit (see Fig. 2.10).(Reprinted from Ref. [560] with permission. © 2017 by WILEY-VCH Verlag GmbH & Co. KGaA, Weinheim).

where we have used Eqs. (2.28), (2.51), and (2.54), and we have introduced (**Exercise 2.10**)

$$p_5(E) \equiv \gamma_A \text{tr}(\mathbf{Rr}) = p_2(E)p_3(E) - E + \epsilon. \qquad (2.56)$$

By comparing Eqs. (2.54) and (2.55), we see that the polynomial p_3 appears in the factorized expression of commutator $[\mathbf{rRr}, \mathbf{RR}]$, but it does not appear in the expressions of both commutators $[\mathbf{RrR}, \mathbf{r}]$ and $[\mathbf{rRr}, \mathbf{R}]$. This is due to the fact that the scale length related to matrix strings involving these conjugate commutators is shorter than that corresponding to the former one. On the other hand, the presence of the polynomial $p_5(E)$ in the three commutators above discloses a new set of resonance energy values, given by the condition $p_5(E) = 0$, which defines the second-order fragmentation pattern in the energy spectrum.

In the weak bonding limit $\gamma_A \to 0$, the roots of the polynomial $p_5(E, \epsilon, 0) = (E - \epsilon)(p_2^2 - 1)$ coincide with the $A - B - A - B - A$ molecule energy levels, namely, $E_0^{(2)} = \epsilon$ (note that the

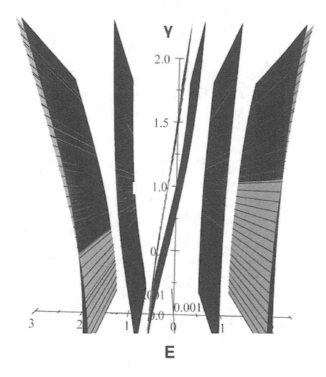

Figure 2.15 Second-order energy spectrum structure determined by the roots of the polynomials $p_3(E, \epsilon, \gamma)$ and $p_5(E, \epsilon, \gamma)$ for a FQC with $N = 34$ atoms, $\epsilon = 1/2$ and $0 \leq \gamma_A \leq 2$. The upper and lower bands resolve into two bands belonging to different spectral polynomials each. (Reprinted from Ref. [560] with permission. © 2017 by WILEY-VCH Verlag GmbH & Co. KGaA, Weinheim).

energy level $\epsilon = E_0^{(2)} = E_0^{(1)} = E_0^{(0)}$ becomes triply degenerated at this stage of the energy spectrum generation), $E_{\pm 1}^{(2)} = \pm\sqrt{\epsilon^2 + 1}$, and $E_{\pm 2}^{(2)} = \pm\sqrt{\epsilon^2 + 3}$ (see Fig. 2.10). In Fig. 2.15, we show the overall structure of the second order hierarchy in the energy spectrum of a FQC with on-site energy $\epsilon = 1/2$ over the range $0 \leq \gamma_A \leq 2$.

2.4.4.3 Higher-order fragmentation patterns

At this stage, it is convenient to introduce a family of polynomials which are related to the traces of the Fibonacci string matrices introduced in Sec. 2.4.2. Indeed, since $\mathbf{F}_n$ matrices are unimodular and they obey the concatenation rule $\mathbf{F}_{n+1} = \mathbf{F}_n \mathbf{F}_{n-1}$, their traces satisfy the recursion relation [431, 432],

$$\mathrm{tr}\mathbf{F}_{n+1} = \mathrm{tr}\mathbf{F}_n \mathrm{tr}\mathbf{F}_{n-1} - \mathrm{tr}\mathbf{F}_{n-2}, \quad n \geq 2, \tag{2.57}$$

with $\text{tr}\mathbf{F}_0 \equiv \text{tr}(\mathbf{r}) = E^2 - \epsilon^2 - 2 \equiv p_2(E)$, $\text{tr}\mathbf{F}_1 \equiv \text{tr}(\mathbf{R}) \equiv \gamma_A^{-1}p_3(E)$, and $\text{tr}\mathbf{F}_2 \equiv \text{tr}(\mathbf{Rr}) = \text{tr}(\mathbf{rR}) = \gamma_A^{-1}p_5(E)$. Making use of Eq. (2.57) recursively, we obtain

$$\text{tr}\mathbf{F}_3 = \text{tr}\mathbf{F}_2\text{tr}\mathbf{F}_1 - \text{tr}\mathbf{F}_0 = \gamma_A^{-2}p_5p_3 - p_2 \equiv \gamma_A^{-2}p_8, \tag{2.58}$$

$$\text{tr}\mathbf{F}_4 = \text{tr}\mathbf{F}_3\text{tr}\mathbf{F}_2 - \text{tr}\mathbf{F}_1 = \gamma_A^{-3}p_8p_5 - \gamma_A^{-1}p_3 \equiv \gamma_A^{-3}p_{13}, \tag{2.59}$$

$$\text{tr}\mathbf{F}_5 = \text{tr}\mathbf{F}_4\text{tr}\mathbf{F}_3 - \text{tr}\mathbf{F}_2 = \gamma_A^{-5}p_{13}p_8 - \gamma_A^{-1}p_5 \equiv \gamma_A^{-5}p_{21}, \tag{2.60}$$

$$\vdots$$

$$\text{tr}\mathbf{F}_j = \text{tr}\mathbf{F}_{j-1}\text{tr}\mathbf{F}_{j-2} - \text{tr}\mathbf{F}_{j-3} = \gamma_A^{-F_{j-1}}p_{F_{j+2}}, \quad j \geq 3, \tag{2.61}$$

where we have defined the energy dependent polynomials

$$p_{F_j}(E) \equiv p_{F_{j-1}}(E)p_{F_{j-2}}(E) - \gamma_A^{2F_j-5}p_{F_{j-3}}(E), \quad j \geq 5, \tag{2.62}$$

whose degree is given by Fibonacci numbers. Henceforth, we will refer to the polynomials introduced in Eq. (2.62) as *Fibonacci spectral polynomials* [560].

The next hierarchies in the energy spectrum fragmentation pattern are obtained by considering blocking schemes based on the quadruplet block $\mathbf{rRRr}$ in combination with: (1) isolated $\mathbf{R}$ matrices, (2) $\mathbf{RR}$ pairs, and (3) $\mathbf{RrR}$ triplets. Accordingly, we must consider the following commutators:

$$[\mathbf{rRRr}, \mathbf{R}] = [\mathbf{r}((\text{tr}\mathbf{F}_1)\mathbf{R} - \mathbf{I})\mathbf{r}, \mathbf{R}] = \text{tr}\mathbf{F}_1[\mathbf{rRr}, \mathbf{R}] - [\mathbf{r}^2, \mathbf{R}] \tag{2.63}$$

$$= -\text{tr}\mathbf{F}_1\text{tr}\mathbf{F}_2[\mathbf{R}, \mathbf{r}] + \text{tr}\mathbf{F}_0[\mathbf{R}, \mathbf{r}] = -\text{tr}\mathbf{F}_3[\mathbf{R}, \mathbf{r}] = -\gamma^{-3}p_1p_8\mathbf{F}, \tag{2.64}$$

where we have taken into account Eqs. (2.51), (2.54), (2.58), (2.28), respectively, and we used Eq. (2.34) to express $\mathbf{r}^2 = (\text{tr}\mathbf{F}_0)\mathbf{r} - \mathbf{I}$.

$$[\mathbf{rRRr}, \mathbf{RR}] = -\text{tr}(\mathbf{RrR})[\mathbf{RR}, \mathbf{r}] = -\text{tr}\mathbf{F}_3\text{tr}\mathbf{F}_1[\mathbf{R}, \mathbf{r}] = -\gamma^{-4}p_1p_3p_8\mathbf{F}, \tag{2.65}$$

where we made use of (2.112) taking $\mathbf{A} \equiv \mathbf{RR}$ and $\mathbf{B} \equiv \mathbf{r}$, as well as the cyclic permutation property of the trace $\text{tr}(\mathbf{rRR}) = \text{tr}(\mathbf{RrR}) = \text{tr}\mathbf{F}_3$, along with Eqs. (2.58) and (2.53).

$$[\mathbf{rRRr}, \mathbf{RrR}] = [\mathbf{r}((\text{tr}\mathbf{F}_1)\mathbf{R} - \mathbf{I})\mathbf{r}, \mathbf{RrR}] = \text{tr}\mathbf{F}_1[\mathbf{rRr}, \mathbf{RrR}] + \text{tr}\mathbf{F}_0[\mathbf{RrR}, \mathbf{r}]$$

$$= (\text{tr}\mathbf{F}_1 - \text{tr}\mathbf{F}_2(\text{tr}\mathbf{F}_2\text{tr}\mathbf{F}_1 - \text{tr}\mathbf{F}_0))[\mathbf{R}, \mathbf{r}] = (\text{tr}\mathbf{F}_1 - \text{tr}\mathbf{F}_3\text{tr}\mathbf{F}_2)[\mathbf{R}, \mathbf{r}]$$

$$= -\text{tr}\mathbf{F}_4[\mathbf{R}, \mathbf{r}] = -\gamma^{-4}p_1p_{13}\mathbf{F}, \tag{2.66}$$

where we have used Eqs. (2.51), (2.54), (2.58), and (2.59), along with Eq. (2.114) (**Exercise 2.9**).

The roots of the Fibonacci spectral polynomials $p_8(E)$ in Eqs. (2.64) and (2.65), and $p_{13}(E)$ in Eq. (2.66), respectively, determine the main features of the third and fourth-order fragmentation patterns of the energy spectrum. The fourth-order fragmentation pattern can also be obtained by considering blocking schemes including longer string matrices. For instance, let us consider the commutator

$$[\mathbf{RRrRrRR}, \mathbf{rRr}] = -\text{tr}(\mathbf{rRrRR})[\mathbf{rRr}, \mathbf{RR}] = -\text{tr}(\mathbf{RrRRr})[\mathbf{rRr}, \mathbf{RR}]$$

$$= \text{tr}\mathbf{F}_1\text{tr}\mathbf{F}_2\text{tr}\mathbf{F}_4[\mathbf{R}, \mathbf{r}] = \gamma^{-6}p_1p_3p_5p_{13}\mathbf{F}, \tag{2.67}$$

where we used Eq. (2.112) taking $\mathbf{A} = \mathbf{rRr}$ and $\mathbf{B} = \mathbf{RR}$ (**Exercise 2.9**), along with Eqs. (2.55), (2.54), and (2.59). As we see, three consecutive fragmentation pattern hierarchies are included via the Fibonacci spectral polynomials p_3, p_5, and p_{13} appearing in the factorized expression of this commutator.

Finally, blocking schemes yielding the fifth-order fragmentation pattern involve strings containing seven elemental matrices $\mathbf{r}$ and $\mathbf{R}$ at least. The simpler commutator representative is given by

$$[\mathbf{rRRrRRr}, \mathbf{RrR}] = [(\mathbf{rR}^2)^2\mathbf{r}, \mathbf{RrR}] = \mathrm{tr}\mathbf{F}_3[\mathbf{rRRr}, \mathbf{RrR}] + [\mathbf{RrR}, \mathbf{r}]$$

$$= (\mathrm{tr}\mathbf{F}_2 - \mathrm{tr}\mathbf{F}_3\mathrm{tr}\mathbf{F}_4)[\mathbf{R}, \mathbf{r}] = -\mathrm{tr}\mathbf{F}_5[\mathbf{R}, \mathbf{r}] = -\gamma^{-6}p_1 p_{21}\mathbf{F}, \qquad (2.68)$$

where we used Eq. (2.34) to get $(\mathbf{rR}^2)^2 = U_1(z)\mathbf{rR}^2 - U_0(z)\mathbf{I} = \mathrm{tr}\mathbf{F}_3(\mathbf{rR}^2) - \mathbf{I}$, as well as Eqs. (2.66), (2.54), and (2.60). The roots of the Fibonacci spectral polynomial $p_{21}(E)$ determine the fifth-order fragmentation pattern of the energy spectrum.

Following a different approach, the hierarchical structure of the energy spectrum in Fibonacci, Thue-Morse, and Rudin-Shapiro BACs has recently been discussed in terms of a combination of different resonator-like atomic clusters by means of a systematic approach based on quantum mechanical perturbation theory, where the system Hamiltonian is expressed in the form

$$H = H_0 + \xi H_I \equiv \begin{pmatrix} v_1 & 0 & \cdots & 0 \\ 0 & \ddots & \ddots & \vdots \\ \vdots & \ddots & \ddots & 0 \\ 0 & \cdots & 0 & v_N \end{pmatrix} + \xi \begin{pmatrix} 0 & t_{12} & \cdots & 0 \\ t_{12} & \ddots & \ddots & \vdots \\ \vdots & \ddots & \ddots & t_{N-1,N} \\ 0 & \cdots & t_{N-1,N} & 0 \end{pmatrix},$$

where ξ is the perturbation parameter [754]. Indeed, the role of local symmetries in both the energy spectrum structure and the localization degree of its related eigenstates has been studied in detail during the last decade [396, 397, 398, 399, 418, 626, 627].

2.5 FREQUENCY SPECTRUM OF ONE-DIMENSIONAL QUASICRYSTALS

The study of phonon propagation through 1D FQCs has been extensively considered in order to understand the role of QPO in their physical properties in terms of the Hamiltonian given by Eq. (2.8), describing a harmonic chain model composed of two kinds of masses, m_A and m_B, arranged according to a given QP sequence, and two kinds of springs, K_{AA} and $K_{AB} = K_{BA}$, whose value will generally depend on the type of joined atoms. Within this framework different kinds of models can be considered. In the on-site model one assumes all the spring constants to be equal. In the transfer model, all the masses are assumed to be identical instead and the springs strength take on two different values [808]. From a physical point of view, one expects that the nature of the chemical bonding between the different atoms (and thereof the value of the spring constant representing the bond) will depend on the nature of the involved atoms. In this case, the aperiodic distribution of masses in the system induces an aperiodic distribution of spring constants throughout, hence leading to the so-called mixed (or tridiagonal) models [560, 562].

In order to study phonon dynamics in BACs, Eq. (2.10) is conveniently cast in the matrix form

$$\begin{pmatrix} u_{k+1} \\ u_k \end{pmatrix} = \begin{pmatrix} \frac{v_k}{K_{k,k+1}} & -\frac{K_{k,k-1}}{K_{k,k+1}} \\ 1 & 0 \end{pmatrix} \begin{pmatrix} u_k \\ u_{k-1} \end{pmatrix} \equiv \mathbf{T}_{k,k\pm1} \begin{pmatrix} u_k \\ u_{k-1} \end{pmatrix}, \qquad (2.69)$$

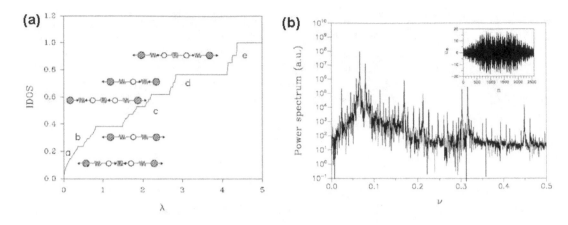

Figure 2.16 (a) Frequency spectrum for a tridiagonal FQC with $N = 610$, $\alpha = \gamma_A = 2$, $m_A = 1$, and $K_{AB} = 1$, as given by the integrated density of states (IDOS). The main subbands (labeled a, b, c, d, and e) are directly related to the normal modes associated to the BAB trimers and BAAB tetramers introduced in a suitable renormalization scheme [530]. (Reprinted figure with permission from Maciá E, Thermal conductivity of one-dimensional Fibonacci quasicrystals, 2000 Physical. Review B **61** 6645. Copyright (2000) by the American Physical Society) (b) Power spectrum of the critical normal mode corresponding to the resonance frequency $\lambda^* = 2$, whose atomic displacements amplitudes are shown in the inset, in a FQC with $N = 2584$, $\gamma_A = 1/2$, and $\alpha = 34/21$. The normal mode amplitudes have been obtained by iterating the dynamical equation (2.69) with the initial conditions $u_0 = 0$ and $u_1 = 1$ [529]. (Reprinted figure with permission from Maciá E, Physical nature of critical modes in Fibonacci quasicrystals, 1999 Physical. Review B **60** 10032. Copyright (1999) by the American Physical Society).

where $v_k \equiv K_{k,k-1} + K_{k,k+1} - m_k \omega^2$ (see Table 2.5), so that the allowed regions of the frequency spectrum are determined from the condition $|\mathrm{tr}\mathcal{M}_N(\omega)| \leq 2$, where $\mathcal{M}_N(\omega) \equiv \prod_{k=N}^{1} \mathbf{T}_{k,k\pm1}$, is the global transfer matrix, and the eight possible local transfer matrices are given by

$$\mathbf{T}_{AAA} = \begin{pmatrix} 2 - \gamma_A^{-1}\lambda & -1 \\ 1 & 0 \end{pmatrix}, \quad \mathbf{T}_{AAB} = \begin{pmatrix} 1 + \gamma_A - \lambda & -\gamma \\ 1 & 0 \end{pmatrix},$$

$$\mathbf{T}_{ABA} = \begin{pmatrix} 2 - \alpha\lambda & -1 \\ 1 & 0 \end{pmatrix}, \quad \mathbf{T}_{BAA} = \begin{pmatrix} 1 + \gamma_A^{-1}(1 - \lambda) & -\gamma_A^{-1} \\ 1 & 0 \end{pmatrix},$$

$$\mathbf{T}_{ABB} = \begin{pmatrix} 1 + \gamma_B^{-1}(1 - \alpha\lambda) & -\gamma_B^{-1} \\ 1 & 0 \end{pmatrix}, \quad \mathbf{T}_{BBA} = \begin{pmatrix} 1 + \gamma_B - \alpha\lambda & -\gamma_B \\ 1 & 0 \end{pmatrix},$$

$$\mathbf{T}_{BAB} = \begin{pmatrix} 2 - \lambda & -1 \\ 1 & 0 \end{pmatrix}, \quad \mathbf{T}_{BBB} = \begin{pmatrix} 2 - \alpha\gamma_B^{-1}\lambda & -1 \\ 1 & 0 \end{pmatrix}, \quad (2.70)$$

where $\gamma_A \equiv K_{AA}/K_{AB}$, $\gamma_B \equiv K_{BB}/K_{AB}$, $\alpha \equiv m_B/m_A$, and $\lambda \equiv m_A\omega^2/K_{AB}$.

The overall structure of the frequency spectrum corresponding to a general FQC is illustrated in Fig. 2.16a in terms of the parametrized frequency λ for a representative choice of the parameters α and γ_A [530]. As we can see, the frequency spectrum shows a splitting scheme, characterized by the presence of five main subbands separated by well-defined gaps. Inside each main subband the fragmentation scheme follows a trifurcation pattern, in which each subband further trifurcates obeying a hierarchy of splitting from one to three subsubbands. Additional physical insight can be gained by analyzing the spectrum within the framework of real-space renormalization techniques [779, 121]. To this end, the original FQC is decomposed into a certain number of atomic clusters, according to a given criteria referred to as blocking scheme. In the study of QP systems the minimization of the information entropy (in Shannon's sense) provides a useful guide to choose appropriate blocking schemes [525]. Using this criterion the original lattice is decomposed in a series of trimers and tetramers of the form $B - A - B$ and $B - A = A - B$. Neglecting the trivial translation modes related to $\lambda_0 = 0$, each trimer contributes with two different normal vibration modes, whose respective frequencies are given by $\lambda_b = \alpha^{-1}$, $\lambda_d = 2 + \alpha^{-1}$, and, analogously, each tetramer contributes with three different normal modes given by $\lambda_c = 1 + \alpha^{-1}$, $\lambda_e = \gamma_A + \eta + \sqrt{(\eta - \gamma_A)^2 + 2\gamma_A}$, and $\lambda_a = \gamma_A + \eta - \sqrt{(\eta - \gamma_A)^2 + 2\gamma_A}$, where $\eta \equiv (1 + \alpha^{-1})/2$ (**Exercise 2.11**). Hence, we have *five* basic normal modes describing the dynamical state of a tridiagonal FQC at a fundamental level. If we assume that these normal modes are coupled via resonance effects, we realize that the number and frequencies of the normal modes appearing at the first stage of the renormalization process respectively determine the number and approximate location of the main subbands in the resulting phonon spectrum [525, 653, 654]. In this way, we can assign each main subband appearing in the frequency spectrum to a specific normal mode. Such a procedure is illustrated in Fig. 2.16a, where we see that the lower frequency region of the spectrum ($\lambda < 1$) contains two main contributions: the lowest frequency contribution ($\lambda \lesssim 0.5$), which is related to the tetramers λ_a (contributing with τ^{-3} states) and the frequency interval $0.5 \lesssim \lambda < 1$, which is related to the trimer's normal mode λ_b (contributing with τ^{-4} states). Therefore, although both subbands are separated by a relatively narrow gap, their origin can be traced back to the dynamics of quite different clusters in the FQC.

In order to gain additional insight on the frequency spectrum main features, we will extend the renormalization algebraic approach discussed in Sec. 2.3.3 in order to study the phonon dynamics as well. To this end, the original FQC is resolved in terms of $B - A$ and $B - A = A$ clusters, respectively related to the renormalized matrices $\mathbf{R}_B^{(1)}$ and $\mathbf{R}_A^{(1)}$ (see Table 2.6), where the related local transfer matrices are respectively given by Eq. (2.70). The corresponding commutator reads

$$[\mathbf{R}_A^{(1)}, \mathbf{R}_B^{(1)}] = \lambda \gamma_A^{-1} p_1(\lambda) \begin{pmatrix} 1 & 0 \\ 2 - \alpha\lambda & -1 \end{pmatrix} \equiv \lambda \gamma_A^{-1} p_1(\lambda)\mathbf{P}, \qquad (2.71)$$

where $p_1(\lambda) \equiv 2\gamma_A - 1 - \alpha[1 + \lambda(\gamma_A - 1)]$. This expression is completely analogous to Eq. (2.28), and the resonance condition $p_1(\lambda) = 0$ determines the zeroth-order frequency spectrum structure in terms of the bi-parametric relation

$$\lambda^*(\alpha, \gamma_A) = \frac{\alpha - 2\gamma_A + 1}{\alpha(1 - \gamma_A)}, \qquad (2.72)$$

which plays the same role as Eq. (2.32) in the analysis of the electronic energy spectrum. By following a completely analogous approach, we can introduce the second order renormalized matrices $\mathbf{R}'_A = \mathbf{R}_B^{(1)}\mathbf{R}_A^{(1)}\mathbf{R}_A^{(1)}\mathbf{R}_B^{(1)}$ and $\mathbf{R}'_B = \mathbf{R}_B^{(1)}\mathbf{R}_A^{(1)}\mathbf{R}_B^{(1)}$, describing the phonon dynamics throughout longer atomic clusters,[21] to get the commutator $[\mathbf{R}'_A, \mathbf{R}'_B] = \lambda \gamma_A^{-1} p_1(\lambda)p_2(\lambda)$ $\mathbf{P} = p_2(\lambda)[\mathbf{R}_A, \mathbf{R}_B]$, where $p_2(\lambda) \equiv \alpha\lambda^2 - 2\lambda(\alpha + 1) + 2$ (**Exercise 2.12**). Quite remarkably, the $p_2(\lambda)$ roots $\lambda_\pm = 1 + \alpha^{-1} \pm \sqrt{1 + \alpha^{-2}}$, are independent of the adopted elastic

---

[21]We note that in this second order renormalization blocking scheme matrices $\mathbf{R}_A^{(1)}$ and $\mathbf{R}_B^{(1)}$ play the analogous role of A and B atoms in the clusters shown in Fig. 2.16a.

constants values. Accordingly, the frequency spectrum of FQCs can be analyzed in terms of a nested sequence of resonance frequencies, which are given by the roots of a series of $p_j(\lambda)$ polynomials, in close analogy with the results obtained in the study of the electronic energy spectrum structure in Sec. 2.4.3.

An interesting property of tridiagonal FQCs is related to the appealing possibility of *modulating* the transport properties of phonons by properly selecting the values of the masses composing the chain (isotopic effect). To illustrate this property, we will consider the particular case given by the condition $K_{AA} = K_{AB}/2$. In this case ($\gamma_A = 1/2$), Eq. (2.72) reduces to $\lambda^* = 2$ for any arbitrary choice of the masses m_A and m_B. The renormalized matrices $\mathbf{R}_A^{(1)}$ and $\mathbf{R}_B^{(1)}$ then adopt the simple form

$$\mathbf{R}_A^{(1)}(\lambda^* = 2) = \begin{pmatrix} 1 & 0 \\ 2(\alpha - 2) & 1 \end{pmatrix} \qquad \mathbf{R}_B^{(1)}(\lambda^* = 2) = \begin{pmatrix} -1 & 0 \\ 2(1 - \alpha) & -1 \end{pmatrix}, \qquad (2.73)$$

and the global transfer matrix reads (**Exercise 2.13a**)

$$\mathcal{M}_N(\lambda^* = 2) \equiv [\mathbf{R}_A^{(1)}(\lambda^* = 2)]^{n_A}[\mathbf{R}_B^{(1)}(\lambda^* = 2)]^{n_B} = (-1)^{F_{n-4}} \begin{pmatrix} 1 & 0 \\ 2(\alpha F_{n-2} - F_{n-1}) & 1 \end{pmatrix}. \qquad (2.74)$$

Thus, we realize that the band edge frequency $\lambda^* = 2$ belongs to the frequency spectrum regardless of the system length, since $|\text{tr}[\mathcal{M}_N(\lambda^*)]| = 2$ in this particular case, describing a band-edge mode. The spatial structure of this critical normal mode is determined by two different contributions, corresponding to two separate scale lengths. Thus, although at long scales (greater than, say, 100 sites) the state shows a long-range modulation given by a broad envelope extending over the entire system's length (as can be seen in the inset to Fig. 2.16b), the atomic amplitudes distribution exhibits a series of QP oscillations at shorter length scales, as it is indicated by the power spectrum shown in Fig. 2.16b main frame. Therefore, the structure of the critical normal mode corresponding to the reduced frequency $\lambda^* = 2$ is QP, since the separation between successive high intensity peaks in the power spectrum of the amplitudes distribution are arranged according to successive powers of the golden mean.

Another interesting particular case is given by the mid-band frequency $\lambda^* = 1$, which according to Eq. (2.72) corresponds to the choice $\gamma_A = (2 - \alpha)^{-1}$. In this case the global transfer matrix reads (**Exercise 2.13b**)

$$\mathcal{M}_N(\lambda^* = 1) \equiv [\mathbf{R}_A^{(1)}(\lambda^*)]^{n_A}[\mathbf{R}_B^{(1)}(\lambda^*)]^{n_B} = (-1)^{n_A} \begin{pmatrix} U_{n_B} + U_{n_B-1} & -U_{n_B-1} \\ (2 - \alpha)U_{n_B-1} & -U_{n_B-1} - U_{n_B-2} \end{pmatrix}, \qquad (2.75)$$

where the second kind Chebyshev polynomials depend on the variable $z = \frac{1}{2}\text{tr}\mathbf{R}_B^{(1)}(\lambda^* = 1) = -\alpha/2$. Therefore, $|\text{tr}[\mathcal{M}_N(\lambda^* = 1)]| = |U_{n_B} - U_{n_B-2}| = |2T_{n_B}| \leq 2$, and we conclude that the frequency $\lambda^* = 1$ belongs to the frequency spectrum for any α value and regardless of the system length.

Finally, it is interesting to illustrate the potential of the commutator based approach to study quite general realizations of order in 1D systems by considering the renormalization of an arbitrary lattice chain in terms of A and B isolated atoms and $A - B - A$ and $B - A - B$ trimers, whose local transfer matrices are respectively given by $\mathbf{T}_{AAA}$ and $\mathbf{T}_{BAA}\mathbf{T}_{ABA}\mathbf{T}_{AAB}$ and the conjugated ones $\mathbf{T}_{BBB}$ and $\mathbf{T}_{ABB}\mathbf{T}_{BAB}\mathbf{T}_{BBA}$, respectively. Making use of Eq. (2.70) the corresponding commutators read

$$[\mathbf{T}_{AAA}, \mathbf{T}_{BAA}\mathbf{T}_{ABA}\mathbf{T}_{AAB}] = \lambda\gamma_A^{-1}q_1(\gamma_A, \lambda)\sigma_x,$$
$$[\mathbf{T}_{BBB}, \mathbf{T}_{ABB}\mathbf{T}_{BAB}\mathbf{T}_{BBA}] = \lambda\gamma_B^{-1}\bar{q}_1(\gamma_B, \lambda)\sigma_x,$$

where $q_1(\gamma_A, \lambda) = \alpha\lambda(\gamma_A - 1) + 1 + \alpha - 2\gamma_A$ and $\bar{q}_1(\gamma_B, \lambda) \equiv \alpha\lambda(\gamma_B - 1) + 1 + \alpha - 2\alpha\gamma_B$. Enforcing $q_1(\gamma_A, \lambda) = 0$ and $\bar{q}_1(\gamma_B, \lambda) = 0$ simultaneously with $\alpha \neq 1$, $\gamma_B \neq 1$, and $\gamma_A \neq 1$, we get $2\gamma_A\gamma_B = \gamma_A + \gamma_B$. This relationship can be interpreted as a correlation in the *parameter space*, thereby leading to the presence of simultaneous resonances among all the involved atomic clusters which lead to the emergence of delocalized phonon modes irrespective of the particular order realization in the original lattice. Thus, the presence of unexpected transparent states recently reported in disordered chains, which do not exhibit atomic correlations in physical space [991], can be properly accounted for in a natural way within the commutator formalism.

2.6 EIGENSTATES IN QUASIPERIODIC POTENTIALS

In previous section, we have highlighted that QCs do not occupy a vague intermediate position between periodic crystal and amorphous matter representatives. However, there exist physical scenarios where one can properly talk about the existence of states describing intermediate orderings. A representative example is provided by a system undergoing a phase transition from solid to liquid states. This situation is referred to as a passage through a "critical point". At the critical temperature various thermodynamic functions develop a singular behavior which is related to long-range correlations and large fluctuations. Actually, the system should appear identical on all length scales at, exactly, the critical temperature and, consequently, it would be scale invariant. This picture, characteristic of thermodynamic phase transitions, was borrowed to the study of QCs, since analogous features can occur in a solid that preserves its structural integrity but experiences a transition from a metallic-like behavior to an insulator-like one. This kind of phase transitions, affecting the transport properties rather than the lattice structure of a given material, played a very important role in the earlier investigations in the field, leading to a somewhat misleading terminology over the course of the years, as we briefly review in Sec. 2.6.2. For the sake of clarity, henceforth, we will refer to any eigenstate belonging to a purely singular continuous spectrum as a *critical* state, whereas the terms Bloch state or exponentially localized state will refer to the eigenstates of systems characterized by absolutely continuous and pure point energy (frequency) spectra, respectively. In adopting this semantics one should keep in mind that any preconception regarding the transport properties of the so-called critical states should be avoided. In fact, as we will discuss in Sec. 2.7, there exist critical states exhibiting transport properties closely resembling those of Bloch states in periodic systems, but one can also find critical states in the same system (even having energy values very close to those of the former states) which behave in a way very similar to that typical of localized states in amorphous systems. Such a diversity of physical behaviors for states belonging to the same energy spectrum makes it necessary to pay a special attention to both conceptual aspects and terminological issues alike.

2.6.1 Wave functions in periodic and random systems

"To make my life easy, I began by considering wave-functions in a one-dimensional periodic potential. By straight Fourier analysis I found to my delight that the wave differed from the plane wave of free electron only by a periodic modulation. This was so simple that I couldn't think it could be much of a discovery, but when I showed it to Heisenberg he said right away: 'That's it!' ". (Felix Bloch, 1976)[77]

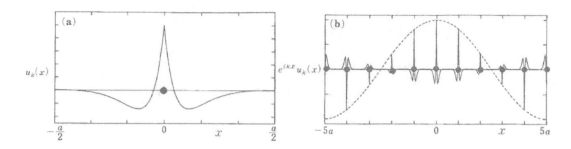

Figure 2.17 Illustration of a Bloch function in one dimension. In (a) we show the periodic function $u_k(x)$ centered at the origin of the unit cell within the range $-a/2 \leq x \leq a/2$, where a is the lattice constant. In (b) the Bloch function is constructed by using the function shown in (a). At every lattice site (solid circles representing atoms) the function u_k is modulated by the plane wave e^{ikx}. (Only the real part is plotted) [608]. (Courtesy of Uichiro Mizutani).

> *"If examined in detail, Bloch waves are found to be more complicated than their simple form may suggest"* (Julian D. Maynard, 2001)[592]

In order to properly appreciate the main characteristic features of critical states, let us recall first the explicit mathematical expressions for extended Bloch functions and exponentially localized states. It is well-known that eigenstates in periodic crystals are described in terms of the so-called *Bloch functions*. The conceptual appeal of Bloch functions in the description of the physical properties of classical crystals is easily grasped by solving the Schrödinger equation describing the motion of an electron with a wave function ψ, energy E and effective mass m, under the action of a potential $V(\mathbf{r})$

$$\frac{\hbar^2}{2m}\nabla^2\psi + [E - V(\mathbf{r})]\psi = 0, \tag{2.76}$$

where $\hbar$ is the reduced Planck's constant. In the absence of any interaction (i.e., $V(\mathbf{r}) \equiv 0$) the solution to Eq. (2.76) for a free electron is readily obtained as a linear combination of plane waves of the form $\psi_\pm(\mathbf{r}) = A_\pm e^{\pm i\mathbf{k}\cdot\mathbf{r}}$, where $k = \sqrt{2mE}/\hbar$ is the wave vector modulus. The next step is to consider the motion of an electron interacting with the atoms forming a periodic crystal lattice with a lattice constant $\mathbf{a}$. Since this lattice is periodic, the resulting interaction potential naturally inherits this periodicity, so that one has $V(\mathbf{r} + \mathbf{a}) = V(\mathbf{r})$. Within this context, the celebrated Bloch's theorem states that the solution to Eq. (2.76) reads [608],

$$\psi_\mathbf{k}(\mathbf{r}) = u_\mathbf{k}(\mathbf{r})e^{i\mathbf{k}\cdot\mathbf{r}}, \tag{2.77}$$

where the function $u_\mathbf{k}(\mathbf{r})$ is real and periodic, with the same period as the underlying lattice, that is, $u_\mathbf{k}(\mathbf{r} + \mathbf{a}) = u_\mathbf{k}(\mathbf{r})$. In this way, the periodicity of function $u_\mathbf{k}(\mathbf{r})$ guarantees the periodicity of the Bloch function *squared modulus* $|\psi_\mathbf{k}(\mathbf{r})|^2 = u_\mathbf{k}^2(\mathbf{r})$, which describes the electronic charge spatial distribution. It is important to note that the $u_\mathbf{k}(\mathbf{r})$ function usually describes the structure of the wave function in the atomic neighborhood, and it is generally relatively localized around each atom (Fig. 2.17a). Thus, the extended nature of Bloch functions ultimately arises from the plane wave modulation, as it is illustrated in Fig. 2.17b. Thus, plane waves can be regarded as special cases of Bloch functions, when the periodic part $u_\mathbf{k}(\mathbf{r})$ is a constant. Note, however, that the wave function given by Eq. (2.77) does not have the same period as the lattice potential $V(\mathbf{r})$, though the charge distribution related to it, namely, $|\psi_\mathbf{k}(\mathbf{r})|^2$ does. This property indicates that the corresponding electron has the same probability of being found in any unit cell, which is usually interpreted by saying that the electron is able to move freely through the lattice.

Now, if examined in detail, Bloch functions are found to be more complicated that their simple form might suggest at first glance. In fact, let us consider the expression

$$\psi_\mathbf{k}(\mathbf{r} + \mathbf{a}) = u_\mathbf{k}(\mathbf{r} + \mathbf{a})e^{i\mathbf{k}\cdot(\mathbf{r}+\mathbf{a})} = u_\mathbf{k}(\mathbf{r})e^{i\mathbf{k}\cdot\mathbf{r}}e^{i\mathbf{k}\cdot\mathbf{a}} = \psi_\mathbf{k}(\mathbf{r})e^{i\mathbf{k}\cdot\mathbf{a}}. \qquad (2.78)$$

Thus, we see that the wavefunctions at positions $\mathbf{r}$ and $\mathbf{r}+\mathbf{a}$ are essentially the same, except for a phase factor $e^{i\mathbf{k}\cdot\mathbf{a}}$. In order to exactly satisfy the periodicity condition $\psi_\mathbf{k}(\mathbf{r} + \mathbf{a}) = \psi_\mathbf{k}(\mathbf{r})$, we must then impose the condition $e^{i\mathbf{k}\cdot\mathbf{a}} = 1 \implies \mathbf{k}\cdot\mathbf{a} = m\pi$, $m \in \mathbb{Z}$. To gain a proper understanding about the physical meaning of the exponent $\mathbf{k}\cdot\mathbf{a}$ above, we obtain the eigenvalue of the momentum operator $\hat{\mathbf{p}} = -i\hbar\nabla$ acting on Eq. (2.77), namely,

$$\hat{\mathbf{p}}\psi_\mathbf{k}(\mathbf{r}) = -i\hbar\left(i\mathbf{k} + \nabla \ln u_\mathbf{k}(\mathbf{r})\right)\psi_\mathbf{k}(\mathbf{r}), \qquad (2.79)$$

which differs from the simple relation $\hat{\mathbf{p}}\psi_\mathbf{k}(\mathbf{r}) = \hbar\mathbf{k}\psi_\mathbf{k}(\mathbf{r})$ corresponding to the free electron case. Therefore, the vector $\mathbf{k}$ in the Bloch function (usually referred to as quasi-momentum) differs from that corresponding to a free electron, because the atomic lattice potential exerts a force on the electron, which can be described in terms of the function $u_\mathbf{k}(\mathbf{r})$ and its first order derivative. Note that, in the case of periodic crystals, the definition of quasi-momentum depends only on the symmetry of the crystal and not on the detail of the structure within the unit cell [401].

Under a closer inspection, we realize that Eq. (2.77) actually describes an *almost periodic* function (see Fig. 1.17), given by the product of two periodic functions, $u_\mathbf{k}(\mathbf{r})$ and $e^{i\mathbf{k}\cdot\mathbf{r}}$, whose respective periods are not commensurate to each other in general (**Exercise 2.14**) [753]. Once this has been realized, we note that when describing the electron propagation through periodic structures it is customary to adopt the Born-von Karman periodic boundary conditions $\psi_\mathbf{k}(\mathbf{r} + m\mathbf{a}) = \psi_\mathbf{k}(\mathbf{r})$, thus ensuring that the reciprocal vector $\mathbf{k}$ fulfills the condition $e^{im\mathbf{k}\cdot\mathbf{a}} \equiv 1$. In this case, Eq. (2.126) reads $|\psi_\mathbf{k}(\mathbf{r} + m\mathbf{a}) - \psi_\mathbf{k}(\mathbf{r})| = 0$, and the originally almost periodic Bloch function reduces to the periodic subclass.

On the other hand, in amorphous materials characterized by a random distribution of atoms through the space, the electronic states are exponentially localized according to an expression of the form

$$\psi(\mathbf{r}) = A(\mathbf{r})e^{-\frac{|\mathbf{r}|}{\ell}}, \qquad (2.80)$$

which shows that at some site the wave function has a maximum amplitude and decreases exponentially away from that site, so that ℓ provides a measure of the spatial extension of the wave function, which is referred to as its *localization length*. In 1D discrete lattices the above expression reads $\psi_n(x) = A_n e^{-\frac{|x|}{\ell_n}}$, and it is important to emphasize that the ultimate reason leading to the localization of electronic states in random 1D chains is not the presence of exponentially decaying modulations in Eq. (2.80), but the fact that both the amplitudes A_n and the reciprocal localization lengths ℓ_n, form an uncorrelated random ensemble [171]. This property guarantees that possible resonances between electronic states belonging to neighboring atoms cannot extend to other atoms located far away along the chain. In fact, as soon as any sort of (even short-range) correlation is present in an otherwise disordered chain, one can observe the emergence of a significant number of states uniformly spreading through the lattice [215, 770].

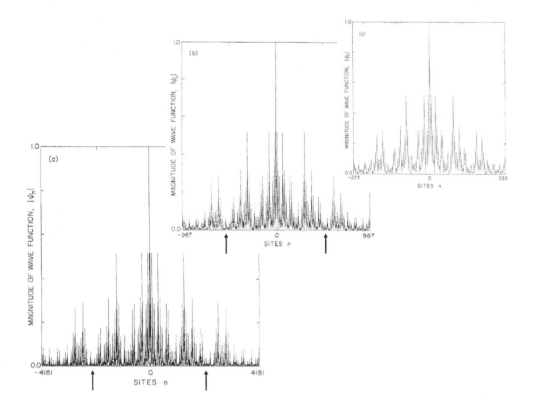

Figure 2.18 Illustration of a self-similar critical wave function. The amplitude distribution shown in (a) for the wave function as a whole is identically reproduced at progressively smaller scales (indicated by the arrows) in (b) and (c) after a proper rescaling. This wave function corresponds to the eigenstate located at the center of the spectrum ($E = 0$) in the Aubry-André model with a potential strength $\lambda = 2$. (Adapted from [665]. With permission from Ostlund S and Pandit R 1984 *Phys. Rev. B* **29** 1394 © 1984 by the American Physical Society).

2.6.2 The notion of critical eigenstates

Following a chronological order the concept of critical wave function was born in the study of the Anderson Hamiltonian, which describes a periodic lattice with site-diagonal disorder. This model is known to have extended states for weak disorder in 3D systems, as well as in 2D samples with a strong magnetic field. For strong disorder, on the other hand, the electronic states are localized. For 1D systems it was proved that localized states decay exponentially in space in most cases [628]. However, this exponential decay relates to the asymptotic evolution of the envelope of the wave function while the short-range behavior is characterized by strong fluctuations. The magnitude of these fluctuations seems to be related to certain physical parameters, such as the degree of disorder which, in turn, controls the appearance of the so-called mobility edges. Approaching a mobility edge, from the insulator regime, the exponential decay constant diverges, so that the wave function amplitudes can be expected to feature fluctuations on all length scales larger than the lattice spacing. This singular fact turns out to be very convenient to explain metal-insulator transitions. Since Bloch states are regarded prototypical states of periodic systems, whereas exponentially

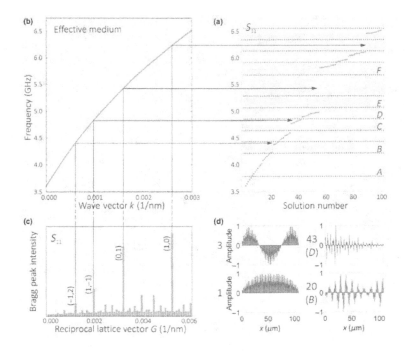

Figure 2.19   a) Numerical spin wave spectrum for a magnonic FQC with $n = 89$ layers compared to the analytical spin wave dispersion in a homogeneous material. The horizontal lines marked from A to F indicate the frequencies at which experimental measurements have been performed. The spatial Fourier spectrum is given in c). The arrows link the reciprocal lattice vectors G, corresponding to the Bragg peaks of higher intensity, to the frequencies of the largest magnonic gaps of the FQC shown in a). The amplitude and the phase distribution of some critical modes are shown in d). (Lisiecki F, Rychly J, Kuswik P, Glowinski H, Klos J, Groß F, Träger N, Bykova I, Weigand M, Zelent M, Goering E J, Schütz G, Krawczyk M, Stobiecki F, Dubowik J, Gräfe J, Phys. Rev. Appl. 11, 054061 (10pp) 2019. Work licensed under a Creative Commons Attribution 4.0 International License).

localized states are the typical states found in uncorrelated random systems, the states occurring at the critical point in a metal-insulator transition, that is, critical states, were originally defined as being neither Bloch functions nor exponentially localized states, but occupying a fuzzy intermediate position between them.

> *These states, which we will term critical, have a maximum at a site (in the lattice) and a series of subsidiary maxima at (a number of other) sites which do not decay to zero* [664].

Thus, the notion of "criticality" can be understood as follows. An extended wave function is expected to extend *homogeneously* over the whole sample. On the other hand, for a wave function localized at a particular site of the sample, one expects its probability density to display a single dominant maximum at, or around, this site, and its envelope function is generally observed to decay exponentially in space. On the contrary, a critical state is characterized by *strong spatial fluctuations* of the wave function amplitudes. This unusual

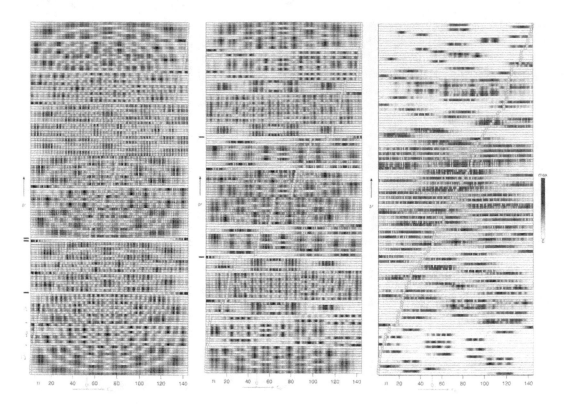

Figure 2.20   Eigenstate maps of Fibonacci (left panel), Thue-Morse (middle panel) and Rudin-Shapiro (right panel) aperiodic lattices with $N = 144$, $\epsilon_A = 0$, $\epsilon_B = 0.15$ and $t_{k,k\pm1} = 0.1$ (on-site model). Each horizontal stripe shows the square root of the wave function amplitude modulus in a grayscale from white (zero) to black (maximum). Superimposed are the eigenvalues in arbitrary units with origin at $\epsilon = 0$. Black horizontal lines on the left indicate band-edge states. (Reprinted figure with permission from Röntgen M, Morfonios C V, Wang R, Dal Negro L, Schmelcher P, Phys. Rev. B 99 214201 (24 pp) 2019. Copyright (2019) by the American Physical Society).

behavior, consisting of an alternatively decaying and recovering of the wave function amplitudes, is illustrated in Fig. 2.18 for a self-similar wave function. Two main features of this wave function amplitude distribution must be highlighted. On the one hand, although the main local maxima are modulated by an overall decaying envelope, this envelope *cannot* be fitted to an exponential function.[22] On the other hand, the subsidiary peaks around the main local maxima display self-similar scaling features. This property is observed in many QP systems eigenstates, but other possible spatial patterns are also frequently observed, including multifractal distributions.

In Fig. 2.19, the main features of the experimental spin wave spectrum of a FQC consisting of alternating strips of magnetic materials separated by air gaps is shown. In the first place, we see that the gap labeling theorem is clearly obeyed, so that the main gaps

---

[22]Though exponentially localized decaying states are considered as representative of most localization processes in solid state physics one must keep in mind that other possible mathematical functions are also possible to this end.

in the spin wave spectrum are directly related to the most intense reflection peaks in the Fourier spectrum of the structure. In Fig. 2.19d, we see the amplitude distribution of two low-frequency critical states belonging to the spin wave spectrum, where a QP pattern of local resonance peaks are modulated in terms of a typical standing wave profile on a large scale. At higher frequencies in the spectrum, however, the envelope of the spin waves becomes more fractal-like, with several maxima and minima located at different locations throughout the structure [506].

In Fig. 2.20, a detailed picture of the wave function amplitudes spatial distribution is shown for the eigenstates belonging to the energy spectrum of FQC, Thue-Morse and Rudin-Shapiro aperiodic lattices. A close inspection reveals that eigenstates tend to localize into locally reflection-symmetric peaks (black regions) which in turn are evenly distributed throughout the lattice. This typical feature of critical eigenstates is clearly appreciated for most states in both Fibonacci and Thue-Morse spectra, while most states belonging to the Rudin-Shapiro lattice are mainly localized, as expected for a system exhibiting an absolutely continuous Fourier spectrum. On the other hand, within the gaps between subbands in both Fibonacci and Thue-Morse lattices there may appear spectrally isolated modes, reminiscent of gap mode localized on defects within a periodic lattice [754].

2.6.3 Generalizing the Bloch function notion

"The electronic wave functions appear to obey a modified Bloch theorem".
(Levine and Steinhardt, 1984)[486]

How do the energy and frequency eigenstates in QCs compare to those observed in periodic crystals? Bloch's theorem was a major milestone that established the presence of allowed bands and forbidden bandgaps in periodic crystals. Although it was once believed that, in virtue of this theorem, bandgaps could form only under periodic conditions, this restriction has been disproved by the discovery of QCs, and even the existence of Bloch-like eigenstates in certain class of disordered potentials has been recently shown by exploiting supersymmetry techniques,[23] within the framework of the isospectral approach[24][983]. Notwithstanding this, since QCs lack translational invariance Bloch's theorem does not longer hold in its standard form. This shortcoming has spurred several attempts to search for an equivalent result valid for QP systems.

Certainly, the more straightforward way to extend Bloch's theorem will be by directly exploiting the new symmetries present in a given aperiodic lattice. A suitable example is provided by fractal lattices, which are characterized by the replacement of the usual translation symmetry by an inflation one. To this end, we apply the inflation operator $\mathbf{S}_\lambda \psi_\mathbf{k}(\mathbf{r}) = \psi_\mathbf{k}(\lambda\mathbf{r})$ acting upon the wavefunction $\psi_\mathbf{k}(\mathbf{r})$, where $\lambda > 0$ is the inflation factor, and $\mathbf{k}$ is a suitable label. Thus, the successive application of two inflation operations can be written

$$\mathbf{S}_\lambda \mathbf{S}_{\lambda'} \psi_\mathbf{k}(\mathbf{r}) = \mathbf{S}_\lambda \psi_\mathbf{k}(\lambda'\mathbf{r}) = \psi_\mathbf{k}(\lambda(\lambda'\mathbf{r})) = \psi_\mathbf{k}(\lambda\lambda'\mathbf{r}). \quad (2.81)$$

On the other hand, the inflation operator eigenvalues can be expressed as $\mathbf{S}_\lambda \psi_\mathbf{k}(\mathbf{r}) = c_\mathbf{k}(\lambda)\psi_\mathbf{k}(\mathbf{r})$, so that the equation corresponding to the successive application of two inflation operations reads

$$\mathbf{S}_\lambda \mathbf{S}_{\lambda'} \psi_\mathbf{k}(\mathbf{r}) = \mathbf{S}_\lambda [c_\mathbf{k}(\lambda')\psi_\mathbf{k}(\mathbf{r})] = c_\mathbf{k}(\lambda)c_\mathbf{k}(\lambda')\psi_\mathbf{k}(\mathbf{r}). \quad (2.82)$$

---

[23]Supersymmetry is a promising postulate in theoretical particle physics that describes the relationship between bosons and fermions.

[24]The so-called isospectral problem addresses the possible existence of different sorts of potential distributions producing the same eigenstates set spectrum.

Now, the eigenvalues corresponding to Eq. (2.81) can also be formally expressed as $\mathbf{S}_\lambda \mathbf{S}_{\lambda'}\psi_\mathbf{k}(\mathbf{r}) = c_\mathbf{k}(\lambda\lambda')\psi_\mathbf{k}(\mathbf{r})$, and comparing with Eq. (2.82) we get $c_\mathbf{k}(\lambda\lambda') = c_\mathbf{k}(\lambda)c_\mathbf{k}(\lambda')$. This relationship is satisfied by functions of the form $c_\mathbf{k}(\lambda) = (e^{i\mathbf{k}\cdot\mathbf{r}})^{\ln\lambda}$, as it can be readily verified. Therefore, the eigenvalue inflation operator equation finally reads

$$\mathbf{S}_\lambda\psi_\mathbf{k}(\mathbf{r}) = (e^{i\mathbf{k}\cdot\mathbf{r}})^{\ln\lambda}\psi_\mathbf{k}(\mathbf{r}). \tag{2.83}$$

Since $\lambda\lambda' = \lambda'\lambda$ after Eq. (2.81) one readily obtains $\mathbf{S}_\lambda\mathbf{S}_{\lambda'}\psi(\mathbf{r}) = \mathbf{S}_{\lambda'}\mathbf{S}_\lambda\psi(\mathbf{r})$, and therefore $[\mathbf{S}_\lambda, \mathbf{S}_{\lambda'}] = \mathbf{0}$. This commutator involving inflation operators plays the same role in self-similar systems than the well-known commutator $[\mathbf{T}_\mathbf{R}, \mathbf{T}_{\mathbf{R}'}] = \mathbf{0}$, involving the translation operator $\mathbf{T}_\mathbf{R}\psi(\mathbf{r}) \equiv \psi(\mathbf{r}+\mathbf{R})$, does in periodic systems. In order to generalize the Bloch theorem let us now consider $\psi_\mathbf{k}(\mathbf{r})$ be an eigenfunction of the Schrödinger equation

$$\mathbf{H}(\mathbf{r})\psi_\mathbf{k}(\mathbf{r}) = E(\mathbf{k})\psi_\mathbf{k}(\mathbf{r}), \tag{2.84}$$

where $\mathbf{H}(\mathbf{r})$ is a Hamiltonian operator satisfying the scale invariance property $\mathbf{H}(\lambda\mathbf{r}) = \mathbf{H}(\mathbf{r})$. Then, we have

$$\begin{aligned}(\mathbf{S}_\lambda\mathbf{H}(\mathbf{r}))\psi_\mathbf{k}(\mathbf{r}) &= \mathbf{S}_\lambda(E(\mathbf{k})\psi_\mathbf{k}(\mathbf{r})) = E(\lambda\mathbf{k})\psi_\mathbf{k}(\lambda\mathbf{r}),\\ (\mathbf{H}(\mathbf{r})\mathbf{S}_\lambda)\psi_\mathbf{k}(\mathbf{r}) &= \mathbf{H}(\mathbf{r})\psi_\mathbf{k}(\lambda\mathbf{r}) \equiv \mathbf{H}(\lambda\mathbf{r})\psi_\mathbf{k}(\lambda\mathbf{r}) = E(\lambda\mathbf{k})\psi_\mathbf{k}(\lambda\mathbf{r}),\end{aligned} \tag{2.85}$$

hence, we obtain $[\mathbf{S}_\lambda, \mathbf{H}] = \mathbf{0}$. This expression generalizes to fractal systems the well-known relationship $[\mathbf{T}_\mathbf{R}, \mathbf{H}] = \mathbf{0}$ for periodic systems. We note that $[\mathbf{S}_\lambda, \mathbf{T}_\mathbf{R}] \neq \mathbf{0}\ \forall\lambda \neq 1$. Making use of Eq. (2.83), we can express the common eigenfunctions of the inflation and Hamiltonian operators in the form

$$\psi_\mathbf{k}(\mathbf{r}) = u_\mathbf{k}(\mathbf{r})(e^{i\mathbf{k}\cdot\mathbf{r}})^{\ln\lambda}, \tag{2.86}$$

where $u_\mathbf{k}(\mathbf{r})$ is a scale invariant function with a suitable scaling factor (see Sec. 1.3.1). By comparing with Eq. (2.77), we see that Eq. (2.86) provides a natural generalization of the usual Bloch functions to the case of a self-similar fractal lattice exhibiting a fractal energy spectrum.

Analogously, one may think of a similar generalization for almost periodic systems which simply consists in replacing the periodic function $u_\mathbf{k}(\mathbf{r})$ in Eq. (2.77) by a suitable QP or AP one, so that the generalized Bloch functions will preserve their original functional form given by Eq. (2.77). In this way, the basic physical picture of a wave function whose amplitude is modulated by the lattice potential remains, albeit the long-range order of the underlying lattice is no longer periodic but QP in the case of currently known QCs.[25] Thus, one cannot exclude the Bloch function *picture* outright solely on the basis of the lack of periodicity. In fact, even in a QP potential the eigenfunctions might happen to be extended throughout the system and if so they could be cast in a form analogous to the Bloch one with the only distinction from the periodic case traced to the fact that the momentum $\mathbf{k}$ cannot be reduced to a single Brillouin zone because of the absence of a smallest reciprocal lattice vector (i.e., $\left|\frac{\pi}{na}\right| \to \infty$) [97], a difficulty which can be conveniently circumvented by the use of an extended-zone scheme [753]. Accordingly, from a fundamental point of view, the absence of translational periodicity *does not* necessarily *preclude* the existence of delocalized electronic states in QCs.

In Sec. 1.1.2, it was noted that the electron diffraction spectrum of QCs is dominated by a few reciprocal lattice points that may be taken to construct a quasi-Brillouin zone [650, 822].

---

[25]Note that the very possibility of considering ordered arrangements of matter where atomic positions are described in terms of the more general class of AP functions should not be excluded at all from a fundamental point of view.

Thus, wave functions of the form given by Eq. (2.77), where $u_{\mathbf{k}}(\mathbf{r})$ is a QP function which is formally defined on a countable set of reciprocal lattice vectors, envisioned as plane waves with QP modulation, could be used to describe the electron dynamics in QCs. In fact, this scenario was experimentally tested to some extent in a study of the valence bands of a dQC belonging to the AlNiCo system [757]. Likewise, Bloch-like surface waves associated with a 2D QP structure were experimentally observed in a classic wave propagation experiment consisting of pulse propagation with a shallow fluid covering a QP drilled bottom (see Sec. 4.5.2) [895].

> *"To construct the potential V, we consider the array of ions, in which the electrons move, as a kind of oblique projection of three n-dimensional periodic lattices on three independent directions of the usual Euclidean space".* (Michel V. Romerio, 1971)[753][26]

A remarkable link between AP functions and periodic structures in higher dimensional spaces was early put forward by Michel V. Romerio in a paper entitled *"Almost periodic functions and the theory of disordered systems"*, published in the *Journal of Mathematical Physics* in 1971. In this work the Bloch theorem was generalized to AP functions and some properties of the DOS and spatial localizations of electrons were obtained. Since this article has remained quite unnoticed in the condensed matter community,[27] it is instructive to consider its main tenets in the light of our current understanding about QCs. To this end, we will start by recalling two theorems which were originally demonstrated in Romerio's work [753].

Theorem 2.1 *Let $V(\mathbf{x})$ be an AP function on $\mathbb{R}^3$ and $W(\mathbf{x}_1, \ldots, \mathbf{x}_n)$ a multiperiodic function on $\mathbb{R}^3 \times \mathbb{R}^3 \times \ldots \times \mathbb{R}^3 = \mathbb{R}^{3n}$, with independent periods for each variable $\mathbf{x}_i \in \mathbb{R}^3$, such that $diag[W] \equiv W(\mathbf{x}, \ldots, \mathbf{x}) = V(\mathbf{x})$. Then, there exists a second order linear differential operator D_n such that all the eigenfunctions $\phi(\mathbf{x}_1, \ldots, \mathbf{x}_n)$ of the generalized Schrödinger equation $(-D_n + W)\phi = E\phi$ in the hyperspace $\mathbb{R}^{3n}$, satisfy the Schrödinger equation in physical space $\mathbb{R}^3$*

$$(-\nabla^2 + V - E) \ diag[\phi] = 0,$$

with $diag[\phi] \neq 0$.

The proof of this theorem relies in the introduction of the differential operator

$$D_n \equiv \sum_{s,t=1}^{n} \nabla_s \cdot \nabla_t = \sum_{s,t=1}^{n} \sum_{i=1}^{3} \frac{\partial^2}{\partial x_s^i \partial x_t^i}, \qquad (2.87)$$

where $\nabla_s = \sum_{i=1}^{3} \frac{\partial}{\partial x_s^i} \mathbf{e}_i$, and $\{\mathbf{e}_1, \mathbf{e}_2, \mathbf{e}_3\}$ is the canonical base of $\mathbb{R}^3$. Let us express the eigenfunctions $\phi(\mathbf{x}_1, \ldots, \mathbf{x}_n)$ in $\mathbb{R}^{3n}$ as a Fourier integral of the form

$$\phi(\mathbf{x}_1, \ldots, \mathbf{x}_n) = \lim_{V \to \infty} \int_V A(\mathbf{k}_1, \ldots, \mathbf{k}_n) e^{i \sum_{j=1}^{n} \mathbf{k}_j \cdot \mathbf{x}_j} d^{3n}k, \qquad (2.88)$$

---

[26]To put the results reported in this work into a proper perspective we should note that they appeared before the notion of superspace were introduced by de Wolff in 1974 to account for the indexation of the diffraction patterns of incommensurately modulated structures (see Sec. 1.4).

[27]According to the ISI Web of Knowledge this paper has received 17 cites to date (June 2020), the more recent one in a paper published in 2007.

where $\mathbf{k}_j$ are the reciprocal vectors of the lattice vectors $\mathbf{x}_j$. After Eqs. (2.87) and (2.88), we obtain

$$-D_n\phi = \lim_{V\to\infty} \int_V \mathbf{k}^2 A(\mathbf{k}_1,\ldots,\mathbf{k}_n)e^{i\sum_{j=1}^n \mathbf{k}_j\cdot\mathbf{x}_j}d^{3n}k. \tag{2.89}$$

where we have introduced the relation

$$\sum_{s,t=1}^n \mathbf{k}_s\cdot\mathbf{k}_t = (\mathbf{k}_1+\mathbf{k}_2+\ldots+\mathbf{k}_n)^2 \equiv \mathbf{k}^2, \tag{2.90}$$

where $\mathbf{k} \equiv \sum_{j=1}^n \mathbf{k}_j \in \mathbb{R}^3$. Thus,

$$(-D_n+W)\phi = \lim_{V\to\infty} \int_V (\mathbf{k}^2+W)A(\mathbf{k}_1,\ldots,\mathbf{k}_n)e^{i\sum_{j=1}^n \mathbf{k}_j\cdot\mathbf{x}_j}d^{3n}k, \tag{2.91}$$

and performing the diag operation, which transforms $\mathbf{x}_j \to \mathbf{x}$ $\forall j$, keeping the reciprocal vectors $\mathbf{k}_j$ unchanged, we have

$$
\begin{aligned}
\mathrm{diag}[(-D_n+W)\phi] &= \lim_{V\to\infty} \int_V (\mathbf{k}^2+\mathrm{diag}[W])A(\mathbf{k}_1,\ldots,\mathbf{k}_n)e^{i(\sum_{j=1}^n \mathbf{k}_j)\cdot\mathbf{x}}d^{3n}k \\
&= \lim_{V\to\infty} \int_V (\mathbf{k}^2+V(\mathbf{x}))A(\mathbf{k}_1,\ldots,\mathbf{k}_n)e^{i\mathbf{k}\cdot\mathbf{x}}d^{3n}k \\
&= \lim_{V\to\infty} \int_V \mathbf{k}^2 A(\mathbf{k}_1,\ldots,\mathbf{k}_n)e^{i\mathbf{k}\cdot\mathbf{x}}d^{3n}k + V(\mathbf{x})\,\mathrm{diag}[\phi],
\end{aligned} \tag{2.92}
$$

where we have taken into account Eq. (2.88).

On the other hand, we calculate the derivative

$$
\begin{aligned}
-\nabla^2(\mathrm{diag}[\phi]) &= -\sum_{i=1}^3 \frac{\partial^2}{\partial x_i^2}\mathrm{diag}[\phi] = -\lim_{V\to\infty}\int_V A(\mathbf{k}_1,\ldots,\mathbf{k}_n)\left(\sum_{i=1}^3 \frac{\partial^2}{\partial x_i^2}e^{i\mathbf{k}\cdot\mathbf{x}}\right)d^{3n}k \\
&= -\lim_{V\to\infty}\int_V A(\mathbf{k}_1,\ldots,\mathbf{k}_n)\left(\sum_{i=1}^3 \frac{\partial^2}{\partial x_i^2}e^{i\sum_{i=1}^3 k_i x_i}\right)d^{3n}k \\
&= \lim_{V\to\infty}\int_V \left(\sum_{i=1}^3 k_i^2\right)A(\mathbf{k}_1,\ldots,\mathbf{k}_n)e^{i\mathbf{k}\cdot\mathbf{x}}d^{3n}k \\
&= \lim_{V\to\infty}\int_V \mathbf{k}^2 A(\mathbf{k}_1,\ldots,\mathbf{k}_n)e^{i\mathbf{k}\cdot\mathbf{x}}d^{3n}k,
\end{aligned} \tag{2.93}
$$

so that, by comparing Eqs. (2.92) and (2.93) we obtain the relationship

$$\mathrm{diag}[(-D_n+W)\phi] = \left(-\nabla^2+V\right)\,\mathrm{diag}[\phi]. \tag{2.94}$$

Therefore, since by assumption ϕ is an eigenfunction of the generalized Schrödinger equation $(-D_n+W)\phi = E\phi$, we have $\mathrm{diag}[(-D_n+W)\phi] = E\,\mathrm{diag}[\phi]$, where we take into account that E in general depends on $\mathbf{k}$ but not on $\mathbf{x}$, so that Eq. (2.94) can finally be expressed as

$$\left(-\nabla^2+V-E\right)\mathrm{diag}[\phi] = 0. \tag{2.95}$$

Q.E.D.

Theorem 2.2 *For every eigenfunction ψ of the AP Schrödinger operator $-\nabla^2+V$ in the physical space there exist an eigenfunction ϕ of the generalized Schrödinger operator $-D_n+W$ in the hyperspace, such that $\mathrm{diag}[\phi] = \psi$.*

Let us express the eigenfunction $\psi(\mathbf{x})$ in $\mathbb{R}^3$ as a Fourier integral of the form

$$\psi(\mathbf{x}) = \lim_{V_k \to \infty} \int_{V_k} a(\mathbf{k}) e^{i\mathbf{k} \cdot \mathbf{x}} d^3 k, \qquad (2.96)$$

where $\mathbf{k}$ is the reciprocal vector of the lattice vector $\mathbf{x}$ given by Eq. (2.90), and let us define the eigenfunction $\phi(\mathbf{x}_1, \ldots, \mathbf{x}_n)$ in $\mathbb{R}^{3n}$ as

$$\begin{aligned}
\phi(\mathbf{x}_1, \ldots, \mathbf{x}_n) &= \lim_{V \to \infty} \int_V a(\mathbf{k}_1 + \ldots + \mathbf{k}_n) e^{i \sum_{j=1}^n \mathbf{k}_j \cdot \mathbf{x}_j} d^{3n} k \\
&= \lim_{V \to \infty} \int_V a(\mathbf{k}) e^{i\mathbf{k} \cdot \mathbf{x}_1} e^{i[\mathbf{k}_2 \cdot (\mathbf{x}_2 - \mathbf{x}_1) + \ldots + \mathbf{k}_n \cdot (\mathbf{x}_n - \mathbf{x}_1)]} d^{3n} k, \qquad (2.97)
\end{aligned}$$

Then, we can write

$$\begin{aligned}
(-D_n + W)\phi &= \lim_{V \to \infty} \int_V \left(\mathbf{k}^2 + W\right) a(\mathbf{k}) e^{i\mathbf{k} \cdot \mathbf{x}_1} e^{i[\mathbf{k}_2 \cdot (\mathbf{x}_2 - \mathbf{x}_1) + \ldots + \mathbf{k}_n \cdot (\mathbf{x}_n - \mathbf{x}_1)]} d^{3n} k \\
&= (2\pi)^{3(n-1)} \prod_{j=2}^n \delta(\mathbf{x}_j - \mathbf{x}_1) \lim_{V_k \to \infty} \int_{V_k} \left(\mathbf{k}^2 + W\right) a(\mathbf{k}) e^{i\mathbf{k} \cdot \mathbf{x}_1} d^3 k, \qquad (2.98)
\end{aligned}$$

where we have taken into account Eq. (2.89), and used the integral expression

$$\lim_{V_j \to \infty} \int_{V_j} e^{i\mathbf{k}_j \cdot (\mathbf{x}_j - \mathbf{x}_1)} d^3 k_j = (2\pi)^3 \delta(\mathbf{x}_j - \mathbf{x}_1).$$

Now, performing the diag operation the prefactor of the integral in Eq. (2.98) reduces to unity, so that we get

$$\begin{aligned}
\text{diag}[(-D_n + W)\phi] &= \lim_{V_k \to \infty} \int_{V_k} \left(\mathbf{k}^2 + V\right) a(\mathbf{k}) e^{i\mathbf{k} \cdot \mathbf{x}} d^3 k \\
&= \lim_{V_k \to \infty} \int_{V_k} \mathbf{k}^2 a(\mathbf{k}) e^{i\mathbf{k} \cdot \mathbf{x}} d^3 k + V\psi = (-\nabla^2 + V)\psi, \qquad (2.99)
\end{aligned}$$

where we have taken into account Eq(2.96) and the expression

$$\begin{aligned}
-\nabla^2 \psi &= -\lim_{V_k \to \infty} \int_{V_k} a(\mathbf{k}) \left(\sum_{i=1}^3 \frac{\partial^2}{\partial x_i^2} e^{i \sum_{i=1}^3 k_i x_i}\right) d^3 k = \lim_{V_k \to \infty} \int_{V_k} \left(\sum_{i=1}^3 k_i^2\right) a(\mathbf{k}) e^{i\mathbf{k} \cdot \mathbf{x}} d^3 k \\
&= \lim_{V_k \to \infty} \int_{V_k} \mathbf{k}^2 a(\mathbf{k}) e^{i\mathbf{k} \cdot \mathbf{x}} d^3 k. \qquad (2.100)
\end{aligned}$$

Therefore, by comparing Eqs. (2.94) and (2.99), we finally obtain $\text{diag}[\phi] = \psi$, Q.E.D.

Corollary *All eigenfunctions ψ of the AP Schrödinger operator $-\nabla^2 + V$ can be expressed in the form*

$$\psi(\mathbf{x}) = e^{i\mathbf{k} \cdot \mathbf{x}} \text{diag}[u_k(\mathbf{x}_1, \ldots, \mathbf{x}_n)], \qquad (2.101)$$

where $u_k(\mathbf{x}_1, \ldots, \mathbf{x}_n)$ is a multiperiodic function in the hyperspace $\mathbb{R}^{3n}$ having the same set of periods as $W(\mathbf{x}_1, \ldots, \mathbf{x}_n)$.

In fact, since $\phi(\mathbf{x}_1, \ldots, \mathbf{x}_n)$ is multiperiodic in $\mathbb{R}^{3n}$ the Bloch theorem applies in each $\mathbb{R}^3$ subspace of $\mathbb{R}^{3n}$, so that we can express

$$\phi(\mathbf{x}_1, \ldots, \mathbf{x}_n) = e^{i(\mathbf{k}_1 \cdot \mathbf{x}_1 + \ldots + \mathbf{k}_n \cdot \mathbf{x}_{n1})} u_{\mathbf{k}_1, \ldots \mathbf{k}_n}(\mathbf{x}_1, \ldots, \mathbf{x}_n),$$

and performing the diag operation we get

$$\psi = \text{diag}[\phi] = e^{i(\sum_{j=1}^n \mathbf{k}_j)\cdot\mathbf{x}}\text{diag}[u_k(\mathbf{x}_1,\ldots,\mathbf{x}_n)] = e^{i\mathbf{k}\cdot\mathbf{x}}\text{diag}[u_k(\mathbf{x}_1,\ldots,\mathbf{x}_n)],$$

where we have taken into account Theorem 2. As we see, this result provides a generalization of the Bloch theorem for periodic potentials to the case of more general almost periodic potentials.

At this point some comments are in order. In the first place, the hyperspace considered in the framework of Romerio's theorems is not as general as that considered in the framework of QCs hyperspace crystallography. Thus, while icosahedral QCs, described in terms of an hypercube in $\mathbb{R}^6$, can fit within the Romerio's approach with $n = 2$, the decagonal QCs described in terms of a $\mathbb{R}^5$ hyperspace, can not. In the second place, the diag operation, involving orthogonal projections of different $\mathbb{R}^3$ subspaces composing $\mathbb{R}^{3n} = (\mathbb{R}^3)^n$, all of them mutually oriented in a parallel manner, is quite different from the projections involving spaces oriented in a nonparallel way, as it occurs in the cut-and-project method. Anyway, it is certainly remarkable that certain characteristic features of the energy spectrum of aperiodic systems, such as its highly fragmented structure

"$E_n(\mathbf{k})$ is a function compounded of a succession of discontinuous small pieces, i.e., $E_n(k)$ has a very great number of gaps (This number becomes infinite for every finite region when the periods of W are independent)", [753]

or the peculiar spatial distribution of the eigenstates

"systems described by almost periodic potentials have electron states which are more localized than in the pure crystal. More precisely (...), the wave function is characterized by a certain number of important peaks where the probability of presence of a localized electron is large", [753]

were explicitly described in Romerio's paper as early as 1971.

By adopting a very different perspective the existence of the so-called approximant phases also suggest a continuity argument based on the existence of well-defined Bloch functions for these approximants. Then, one may think of a relatively smooth transition from strict Bloch functions eigenstates to some form of generalized Bloch functions which will be attained when the QP limit is approached by systematically increasing the order of the approximant. This approach was successfully undertaken by C. de Lange and Ted Janssen (1936–2017) in 1983 [168]. In their study they focused on incommensurate crystals whose structure can be described in terms of a 4D hypercrystal, so that the electrons dynamics can be described in terms of a periodic potential in the 4D superspace of the form

$$U(\mathbf{x}, x_4) = \sum_{\mathbf{k}} V(\mathbf{k})e^{i(\mathbf{k}\cdot\mathbf{x}+k_4 x_4)}, \tag{2.102}$$

where $\mathbf{x}$ and $\mathbf{k}$ are the direct and reciprocal space vectors in the 3D physical space, respectively, and the $k_4 x_4$ term in Eq. (2.102) can be interpreted as a phase modulation. The solution to the Schrödinger equation in 4D can then be obtained in terms of the 4D version of Bloch's theorem, that according to Eq. (2.77) reads

$$\psi(\mathbf{x}, x_4) = e^{i(\mathbf{k}\cdot\mathbf{x}+k_4 x_4)}u_{\mathbf{k},k_4}(\mathbf{x}, x_4), \tag{2.103}$$

where $u_{\mathbf{k},k_4}(\mathbf{x}, x_4)$ is periodic in the 4D hyperspace. Then, the crucial point consists in defining a mathematical procedure to relate this Bloch function in the hyperspace to the actual wave function in the 3D aperiodic system. In the Romerio's approach this was attained through the diag operation. A more sophisticated and accurate way was introduced by J. P. Liu and L. Birman in 1987 in terms of a boost technique transforming the general problem of a Schrödinger equation with a QP potential in D dimensions to the solution of a periodic effective Schrödinger equation in $N > D$ dimensions, where the Bloch theorem could be applied [514]. In so doing, the eigenfunctions of the original problem and the boosted problem were related to each other by a Radon transform (see Sec. 1.4.2), in such a way that the corresponding eigenvalues exactly coincide with those obtained for the hyperspace wave function (a feature this approach shares with that earlier introduced by Romerio). This approach was introduced as a

> *"'quasi-Bloch' theory which will enable one to analyze the analytical properties of electronic eigenvalues and eigenfunctions of a QP system."* [515]

Nevertheless, to the best of my knowledge, this approach has not subsequently been considered in the literature.[28]

Certainly, the very possibility of obtaining a proper generalization of the Bloch's theorem by exploiting the underlying higher-dimensional lattice via the so-called "cut-and-project" scheme seems appealing, and it has been more or less openly suggested in the literature, although the higher-dimensional description cannot be exploited directly because the cut corresponding to the "physical" space imposes boundary conditions that break the translational symmetry of the hyperlattice [310]. Interestingly enough, however, a method allowing to obtain the energy spectrum and eigenstates of photonic QCs (see Sec. 4.3) by solving Maxwell's equations in higher dimensions was reported. In this approach the important physical quantity is not so much the band structure, since the reciprocal vector has no clear physical meaning once the solution is projected into physical space, but rather the DOS, formed by projecting the band structure onto the frequency axis instead [749].

2.7 TRANSPORT PROPERTIES OF CRITICAL STATES

The notion of critical wave function has evolved continuously since its introduction in the study of QP systems, leading to a somewhat confusing situation. For instance, references to self-similar, chaotic, lattice-like, or quasilocalized wave functions can be found in the literature depending on the different criteria adopted to characterize them[276, 436, 761, 797]. Generally speaking, critical states exhibit a rather involved oscillatory behavior, displaying strong spatial fluctuations which show distinctive self-similar features (Fig. 2.18). Thus, the wave function is peaked on short chain sequences but peaks reappear far away on chain sequences showing the same local lattice ordering. As a consequence, the notion of an envelope function, which has been most fruitful in the study of both extended and localized states, is mathematically ill-defined in the case of critical states, and other approaches are required to properly describe them and to understand their structure.

As we have seen in Sec. 2.2, from a rigorous mathematical point of view the nature of a state is uniquely determined by the measure of the spectrum to which it belongs. In this way, since it has been proven that a number of QCs model Hamiltonians have purely singular continuous energy spectra, we must conclude that the associated electronic states cannot be, strictly speaking, extended in Bloch's sense. However, this fact does not necessarily imply that states belonging to singular continuous spectra of *different* QCs will behave in exactly the same way from a physical viewpoint. Even in the case of eigenstates belonging

---

[28]"A theory has been proposed by Lu and Birman [514], but D. Thouless and D. P. DiVincenzo (private communication) have indicated that the theory may not be generally applicable" [330].

to a given system the coexistence of extended, localized and some sort of intermediate localization degree states was reported as earlier as in 1982 for incommensurate systems [168].

In fact, physically states can be properly classified according to their related transport properties. Thus, conducting states in crystalline systems are described by Bloch functions, whereas insulating systems exhibit exponentially decaying functions corresponding to localized states. Within this scheme the notion of critical states is somewhat imprecise, because critical states exhibit strong spatial fluctuations at different scales. In this regard, a first step towards a better understanding of critical states was provided by the demonstration that the amplitudes of critical states in on-site FQC models do not tend to zero at infinity, but are bounded below through the system [363]. This result suggested that the physical behavior of critical states is more similar to that corresponding to extended states than to localized ones. Indeed, the possible existence of extended states in several kinds of aperiodic systems has been discussed during the last decades [455, 456, 527, 528, 797, 807, 969], and arguments supporting the convenience of widening the very notion of extended state in aperiodic systems to include extended critical states which are not Bloch functions have been put forward [527, 528].

The transport properties of the electronic states in 1D systems can be studied by means of the Landauer conductance, given by the expression $g_N(E) = g_0 T_N(E)$, where $g_0 = 2e^2/h$ is the quantum of conductance, and

$$T_N(E) = \frac{4(\det \mathcal{M}_N)^2 \sin^2 \kappa}{[M_{12} - M_{21} + (M_{11} - M_{22}) \cos \kappa]^2 + (M_{11} + M_{22})^2 \sin^2 \kappa}, \qquad (2.104)$$

is the transmission coefficient for a chain containing N atoms, where M_{ij} are the global transfer matrix elements [548]. A completely analogous expression holds in the case of phonon dynamics. In obtaining Eq. (2.104) one assumes the system is sandwiched between two identical periodic chains (playing the role of contacts), each one with on-site energy ϵ' and transfer integral t', so that their dispersion relation is given by $E = \epsilon' + 2t' \cos \kappa$.[29]

For the sake of illustration, in the case of the resonance energy given by Eq. (2.32) for the FQC we considered in Sec. 2.4.2, we have (**Exercise 2.8**)[559]

$$T_{N_1}(E_*) = \left[1 + \frac{(1 - \gamma_A^2)^2}{(4 - E_*^2)\gamma_A^2} \sin^2(N_1 \varphi) \right]^{-1}, \quad |E_*| < 2, \qquad (2.105)$$

where $\varphi \equiv \cos^{-1}\left(\frac{\epsilon \gamma_A}{1 - \gamma_A^2}\right)$, N_1 is a Fibonacci number, and we have assumed $\epsilon' = 0$ and $t' = 1$ for simplicity. Proceeding in a similar way, one can get the resonance energy transmission coefficient for the period doubling based BAC, which has the same functional form as that given by Eq. (2.105) for the FQC, just replacing the Fibonacci number N_1 by the corresponding number of atoms in the considered period doubling lattice [559]. As it can be readily seen, the transmission coefficient given by Eq. (2.105) is always bounded below (i.e., $T_{N_1}(E_*) \neq 0$) for *any* lattice length, which proves the physically extended character of the E_* resonant states. Notwithstanding this, we must highlight that by extended we mean non-localized, and we do not necessarily have in mind a fully transparent, highly conductive state, as it occurs for Bloch states in periodic lattices. In fact, by systematically varying the FQC size, the resulting transmission coefficients can take on values within the range

---

[29]More generally, one may consider different materials on the left and right leads, characterized by different $\epsilon'_{L,R}$ and $t'_{L,R}$ values, respectively. The role of such asymmetric contacts in the resulting transport properties of atomic carbon wires was recently discussed in Ref. [468].

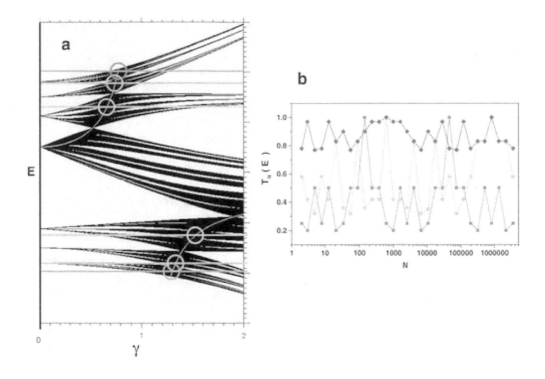

Figure 2.21 (a) Numerically obtained phase diagram for the FQC shown in Fig. 2.10, for the model parameters choice $\epsilon = 1/2$ and $N_1 = 34$, including the resonance energies $E_{\pm 2}^{(2)} = \pm\sqrt{13}/2$, $E_{\pm}^{+} = \pm\frac{\sqrt{11+2\sqrt{5}}}{2}$, and $E_{\pm}^{-} = \pm\frac{\sqrt{11-2\sqrt{5}}}{2}$, (encircled). (b) Variation of the transport coefficients corresponding to the resonance energies $E_{\pm 2}^{(2)}$ (diamonds), $E_{\pm}^{+}$ (dots), and $E_{\pm}^{-}$ (squares), encircled in (a), as a function of the system size (ranging from $N_1 = F_2 = 2$ to $N_1 = F_{32} = 3524\,578$).(Reprinted from Ref. [560] with permission. © 2017 by WILEY-VCH Verlag GmbH & Co. KGaA, Weinheim).

$T_{N_1}^{\min} \leq T_{N_1} \leq 1$, where full transmission states with $T_{N_1}(E_*) = 1$ are obtained by imposing the condition $\sin(N_1\varphi) = 0$ in Eq. (2.105), and the minima

$$T_{N_1}^{\min} = \left[1 + \frac{(1-\gamma_A^2)^2}{(4-E_*^2)\gamma_A^2}\right]^{-1}, \tag{2.106}$$

are determined by those model parameters satisfying the relationship $\varphi_m = (\frac{1}{2} + m)\pi/N_1$, $m \in \mathbb{Z}$. Thus, depending on both the adopted model parameters and the system length, among the E_* critical states, we generally find full transparent states interspersed with states exhibiting a broad palette of possible Landauer conductance values, ranging from highly conductive to highly resistive ones. Such a behavior has been reported for other kinds of aperiodic lattices as well, including Cantor, Kolakoski, Thue-Morse, and Rudin-Shapiro representatives [446, 445, 470].

For the sake of illustration in Fig. 2.21b, we plot the transmission coefficients corresponding to the resonance energies encircled in Fig. 2.21a as a function of the number of atoms in a FQC. These energies correspond to the crossing points between the zeroth-order energy curve and the first- and second-order energy spectra bands, respectively (see Fig. 2.10). Two main conclusions can be drawn from Fig. 2.21. First, the Landauer conductance of these resonant states sensitively depends on the number of atoms in the considered FQC, exhibiting a remarkable log-periodic pattern which is characteristic of each energy value. This log-periodic feature stems from an interesting relationship between the Fibonacci number $N_1 = F_j$ value and its related divisors, namely that F_k is a divisor of F_{pk} for all $p > 0$ [395]. Thus, for a given system size, the smaller k is, the more periods the envelope of the wave functions has, and the period of this envelope is exactly N_1/k. Thus, the wave function amplitudes distribution becomes progressively more homogeneous as the value of k gets smaller. In that case, the wave functions exhibit a remarkable spatial distribution, where the QP component (characteristic of short spatial scales) is nicely modulated by a long scale periodic component [351]. Second, the Landauer conductance variation range depends on the relative location of the considered resonant state in the overall energy spectrum, taking on significantly higher (lower) $T_{N_1}^{\min}$ values for eigenvalues located closer to the less (more) fragmented regions of the spectrum, respectively. In addition, we note that the resonance energies $E_{\pm}^{+} \simeq \pm 1.966\,731\dots$ (which are close to the energy spectrum band-edges located

at $E_*^{\mathrm{b}} = \pm\sqrt{\epsilon^2 + 4\cos^2\left(\frac{\pi}{2N_1}\right)} = \pm 2.059482\dots$ for the model parameters choice $\epsilon = 1/2$

and $N_1 = 34$) exhibit relatively high conductance values (including full transmission ones for some specific N_1 values), in agreement with the experimental results obtained in the study of light transport through the band edge states of photonic FQCs [149, 284].

By following an analogous procedure, we can obtain the resonance energy transmission coefficient for any precious mean based BAC. For instance, for the silver mean based QC, we have (**Exercise 2.15**)

$$T_{N_2}^{(2)}(E_*) = \left[1 + \frac{(\gamma_A - 1)^4 E_*^2}{(4 - E_*^2)\left[(1 + \gamma_A^2)^2 - \gamma_A^2 E_*^2\right]} \sin^2\left(N_2\phi\right)\right]^{-1}, \qquad (2.107)$$

where N_2 is the number of atoms in the silver mean QC. For QCs based on precious means with $n > 2$ in the corresponding substitution rule (see Table 2.6), one obtains completely similar expressions by just replacing N_2 by the corresponding number of atoms in the considered lattice into Eq. (2.107). Accordingly, we conclude that the transmission coefficient is always bounded below (i.e., $T_N^{(n)}(E_*) \neq 0$) for *any* lattice length for the resonant state E_* shared by all these QCs representatives, thereby analytically confirming the physically extended character of this resonance state in a broad class of aperiodic systems. Thus, resonant states corresponding to Fibonacci and period doubling BACs exhibit very similar trends, characterized by the presence of certain γ_A values supporting full transparent states, alternating with γ_A values where the transmission coefficients take on significantly small values. In addition, the transmission coefficient minima exhibit progressively decreasing values as we approach the limits $\gamma_A \to 0$ and $\gamma_A \to \infty$, describing the physical situation when one of the two types of chemical bonds, that is, $A - A$ or $A - B$, has a bonding strength which is negligible with respect to the other, so that the original lattice can be regarded as effectively decoupled in terms of a series of atomic clusters. A similar alternating behavior can be also appreciated in the transmission coefficient corresponding to precious means based BACs, such as the silver and bronze means, but in these systems the minima take on significantly higher values. In fact, after Eq. (2.107) we get the limits

$$\lim_{\gamma \to 0} T_N^{\min}(\epsilon, \gamma) = \lim_{\gamma \to \infty} T_N^{\min}(\epsilon, \gamma) = 1 - \left(\frac{\epsilon}{2}\right)^2,$$

which are independent of the chain length and, at variance with the limits previously obtained for the Fibonacci and period doubling BACs, can take on values significantly close to the unity, hence indicating a much more conductive nature for the resonant state E_* in this case. This physical property can be traced back to the different atomic cluster structure underlying both kinds of BACs. In fact, whereas precious mean based BACs display progressively longer A^n clusters as we increase the substitution sequence generation index n (see Fig. 2.7 and Table 2.6), the Fibonacci and period doubling BACs only have AA dimers and AAA trimers, respectively (see Figs. 2.7–2.8 and Table 2.6). Accordingly, resonance effects stemming from the presence of T_{AAA} local transfer matrices in the silver and bronze BACs accounts for their significantly improved transport properties.

In a similar vein, the electronic transport at zero temperature in BAC based on mixed mean sequences achieving macroscopic lengths of up to 10^8 atoms by using a real-space renormalization method developed for the Kubo-Greenwood formula within the tight-binding formalism. In so doing, the existence of transparent states in several mixed mean BACs was disclosed, as well as the presence of a large number of nearly transparent states (with $1 - T_N \simeq 10^{-12}$ or below) located close to the full transparent states in the parameter phase diagram [943]. Furthermore, a significant number of electronic states exhibiting transmission coefficients very close to unity can be found close to the $E_*(\epsilon, \gamma_A)$ curve given by Eq. (2.32) in the energy phase diagram (see Fig. 2.10) [775]. In order to analyze the role of these critical states in the resulting transport properties the study of the AC conductivity at zero temperature is very convenient, since it is very sensitive to the distribution nature of eigenvalues and the localization properties of the wave function close to the Fermi energy. In this way, by comparing the AC conductivities corresponding to both periodic and FQCs it was concluded that the value of the AC conductivity takes on systematically smaller values in the FQC case, due to the fact that the AC conductivity involves the contribution of non-transparent states within an interval of $\hbar\omega$ around the Fermi level in this case [667, 774, 775, 776].

Analogous results are obtained in the case of phonons propagating through FQCs. For instance, from the knowledge of the global transfer matrix given by Eq. (2.75), we can obtain the transmission coefficient (**Exercise 2.13c**)

$$T_N(\lambda^* = 1) = \frac{1}{1 + \frac{q(\alpha)}{4} U_{n_B-1}^2(\alpha/2)}, \qquad (2.108)$$

where $q(\alpha) \equiv 2\alpha^2 - 6\alpha + 5$ has no real roots. By inspecting Eq. (2.108), we see that, depending on the adopted α value, the transmission coefficient can range from a transparent state with $T_N(1) = 1$ (when $U_{n_B-1} = 0 \Rightarrow \alpha = 2\cos\left(\frac{\ell\pi}{n_B}\right)$) to the lowest value $T_N^*(1) = 4(2\alpha^2 - 6\alpha + 9)^{-1}$ (when $U_{n_B-1} = 1 \Rightarrow \alpha = 2\cos\left(\frac{\ell\pi}{2n_B}\right)$, with $\ell \in \mathbb{N}$). This curve exhibits a maximum ($T_N^*(1) = 8/9$) for the mass ratio value $\alpha_m = 3/2$, so that we have $4/9 \leq T_N^*(1) \leq 8/9$ within the interval $0 \leq \alpha \leq 3/2$, whereas for mass ratio values $\alpha > \alpha_m$ the transmission coefficient lower value $T_N^*(1)$ progressively decreases approaching zero as α increases. This result properly illustrates the rich physical behavior of critical states in terms of their related transport properties. In fact, studies of band structure effects in the thermal conductivity of FQCs indicated that the transmission properties of states with different frequencies play a significant role in the resulting thermal conductivity values [528, 530]. Indeed, the existence of tunable phonon resonances could be exploited in order to design novel thermal-conducting devices [110].

Similar results concerning the nature of critical modes in other kind of self-similar structures, such as Thue-Morse chains and hierarchical lattices, have been reported in the literature [120, 512], and its role in the transport properties has been analyzed in detail in terms of multifractal formalism on the basis that fractal dimension is directly associated to the localization degree of the eigenstates [15, 637]. More precisely, the phonon diffusivity, D, is related to the spectrum effective bandwidth $S \equiv \bigcup_i \Delta\omega_i$, where $\Delta\omega_i$ denotes the length of each subband in the spectrum, by a power law of the form $S \sim L^{-D}$, where $L = F_n$ for FQCs and $L = 2^n$ for Thue-Morse and period-doubling lattices, respectively, with n denoting the generation index of the sequences.

In summary, although when considering Bloch functions in periodic systems the notion of extended wave functions coincides with transparent ones, this is no longer true in the case of fractal and QPS. In particular, for general Fibonacci systems in which both diagonal and off-diagonal QP order are present in their model Hamiltonian, we have critical states which are not localized (i.e., $T_N \neq 0 \ \forall N$, when $N \to \infty$). For finite Fibonacci chains, one can find transparent states exhibiting a physical behavior completely analogous to that corresponding to usual Bloch states in periodic systems (i.e., $T_N = 1$) for a given choice of the model parameters. There exists a second class of critical states, those located close to the transparent ones, which are not strictly transparent (i.e., $T_N \lesssim 1$), but exhibit transmission coefficient values very close to unity. Finally, the remaining states in the spectrum show a broad diversity of possible values of the transmission coefficient (i.e., $0 < T_N \ll 1$), in agreement with the earlier view of critical states as intermediate between periodic Bloch wave functions ($T_N = 1$) and fully localized states ($T_N = 0$). Accordingly, most resonant states in the energy (frequency) spectrum show a great diversity of possible electrical (thermal) conductance values, which can take on either low or high values depending on the adopted model parameters, at variance with the highly conductive Bloch states found in periodic systems. In particular, the variation of both the electrical and thermal conductance values as a function of the chemical bonding strength parameter $\gamma_{A(B)}$ can help to understand the reported sensitivity of the electrical and thermal transport properties of most iQCs to different stoichiometric compositions (see Sec. 3.2.1).

2.8 TWO-DIMENSIONAL MODEL HAMILTONIANS

> *"More than 30 years after the discovery of quasicrystals a satisfactory quantum theory of quasicrystalline solids is still far from being constructed. (...) Moreover, there is a striking difference between the one-dimensional case, where progress has been achieved and the more physical case of higher dimensions, where the results are quite scant."* (Pavel Kalugin and André Katz, 2014)[401]

From the 1D Hamiltonian given by Eq. (2.7), one can construct 2D related ones in several ways, such that the corresponding wave functions are given by products of 1D wave functions. The first one relies on the Euclidean product of two 1D lattices, which results in a rectangular grid. In this case, the eigenvalues are sums of 1D eigenvalues [579]. For instance, the Hamiltonian of a 2D aperiodic square lattice is obtained from the Euclidean products $H_{sq}^{2D} = H_x^{1D} \otimes \mathbf{I}_y + H_y^{1D} \otimes \mathbf{I}_x$, where $\mathbf{I}_{x(y)}$ are the 1D identity operators. The energy spectrum of H_{sq}^{2D} is given by $E_{sq}^{2D} = E_x^{1D} + E_y^{1D}$, where $E_{x(y)}^{1D}$ is the spectrum of Eq. (2.7) in the X or Y space, respectively. More general 2D Hamiltonians involving Euclidean products between two 1D Hamiltonians can be written as

$$H_{\otimes}^{2D} = a(H_{\parallel} + \epsilon_{\parallel}\mathbf{I}_{\parallel}) \otimes (H_{\perp} + \epsilon_{\perp}\mathbf{I}_{\perp}), \qquad (2.109)$$

where a is a suitable scale factor, $H_{\parallel}$ and $H_{\perp}$ are 1D Hamiltonians given by Eq. (2.7) along two orthogonal directions, and $\epsilon_{\parallel}$ and $\epsilon_{\perp}$ are the on-site energies of atoms along parallel and perpendicular subspaces, respectively. The corresponding eigenspectrum $H_{\otimes}^{2D} |\alpha, \beta\rangle = E_{\alpha\beta}^{2D}$ is given by $E_{\alpha\beta}^{2D} = a(E_{\alpha}^{\parallel} + \epsilon_{\parallel})(E_{\beta}^{\perp} + \epsilon_{\perp})$, where $H^{\parallel} |\alpha\rangle = E_{\alpha}^{\parallel} |\alpha\rangle$ and $H^{\perp} |\beta\rangle = E_{\beta}^{\perp} |\beta\rangle$,

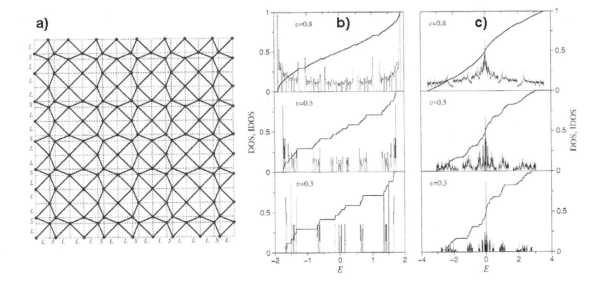

Figure 2.22 (a) Labyrinth tiling (thick lines) based on the QP square lattice obtained from the Euclidean product of two perpendicular 1D silver mean lattices (dashed lines grid). (b) The DOS and the IDOS (bold line) for various hopping parmeters v for a 1D silver mean lattice, (c) the same as (b) for the 2D Labyrinth tiling shown in (a). (Reprinted figure with permission from Yuan H Q, Grimm U, Repetowicz P, Schreiber M, *Phys. Rev. B* 62 15569–15578 (2000). Copyright (2000) by the American Physical Society).

so that the 2D energy spectrum is obtained from the product of the 1D eigenenergies. In the limiting cases $\epsilon_\parallel = \epsilon_\perp = a^{-1}$ and $\epsilon_\parallel = \epsilon_\perp = 0$, Eq. (2.109) reduces to H_{sq}^{2D} and H_{Lb}^{2D}, respectively, where H_{Lb}^{2D} describe the so-called Labyrinth tiling, first introduced by Clement Sire in 1989 [816]. This 2D rhombic lattice with non-constant bond lengths can be obtained by connecting second neighbors in a 2D aperiodic lattice based on the precious means (Fig. 2.22a) [773]. Its energy spectrum sensitively depends on the values of the parameters $\lambda_{x,y} \equiv |t_A^2 - t_A^2|(t_A t_B)^{-1}$ along the X and Y directions, respectively. If $\lambda_{x,y}$ are close to zero the spectrum is absolutely continuous, but it is a Cantor set of zero Lebesgue measure if λ_x and λ_y are large enough [858].

It is interesting to note that the gap labeling theorem is not limited to 1D systems, but it specifies the possible locations of gaps in the IDOS in higher-dimensional systems as well. In that case, however, several studies have shown that gaps tend to be generically closed in more than one dimension [310]. For instance, it has been reported that for small enough transfer integral values there are not gaps in the energy spectrum of 2D QP systems, such as the square Fibonacci lattices [33], or the Labyrinth tiling. This is illustrated in Fig. 2.22c, which shows the DOS and the IDOS for the Labyrinth tiling based on the silver mean lattice drawn in Fig. 2.22a. This tiling has three different hopping terms, taking on the values 1, v, and v^2, whereas the silver mean lattice has two different values, namely, 1 and v. By comparing Figs. 2.22b and c, we see the typical fingerprints indicating that the IDOS for the 1D silver mean lattice is a devil's staircase and its DOS is singular continuous with zero Lebesgue measure, for any v value, whereas for large enough v values the 2D Labyrinth spectrum progressively fills the gaps originally present for small v values. Indeed, the DOS is supported in a continuous interval for almost all values of hopping integral in the small coupling regime [816, 858]. Extensions of these results to general tridiagonal Hamiltonians have been reported as well [596].

Other popular representatives of 2D QPS are provided by the pentagonal Penrose and the octagonal Ammann-Beenker tilings (see Fig. 4.7 in Chapter 4). The electronic energy spectra of these QP tilings can be studied in terms of tight-binding Hamiltonians given by (an analogous expression can be written down for the phonon problem),

$$H_e = \sum_{n,m} \epsilon_{n,m} \, |n,m\rangle \, \langle n,m| + \sum_{\langle nm,jk\rangle} t_{nm,jk} \, |n,m\rangle \, \langle j,k| , \qquad (2.110)$$

where $|n,m\rangle$ is a Wannier state located at site (n,m), and the hopping terms include nearest neighbors only. It has been shown that the ground state wave functions can be expressed in the form $\psi_{n,m} = C_{n,m} e^{i\varkappa h_{n,m}}$, where the pre-exponential factors $C_{n,m}$ depend on the local environment at site (n,m) and they are related to a modulating QP function, $\varkappa$ is a constant playing the role of a suitable scale factor, and $h_{n,m}$ describes a height field defined on the vertices of the tiling, whose spatial pattern exhibits the long-range self-similar symmetry of the underlying QP tiling, which can be determined by using the related substitution rule (**Exercise 2.7**) [523]. We note that this ansatz agrees with the general expression given by Eq. (2.86) for generalized Bloch functions in QP potentials [401].

Regarding the Penrose tiling, its IDOS presents a dense central subband amounting about 10% of the total number of states in the spectrum. For an infinite Penrose tiling the gap labeling theorem established a tetrafurcation hierarchy of main gaps indicated in the ordinate axis of Fig. 2.23a [936]. As we see, the frequency spectrum is dominated by the couplings along the short diagonals of thin rhombi in the Penrose tile, in close analogy with the splitting of the frequency spectrum of the 1D FQC shown in Fig. 2.16a. In Fig. 2.23b the visualization of a representative electronic state in a synthetic Penrose molecular QC is done through the measurement of the differential tunneling conductance, normalized by the spatially averaged tunneling conductance measured on the bare Cu substrate surface. This conductance map reveals a pattern where electrons form a standing wave exhibiting high charge concentrations at particular sites, where electrons resonate at a particular local vertex structure of the Penrose QC [138], in close analogy with the physical origin of the frequency energy spectrum plotted in Fig. 2.23a. Both the local and global electronic DOS in Penrose lattices have been studied by considering a renormalization method which neglects the relatively small hopping integrals corresponding to the long diagonal of kites.

The electron dynamics on a square lattice in which the hopping terms follow an aperiodic distribution under the presence of a static electric field parallel to the lattice plane has been recently reported. The hopping integrals network is given by the distribution $t_{nm,jk} = e^{-|\zeta_{nm}-\zeta_{jk}|}$, with $\zeta_{nm} = W \cos(\pi a n^\nu) \cos(\pi a m^\nu)$, where W is a constant, a is an irrational number, and the exponent ν controls the degree of aperiodicity in the lattice. In the absence of external electrical field transport properties typical of ballistic propagation of the states are obtained depending on the adopted ν value. When the electrical field is turned on these states exhibit an oscillatory behavior similar to that observed in periodic systems (Bloch oscillations, see Sec. 3.4.4) [216].

Recent advances in controlling the rotation angle of graphene over graphene layers allows the use of the resulting bilayer structure as an effective knob for tuning its electronic properties. In this regard, it has been recently reported that an incommensurate structure analogue to a 12-fold QC [17], can be obtained by epitaxial growth of a 30° twisted bilayer graphene on top of a suitable stabilizing substrate [5]. This ddQC can be regarded as a material realization of the algorithmic 2D QC construction process originally proposed by

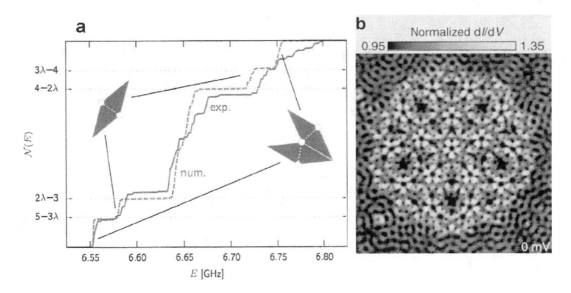

Figure 2.23   a) Solid line: normalized IDOS $\mathcal{N}(E)$ for a Penrose lattice whose $N = 164$ vertices are occupied by dielectric resonators (ceramic cylinders of 5 mm height and 8 nm diameter) with a high index of refraction ($n = 6$). The dashed line is obtained by direct diagonalization of the Hamiltonian given by Eq. (2.110) with $N = 2665$ sites. The horizontal dotted lines mark the gap labeling values defining the main gaps locations for an infine tile ($\lambda \equiv \tau$). The origin of the main four subbands can be related to dimer and trimer vertex resonances. (Reprinted figure with permission from Vignolo P, Bellec M, Böhm J, Camara A, Gambaudo J M, Kuhl U, *Phys. Rev. B* 93, 075141 (7pp) 2016. Copyright (2016) by the American Physical Society). b) Normalized differential conductance map over a 42 × 42 nm field of a molecular Penrose QC formed by 460 CO molecules arranged on the surface of Cu(111) one at a time, measured at zero bias voltage. (Collins L C, Witte T G, Silverman R, Green D B, Gomes K K, Imaging quasiperiodic electronic states in a synthetic Penrose tiling, *Nature Commun.* **8** 15961 (2017). DOI:10.1038/ncomms15961. Work licensed under a Creative Commons Attribution 4.0 International License).

Stampfli using two ideal hexagonal grids [834]. Multiple Dirac cones replicated with the 12-fold rotational symmetry were experimentally observed in angle-resolved photoemission spectra, thereby signaling a strong interlayer coupling. This platform hence provides a way to explore physical properties of relativistic electrons moving in 2D QP structures. In this way, the emergence of electronic localization as an intrinsic effect due to long-range QPO has been disclosed in this sort of synthetic ddQCs [673].

2.9   EXERCISES WITH SOLUTIONS

2.1 *Let us consider the general substitution sequence $g(A) = A^\alpha B^\beta$, $g(B) = A^\gamma B^\delta$, where α, β, γ and δ are natural numbers. Show that the substitution sequence $g(A) = B^\gamma A^\delta$, $g(B) = B^\alpha A^\beta$, obtained from the former by the letter exchange $A \leftrightarrow B$, belongs to the same spectral class.*

The substitution matrices of the original and transformed sequences are

$$\mathbf{S} = \begin{pmatrix} \alpha & \gamma \\ \beta & \delta \end{pmatrix}, \quad \tilde{\mathbf{S}} = \begin{pmatrix} \delta & \beta \\ \gamma & \alpha \end{pmatrix}.$$

We have $\det \mathbf{S} = \alpha\delta - \beta\gamma = \det \tilde{\mathbf{S}}$ and $\mathrm{tr}\mathbf{S} = \alpha+\delta = \mathrm{tr}\tilde{\mathbf{S}}$. This guarantees that their characteristic polynomials $P_\lambda = \lambda^2 - \lambda\mathrm{tr}\mathbf{S} + \det \mathbf{S}$ are invariant under the letter exchange $A \leftrightarrow B$.[30] Therefore, since both substitution rules have the same eigenvalues and determinant values, they belong to the same spectral class.

2.2 *The following substitution rules have been discussed in the literature as possible representatives of a ternary Fibonacci lattice: (a) $g(A) = AB$, $g(B) = C$, $g(C) = A$ [283], (b) $g(A) = ABC$, $g(B) = A$, $g(C) = B$ [358, 933], (c) $g(A) = B$, $g(B) = ACA$, $g(C) = A$ [543], (d) $g(A) = AB$, $g(B) = BC$, $g(C) = AC$. Determine if they satisfy the PV property and classify them. Compare the obtained results with those corresponding to the substitution rule (e) $g(A) = B$, $g(B) = C$, $g(C) = A$.*

TABLE 2.7   Substitution matrix **S**, its determinant value, and their eigenvalues for the substitution sequences listed above. The corresponding spectral class are included in the last column. Those sequences satisfying the Pisot property are highlighted (bold face) in the first column.

SEQUENCE	S	detS	EIGENVALUES	CLASS
(a)	$\begin{pmatrix} 1 & 0 & 1 \\ 1 & 0 & 0 \\ 0 & 1 & 0 \end{pmatrix}$	1	$\lambda_+ \simeq 1.465\ldots,\ |\lambda_{2,3}| = \lambda_+^{-1/2}$	Π_{I}
(b)	$\begin{pmatrix} 1 & 1 & 0 \\ 1 & 0 & 1 \\ 1 & 0 & 0 \end{pmatrix}$	1	$\lambda_+ \simeq 1.839\ldots,\ |\lambda_{2,3}| = \lambda_+^{-1/2}$	Π_{I}
(c)	$\begin{pmatrix} 0 & 2 & 1 \\ 1 & 0 & 0 \\ 0 & 1 & 0 \end{pmatrix}$	1	$\lambda_+ = \tau,\ \lambda_2 = -1,\ \lambda_3 = -\tau^{-1}$	Π_{III}
(d)	$\begin{pmatrix} 1 & 0 & 1 \\ 1 & 1 & 0 \\ 0 & 1 & 1 \end{pmatrix}$	2	$\lambda_+ = 2,\ \lambda_{2,3} = \frac{1 \pm i\sqrt{3}}{2}$	Π_{III}
(e)	$\begin{pmatrix} 0 & 0 & 1 \\ 1 & 0 & 0 \\ 0 & 1 & 0 \end{pmatrix}$	1	$\lambda_+ = 1,\quad \lambda_{2,3} = \frac{-1 \pm i\sqrt{3}}{2}$	Π_0

Making use of Eq. (2.1) in Table 2.7, we list the substitution matrices **S**, their determinant value, and their eigenvalues for the sequences given in the enunciate. By inspecting this Table several conclusions can be drawn:

- The sequences (a) and (b) have the same characteristic polynomial than the ternary Fibonacci and Tribonacci sequences listed in Table 2.4, respectively, since the corresponding **S** matrices are mutually transposed to each other.

- Both sequences (a) and (b) satisfy the PV property and, since their determinants equal unity, they belong to the Π_{I} class.

---

[30]We note that $\tilde{\mathbf{S}} = \sigma\mathbf{S}\sigma^{-1}$, where $\sigma = \begin{pmatrix} 0 & 1 \\ 1 & 0 \end{pmatrix}$.

- The leading eigenvalue of the characteristic polynomial corresponding to the sequence (c) is larger than one but one of the conjugate roots has a unity modulus. Therefore, λ_+ is a Salem number and the substitution rule does not satisfy the PV property, so that this sequence belongs to the Π_{III} class.

- The leading eigenvalue of the characteristic polynomial corresponding to the sequence (d) is larger than one but the moduli of the two conjugate roots equal unity. Therefore, λ_+ is also a Salem number and the substitution rule does not satisfy the PV property, so that this sequence belongs to the Π_{III} class as well.

- The leading eigenvalue of the characteristic polynomial corresponding to the sequence (c) is one, and the moduli of its conjugate roots also equal the unity. Therefore, λ_+ is neither a PV nor a Salem number. One can readily check that the substitution rule $g(A) = B$, $g(B) = C$, $g(C) = A$, describes cyclic permutations in the letter set $\mathcal{A} = \{A, B, C\}$.

2.3 *The quaternary substitution rule given by $g(A) = ABDB$, $g(B) = ABAC$, $g(C) = DCDB$, $g(D) = DCAC$, describes a generalization of the Rudin-Shapiro sequence exhibiting a purely absolute continuous spectrum* [122]. *(a) Show that it is primitive. (b) Determine if it satisfies the Pisot property. (c) Calculate the frequency of apparition of each letter in the limit $N \to \infty$.*

(a) The corresponding substitution matrix reads

$$\mathbf{S} = \begin{pmatrix} 1 & 2 & 0 & 1 \\ 2 & 1 & 1 & 0 \\ 0 & 1 & 1 & 2 \\ 1 & 0 & 2 & 1 \end{pmatrix},$$

so that

$$\mathbf{S}^2 = \begin{pmatrix} 1 & 2 & 0 & 1 \\ 2 & 1 & 1 & 0 \\ 0 & 1 & 1 & 2 \\ 1 & 0 & 2 & 1 \end{pmatrix} \begin{pmatrix} 1 & 2 & 0 & 1 \\ 2 & 1 & 1 & 0 \\ 0 & 1 & 1 & 2 \\ 1 & 0 & 2 & 1 \end{pmatrix} = \begin{pmatrix} 6 & 4 & 4 & 2 \\ 4 & 6 & 2 & 4 \\ 4 & 2 & 6 & 4 \\ 2 & 4 & 4 & 6 \end{pmatrix}.$$

Since all entries in $\mathbf{S}^2$ are positive and non-zero, none of the matrix elements of successive power matrices can vanish, which guarantees that this substitution rule is primitive.

(b) The $\mathbf{S}$ matrix characteristic polynomial can be factorized as

$$P_\lambda = \lambda \, (\lambda - 2) \, (\lambda + 2) \, (\lambda - 4),$$

so that its eigenvalues read $\lambda_1 = 0$, $\lambda_2 = 2 = -\lambda_3$, and $\lambda_+ = 4$. Thus, since $|\lambda_2| = |\lambda_3| > 1$, this substitution rule does not satisfy the Pisot property.

(c) The $\mathbf{S}$ matrix normalized leading eigenvector reads $\mathbf{v}_+ = \frac{1}{4}(1, 1, 1, 1)$, hence indicating that all letters are equiprobable (25%) in the $N \to \infty$ limit.

2.4 *The bronze-mean hexagonal QC tiling is generated by the substitution rules* $g(A) = AABA$, $g(B) = A$, $g(C) = CDC$, *and* $g(D) = CDCC$ [196, 634]. *Show that: (a) the substitution sequence is non-primitive; (b) the structure belongs to the* Π_{I} *spectral class.*

(a) The corresponding substitution matrix reads

$$\mathbf{S} = \begin{pmatrix} 3 & 1 & 0 & 0 \\ 1 & 0 & 0 & 0 \\ 0 & 0 & 2 & 3 \\ 0 & 0 & 1 & 1 \end{pmatrix} \equiv \begin{pmatrix} \mathbf{S}_\mathrm{B} & \mathbf{0} \\ \mathbf{0} & \mathbf{C} \end{pmatrix},$$

where $\mathbf{S}_\mathrm{B} = \begin{pmatrix} 3 & 1 \\ 1 & 0 \end{pmatrix}$ is the bronze mean substitution matrix (see Table 2.3), with characteristic polynomial $P_\lambda^\mathrm{B} = \lambda^2 - 3\lambda - 1$, and $\mathbf{C} \equiv \begin{pmatrix} 2 & 3 \\ 1 & 1 \end{pmatrix}$, with characteristic polynomial $P_\lambda^\mathrm{C} = P_\lambda^\mathrm{B}$. Therefore,

$$\mathbf{S}^N = \begin{pmatrix} \mathbf{S}_\mathrm{B} & \mathbf{0} \\ \mathbf{0} & \mathbf{C} \end{pmatrix}^N = \begin{pmatrix} \mathbf{S}_B^N & \mathbf{0} \\ \mathbf{0} & \mathbf{C}^N \end{pmatrix},$$

and we see that this substitution sequence is non-primitive, since the anti-diagonal block matrices of $\mathbf{S}^N$ equal zero $\forall N \in \mathbb{N}$.

(b) Since $P_\lambda^\mathrm{C} = P_\lambda^\mathrm{B}$, the characteristic polynomial of the diagonal block matrix $\mathbf{S}$ can be factorized in the form $P_\lambda = (P_\lambda^\mathrm{B})^2 = (\lambda^2 - 3\lambda - 1)^2$. In this case, the degenerate Frobenius eigenvalue, $\lambda_+ = \frac{1}{2}\sqrt{13} + \frac{3}{2} \simeq 3.302\ldots$, satisfy the PV property, for $\lambda_- = \frac{3}{2} - \frac{1}{2}\sqrt{13} \simeq -0.302\ldots$. On the other hand, $\det \mathbf{S} = 1$, so that this hexagonal QC belongs to the Π_{I} spectral class, albeit its substitution matrix is non-primitive.

2.5 *Demonstrate that the substitution rule related to the general binary Cantor lattice given by Eq. (2.3) is non-primitive.*

The eigenvalues and eigenvectors of the substitution matrix

$$\mathbf{S} = \begin{pmatrix} r & 0 \\ s - r & s \end{pmatrix},$$

are given by $\lambda_+ = r$, $\lambda_- = s$, and $\mathbf{u}_+ = (-1, 1)$, $\mathbf{u}_- = (0, 1)$, respectively. Thus, the N-th power of the substitution matrix can be obtained as

$$\mathbf{S}^N = \begin{pmatrix} -1 & 0 \\ 1 & 1 \end{pmatrix} \begin{pmatrix} r^N & 0 \\ 0 & s^N \end{pmatrix} \begin{pmatrix} -1 & 0 \\ 1 & 1 \end{pmatrix}^{-1} = \begin{pmatrix} r^N & 0 \\ s^N - r^N & s^N \end{pmatrix}.$$

As we see, there exist one matrix element in $\mathbf{S}^N$ which is equal to zero $\forall N \in \mathbb{N}$, and the related substitution sequence is therefore non-primitive.

2.6 *Classify the generalized Fibonacci sequences given by the substitution rule family* $g(A) = A^{F_n} B^{F_{n-1}}$, $g(B) = A^{F_{n-1}} B^{F_{n-2}}$, *where* F_m *are Fibonacci numbers.*

The corresponding substitution matrix is given by Eq. (2.5), whose determinant reads $\det \mathbf{S} = F_n F_{n-2} - F_{n-1}^2 = -1$. The characteristic polynomial can be written as

$$P_\lambda = F_n F_{n-2} - F_{n-1}^2 - \lambda(F_n + F_{n-2}) + \lambda^2 = -1 - \lambda \operatorname{tr} \mathbf{S} + \lambda^2$$

The eigenvalues are $\lambda_\pm = \left(\mathrm{tr}\mathbf{S} \pm \sqrt{\mathrm{tr}^2\mathbf{S} + 4}\right)/2$, and since $\mathrm{tr}\mathbf{S} \geq 2$ we get $\lambda_+ > 1$ and $|\lambda_-| < 1$. Therefore, the leading eigenvalue satisfies the Pisot property and $|\det \mathbf{S}| = 1$, so that the generalized Fibonacci sequences belong to the Π_I class.

2.7 *Determine the inflation factors and the relative frequency of the different tiles for: (a) the 2D Fibonacci tiling given by Eq. (2.6); (b) the two-dimensional octagonal (Ammann-Beenker) tiling given by the substitution rule $A \to 3A + 2B$, $B \to 4A + 3B$ [840].*

(a) The characteristic polynomial of the substitution matrix $\mathbf{S}_{2D}$ can be factorized in the form $P_\lambda = \lambda^4 - \lambda^3 - 4\lambda^2 - \lambda + 1 = (\lambda + 1)^2 \left(\lambda^2 - 3\lambda + 1\right) \equiv (\lambda + 1)^2 \tilde{P}_\lambda$, so that the roots of the irreducible component $\tilde{P}_\lambda$ read $\lambda_\pm = \tau^{\pm 2}$, and the scaling factors are $s_\pm \equiv \sqrt{\lambda_\pm} = \tau^{\pm 1}$. The leading vector normalized components are given by $\mathbf{v}_+ = (\tau^{-4}, \tau^{-3}, \tau^{-3}, \tau^{-2})$ [35].

(b) The corresponding substitution matrix is given by[31]

$$\mathbf{S}_{AB} = \begin{pmatrix} 3 & 4 \\ 2 & 3 \end{pmatrix},$$

with determinant $\det \mathbf{S} = 1$. The eigenvalues are $\lambda_\pm = 3 \pm 2\sqrt{2}$, so that the scaling factors are given by $s_\pm \equiv \sqrt{\lambda_\pm} = \sqrt{3 \pm 2\sqrt{2}} = \sqrt{2} \pm 1$. On the other hand, making use of Eq. (2.2) we get,

$$v_A = 2 - \sqrt{2} \simeq 0.586\ldots, \qquad v_B = \sqrt{2} - 1 \simeq 0.414\ldots.$$

2.8 *Derive Eq. (2.37).*

When evaluated at the resonance energy E_* given by Eq. (2.32) the elemental cluster matrices given by Eq. (2.24) read

$$\mathbf{R}(E_*) = \begin{pmatrix} q(q^2 - 2) & \gamma_A(1 - q^2) \\ (q^2 - 1)\gamma_A^{-1} & -q \end{pmatrix}, \, \mathbf{r}(E_*) = \begin{pmatrix} q^2 - 1 & -q\gamma_A \\ q\gamma_A^{-1} & -1 \end{pmatrix},$$

where $q \equiv 2\epsilon\gamma_A(1 - \gamma_A^2)^{-1}$. By introducing the auxiliary unimodular matrix

$$\mathbf{Q} \equiv \begin{pmatrix} q & -\gamma_A \\ \gamma_A^{-1} & 0 \end{pmatrix},$$

we readily check that $\mathbf{r}(E_*) = \mathbf{Q}^2$ and $\mathbf{R}(E_*) = \mathbf{Q}^3$, so that $\mathcal{M}_N^{(1)}(E_*) \equiv \mathbf{R}(E_*)^{n_A} \mathbf{r}(E_*)^{n_B} = \mathbf{Q}^{3n_A + 2n_B} = \mathbf{Q}^N$, for $3n_A + 2n_B = 3F_{j-3} + 2F_{j-4} = F_{j-3} + 2F_{j-2} = F_{j-1} + F_{j-2} = F_j \equiv N$. Now, since $\det \mathbf{Q} = 1$ we can exploit the Cayley-Hamilton theorem to calculate the required matrix power. Accordingly, making use of Eq. (2.34), we obtain

$$\begin{aligned} \mathbf{Q}^N &= U_{N-1}(z)\mathbf{Q} - U_{N-2}(z)\mathbf{I} = U_{N-1}(z)\begin{pmatrix} 2z & -\gamma_A \\ \gamma_A^{-1} & 0 \end{pmatrix} - U_{N-2}(z)\begin{pmatrix} 1 & 0 \\ 0 & 1 \end{pmatrix} \\ &= \begin{pmatrix} U_N & -\gamma_A U_{N-1} \\ \gamma_A^{-1}U_{N-1} & -U_{N-2} \end{pmatrix}, \end{aligned}$$

where $z \equiv \mathrm{tr}\mathbf{Q}/2 = q/2$, and we have used Eq. (2.36).

---

[31]Note that $\mathbf{S}_{AB} = \begin{pmatrix} 1 & 2 \\ 1 & 1 \end{pmatrix}^2$.

2.9 *Derive the commutator relationships* $[\mathbf{ABA}, \mathbf{B}] = \mathrm{tr}(\mathbf{AB})[\mathbf{A}, \mathbf{B}]$ *and* $[\mathbf{ABA}, \mathbf{BAB}] = (\mathrm{tr}^2(\mathbf{ABA}) - 1)[\mathbf{A}, \mathbf{B}]$, *for unimodular matrices.*

Let $\mathbf{A}$ and $\mathbf{B}$ be 2×2 unimodular matrices. According to the Cayley-Hamilton theorem we have (see Eq. (2.34))

$$(\mathbf{AB})^2 = U_1(z)\mathbf{AB} - U_0(z)\mathbf{I}, \quad (\mathbf{BA})^2 = U_1(z)\mathbf{BA} - U_0(z)\mathbf{I}, \tag{2.111}$$

with $U_1(z) = 2z = \mathrm{tr}(\mathbf{AB}) = \mathrm{tr}(\mathbf{BA})$. Making use of (2.111) we readily obtain the useful relation

$$[\mathbf{ABA}, \mathbf{B}] = (\mathbf{AB})^2 - (\mathbf{BA})^2 = U_1(z)[\mathbf{A}, \mathbf{B}] = \mathrm{tr}(\mathbf{AB})[\mathbf{A}, \mathbf{B}]. \tag{2.112}$$

which is used to derive Eqs. (2.54), (2.55), (2.65), (2.67), and (2.68).

Analogously, after

$$(\mathbf{AB})^3 = U_2(z)\mathbf{AB} - U_1(z)\mathbf{I}, \quad (\mathbf{BA})^3 = U_2(z)\mathbf{BA} - U_1(z)\mathbf{I}, \tag{2.113}$$

we get

$$[\mathbf{ABA}, \mathbf{BAB}] \equiv (\mathbf{AB})^3 - (\mathbf{BA})^3 = U_2(z)[\mathbf{A}, \mathbf{B}] = (\mathrm{tr}^2\mathbf{F}_2 - 1)[\mathbf{R}, \mathbf{r}]. \tag{2.114}$$

which is used to derive Eq. (2.66).

2.10 *Derive Eq. (2.56).*

Making use of $p_2(E) = E^2 - \epsilon^2 - 2$, and $p_3(E) \equiv R_{11}(E) - v_A$ (see Eq. (2.52)) into Eq. (2.28), we can express the elemental cluster matrices in the form

$$\mathbf{R} = \gamma_A^{-1} \begin{pmatrix} v_A + p_3 & \gamma_A^2 - v_A^2 \\ 1 + p_2 & -v_A \end{pmatrix}, \quad \mathbf{r} = \begin{pmatrix} 1 + p_2 & -v_A \\ v_B & -1 \end{pmatrix},$$

so that

$$\mathbf{Rr} = \gamma_A^{-1} \begin{pmatrix} (1+p_2)(v_A + p_3) + v_B(\gamma_A^2 - v_A^2) & -\gamma_A^2 - v_A p_3 \\ (1+p_2)^2 - v_A v_B & -v_A p_2 \end{pmatrix},$$

and we have $\mathrm{tr}(\mathbf{Rr}) = \gamma_A^{-1}(p_2 p_3 - E + \epsilon)$, where we have used Eq. (2.52).

2.11 *Obtain the normal frequencies of the atomic clusters* $B - A - B$ *and* $B - A = A - B$.

The normal modes of harmonic chains composed of two types of atoms are derived in many textbooks [27, 299, 881], where it is shown that the corresponding normal frequencies are obtained as the roots of the characteristic polynomial associated to the determinant $|\mathbf{M}\omega^2 - \mathbf{K}|$, where $\mathbf{K} \equiv \left(\frac{\partial^2 U}{\partial \eta_j^2}\right)_0$ is the Hessian matrix related to the potential energy function $U(\eta_j)$, where η_j measures the elongation of the j-th particle with respect to the equilibrium position, and $\mathbf{M} \equiv \left(\frac{\partial^2 T}{\partial \dot{\eta}_j^2}\right)_0$ is the Hessian matrix related to the kinetic energy function $T(\eta_j, \dot{\eta}_j)$, where $\dot{\eta}_j$ measures the velocity of the j-th particle. Both Hessian matrices are evaluated with respect to the equilibrium configuration $\{\eta_j = 0\} \; \forall j$ (labeled by the subscript **0**).

(a) In the case of the triatomic molecule $B - A - B$, where atoms of masses m_A and m_B are respectively coupled with springs of elastic constant $K_{AB} \equiv k$, we have

$$T = \frac{1}{2}m_B(\dot{\eta}_1^2 + \dot{\eta}_3^2) + \frac{1}{2}m_A\dot{\eta}_2^2, \quad U = \frac{1}{2}k(\eta_1 - \eta_2)^2 + \frac{1}{2}k(\eta_2 - \eta_3)^2, \tag{2.115}$$

so that the Hessian matrices are given by

$$\mathbf{M} = \begin{pmatrix} m_B & 0 & 0 \\ 0 & m_A & 0 \\ 0 & 0 & m_B \end{pmatrix}, \quad \mathbf{K} = \begin{pmatrix} k & -k & 0 \\ -k & 2k & -k \\ 0 & -k & k \end{pmatrix}, \qquad (2.116)$$

and the normal frequencies are obtained by solving

$$|\mathbf{M}\omega^2 - \mathbf{K}| = \omega^2 \left(m_B \omega^2 - k \right) \left(m_B m_A \omega^2 - 2km_B - km_A \right) = 0,$$

to get

$$\omega_1^2 = 0, \quad \omega_2^2 = \frac{k}{m_B}, \quad \omega_3^2 = \frac{k}{m_B} \left(1 + \frac{2m_B}{m_A} \right),$$

which can be expressed in the form $\lambda_1 = 0$, $\lambda_2 = \alpha^{-1}$ and $\lambda_3 = 2 + \alpha^{-1}$ in terms of the reduced frequency $\lambda = m_A \omega^2 / k$.

(b) In the case of the tetramer molecule $B - A = A - B$ we have

$$T = \frac{m_B}{2}(\dot{\eta}_1^2 + \dot{\eta}_4^2) + \frac{m_A}{2}(\dot{\eta}_2^2 + \dot{\eta}_3^2), \quad U = \frac{k}{2}(\eta_1 - \eta_2)^2 + \frac{K}{2}(\eta_2 - \eta_3)^2 + \frac{k}{2}(\eta_3 - \eta_4)^2, \quad (2.117)$$

where K is the elastic constant coupling consecutive $A = A$ atoms. The Hessian matrices are given by

$$\mathbf{M} = \begin{pmatrix} m_B & 0 & 0 & 0 \\ 0 & m_A & 0 & 0 \\ 0 & 0 & m_A & 0 \\ 0 & 0 & 0 & m_B \end{pmatrix}, \quad \mathbf{K} = \begin{pmatrix} k & -k & 0 & 0 \\ -k & k+K & -K & 0 \\ 0 & -K & k+K & -k \\ 0 & 0 & -k & k \end{pmatrix}, \qquad (2.118)$$

and the normal frequencies are obtained by solving

$$|\mathbf{M}\omega^2 - \mathbf{K}| = \omega^2 \left(m_A k + k m_B - m_A m_B \omega^2 \right) \left(m_A m_B \omega^4 - \omega^2 \left(2m_B K + k m_A + k m_B \right) + 2Kk \right) = 0,$$

to get

$$\omega_4^2 = 0, \quad \omega_5^2 = \frac{m_A + m_B}{m_A m_B} k,$$

$$\omega_\pm^2 = \frac{(k + 2K)m_B + m_A k \pm \sqrt{m_B^2 (2K + k)^2 + m_A^2 k^2 - 2km_A m_B (2K - k)}}{2m_A m_B},$$

which can be expressed in the form $\lambda_4 = 0$, $\lambda_5 = 1 + \alpha^{-1}$ and $\lambda_\pm = \gamma + \eta \pm \sqrt{(\eta - \gamma)^2 + 2\gamma}$ in terms of the reduced frequency $\lambda = m_A \omega^2 / k$, where $\gamma = K/k$ and $\eta = (1 + \alpha^{-1})/2$.

2.12 *Derive the relationship* $[\mathbf{R}'_A, \mathbf{R}'_B] = p_2(\lambda)[\mathbf{R}_A, \mathbf{R}_B]$.

Defining $\mathbf{R}'_A = \mathbf{R}_B^{(1)} \mathbf{R}_A^{(1)} \mathbf{R}_A^{(1)} \mathbf{R}_B^{(1)}$ and $\mathbf{R}'_B = \mathbf{R}_B^{(1)} \mathbf{R}_A^{(1)} \mathbf{R}_B^{(1)}$, and making use of Eq. (2.70) for the related local transfer matrices $\mathbf{T}_{AAB}$, $\mathbf{T}_{BAA}$, $\mathbf{T}_{ABA}$, and $\mathbf{T}_{BAB}$, after some algebra, we obtain

$$\begin{aligned} [\mathbf{R}'_A, \mathbf{R}'_B] &= \mathbf{R}_B \mathbf{R}_A \mathbf{R}_A \mathbf{R}_B \mathbf{R}_B \mathbf{R}_A \mathbf{R}_B - \mathbf{R}_B \mathbf{R}_A \mathbf{R}_B \mathbf{R}_B \mathbf{R}_A \mathbf{R}_A \mathbf{R}_B \\ &= \frac{\lambda}{\gamma_A} \left(2\gamma_A - 1 - \alpha \left(1 + (\gamma_A - 1)\lambda \right) \right) \left(2 - 2\lambda - 2\alpha\lambda + \alpha\lambda^2 \right) \begin{pmatrix} 1 & 0 \\ 2 - \alpha\lambda & -1 \end{pmatrix} \\ &= \frac{\lambda}{\gamma_A} p_1(\lambda) p_2(\lambda) \begin{pmatrix} 1 & 0 \\ 2 - \alpha\lambda & -1 \end{pmatrix} = p_2(\lambda)[\mathbf{R}_A, \mathbf{R}_B] \end{aligned}$$

where $p_2(\lambda) \equiv \alpha\lambda^2 - 2\lambda(\alpha + 1) + 2$, and we have taken into account Eq. (2.71).

2.13 *Obtain (a) Eq. (2.74), (b) Eq. (2.75), and Eq. (2.108).*

(a) For any resonance frequency λ_* satisfying Eq. (2.72) the renormalized matrices $\mathbf{R}_A^{(1)}(\lambda_*)$ and $\mathbf{R}_B^{(1)}(\lambda_*)$ commute, so that the global transfer matrix can be rearranged in the convenient form $\mathcal{M}_N(\lambda_*) \equiv \mathbf{R}_A^{n_A} \mathbf{R}_B^{n_B}$, where $N = F_n$, $n_A = F_{n-3}$ and $n_B = F_{n-4}$ are Fibonacci numbers. In the particular case $\lambda_* = 2$, the renormalized matrices adopt the simple form given by Eq. (2.73), so that they are both unimodular. Therefore, we can use the Cayley-Hamilton theorem to obtain the required power matrices. Thus, making use of Eq. (2.34), we have

$$
\begin{aligned}
[\mathbf{R}_A^{(1)}(\lambda_* = 2)]^{n_A} &= U_{n_A-1}(1) \begin{pmatrix} 1 & 0 \\ 2(\alpha - 2) & 1 \end{pmatrix} - U_{n_A-2}(1) \begin{pmatrix} 1 & 0 \\ 0 & 1 \end{pmatrix} \\
&= \begin{pmatrix} 1 & 0 \\ 2n_A(\alpha - 2) & 1 \end{pmatrix}
\end{aligned} \tag{2.119}
$$

$$
\begin{aligned}
[\mathbf{R}_B^{(1)}(\lambda_* = 2)]^{n_B} &= U_{n_B-1}(-1) \begin{pmatrix} -1 & 0 \\ 2(1 - \alpha) & -1 \end{pmatrix} - U_{n_B-2}(-1) \begin{pmatrix} 1 & 0 \\ 0 & 1 \end{pmatrix} \\
&= (-1)^{n_B} \begin{pmatrix} 1 & 0 \\ 2n_B(\alpha - 1) & 1 \end{pmatrix},
\end{aligned} \tag{2.120}
$$

where we have used the relations $U_n(1) = n + 1$ and $U_n(-1) = (-1)^n(n + 1)$. Making use of Eqs. (2.119)–(2.120) along with the relation $F_n = F_{n-1} + F_{n-2}$, we finally obtain

$$
\mathcal{M}_N(\lambda_* = 2) \equiv [\mathbf{R}_A^{(1)}(\lambda_* = 2)]^{n_A} [\mathbf{R}_B^{(1)}(\lambda_* = 2)]^{n_B} = (-1)^{F_{n-4}} \begin{pmatrix} 1 & 0 \\ 2(\alpha F_{n-2} - F_{n-1}) & 1 \end{pmatrix}.
$$

(b) In the case $\lambda_* = 1$ ($\Rightarrow \gamma_A = (2 - \alpha)^{-1}$) the local transfer matrices given by Eq. (2.70) take the form

$$
\mathbf{T}_{ABA} = \begin{pmatrix} 2 - \alpha & -1 \\ 1 & 0 \end{pmatrix}, \quad \mathbf{T}_{BAA} = \begin{pmatrix} 1 & \alpha - 2 \\ 1 & 0 \end{pmatrix},
$$

$$
\mathbf{T}_{AAB} = \begin{pmatrix} (2 - \alpha)^{-1} & -(2 - \alpha)^{-1} \\ 1 & 0 \end{pmatrix}, \quad \mathbf{T}_{BAB} = \begin{pmatrix} 1 & -1 \\ 1 & 0 \end{pmatrix}, \tag{2.121}
$$

so that the renormalized matrices $\mathbf{R}_A^{(1)}(\lambda_*)$ and $\mathbf{R}_B^{(1)}(\lambda_*)$ are given by

$$
\mathbf{R}_A^{(1)}(\lambda_* = 1) = \begin{pmatrix} -1 & 0 \\ 0 & -1 \end{pmatrix} = -\mathbf{I} \qquad \mathbf{R}_B^{(1)}(\lambda_* = 1) = \begin{pmatrix} 1 - \alpha & -1 \\ 2 - \alpha & -1 \end{pmatrix}, \tag{2.122}
$$

and $\mathcal{M}_N(\lambda_* = 1) \equiv [\mathbf{R}_A^{(1)}(\lambda_* = 1)]^{n_A}[\mathbf{R}_B^{(1)}(\lambda_* = 1)]^{n_B} = (-1)^{n_A}[\mathbf{R}_B^{(1)}(\lambda_* = 1)]^{n_B}$. Since $\mathbf{R}_B^{(1)}(\lambda_* = 1) \in SL(2, \mathbb{R})$ we can use the Cayley-Hamilton theorem to obtain

$$
\begin{aligned}
[\mathbf{R}_B^{(1)}(\lambda_* = 1)]^{n_B} &= U_{n_B-1}(-\alpha/2) \begin{pmatrix} 2z + 1 & -1 \\ 2 - \alpha & -1 \end{pmatrix} - U_{n_B-2}(-\alpha/2) \begin{pmatrix} 1 & 0 \\ 0 & 1 \end{pmatrix} \\
&= \begin{pmatrix} U_{n_B} + U_{n_B-1} & -U_{n_B-1} \\ (2 - \alpha)U_{n_B-1} & -U_{n_B-1} - U_{n_B-2} \end{pmatrix},
\end{aligned}
$$

where we made use of Eq. (2.36). Therefore

$$
\mathcal{M}_N(\lambda_* = 1) \equiv (-1)^{n_A} \begin{pmatrix} U_{n_B} + U_{n_B-1} & -U_{n_B-1} \\ (2 - \alpha)U_{n_B-1} & -U_{n_B-1} - U_{n_B-2} \end{pmatrix}. \tag{2.123}
$$

We note that, since $\det \mathbf{R}_B^{(1)}(\lambda_* = 1) = 1$, then $\det \mathbf{M}_N(\lambda_* = 1) = 1$.

(c) From the knowledge of the global transfer matrix the transmission coefficient is obtained from Eq. (2.104), where the dispersion relation for the frequency $\lambda_* = 1$ reads $\sin(qa_A) = 1$ (we assume the Fibonacci superlattice (FSL) is embedded between A layers with $a_A = 1$). Then, plugging the global transfer matrix elements given by Eq. (2.123) into Eq. (2.104) we get

$$T_N(\lambda_* = 1) = \frac{1}{1 + \frac{2\alpha^2 - 6\alpha + 5}{4} U_{n_B-1}^2(\alpha/2)}, \qquad (2.124)$$

where we have used the relationships $U_{n_B} - U_{n_B-2} = 2T_{n_B}$ and $T_{n_B}^2 = 1 + (z^2 - 1)U_{n_D-1}^2$ involving Chebyshev polynomials of first and second kind.

2.14 *Show that the Bloch function given by Eq. (2.77) describes an almost periodic function.*

After Eq. (2.77) we have

$$|\psi_\mathbf{k}(\mathbf{r} + m\mathbf{a}) - \psi_\mathbf{k}(\mathbf{r})| = |e^{i\mathbf{k}\cdot\mathbf{r}}||u_\mathbf{k}(\mathbf{r})||e^{im\mathbf{k}\cdot\mathbf{a}} - 1|, \qquad (2.125)$$

where $m \in \mathbb{Z}$. Therefore,

$$|\psi_\mathbf{k}(\mathbf{r} + m\mathbf{a}) - \psi_\mathbf{k}(\mathbf{r})| \leq \sup |u_\mathbf{k}(\mathbf{r})||e^{im\mathbf{k}\cdot\mathbf{a}} - 1|. \qquad (2.126)$$

Now, for any value of the electron energy fixing the $\mathbf{k}$ value, m can be chosen in such a way that $|e^{im\mathbf{k}\cdot\mathbf{a}} - 1|$ is arbitrarily small, so that $\sup |u_\mathbf{k}(\mathbf{r})||e^{im\mathbf{k}\cdot\mathbf{a}} - 1| < \epsilon$, for any given $\epsilon > 0$, in agreement with the definition of an almost periodic function given in Sec. 1.3.2.

2.15 *Derive Eq. (2.107).*

In the case of the silver mean based QC the global transfer matrix, evaluated at the resonance energy E_*, can be expressed as $\mathcal{M}_{N_2}^{(2)}(E_*) \equiv \left[\mathbf{R}_A^{(2)}(E_*)\right]^l \left[\mathbf{R}_B^{(2)}(E_*)\right]^k$, where l and k are the numbers of $\mathbf{R}_A^{(2)}$ and $\mathbf{R}_B^{(2)}$ blocks, respectively. In order to explicitly calculate these power matrices we will rewrite $\mathbf{R}_A^{(2)}$ in the general form $\mathbf{R}_A^{(2)} = \mathbf{T}_{AAB}^{-1} \mathbf{R}_A^{(1)} \mathbf{T}_{AAB}$, relating the corresponding elemental block matrices in the silver mean and FQCs (compare the first two rows in Table 2.6). On the other hand, we can express (see **Exercise 2.8**) [390],

$$\mathbf{R}_A^{(1)}(E_*) = \begin{pmatrix} q & -\gamma_A \\ \gamma_A^{-1} & 0 \end{pmatrix}^3 = \begin{pmatrix} U_3(x) & -\gamma_A U_2(x) \\ \gamma_A^{-1} U_2(x) & -U_1(x) \end{pmatrix}, \qquad (2.127)$$

where $x \equiv q/2 = \epsilon\gamma_A(1 - \gamma_A^2)^{-1}$, and we have used the Cayley-Hamilton theorem. Now, we can use Eq. (2.21) in $\mathbf{R}_A^{(2)} = \mathbf{T}_{AAB}^{-1} \mathbf{R}_A^{(1)} \mathbf{T}_{AAB}$, to obtain

$$\begin{aligned}
\mathbf{R}_A^{(2)}(E_*) &= \begin{pmatrix} 0 & 1 \\ -\gamma_A^{-1} & 2x \end{pmatrix} \begin{pmatrix} U_3(x) & -\gamma_A U_2(x) \\ \gamma_A^{-1} U_2(x) & -U_1(x) \end{pmatrix} \begin{pmatrix} 2x\gamma_A & -\gamma_A \\ 1 & 0 \end{pmatrix} \\
&= \begin{pmatrix} U_3(x) & -U_2(x) \\ U_2(x) & -U_1(x) \end{pmatrix}.
\end{aligned}$$

Since $\det\left[\mathbf{R}_A^{(2)}(E_*)\right] = U_2^2(x) - U_1(x)U_3(x) = 1$, we can make use of Eq. (2.34) to get

$$\begin{aligned}
\left[\mathbf{R}_A^{(2)}(E_*)\right]^l &= \begin{pmatrix} U_l(y) + U_1(x)U_{l-1}(y) & -U_2(x)U_{l-1}(y) \\ U_2(x)U_{l-1}(y) & -U_1(x)U_{l-1}(y) - U_{l-2}(y) \end{pmatrix} \\
&= \begin{pmatrix} U_{3l}(x) & -U_{3l-1}(x) \\ U_{3l-1}(x) & -U_{3l-2}(x) \end{pmatrix}, \qquad (2.128)
\end{aligned}$$

where $y \equiv \frac{1}{2}\mathrm{tr}(\mathbf{R}_A^{(2)}(E_*)) = (U_3(x) - U_1(x))/2 = T_3(x)$, and we have used the functional relations $U_{n-1}[T_m(x)]\,U_{m-1}(x) = U_{mn-1}(x)$ and $T_m[T_n(x)] = T_{mn}(x)$, along with the relationship $T_n = xU_n - U_{n-2} = U_n - xU_{n-1}$. On the other hand, we have

$$\left[\mathbf{R}_B^{(2)}(E_*)\right]^k = [\mathbf{T}_{AAA}(E_*)]^k = \begin{pmatrix} U_k(x) & -U_{k-1}(x) \\ U_{k-1}(x) & -U_{k-2}(x) \end{pmatrix}. \tag{2.129}$$

Thus, multiplying Eqs. (2.128) and (2.129), and then plugging the resulting $\mathcal{M}_N^{(2)}(E_*)$ matrix elements into Eq. (2.104), using the multiplication formula $U_{m-1}U_{n-1} = (1 - x^2)^{-1}(T_{m-n} - T_{m+n})/2$, after some algebra we finally obtain the closed expression given by Eq. (2.107).

CHAPTER 3

Physical Properties of Intermetallic Quasicrystals

3.1 UNCONVENTIONAL SOLIDS MADE OF METALLIC ATOMS

"If real quasicrystals exist, as suggested by Shechtman et al., they are sure to possess a wealth of remarkable new structural and electronic properties". (Dov Levine and Paul J. Steinhardt, 1984) [486]

"I point out that there is no reason to expect these alloys to have unusual physical properties". (Linus Pauling, 1987) [679]

The discovery, in 1987, of the first thermodynamically stable QCs belonging to the Al-Li-Cu and Zn-Mg-Ga alloy systems made it possible to grow relatively large grains (millimeter size) by conventional solidification techniques in close to equilibrium conditions (Sec. 1.1.4). However, these QCs did not show good enough structural quality to properly disclose the possible existence of quasiperiodicity related effects in their physical properties, since they were usually contaminated with small inclusions of periodic crystal phases, and exhibited a relatively large number of structural imperfections as well. To this end, one should be able to study single-grain samples completely free of secondary phases, without phasonic defects or chemical disorder, with no extensive crack networks nor great voids, and exhibiting a robust enough long-range QPO.

The discovery of second generation thermodynamically stable QCs of *high structural quality* in the Al-Cu-(Fe,Ru,Os), Al-Pd-(Mn,Re), Zn-Mg-(Y,Gd,Tb,Dy,Ho,Er), and Cd-(Yb,Ca) icosahedral alloy systems, as well as in the Al-Co-(Cu,Ni) decagonal alloy system, during the late 1980s and the 1990s, made it possible to perform accurate enough experimental measurements of *intrinsic* transport properties of QCs. Accordingly, detailed studies of the temperature dependence of specific heat, electrical, optical, and thermal conductivities, Hall and Seebeck coefficients, along with magnetic, mechanical and tribological properties, were progressively reported. The obtained results clearly indicated that these materials occupy an odd position among the well-ordered condensed matter phases. In fact, since QCs consist of metallic elements one would reasonably expect they should behave as metals. Nonetheless, as we will describe in this Chapter, it is now well established that most physical properties of stable intermetallic QCs are quite unusual by the standard of common metallic alloys. This leads us to face a *second paradox*,[1] namely, that QCs are a

---

[1]The first one was that posed by the very existence of intermetallic QCs exhibiting symmetries not compatible with lattice periodicity.

TABLE 3.1 Comparison between the physical properties of intermetallic QCs versus typical metallic materials. I (S) stands for ionic (semiconducting) materials typical properties. (*) Reported for approximant phases only to date

PROPERTY	METALS	QUASICRYSTALS
MECHANICAL	Ductility, malleability	Brittle (I)
TRIBOLOGICAL	Relatively soft	Very hard (I)
	Moderate friction	Low friction coefficient
	Easy corrosion	Corrosion resistant
ELECTRICAL	High conductivity	Low conductivity (S)
	Resistivity increases with T	Resistivity decreases with T (S)
	Small thermopower	Moderate thermopower (S)
MAGNETIC	Paramagnetic	Diamagnetic, antiferromagnetic
	Ferromagnetic	Ferri- or ferromagnetic (*)
THERMAL	High conductivity	Very low conductivity (I)
	Large specific heat	Small specific heat
OPTICAL	Drude peak	No Drude peak, IR absorption (S)

peculiar class of intermetallic compounds which do not behave as standard metallic alloys usually do.

In fact, by the light of the experimental knowledge gained about the intrinsic properties of stable QCs during the last three decades, it is currently well documented that they exhibit a significant number of unique physical properties, which significantly differ from those observed in their periodic counterparts within the same alloy phase. For the sake of comparison, in Table 3.1, we list several characteristic physical properties of usual metallic alloys and intermetallic QCs. By inspecting this table one realizes that QCs remarkably depart from standard metallic behavior, resembling either ionic or semiconducting materials, depending on the considered physical property. In particular, the electrical transport properties of thermodynamically stable QCs of high structural quality resemble a more semiconductor-like than metallic character, whereas their thermal transport properties are similar to those observed in insulating materials. Therefore, QCs provide an intriguing example of well ordered solids made of typical metal atoms which do not exhibit most of the physical properties usually signaling the presence of metallic bonding. In fact, many physical properties of QCs seem to be just *opposite* to the typical behavior of metallic materials.[2] Thus, albeit the nature of the structural order present in intermetallic QCs was understood relatively soon, the role that this order plays in their physical properties still remains an open question in the field [210, 548, 562].

3.2 ELECTRONIC STRUCTURE

3.2.1 Hume-Rothery electron concentration rule

The fundamental reason for the existence of the metallic state is that in an isolated metal atom the valence electrons occupy positions close to the upper edge of the potential well, so that the perturbations introduced during the condensation processes by neighboring metal atoms lead to delocalization of the valence electrons [273]. Thus, metallic bonding is described by considering that free electrons negative charge is uniformly distributed

---

[2]We should note here that intermetallic QCs also exhibit a few properties typically associated to metallic behavior, such as a gray metallic luster in the visible spectral window, or a linear current–voltage characteristic curve (Ohm's law, see Sec. 3.4.4).

throughout an ordered array of potentials due to positive ions, and that the total charge of electrons exactly balances that of the ions. Along with the electrostatic energy among electrons and ions one must include in the total energy budget the potential energy of electrons moving in the ionic lattice, the electron–electron interactions and the kinetic energy of electrons. Since no more than two electrons can occupy the same quantum state,[3] the available electronic states are progressively filled up to the so-called Fermi energy level E_F, and the resulting kinetic energy of the electronic system systematically increases with the electron concentration. Accordingly, a variety of metallic phases are stabilized in Nature by mechanisms able to lower the kinetic energy of electrons as much as possible.

Of particular interest to us is the mechanism referred to as the Hume-Rothery concentration rule, since it was systematically used by Tsai and co-workers as an useful guide to discover the second generation of high quality, thermodynamically stable QCs mentioned in Sec. 3.1. This rule refers to the stabilization of certain isostructural alloys at a specific composition-averaged valence value, and it was introduced by William Hume-Rothery (1899–1968) in 1926, when he reported that certain metallic compounds with closely related structures (but apparently unrelated stoichiometries) exhibit the same ratio of number of valence electrons to number of atoms, namely, the so-called electron–per-atom ratio, e/a [610, 611]. For example, the isostructural compounds with compositions CuZn, Cu_3Al, and Cu_5Sn share the ratio $e/a = 3/2$, if one considers the valence values Cu = 1, Zn = 2, Al = 3, and Sn = 4.[4] A more striking example is provided by the so-called γ-alloys, such as Cu_5Zn_8, Cu_9Al_4, or $Cu_{31}Sn_8$, sharing the ratio $e/a = 21/13$. We note that the numerator and denominator of this e/a ratio are successive terms in the Fibonacci sequence, so that the electron per atom value of γ-alloys is relatively close to the golden mean value $(1+\sqrt{5})/2 \simeq 1.6180$, as expected after Eq. (1.4).

Hume-Rothery rule can be explained as resulting from a perturbation of the kinetic energy of the valence electrons due to their diffraction by the crystal lattice when an electron has such a de Broglie wavelength[5] and direction as to fit the Bragg reflection condition $n\lambda = 2d\sin\theta$, $n \in \mathbb{N}$, where d measures the distance between suitable crystallographic planes, and 2θ is the angle between the incident and the reflected electron. The perturbation is of such nature as to stabilize electrons with energies close to that corresponding to Bragg reflection and to destabilize electrons with a larger energy. Hence, special stability would be expected for metals with just the right number of electrons. This number is proportional to the volume of a polyhedron in reciprocal space (the so-called Brillouin zone), corresponding to the crystallographic planes giving rise to the perturbation. For instance, the Brillouin polyhedron for the γ-alloys is bounded by twelve {330} and twenty-four {411} planes, as derived from X-ray diffraction data. Due to their great symmetry, this zone is quite close to spherical shape, so that the diffraction condition can be expressed in the form (often referred to as the Hume-Rothery matching condition)

$$K_{hkl} = 2k_F, \qquad (3.1)$$

where K_{hkl} is the reciprocal vector of the considered diffraction plane, $k_F = \sqrt[3]{3\pi^2 n}/a_0$ is the radius of the Fermi sphere, a_0 is the lattice parameter, and $n = (e/a)N$ is the average electron number per unit volume, where N is the number of atoms in the unit cell (**Exercise 3.1**). Therefore in Hume-Rothery alloys the redistribution of electronic states due to the Fermi sphere-Brillouin zone interaction gives rise to a significant reduction of the density of states (pseudogap) close to the Fermi energy, thereby stabilizing the resulting compound as

---

[3] Due to Pauli's exclusion principle.

[4] Other alloys which may be placed in the same class are CuBe, AgZn, AgCd, AgMg, Ag_3Al, and AuZn.

[5] That is, $\lambda = h/\sqrt{2mT}$, where h is the Plack's constant, m is the electron mass, and T its kinetic energy.

1	2											13	14	15
Li 1.02	Be 2.00											B 2.98	C 3.92	N
Na 1.01	Mg 2.01	3	4	5	6	7	8	9	10	11	12	Al 3.01	Si 4.00	P 4.97
K 1.01	Ca 2.00 1.56	Sc 2.94 1.33	Ti 1.14	V 0.90	Cr 0.92	Mn 1.05	Fe 1.05	Co 1.03	Ni 1.16	Cu 1.00	Zn 2.04	Ga 3.00	Ge 4.05	As 4.92
Rb 1.01	Sr 1.96	Y 3.15 1.87	Zr 1.49	Nb 1.32	Mo 1.39	Tc 0.95	Ru 1.04	Rh 1.00	Pd 0.96	Ag 1.01	Cd 2.03	In 3.03	Sn 3.97	Sb 4.99
Cs 1.04	Ba 2.03	La 3.00	Hf 1.76	Ta 1.57	W 1.43	Re 1.40	Os 1.55	Ir 1.60	Pt 1.63	Au 1.00	Hg 2.03	Tl 3.03	Pb 4.00	Bi 4.94

Figure 3.1 Electron per atom concentration ratio e/a for 54 elements in the periodic table. The value in the top (bottom) level for Ca, Sc, and Y is used when these atoms are alloyed with non-transition metal (transition metal) elements, respectively. (Reprinted from Mizutani U and Sato H, The physics of the Hume-Rothery electron concentration rule, *Crystals* **7**, 9 (2017); doi:10.3390/cryst7010009, Creative Commons Atribution License CC BY 4.0).

a result of transferring electrons with the highest kinetic energies into deeper, low energy states in the energy spectrum. For instance, it has been estimated that the formation of a pseudogap at the Fermi level with a width of 0.5 to 1 eV and a height 0.2 to 0.6 times as high as the typical free electron DOS can lower the electronic energy by 30 to 50 kJmol^{-1} relative to the free electron value (**Exercise 3.2**) [608, 611].

Although ideal QCs have a dense reciprocal space, only a relatively small number of Bragg reflections have very strong intensities in actual diffraction patterns. The Hume-Rothery criterion can then be applied to QCs by introducing a pseudo-Brillouin zone related to this set of intense enough reflections. Indeed, Eq. (3.1) can be properly used to explain the stability of iQCs containing elements with a full d-band, such as $Al_{56}Li_{33}Cu_{11}$ ($e/a = 2.129$), $Zn_{43}Mg_{37}Ga_{20}$ ($e/a = 2.221$), $Zn_{60}Mg_{30}(RE)_{10}$ ($e/a = 2.127$), or $Zn_{80}Sc_{15}Mg_5$ ($e/a = 2.039$), by adopting the valence values given in Fig. 3.1, along with RE = 3.[6] For alloys containing transition elements the Hume-Rothery concentration rule was tentatively applied by assuming that transition atoms adopted a negative effective valence, following an approach earlier introduced by Geoffrey V. Raynor (1913–1983).[7] In so doing, the values $e/a = 1.751$ and $e/a = 1.734$ were obtained for i-$Al_{65}Cu_{20}Fe_{15}$ and i-$Al_{70}Pd_{20}Mn_{10}$, respectively, which are significantly lower than the electron per atom ratios derived before for stable QCs not containing transition metal atoms (i.e., $e/a = 2.04 - 2.22$). Nonetheless, recent studies by U. Mizutani and H. Sato clearly indicate the convenience of abandoning the use of Raynor's valence values in the study of Hume-Rothery compounds,

---

[6] We note that the e/a values obtained for these QCs are significantly higher than those obtained for $\alpha-$, $\beta-$, and $\gamma-$brass alloys (cf. **Exercise 3.1**).

[7] The Raynor's valence values for some typical transition metal atoms read Cr = −4.66, Mn = −3.66, Fe = −2.66, Co = −1.61, Ni = −0.71, and Pd = 0.0.

and to properly distinguish the electron concentration ratio e/a, derived from the Fermi sphere radius in reciprocal space, from the chemical valence of the element, which is defined in real space to measure its bonding power with neighboring atoms when it forms chemical compounds (**Exercise 3.1**). Accordingly, the e/a values for 54 elements covering from Group 1 up to Group 16 in the periodic table (including 3d-, 4d-, and 5d-elements) have been determined in a self-consistent manner by performing detailed band structure calculations. In this way, the origin of a pseudogap at the Fermi level for a large number of compounds has been successfully interpreted in terms of the Fermi-surface-Brillouin zone interference condition, regardless of the chemical bond-types involved [611]. Taking into account the updated e/a values given in Fig. 3.1, one gets $e/a = 2.314$ and $e/a = 2.404$ for i-$Al_{65}Cu_{20}Fe_{15}$ and i-$Al_{70}Pd_{20}Mn_{10}$, respectively, figures which come closer to those previously obtained for iQCs bearing elements with a full d-band. It is interesting to note that the electron per atom ratio of mineral icosahedrite (i-$Al_{62}Cu_{31}Fe_7$, $e/a = 2.250$) is not so different to that of i-$Al_{65}Cu_{20}Fe_{15}$ synthetic alloy, albeit the former contains far less iron content [345].

In general the presence of a pseudogap at the Fermi level can be due to two different contributions. For one thing, we have the interference phenomenon of electrons with certain set of lattice planes, which is very sensitive to the long-range order of the considered solid (either periodic or QP). On the other hand, we have the orbital hybridization effect between the unlike constituent atoms in a given solid, which stems from the local atomic distribution, and it is not significantly dependent on the long-range order [611]. Thus, in QCs bearing metal atoms contributing d orbitals to the electronic structure, such as i-AlCu(Fe,Ru,Os) or i-AlPd(Mn,Re), the presence of hybridization effects between sp aluminium states and these d states adds up to the structure related Fermi surface-Brillouin zone diffraction mechanism, further deepening the pseudogap close to the Fermi level. In fact, the role of sp-d hybridization effects in both cohesion energy and transport properties has been demonstrated for a series of QCs belonging to the AlCu(Fe,Ru) and AlPd(Mn,Re) systems [259, 448, 607, 897, 898]. On the other hand, it was reported that the role played by hybridization effects in the stability of the i-Cd(Yb,Ca) phase is significantly larger than that resulting from the Fermi-surface-Brillouin zone mechanism in this binary QC [364, 871]. In this case, the orbitals involved in the hybridization process come from occupied Cd-5p and unoccupied Yb-5d (or Ca-3d) orbitals, which highlights the importance of chemical bonding related effects in these quasicrystalline compounds. Indeed, the influence of sp-d hybridization on the electronic structure of different Al-Mn alloys has been recently studied by photoelectron spectroscopy, and it has been confirmed that these hybridization effects alone suffice to give rise to a pseudogap, even in the absence of a Fermi surface-Brillouin zone mechanism [804].

These results inspired a chemical synthesis exploration project aimed at obtaining new QCs via pseudogap electronic tuning in the Zn-Sc-Cu, Ca-Au-In, and Mg-Cu-Ga systems[499, 500–502]. As a working hypothesis it was assumed that, for a given alloy composition, the Fermi surface-Brillouin zone mechanism give rise to the formation of pseudogaps at energy values which are *not close enough* to the Fermi energy to significantly reduce its electronic energy. Now, one can manage to shift E_F close to that DOS minimum by properly alloying the original compound with electron rich (alternatively, poor) atoms in order to properly match E_F inside the deeper region of the pseudogap. This strategy was first applied to $Zn_{17}Sc_3$ alloy, whose structure is isotypic with that of the prototypical Tsai-type 1/1 approximant. The $Zn_{17}Sc_3$ alloy has an electron concentration ratio $(e/a = 2.175)$[8] somewhat larger than that observed in $Cd_{85}(Ca,Yb)_{15}$ iQCs $(e/a = 2.026)$, but substitution

---

[8]If not otherwise stated all the e/a values reported in this book are obtained making use of the data given in Fig. 3.1.

of some Zn by Cu allowed to tune the e/a value to obtain a novel icosahedral phase with the stoichiometry $Zn_{71.5}Sc_{16.2}Cu_{12.3}$ ($e/a = 2.058$). The electronic tuning via compositional change route was subsequently applied to $Zn_{11}Mg_2$ ($e/a = 2.035$) alloy precursors to get the quaternary i-$Cu_{48}Ga_{34}Sc_{15}Mg_3$ ($e/a = 2.001$) [500], and the ternary i-$Zn_{82.1}Sc_{14.6}Mg_{3.3}$ ($e/a = 2.170$) and i-$Au_{44.2}In_{41.7}Ca_{14.1}$ ($e/a = 1.988$) compounds [501, 502]. In the same vein, the systematic exploration of the Na-Au-Ga system led to the discovery of the first Na-containing QC, i-$Ga_{37.5}Na_{32.5}Au_{30}$, which has a lower $e/a = 1.753$ value [820]. Very recently, the existence of the stable QC i-$Au_{62.7}Al_{23.0}Ca_{14.3}$, with the even lower $e/a = 1.605$ value has been reported [690].

Another route to obtain new QCs is by exploring the confluence zone between icosahedral and decagonal phases in the phase diagram of suitable systems, such as Al-Cu-Fe (for stable iQCs) and Al-Cu-Co (for stable dQCs). In this way, the i-$Al_{64.4}Cu_{22.2}Fe_{10.4}Co_{3.0}$ ($e/a = 2.230$) and $Al_{64.3}Co_{18.3}Cu_{17.3}Fe_{0.1}$ ($e/a = 2.402$) thermodynamically stable, single grain (~ 1 mm size) QCs, have been reported [430]. Thus, at high level of Co substitution with Fe in the ternary d-AlCuCo a quaternary i-AlCu(Co,Fe) QC is formed. Following a completely different approach, aimed at understanding the possible origin of the mineral QCs found in Khatyrka meteorite (see Sec. 1.2), several Al-based alloys were stacked together and shocked in a stainless steel chamber intended to replicate a natural shock that affected the Khatyrka meteorite. Abundant dQCs of composition $Al_{73}Ni_{19}Fe_4Cu_2Mg_{0.6}Mo_{0.4}Mn_{0.3}$ ($e/a = 2.500$) were obtained, showing essentially no measurable phason defects, thereby suggesting that shock synthesis grows essentially strain-free, highly perfect QCs from the onset [661]. This composition is close to that of mineral decagonite, d-$Al_{71}Ni_{24}Fe_5$ ($e/a = 2.468$, see Sec. 1.2), although the synthetic QCs include a significant number of minor components.[9] In a similar way, iQCs with an average composition $Al_{72}Cu_{12}Fe_{12}Cr_3Ni_1$ ($e/a = 2.452$) were found after shocking metallic $CuAl_5$ and $(Mg_{0.75}Fe_{0.25})_2SiO_4$ olivine in the stainless steel chamber [662]. Subsequent shock-recovery experiments, that used AlCuW graded density impactors as starting materials in the targets produced an assemblage of co-existing i-$Al_{61.5}Cu_{30.3}Fe_{6.8}Cr_{1.4}$ ($e/a = 2.238$) or i-$Al_{68.6}Fe_{14.5}Cu_{11.2}Cr_{8.0}Ni_{1.8}$ ($e/a = 2.387$) QCs, together with periodic crystalline phases [345]. These results suggest that even small amounts of additional TM atoms may significantly expand the stability range of both iQC and dQC alloy forming systems, under conditions very different from those of conventional metallurgical processing.

3.2.2 Density of states around the Fermi energy

The suitability of the Hume-Rothery mechanism to account for QCs stability was confirmed by earlier electronic band structure calculations for several crystalline approximants, including R-AlCuLi, α-AlMnSi, 1/1-AlCuFe, or $Al_{13}Fe_4$ [267]. These numerical studies also predicted the existence of a number of very narrow peaks, over an energy scale of about $1 - 10$ meV, in the electronic DOS close to the Fermi level, which was referred to as a spiky feature component of the energy spectrum [268], and it was argued that these peaks may stem from the structural quasiperiodicity of the substrate due to cluster aggregation [375], along with d-orbital resonance effects [899]. Later on, on the basis of extensive *ab-initio* calculations of several QC approximants, it was claimed that the spiky DOS component may probably stem from numerical artifacts [186, 1005–1008]. In any event, the crucial point was to ascertain as to whether the pseudogap at the Fermi energy and the spiky features derived from numerical calculations for crystalline approximant models really existed in actual QCs or not.

---

[9]We note that this compound may be considered as a possible high-entropy alloy representative.

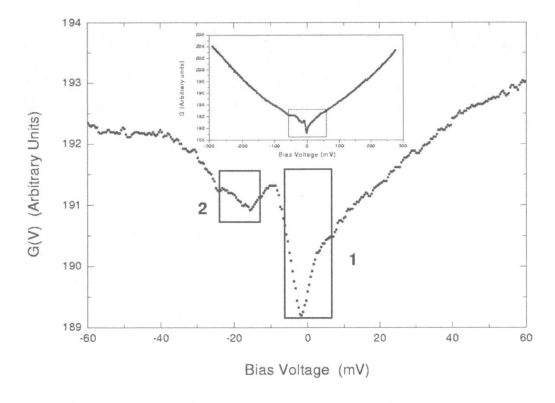

Figure 3.2 Differential conductance for an i-$Al_{63}Cu_{25}Fe_{12}$-Al tunnel junction at a temperature of $T = 2$ K over two different energy scales: ± 60 meV (main frame) and ± 300 meV (inset). Experimental data by courtesy of Roberto Escudero. (Reprinted figure with permission from Maciá E, Modeling the thermopower of icosahedral AlCuFe quasicrystal: Spectral fine structure, *Phys. Rev. B* **69** 132201 (2004). Copyright (2004) by the American Physical Society).

The existence of a pseudogap in the i-AlMn metastable QC was first reported by Esther Belin-Ferré (1939–2018) and her associates in 1988 from soft X-ray emission spectroscopy measurements indicating that the DOS at E_F was substantially depressed relative to its amorphous and crystalline counterparts. Subsequently, the presence of pseudogaps close to E_F in all the studied stable QCs was well established by using different experimental techniques, including high energy photoemission [648], soft X-ray [44, 45], scanning tunneling [162, 426], and point contact spectroscopies [238], along with magnetic susceptibility and nuclear magnetic resonance (NMR) measurements. In this way, the size and shape of the electronic DOS around the Fermi energy was determined in detail, revealing the presence of a relatively broad and shallow reduction of the DOS extending over about 1 eV around the Fermi energy in i-AlCuFe and i-AlPdRe QCs.[10] Indeed, tunneling spectroscopy at low temperatures gave evidence for a symmetric narrow dip of about 50–60 meV wide located very close to the Fermi level [162, 426]. For the sake of illustration in Fig. 3.2, we show low temperature tunneling spectroscopy measurements corresponding to

---

[10]It is worth noticing that the presence of a pseudogap is a generic feature of QCs, but it is not a specific one, since certain periodic crystals can also exhibit a substantial depletion of the electronic DOS close to the Fermi level [609]

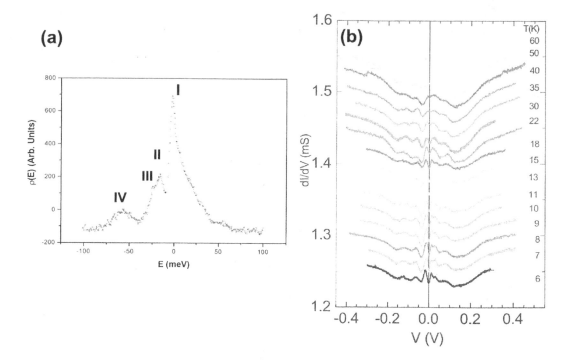

Figure 3.3 (a) Detailed view of the fine structure in the spectral resistivity curve corresponding to the sample shown in Fig. 3.2. To this end, the background DOS has been fitted to a parabolic curve ($r = 0.9649$) and it has been substracted to the original tunnel data (kindly provided by Roberto Escudero). Four Lorentzian peaks can be fitted to the reduced data ($r = 0.9672$), which are labeled with roman numerals from right to left. (b) Differential conductance curve obtained by point contact tunneling spectroscopy on a 5-fold surface of an i-ZnMgTb single grain QC at different temperatures [239]. (Courtesy of Roberto Escudero).

an i-$Al_{63}Cu_{25}Fe_{12}$ sample [238]. These measurements reveal a broad pseudogap extending over an energy scale of about 0.6 eV (shown in the inset) along with some fine structure close to the Fermi level (labeled 1 and 2 in the main frame). In Fig. 3.3a, we display a more detailed view of these fine spectral features, where the background DOS contribution has been removed. In this way, we can identify four Lorentzian peaks in the spectral resistivity curve located at energies $E_I = -2.00$, $E_{II} = -16.11$, $E_{III} = -24.37$, and $E_{IV} = -56.59$ meV with respect to the Fermi level. The existence of a sharp DOS dip of about 20 meV wide at the Fermi level in both quasicrystalline and approximant phases have been confirmed by nuclear magnetic resonance studies, which probe the bulk properties of the considered samples [877]. These observations then confirm that the dip centered at the pseudogap is not a surface feature and that both its width and depth are sample dependent. On the other hand, the dependence of the pseudogap structure with the temperature was investigated by means of tunnelling and point contact spectroscopy, and it was reported that the width of the broad pseudogap component remains essentially unmodified as the temperature is increased from 4 K to 77 K. On the contrary, the dip feature centered at the Fermi level exhibits a significant modification, progressively deepening and narrowing as the temperature is decreased. In Fig. 3.3b, we show the variation of the differential conductance curve around the Fermi energy as a function of the temperature for a i-ZnMgTb QC. At low temperatures several peaks and dips can be seen close to E_F along with a narrow dip centered at the Fermi energy. As the temperature is increased the central dip progressively fills until it completely disappears for $T \gtrsim 35$ K. Quite interestingly the dip located on the left of the central dip (at about -0.04 eV) is quite robust and remains almost unchanged upon temperature variation, hence suggesting this feature may be related to localized electronic states stemming from bonding hybridization effects [238].

At the time being, however, the possible existence of the DOS spiky component is still awaiting for a definitive experimental confirmation [505]. In fact, difficulties in the experimental investigation of fine structure in the DOS arise form the requirement of a high energy resolution, as the peaks and gaps to be observed are only a few meV wide. Thus, both high resolution photoemission and tunneling spectroscopies have failed to detect the theoretically predicted dense distribution of spiky features around the Fermi level. Several reasons have been invoked in order to explain these unsuccessful results. Among them the existence of some residual disorder present even in samples of high structural quality has been invoked as a plausible agent to smear out the finer details of the DOS [830]. It has also been argued that photoemission and STM techniques probe the near surface layers, so that sharp features close to the pseudogap could be removed by subtle structural deviations near the surface from that of the bulk [219].

Notwithstanding this, some tunnelling spectroscopy measurements performed in iQCs at low temperatures (2–10 K) have provided experimental support for the existence of a number of energetically localized features close to the Fermi level in the electronic structure of i-AlPdMn, i-AlPdRe, and i-ZnMgTb samples at certain *local regions* (with a spatial extent of about 0.5–1 nm) [238, 239, 958], though when the observations are performed at larger spatial scales the local DOS is averaged and the finer structure is smeared out [483, 573]. In addition, it has been reported that, in single grain samples, the DOS structure depends on the symmetry (2-, 3-, or 5-fold) of the considered surface, so that the number of observed spectral features increases as the rotation axis order increases [239].

By collecting all the relevant information provided by both experimental measurements and numerical calculations, one can devise a DOS structure model near the Fermi level including three different contributions at $T = 0$, as it is sketched in Fig. 3.4a [531]. Firstly, we have the contribution due to a relatively broad pseudogap (~ 1 eV width) which, according to high resolution STM and NMR measurements, contains two main features: (1) a narrow ($\sim 0.06 - 0.02$ eV) and symmetric parabolic dip, $N_d(E)$, located very close to the Fermi energy and, (2) a square root term, $N_{sr}(E)$, beyond the narrow dip region. Thus, the DOS around the Fermi energy will be given by the function

$$\begin{cases} N_d(E) = N_a + \Omega E^2, & |E| \le b_W/2 \\ N_{sr}(E) = d + c\sqrt{|E|}, & |E| > b_W/2 \end{cases} \tag{3.2}$$

where N_a is the DOS value at the origin of the energy scale, $\Omega \equiv \frac{1}{2}\left(\frac{d^2N}{dE^2}\right)$ measures the curvature of the dip, b_W is the dip width, and the constants $c = \Omega b\sqrt{2b_W}$ and $d = a - 3\Omega b_W^2/4$ guarantee the derivability and continuity of the DOS at $E = b_W/2$. The importance of the DOS local curvature close to E_F can be realized by comparing the electronic energy gain derived from the electronic structure model given by Eq. (3.2) (**Exercise 3.3**), with that corresponding to the rectangular pseudogap model shown in Fig. 3.31 (**Exercise 3.2**).

We note that the model function given by Eq. (3.2) displays even symmetry with respect to the DOS minimum, a feature that is not supported by tunnel spectroscopy

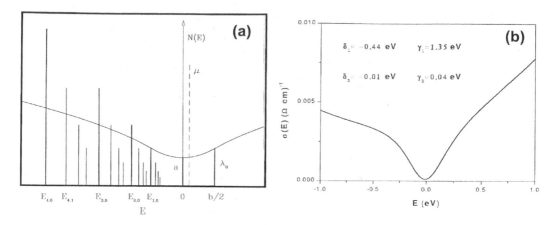

Figure 3.4 (a) Diagram showing the main contributions to the electronic structure of iQCs close to the Fermi energy μ. (Reprinted figure with permission from Maciá E, Modeling the electrical conductivity of icosahedral quasicrystals, Physical Review B **61** 8771-8777 (2000) Copyright (2000) by the American Physical Society). (b) Spectral conductivity curve in the energy interval ± 1 eV around the Fermi level as obtained from Eq. (3.6) for the electronic model parameter values γ_i and δ_i indicated in the frame.

obtained differential conductance curves (see Figs.3.2 and 3.3), nor hard X-ray photoemission measurements [648]. In order to break this symmetry, we may consider the linearly modulated Lorentzian curve [648]

$$N_{ps}(E) = (a'E + b')\left(1 - \frac{c'\Gamma^2}{E^2 + \Gamma^2}\right), \qquad (3.3)$$

which is usually used to reduce the data obtained from photoemission experiments, where a', b', and c' are suitable parameters, and 2Γ is the Lorentzian's full width at the half-maximum. Alternatively, on the basis of the plot shown in Fig. 3.3a we may consider an additional contribution to the DOS due to a self-similar distribution of spiky features given by the expression

$$N_{ss}(E) = \sum_{n=1}^{M} \sum_{j=0}^{M-1} \lambda_{n,j}\,\delta(E - E_{n,j}), \qquad (3.4)$$

where $\lambda_{n,j} \equiv \eta^{n-j-1}\lambda_0$, with $\lambda_0 \equiv N_d(b_W/2)$, measures the strength of the self-similar peaks, and the series $E_{n,j} \equiv -\eta^{n-2}\left[1 + \eta^{-j}(\eta - 1)\right]b_W/2$ determines their positions (see Fig. 3.4a), where the inflation factor $\eta > 1$ is related to the QC structure [375, 376, 377]. This self-similar structure includes M main peaks, labeled by the integer n, and $M(M-1)$ subsidiary peaks, labeled by pairs (n, j). However, we must keep in mind that, albeit some fine spectral features have been identified in the DOS of some iQCs, the number of observed features does not allow for any reliable assessment on possible self-similar fingerprints in their energy distribution and/or depth values of the dips.

3.2.3   Spectral conductivity function

The transport properties of a given material greatly depend on the scattering processes determined by the mutual interactions between the carriers and the lattice, as well as by the presence of impurities and other structural defects, which often vary themselves with both temperature and external force fields. In order to properly include the physical processes determining the temperature dependence of different transport coefficients, such as the electrical and thermal conductivities or the thermoelectric Seebeck coefficient, it is convenient to introduce a characteristic function, which entails detailed information on the electronic structure of the material. This function is referred to as the *spectral conductivity function*, $\sigma(E) > 0$ (measured in $\Omega^{-1}\mathrm{cm}^{-1}$ units), which is defined as the $T \to 0$ conductivity with the Fermi level at energy E. For systems which can be described within the Boltzmann approach, one has $\sigma(E) = e^2 \tau(E) n(E) v^2(E)$, where $\tau(E)$ is the relaxation time, $n(E)$ measures the charge carriers density (in cm^{-3}), and $v(E)$ is the group velocity of the carriers. In the case of systems for which the applicability of the Boltzmann approach is not guaranteed, as it seems to be the case for the more resistive QCs, one can consider the more general relationship

$$\sigma(E) = \frac{e^2}{V} N(E) D(E), \tag{3.5}$$

where $N(E)$ is the DOS, $D(E)$ measures the diffusivity of the states (in $\mathrm{cm}^2\mathrm{s}^{-1}$) and V is the system volume. Although it may be tempting to assume that the $\sigma(E)$ function should closely resemble the overall structure of the DOS, it has been shown that dips in the $\sigma(E)$ curve can correspond to peaks in the DOS at certain energies in QCs [747]. This behavior is likely to be related to the peculiar nature of critical electronic states close to the Fermi level (see Sec. 2.6).

From systematic studies of the electrical conductivity $\sigma(T)$, thermoelectric Seebeck coefficient $S(T)$, and Hall coefficient $R_H(T)$ curves of Al-based iQCs, it has been concluded that the main qualitative features of these transport coefficients can be accounted for by considering an *asymmetric* spectral conductivity function characterized by a broad minimum exhibiting a pronounced dip within it [826, 827, 847]. In fact, a series of *ab-initio* studies have shown that the electronic structure of both QCs and approximant phases belonging to the i-AlCu(Fe,Ru) and i-AlPd(Mn,Re) representatives can be satisfactorily described in terms of a spectral conductivity function exhibiting two basic spectral features close to the Fermi level,[11] namely, a wide and a narrow Lorentzian peaks, according to the expression [473, 474, 826, 827],

$$\sigma(E) = \bar{\sigma} \left\{ \frac{\gamma_1}{(E - \delta_1)^2 + \gamma_1^2} + \frac{\alpha \gamma_2}{(E - \delta_2)^2 + \gamma_2^2} \right\}^{-1}. \tag{3.6}$$

This model includes six parameters, determining the Lorentzian's heights ($\bar{\sigma}/\gamma_i$) and widths ($\sim \gamma_i$), their positions with respect to the Fermi level, δ_i, and their relative weight in the overall structure, $\alpha > 0$. The parameter $\bar{\sigma}$ is a scale factor measured in $(\Omega\mathrm{cm}\,\mathrm{eV})^{-1}$ units. Suitable values for these electronic model parameters can be obtained by properly combining *ab-initio* calculations of approximant phases with experimental transport data of icosahedral samples within a phenomenological approach [539, 540, 541, 546]. In Fig. 3.4b, we depict the $\sigma(E)$ curve for a suitable choice of the model parameters. By comparing this figure with Fig. 3.2, we see that Eq. (3.6) properly captures the main spectral features of realistic samples.

From the knowledge of the spectral conductivity function the temperature dependent transport coefficients can be obtained within the framework of the linear response theory.

---

[11] This spectral conductivity model does not hold for the description of either ZnMgRE and Tsai-type cluster based iQCs.

The central information quantities are the kinetic coefficients,[12]

$$\mathcal{L}_{ij}(T) = (-1)^{i+j} \int_{-\infty}^{+\infty} \sigma(E) \, (E-\mu)^{i+j-2} \left(-\frac{\partial f}{\partial E}\right) dE, \qquad (3.7)$$

where $i,j = 1,2$, $\mu(T)$ is the chemical potential (which equates the Fermi energy, E_F^0, at $T = 0$), and $f(E,T)$ is the Fermi-Dirac distribution function given

$$f(E,T) = \frac{1}{1 + e^{(E-E_F(T))\beta}} = \frac{1}{2}\left[1 - \tanh\left(\frac{E - E_F(T)}{2k_B T}\right)\right], \qquad (3.8)$$

where $\beta \equiv (k_B T)^{-1}$, k_B is the Boltzmann constant, and $E_F(T) = E_F^0 - AT^2$, with $A \equiv \frac{\pi^2 k_B^2}{6}\left(\frac{d(\ln N)}{dE}\right)_{E_F^0}$, describes the variation of the Fermi energy with the temperature [27, 608].

In this formulation all the microscopic details of the system are included in the spectral conductivity function $\sigma(E)$. Therefore, the temperature dependence of the transport coefficients appears in the Fermi-Dirac distribution, whereas all peculiarities of the scattering processes are incorporated in $\sigma(E)$. The kinetic coefficients derived from Eq. (3.7) are valid for both extended and localized states. In fact, this formalism has been applied to study the transport properties in QCs and related approximants characterized by critical wavefunctions exhibiting spatial fluctuations at all scales.

The electrical conductivity $\sigma(T)$, electronic thermal conductivity $\kappa_e(T)$, and Seebeck $S(T)$, transport coefficients are related to the kinetic coefficients given by Eq. (3.7) through the expressions[608],

$$\sigma(T) = \mathcal{L}_{11}(T), \quad S(T) = \frac{1}{|e|T}\frac{\mathcal{L}_{12}(T)}{\mathcal{L}_{11}(T)}, \quad \kappa_e(T) = \frac{1}{e^2 T}\left(\mathcal{L}_{22}(T) - \frac{\mathcal{L}_{12}^2(T)}{\mathcal{L}_{11}(T)}\right), \qquad (3.9)$$

and the so-called Lorenz function

$$L(T) \equiv \frac{\kappa_e(T)}{T\sigma(T)}, \qquad (3.10)$$

in a unified way. As a first approximation one generally assumes $E_F(T) \approx E_F^0$. Then, by expressing Eqs.(3.9) in terms of the scaled variable $x \equiv (E - \mu)\beta$, the transport coefficients can be rewritten as [533, 538, 545],

$$\sigma(T) = \frac{J_0}{4}, \quad S(T) = -\frac{k_B}{|e|}\frac{J_1}{J_0}, \quad \kappa_e(T) = \frac{k_B^2 T}{4e^2 J_0}\begin{vmatrix} J_0 & J_1 \\ J_1 & J_2 \end{vmatrix}, \quad L(T) = \frac{k_B^2}{e^2 J_0^2}\begin{vmatrix} J_0 & J_1 \\ J_1 & J_2 \end{vmatrix}, \qquad (3.11)$$

in terms of the reduced kinetic coefficients

$$J_n(T) = \int_{-\infty}^{+\infty} x^n \sigma(x)\,\mathrm{sech}^2(x/2)dx. \qquad (3.12)$$

Analogously, the Hall coefficient can be expressed in the form [695],

$$R_H(T) \simeq -\frac{|e|D_0}{\sigma^2(T)}\int_{-\infty}^{+\infty}\frac{d\sigma}{dE}\left(-\frac{\partial f}{\partial E}\right)dE \simeq -\frac{12|e|D_0}{\pi^2 k_B T}\frac{J_1}{J_0^2}, \qquad (3.13)$$

---

[12]This approach is known as the Chester-Thellung-Kubo-Greenwood formulation [575, 629]. This description is valid provided that the charge carriers are noninteracting among them and the scattering with impurities and lattice phonons is elastic. No assumption is made about the strength of disorder or the nature of the states.

where we have assumed a weak energy dependence of the diffusivity, so that $D(E) \simeq D_0$. In this way, the temperature dependent transport coefficients of QCs can be consistently obtained by employing a realistic model for the spectral conductivity function [66, 475, 541, 546].

3.3 SPECIFIC HEAT MEASUREMENTS

3.3.1 General considerations

Measurements of the low temperature specific heat of solids allow us to determine two important physical magnitudes of the considered sample, namely, the electron density at the Fermi level, $N(E_F)$, and the Debye temperature, Θ_D. To this end, the measured specific heat at constant pressure curve, $C_P(T)$,[13] is divided by the temperature and the resulting $C_P(T)/T$ values are plotted against T^2 to fit the curve

$$\frac{C_P(T)}{T} = \gamma_e + \alpha T^2 + \delta T^4, \tag{3.14}$$

which is interpreted as a sum of two different contributions: the first term, giving the intercept value γ_e, is due to the conduction electrons and the latter two terms stem from lattice dynamics, where αT^2 is related to the linear term in the dispersion relation phonon spectrum, and δT^4 accounts for deviations from this linear behavior due to anharmonic effects. The physical meaning of the empirical fitting coefficients γ_e, α and δ introduced in Eq. (3.14) is closely related to both the electronic and vibrational DOS. Indeed, the internal energy of electrons in a solid at a given temperature reads [27, 608],

$$U_e(T) = \int_{-\infty}^{+\infty} E N(E) f(E, T) dE, \tag{3.15}$$

where E is the electron energy, $N(E)$ is the DOS, and $f(E, T)$ is the Fermi-Dirac distribution function given by Eq. (3.8). The electronic contribution to the specific heat at constant volume is then given by

$$c_e(T) \equiv \left(\frac{\partial U_e}{\partial T} \right)_V = \int_{-\infty}^{+\infty} E N(E) \left(\frac{\partial f}{\partial T} \right) dE. \tag{3.16}$$

In order to simplify the mathematical treatment, it is convenient to introduce the dimensionless scaled energy variable $x \equiv (E - E_F(T))\beta = (E - E_F^0 + AT^2)\beta$, so that

$$\frac{\partial f}{\partial T} = \frac{1}{4T}(x - 2\beta AT^2) \operatorname{sech}^2 \left(\frac{x}{2} \right), \tag{3.17}$$

and Eq. (3.16) adopts the form

$$c_e(T) = \frac{k_B}{4} \int_{-\infty}^{+\infty} \left[\beta^{-1} x^2 + a_1(T)x - \beta a_0(T) \right] N(x) \operatorname{sech}^2 \left(\frac{x}{2} \right) dx, \tag{3.18}$$

where $a_1(T) \equiv E_F^0 - 3AT^2$, and $a_0(T) \equiv 2AT^2 E_F(T)$. The hyperbolic function $\operatorname{sech}^2 (x/2)$

---

[13]In order to compare with theoretical models one must consider the specific heat at constant volume, which is obtained from the experimentally measured $C_P(T)$ data using the relation $C_V(T) = C_P(T) - 9V_a B\bar{\alpha}_L^2 T$, where V_a is the atomic volume, B is the bulk modulus, and $\bar{\alpha}_L$ is the linear thermal expansion coefficient. Fortunately, at low enough temperatures the approximation $C_V(T) \simeq C_P(T)$ can be safely assumed (see Fig. 3.6 inset).

is an even function with respect to the variable x exhibiting a narrow, pronounced peak around the Fermi energy $E = E_F^0$ value. Accordingly, one can expand the DOS in a Taylor series about E_F^0 in the form

$$
\begin{aligned}
N(E) &= N(E_F^0) + (E - E_F^0)\left(\frac{dN}{dE}\right)_{E_F^0} + \frac{1}{2}(E - E_F^0)^2\left(\frac{d^2N}{dE^2}\right)_{E_F^0} + \cdots \\
N(x) &= N(E_F^0) + (x\beta^{-1} - AT^2)\left(\frac{dN}{dE}\right)_{E_F^0} + \frac{1}{2}(x\beta^{-1} - AT^2)^2\left(\frac{d^2N}{dE^2}\right)_{E_F^0} + \cdots \\
&\equiv N_0 + (x\beta^{-1} - AT^2)N_0' + \frac{1}{2}(x\beta^{-1} - AT^2)^2 N_0'' + \cdots
\end{aligned}
\tag{3.19}
$$

Thus, by plugging Eq. (3.19) into Eq. (3.18), keeping terms up to cubic order in the temperature, one gets

$$
\begin{aligned}
c_e(T) &= \frac{\pi^2 k_B^2}{3}N_0 T + \frac{\pi^4 k_B^4}{30}\left(7N_0'' - \frac{10}{3}E_F^0\frac{N_0'N_0''}{N_0} - 5\frac{N_0'^2}{N_0}\right)T^3, \\
&= e^2 L_0 N_0 T + \frac{3}{10}(e^2 L_0)^2\left[7N_0'' - \frac{5N_0'}{3N_0}(3N_0' + 2E_F^0 N_0'')\right]T^3, \\
&\equiv \gamma_e T + \alpha_e T^3,
\end{aligned}
\tag{3.20}
$$

where e is the electron charge and $L_0 \equiv \frac{\pi^2 k_B^2}{3e^2} \simeq 2.44 \times 10^{-8}$ V^{2}K^{-2} is the Sommerfeld value of the so-called Lorenz number. To obtain Eq. (3.20) we have used the integrals $\int_{-\infty}^{+\infty}\text{sech}^2\left(\frac{x}{2}\right)dx = 4$,

$$
\int_{-\infty}^{+\infty} x^{2m}\,\text{sech}^2\left(\frac{x}{2}\right)dx = 4(2^{2m} - 2)\pi^{2m}B_m, \qquad \int_{-\infty}^{+\infty} x^{2m+1}\,\text{sech}^2\left(\frac{x}{2}\right)dx = 0, \tag{3.21}
$$

where B_m are Bernoulli numbers.[14] Thus, conduction electrons will contribute a T-linear (γ_e) and a T-cubic (α_e) term to the specific heat. As we see, this result applies to any solid at low temperatures, and it is not required the free-electron model to hold, although it is explicitly assumed that $N(E)$ is a smooth function and it does not contain singular components [556, 608]. In conventional metals solids the DOS is a slowly varying function of the energy near the Fermi energy, so that the derivatives N_0' and N_0'' present in the α_e coefficient can be neglected, so that $\alpha_e \simeq 0$. In this case Eq. (3.20) reduces to $c_e(T) = \gamma_e T$, with

$$
\gamma_e = e^2 L_0 N(E_F^0) = \frac{\pi^2 k_B^2}{3}N(E_F^0). \tag{3.22}
$$

The lattice contribution to the specific heat can be obtained by considering that the dynamics of atoms in the crystal lattice can be properly described in terms of a number of collective oscillation modes characterized by their frequency values and their specific pattern of oscillation amplitudes. Within the framework of quantum mechanics these oscillations are described in terms of the so-called phonons, which are elementary excitations characterized by an energy $\hbar\omega$, where $\hbar$ is the reduced Planck constant and ω is the mode frequency. By arranging the available phonons according to their energy value one obtains the vibrational density of states (VDOS) $D(\omega)$. For most solids the VDOS rises quadratically with the

---

[14]The first members of this set are given by $B_1 = 1/6$, $B_2 = 1/30$, $B_3 = 1/42$, $B_4 = 1/30$, $B_5 = 5/66$, $B_6 = 691/2730$, and $B_7 = 7/6$.

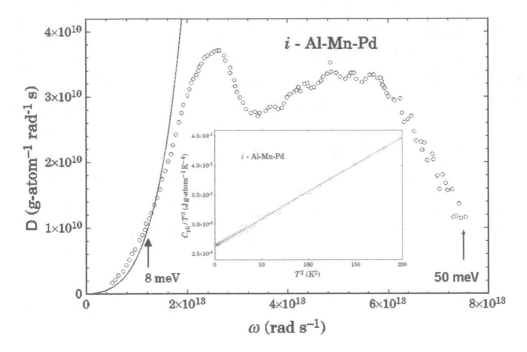

Figure 3.5 VDOS experimentally measured for a single grain i-$Al_{68.2}Pd_{22.8}Mn_9$ QC (circles). The solid curve indicates the low-frequency VDOS given by Eq. (3.26), where the corresponding a_v and b_v parameters are obtained from a low-temperature lattice specific heat $C_{ph} \equiv C_P - \gamma T = \alpha'T^3 + \delta T^5$ fit (see Eq. (3.40)), as it is shown in the inset, to get $\alpha' = 2.63 \times 10^{-5}$ $Jmol^{-1}K^{-4}$ and $\delta = 9.21 \times 10^{-8}$ $Jmol^{-1}K^{-6}$. (Reprinted figure with permission from Wälti Ch, Felder E, Chernikov M A, Ott O R, de Boissieu M, and Janot C, Physical Review B **57** 10504-10511 (1998). Copyright (1998) by the American Physical Society).

frequency for relatively small frequencies, then displays a series of alternating maxima and minima for intermediate frequencies and finally decreases approaching zero at the upper limit cut-off frequency ω_D, referred to as the *Debye frequency* (see Fig. 3.5). The VDOS is related to the phonon dispersion relation, $\omega(q)$, through the relationship [608]

$$D(\omega) = \frac{3V}{2\pi^2}q^2\frac{dq}{d\omega}, \tag{3.23}$$

where V is the sample's volume and q is the phonon's wave vector. In the Debye model a linear dispersion relation of the form $\omega = vq$ is assumed for small enough frequencies, where v is the mean sound velocity of the considered material, $v = 3^{1/3}\left(v_l^{-3} + 2v_t^{-3}\right)^{-1/3}$, where v_l and v_t are the longitudinal and transversal sound speed components, respectively. In that case, the VDOS adopts the parabolic form [608]

$$D(\omega) = \frac{3V}{2\pi^2v^3}\omega^2 \equiv a_v\omega^2. \tag{3.24}$$

At higher frequencies the dispersion relation deviates from linearity, generally with a negative curvature, taking the form $\omega = vq + Aq^3$, with $A < 0$. Solving the cubic equation we get

$$q(\omega) = \Lambda^{-1/3}\left(\Lambda^{2/3} - \frac{v}{3A}\right), \quad \Lambda \equiv \frac{\omega}{2A} + \sqrt{\left(\frac{v}{3A}\right)^3 + \left(\frac{\omega}{2A}\right)^2}, \tag{3.25}$$

and the VDOS can be expressed as

$$D(\omega) \simeq a_v\omega^2\left(1 - \frac{35A}{16v^3}\omega^2\right) \equiv a_v\omega^2 + b_v\omega^4, \tag{3.26}$$

where, making use of Eq. (3.25), we have expanded the function $q^2\frac{dq}{d\omega}$ in Eq. (3.23) in Taylor series keeping terms up to ω^4.

At any given temperature the probability distribution of phonons able to contribute to heat transport is given by the Planck distribution function

$$p(\omega, T) = \frac{1}{e^{\beta\hbar\omega} - 1} = \frac{1}{2}\left[\coth\left(\frac{\hbar\omega\beta}{2}\right) - 1\right]. \tag{3.27}$$

Thus, the internal energy contribution due to the lattice phonons can be expressed as

$$U_k(T) = \int_0^{\omega_D} \hbar\omega D(\omega)p(\omega, T)d\omega,$$

so that we have

$$c_k(T) \equiv \left(\frac{\partial U_k}{\partial T}\right)_V = \int_0^{\omega_D} \hbar\omega D(\omega)\left(\frac{\partial p}{\partial T}\right)d\omega. \tag{3.28}$$

By introducing the dimensionless scaled energy variable $y \equiv \beta\hbar\omega$, Eq. (3.28) can be written as

$$c_k(T) = \frac{k_B^2 T}{4\hbar}\int_0^{\Theta_D/T} y^2 D(y)\operatorname{csch}^2\left(\frac{y}{2}\right)dy, \tag{3.29}$$

where the Debye temperature Θ_D is determined by the highest normal mode of vibration through the relationship $\hbar\omega_D \equiv k_B\Theta_D$. The Debye temperature can be expressed in terms of the sound velocity making use of the normalization condition

$$3N_A \equiv \int_0^{\omega_D} D(\omega)d\omega = \frac{a_v}{3}\omega_D^3, \tag{3.30}$$

where N_A is the Avogadro's constant and we have used Eq. (3.24). In this way, we get the relationship [27, 608]

$$\Theta_D = \frac{\hbar v}{k_B}\left(\frac{6\pi^2 N_A}{V}\right)^{1/3} = \frac{\hbar v}{k_B}\sqrt[3]{6\pi^2 n_a} = \frac{\hbar v}{k_B}\left(\frac{3\rho_m}{4\pi\bar{m}}\right)^{1/3}, \tag{3.31}$$

where $n_a \equiv N_A/V$ is the atomic density, ρ_m is the mass density and $\bar{m}$ the average atomic mass. Taking into account Eqs.(3.24) and (3.31), Eq. (3.29) adopts the form

$$c_k(T) = \frac{9R}{4}\left(\frac{T}{\Theta_D}\right)^3\int_0^{\Theta_D/T} y^4\operatorname{csch}^2\left(\frac{y}{2}\right)dy = 9R\left(\frac{2T}{\Theta_D}\right)^3\int_0^{\frac{\Theta_D}{2T}} z^2 G(z)dz, \tag{3.32}$$

where $R = N_A k_B$ is the perfect gas constant, $z \equiv y/2$ and $G(z) \equiv z^2\operatorname{csch}^2 z$. In the low temperature limit $T \to 0$ we have $\Theta_D/T \to \infty$, and the integral in Eq. (3.32) can be explicitly evaluated by means of the Mellini's transform [414]

$$\int_0^\infty \frac{x^2}{\sinh^2 x}x^{s-1}dx = 2^{-s}\Gamma(s+2)\zeta(s+1). \tag{3.33}$$

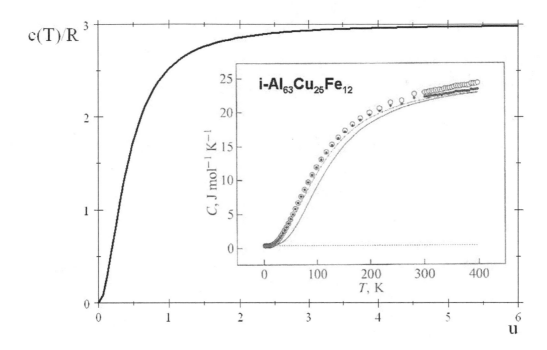

Figure 3.6 (Main frame) Lattice contribution to the specific heat as a function of temperature expressed in terms of the variable $u = 2T/\Theta_D$ in the Debye model. (Inset) Specific heat temperature dependence for the i-Al$_{63}$Pd$_{25}$Fe$_{12}$ QC. Open (closed) circles indicate the specific heat at constant pressure $C_P(T)$ and volume $C_V(T)$, respectively. The solid line corresponds to the lattice specific heat as given by the Debye model, the dot-dashed line shows the lattice specific heat according to the Grüneisen law, and the dotted line denotes the electronic linear contribution to the specific heat. (Inset reprinted by permission from Springer Nature, Physics of the Solid State, Prekul A F, Shalaeva E V, and Shchegolikhina N, Calorimetric investigation of electronic and lattice excitations of the icosahedral quasicrystals in the range of moderate temperatures, 52, 1797-1802 (2010), Copyright (2010)).

where $\Gamma(s)$ and $\zeta(s)$ are the Gamma and Riemann zeta functions, respectively, to obtain[15] [27, 608]

$$c_k(T) = \frac{12\pi^4 R}{5\Theta_D^3}T^3 \equiv \alpha_k T^3. \tag{3.34}$$

Thus, in the low temperature regime the lattice contribution to the specific heat shows a cubic dependence with the temperature, as prescribed by the αT^2 factor in Eq. (3.14). By comparing Eqs.(3.20) and (3.34), we have $\alpha \equiv \alpha_e + \alpha_k$, so that the experimentally obtained α parameter blends contributions due to both electrons and phonons. As we mentioned previously, as far as the DOS is a slowly varying function of E close to the Fermi energy the electronic contribution α_e becomes negligible. Accordingly, in many solids one can safely consider $\alpha_k \simeq \alpha$ in Eq. (3.34) in order to experimentally derive the Debye temperature from

---

[15]We recall that $\Gamma(n) = (n-1)!$ and $\zeta(4) = \pi^4/90$.

the expression

$$\Theta_D^\alpha = \sqrt[3]{\frac{12\pi^4 R}{5\alpha}}. \tag{3.35}$$

In the high temperature limit ($T \gg \Theta_D$) one gets $z \to 0$ and $G(z) = 4z^2(e^z - e^{-z})^{-2} \simeq 1$, so that Eq. (3.32) reduces to

$$c_k(T) \simeq 9R \left(\frac{2T}{\Theta_D}\right)^3 \int_0^{\frac{\Theta_D}{2T}} z^2 dz = 3R, \tag{3.36}$$

in agreement with the so-called Dulong-Petit empirical law. At intermediate temperatures the $G(z)$ function can be approximated as

$$G(z) = \frac{4z^2}{(e^z - e^{-z})^2} \simeq \frac{3}{3 + z^2}, \tag{3.37}$$

where we have kept terms up to the third order in the Taylor expansion of the exponentials. By plugging Eq. (3.37) into Eq. (3.32), we obtain

$$c_k(T) \simeq 27R \left(\frac{2T}{\Theta_D}\right)^3 \int_0^{\frac{\Theta_D}{2T}} \frac{z^2 dz}{3 + z^2} = 27Ru^3 \left[\frac{1}{u} - \sqrt{3}\arctan\left(\frac{\sqrt{3}}{3u}\right)\right], \tag{3.38}$$

where $u \equiv 2T/\Theta_D$. A plot of the temperature dependence of the lattice specific heat as given by Eq. (3.38) is shown in Fig. 3.6 main frame, where we can clearly appreciate the T^3 dependence at $T \ll \Theta_D$, as well as the asymptotic Dulong-Petit limit for $T \gtrsim \Theta_D$.

3.3.2 Quasicrystals and approximant phases

The specific heat curves of most thermodynamically stable iQCs and dQCs have been reported so far, including both polygrained and single-grained samples, spanning over a broad temperature interval ranging from below 1 K up to their melting temperatures. To start with, we will consider the information gained from the study of low-temperature specific heat measurements making use of Eq. (3.14). In so doing, we will tentatively assume that this expression can be physically interpreted for QCs as it is usually done for common metallic alloys, that is, in terms of Eqs. (3.22) and (3.35), and by the light of the obtained results we will subsequently discuss the reliability of such an assumption. In the 3rd and 4th columns of Tables 3.2–3.4, we list the obtained Θ_D^α and γ_e values **(Exercise 3.4)**, for a representative sample of iQCs, classified attending to their electron per atom ratio values (see Sec. 3.2.1). By inspecting the electronic γ_e coefficient data we note a significant reduction of its value, as compared to that typical of conventional metallic elements.[16] We note that such a reduction is more pronounced in the case of Mackay-type iQCs, followed by Bergman-type ones, whereas in the case of Tsai-type iQCs one obtains γ_e values closer to conventional metals. After Eq. (3.22) this reduction has been interpreted as indicative of a significant DOS depletion at the Fermi level $N(E_F)$ (listed in the 5th column of Tables 3.2–3.4), as expected from the Hume-Rothery mechanism induced pseudogap. It was reported that the electronic γ_e coefficient remains the same (or only changes slightly) before and after annealing treatments for i-AlLiCu (i-AlCuFe) QCs, respectively, hence indicating that $N(E_F)$ is scarcely sensitive to the related phason strain elimination in these two representative Bergman-type and Mackay-type systems. The insensitivity of the electronic

---

[16]For the sake of illustration the γ_e values (in mJ mol^{-1} K^{-2} units) of typical QC bearing elements read: 16.74 (Mn), 5.02 (Fe), 1.76 (Li), 1.35 (Al), 1.30 (Mg), 0.69 (Cd), 0.69 (Cu), 0.64 (Ga), 0.64 (Zn) [27, 608].

TABLE 3.2 Debye temperature Θ_D^α (in K), electronic specific heat coefficient γ_e (in mJmol^{-1}K^{-2}), DOS at the Fermi level (in state eV^{-1} atom^{-1}), low temperature electrical conductivity σ (in $(\Omega \text{ cm})^{-1}$), and electronic difussivity (in cm^{2}s^{-1}) for iQCs belonging to the Bergman-type arranged according to their electron per atom ratio e/a.

Sample	e/a	Θ_D^α	γ_e	$N(E_F)$	$\sigma(4.2 \text{ K})$	$D(E_F)$	Ref.
Ti$_{45}$Zr$_{38}$Ni$_{17}$	1.276	411			4807		[31]
Ti$_{41.5}$Zr$_{41.5}$Ni$_{17}$	1.289	270			3676		[31]
Al$_{54.8}$Li$_{33.9}$Cu$_{11.3}$	2.108	435	0.17	0.07	1250	2.09	[938]
Al$_{55.0}$Li$_{35.8}$Cu$_{9.2}$	2.113	359	0.318	0.13	1140	1.02	[417] [441]
Al$_{56.1}$Li$_{33.7}$Cu$_{10.2}$	2.134	346	0.35	0.11	1500	1.22	[938]
Al$_{60}$Li$_{30}$Cu$_{10}$	2.212	465	0.33	0.14			[945]
Zn$_{50}$Mg$_{42}$Y$_8$	2.116	326	0.63	0.27			[325]
Zn$_{62}$Mg$_{30}$Y$_8$	2.120	319	0.796	0.33	3570	1.27	[441]
Zn$_{56.8}$Mg$_{34.6}$Y$_{8.6}$	2.125	348	0.623	0.26	4910	2.24	[360] [131]
Zn$_{59}$Mg$_{29}$Y$_{12}$	2.165	350	0.6	0.25	3571	1.69	[705]
Mg$_{44.1}$Zn$_{41.0}$Al$_{14.9}$	2.171	310	0.73	0.30	6667	2.60	[405]
Zn$_{52}$Mg$_{32}$Ga$_{16}$	2.184	243	0.18	0.08	10200	16.1	[939]
Zn$_{46}$Mg$_{37}$Ga$_{17}$	2.192		0.48	0.20			[940]
Zn$_{40}$Mg$_{39.5}$Ga$_{20.5}$	2.225		0.9	0.38			[606]

structure to phason strain can be understood because those reciprocal space vectors which more contribute to the Fermi-surface Brillouin-zone interaction (see Sec. 3.2.2) are hardly sensitive to phason strain, since the related atomic domains are projected onto parallel physical space from close regions in the perpendicular space (see Sec. 1.4.5).

The three main different types of iQCs also exhibit different Debye temperature values, broadly ranging from 235 − 600 K (Mackay-type), 240 − 460 K (Bergman-type), and 140–240 K (Tsai-type). Albeit the Θ_D^α value of i-AlLiCu QCs remained essentially unchanged under varying annealing conditions, the Debye temperature was observed to significantly increase from $\Theta_D^\alpha = 273 \pm 5$ K to $\Theta_D^\alpha = 350 \pm 5$ K after a 812°C annealing treatment in the i-AlCuFe system [945]. This significant change cast some doubts on the applicability of Eq. (3.35) in this case. Additional concerns on the applicability of the standard Debye model to iQCs arose from studies regarding the elastic properties of i-AlPdMn, i-AlCuFe, i-MgZnY and i-AlLiCu representatives, leading to Debye temperature values (obtained from sound velocity measurements via Eq. (3.31)) that generally do not agree with those derived from low-temperature specific heat fits via Eq. (3.35), as it is summarized in Tables 3.5 and 3.6.

At this point, we must keep in mind that, within the Debye model framework, Θ_D enters in various physical magnitudes, such as the lattice specific heat, the VDOS, the average speed of sound, the Debye-Waller factor affecting the intensities of diffraction peaks upon temperature, or the thermal conductivity coefficient (see Sec. 3.8.2). The Θ_D values calculated from these different empirical data are not necessarily the same since they are sensitive to *different regions* of the frequency spectrum. An illustrative example of the temperature dependence of the specific heat curve in the low to intermediate range temperature is shown in the inset of Fig. 3.6 for an i-AlCuFe representative. The overall behavior of the $C_P(T)$ and $C_V(T)$ curves closely resemble that prescribed by the Debye model, albeit both curves exhibit an enhanced value extending over the entire considered temperature range. A significant excess contribution to the specific heat curve was observed in the high-temperature regime as well, where it has been reported that the $C_P(T)$ and $C_V(T)$ curves of QCs belonging to different alloy systems and symmetries significantly rises over the Dulong-Petit law value $C_V^\infty = 3R = 24.9$ Jmol^{-1}K^{-1}, reaching values up to $5R$−$6R$ ($5R$−$9R$) near

TABLE 3.3   Debye temperature Θ_D^α (in K), electronic specific heat coefficient γ_e (in mJmol^{-1}K^{-2}), DOS at the Fermi level (in state eV^{-1} atom^{-1}), low temperature electrical conductivity σ (in $(\Omega$ cm)$^{-1}$), and electronic difussivity (in cm^{2}s^{-1}) for iQCs belonging to the Mackay-type arranged according to their electron per atom ratio e/a.

Sample	e/a	Θ_D^α	γ_e	$N(E_F)$	$\sigma(4.2$ K$)$	$D(E_F)$	Ref.
Al$_{61.4}$Cu$_{25.4}$Fe$_{13.2}$	2.241	536	0.285	0.12			[852]
Al$_{62}$Cu$_{25.5}$Fe$_{12.5}$	2.253	425	0.40	0.17			[670]
Al$_{62}$Cu$_{25.5}$Fe$_{12.5}$	2.253	350	0.30	0.13			[945]
Al$_{62.5}$Cu$_{24.5}$Fe$_{13}$	2.263	560	0.32	0.14	270	0.24	[695]
Al$_{63}$Cu$_{25}$Fe$_{12}$	2.272	370	0.26	0.11	220	0.24	[424]
Al$_{63}$Cu$_{25}$Fe$_{12}$	2.272	509	0.293	0.12			[710]
Al$_{63.5}$Cu$_{24.5}$Fe$_{12}$	2.282	539	0.31	0.13	230	0.21	[695] [765]
Al$_{65}$Cu$_{20}$Ru$_{15}$	2.313	500	0.11	0.05	75	0.19	[695]
Al$_{65}$Cu$_{20}$Ru$_{15}$	2.313	496	0.05	0.02			[359]
Al$_{64.5}$Cu$_{20}$Ru$_{15}$Si$_{0.5}$	2.317	445	0.27	0.11	180	0.19	[695]
Al$_{64}$Cu$_{20}$Ru$_{15}$Si$_1$	2.322	485	0.21	0.09	280	0.38	[695]
Al$_{68}$Cu$_{17}$Ru$_{15}$	2.377	527	0.23	0.10	180	0.22	[695]
Al$_{70}$Cu$_{15}$Ru$_{15}$	2.413	500	0.20	0.08	100	0.14	[695]
Al$_{70}$Pd$_{21}$Mn$_9$	2.403	467	0.009	0.004			[129]
Al$_{70.5}$Pd$_{21}$Re$_{6.5}$Mn$_2$	2.436	440	0.38	0.16	5	0.004	[314]
Al$_{70}$Pd$_{20}$Re$_{10}$	2.439	460	0.22	0.09	100	0.13	[697]
Al$_{70.5}$Pd$_{21}$Re$_{8.5}$	2.443	425	0.10	0.04	63	0.18	[701]
Al$_{70.5}$Pd$_{21}$Re$_{8.5}$	2.443	450	0.11	0.05	1.5	0.004	[698]
Al$_{70.5}$Pd$_{21}$Re$_{8.5}$	2.443	235	0.11	0.05	2	0.005	[707]

the melting point for iQCs (dQCs), respectively, with no indication of saturation (Fig. 3.7). In summary, the existence of an excess contribution to the specific heat over a broad temperature interval ranging from below 1 K up to the sample's melting temperature clearly indicates that the Debye model used to obtain Eqs.(3.34), (3.36), and (3.38) is not entirely reliable in the case of QCs.

In order to properly discuss this excess contribution one must keep in mind that the presence of different relevant energy scales related to different elementary excitations in QCs (a feature stemming from the fractal nature of both electron and phonon energy spectra) complicates the proper assignation of different contributions to the specific heat in terms of Eq. (3.14). For instance, in amorphous solids a linear increase of the specific heat with the temperature has been reported at very low-temperature (say, below 1 K). This term is associated with the presence of the so-called tunneling states, which related to two-level systems arising from the tunneling of atoms between nearby degenerate configurations. These states have been invoked to account for the VDOS enhancement at temperatures below 4 K, whose density determines the slope of the specific heat curve. This contribution cannot be easily separated to that coming from the very low-temperature limit linear contributions to the specific heat curve given by the electronic γ_e term in Eq. (3.22), which will be partly masked by the possible presence of vibrational tunneling states. Several studies of acoustic properties of high- quality i-AlPdMn, i-AlCuFe, and i-ZnMgY QCs showed the presence of tunneling states with a density of states similar to what is found in amorphous metals, hence indicating that a disordered atomic network is not strictly necessary for tunneling states to exist, but it suffices the local atomic order to be complex enough so that one can find some neighboring configurations for resonant tunneling to occur [56]. This scenario is properly accomplished by the large unit cells of approximant crystals and the resonance conditions

TABLE 3.4 Debye temperature Θ_D^α (in K), electronic specific heat coefficient γ_e (in mJmol^{-1}K^{-2}), DOS at the Fermi level (in state eV^{-1} atom^{-1}), low temperature electrical conductivity σ (in $(\Omega$ cm$)^{-1}$), and electronic difussivity (in cm^{2}s^{-1}) for several thermodynamically stable iQCs belonging to the Tsai-type arranged according to their electron per atom ratio e/a. We have used the mixed-valence value 8/3 for Yb atoms in Au based QCs, [383] and the valence 2 in the other ones. The large γ values may be attributed to the Yb-derived states at the Fermi level [870].

Sample	e/a	Θ_D^α	γ_e	$N(E_F)$	$\sigma(4.2$ K$)$	$D(E_F)$	Ref.
Au$_{51}$Al$_{34}$Yb$_{15}$	1.933	168	6.5	2.77	5050	0.22	[383]
Ag$_{42}$In$_{42}$Yb$_{16}$	2.017	208	1.27	0.54	4878	1.09	[79]
Cd$_{85.1}$Yb$_{14.9}$	2.026	140	1.1	0.47	6757	1.74	[705]
Cd$_{85}$Yb$_{15}$	2.026	138	7.5	3.18	5714	0.22	[148]
Cd$_{84.6}$Yb$_{15.4}$	2.025	140	2.87	1.23	2029	0.20	[867, 631]
Zn$_{81}$Sc$_{13}$Mg$_6$	2.155	240	0.5	0.21	2670	1.52	[324]

further improve for the related QCs due to their local isomorphism property. This QPO related feature can explain the strong enhancement of the tunneling strength observed in the icosahedral phase upon annealing (at variance with the behavior observed in amorphous solids), which apparently correlates with the progressive decrease of the amplitude of the linear phason strain in the sample [56].

As the temperature window is increased one must consider the contribution of electronic states located close to the Fermi energy, within the interval $\Delta E \simeq 2k_B T$. As we have described in Sec. 3.2.2, the electronic structure of iQCs exhibits a characteristic pronounced pseudogap close to the Fermi level (see Fig. 3.4), which can be described by means of a parabolic dip whose width determines the local curvature of the DOS at the Fermi energy. Therefore, the electronic structure of iQCs around Fermi level strongly depends on the energy, as prescribed by Eqs.(3.2) or (3.3), and the condition $\alpha_e \simeq 0$ cannot be taken for granted in Eq. (3.20). Taking the pertinent derivatives in Eq. (3.2) and plugging the obtained results in Eq. (3.20) we obtain

$$\alpha_e = \frac{\Omega}{5}(e^2 L_0)^2 \left(\frac{50 N_a}{N_0} - 29\right). \tag{3.39}$$

When E_F^0 is located close to the DOS minimum (as expected in more stable QCs), we have $N_a \approx N_0$, so that $\alpha_e \simeq 21(e^2 L_0)^2 \Omega/5$. Dip's curvature values within the interval $\Omega = 23 - 45$ states (eV)$^{-3}$ per atom have been reported from NMR measurements for thermodynamically stable TM-bearing iQCs [335], yielding $2.4 \times 10^{-7} \leq \alpha_e \leq 1.1 \times 10^{-5}$ mJmol^{-1}K^{-4}. These values are about 3 to 5 orders of magnitude smaller than those obtained from the specific heat curves making use of Eq. (3.14) (see **Exercise 3.4,** and the 2nd column in Table 3.18 of **Exercise 3.6**). If the Fermi energy is shifted from the dip's minimum then we get values within the range $1.8 \times 10^{-5} \lesssim \alpha_e \lesssim 1.1 \times 10^{-4}$ mJmol^{-1}K^{-4}.[17] Therefore, the electronic contribution to the T^3 term of the specific heat can be safely ignored [941], even when one explicitly takes into account the presence of a parabolic dip close to Fermi level (the possible contribution due to spiky features will be discussed below). Thus, we should focus on the lattice contribution term $\alpha_k(T)$ in order to find a possible origin for the specific heat enhancement.

---

[17]Obtained by plugging into Eq. (3.39) the values $N_0 = 0.05$ states (eV)$^{-1}$mol^{-1} [695], $\Omega = 45$ states (eV)$^{-3}$mol^{-1}, and $N_a = 0.064$ states (eV)$^{-1}$mol^{-1}, or $N_a = 0.24$ states (eV)$^{-1}$mol^{-1}, respectively [531].

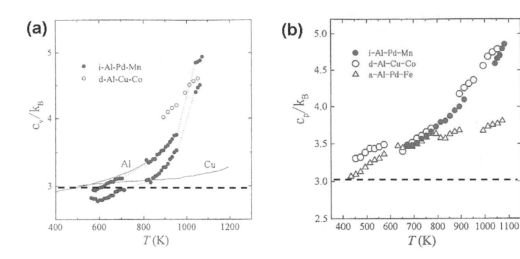

Figure 3.7   Temperature dependence of specific heat per atom at constant volume (a) and pressure (b) for i-Al$_{71}$Pd$_{20}$Mn$_9$ and d-Al$_{64.5}$Cu$_{18.5}$Co$_{17}$ QCs. The data for elemental aluminium and copper are presented for the sake of comparison. In (b) the specific heat curve of the 1/0 approximant crystal Al$_{70}$Pd$_{20}$Fe$_{10}$ is also shown. The dashed horizontal line indicates the Dulong-Petit asymptotic limit. (Reprinted from Edagawa K, Kajiyama K, Tamura R, Takeuchi S, High-temperature specific heat of quasicrystals and a crystal approximant, Material Science Engineering. A 312 293-298 Copyright (2001), and Edagawa K and Kajiyama K, High temperature specific heat of Al-Pd-Mn and Al-Cu-Co quasicrystals, Material Science Engineering A 294–296 646–649, Copyright (2000), with permission from Elsevier).

Possible modifications to the standard Debye model usually involve effects related to a non-linear dispersion relation of phonons. In order to shed some light onto this question coherent inelastic neutron scattering was used to study the phonon dispersion relations in iQCs belonging to the AlPdMn, AlLiCu, AlCuFe, MgZnY, and ZnScMg alloy systems, as well as to the d-AlNiCo system. In all the considered systems well-defined low-energy acoustic phonon modes are found below a certain wave vector of about 3 nm^{-1}. Above this value excitations broaden out into a dispersionless band which may be signaling the presence of mixed acoustic and optical modes at energies extending from about 8 meV up to 12 meV. At higher energies the phonon spectrum is dominated by dispersionless optical modes (see Fig. 3.9) [164, 551]. From these scattering measurements the VDOS can be experimentally obtained for these samples. An example is shown in Fig. 3.5, where the obtained data are compared, in low frequency region, with a biquadratic VDOS curve of the form given by Eq. (3.26), derived from low-temperature specific heat fits making use of Eq. (3.40). By plugging Eq. (3.26) into Eq. (3.29), in the low temperature limit $T \to 0$ we get [941] (**Exercise 3.5**)

$$c_k(T) = \frac{4\pi^4 k_B}{5}\left(\frac{3N_A}{\tilde{\Theta}_D^3} - \frac{b_v}{5}\left(\frac{k_B}{\hbar}\right)^5 \tilde{\Theta}_D^2\right)T^3 + \frac{16\pi^6 k_B}{21}b_v\left(\frac{k_B}{\hbar}\right)^5 T^5 \equiv \alpha' T^3 + \delta T^5, \quad (3.40)$$

TABLE 3.5 Elastic Debye temperatures, Θ_D^S, of iQCs belonging to different alloy systems (arranged by their average atomic mass m) obtained from the mean sound velocity v and mass density ρ_m after Eq. (3.31).

compound	m	ρ_m	v_l	v_t	v	Θ_D^S	Ref
	(amu)	(kgm^{-3})	(ms^{-1})	(ms^{-1})	(ms^{-1})	(K)	
$Ti_{41.5}Zr_{41.5}Ni_{17}$	67.716	6081			2890	325	[258]
$Ti_{39.5}Zr_{39.5}Ni_{21}$	67.281	6207	5230	2480	2790	317	[4]
$Al_{68.7}Pd_{21.7}Mn_{9.6}$	46.817	5088	6512 ± 10	3595 ± 5	4006	481	[12]
$Al_{68.7}Pd_{21.7}Mn_{9.6}$	46.817	5088	6530 ± 10	3590 ± 5	4002	480	[12]
$Al_{70.4}Pd_{21.2}Mn_{8.4}$	46.122	5080	6195 ± 235	3550 ± 30	3944	475	[185]
$Al_{72}Pd_{19.5}Mn_{8.5}$	44.772	4800	6630	3800	4221	504	[879]
$Al_{63}Cu_{25}Fe_{12}$	39.587	4363	7700	3650	4106	495	[92]
$Al_{63.5}Cu_{24.5}Fe_{12}$	39.404	4363	7191	3809	4257	514	[92]
$Al_{55}Li_{35.8}Cu_{9.2}$	23.171	2464	6400 ± 10	3800 ± 10	4208	501	[740, 56]
$Al_{55}Li_{35.8}Cu_{9.2}$	23.171	2464	6500 ± 20	3700 ± 20	4113	490	[740, 879]

TABLE 3.6 Elastic Debye temperature derived from mean sound velocity measurements (Θ_D^S), versus that obtained from specific heat fits assuming a quadratic Debye VDOS (Θ_D^α, see Tables 3.2 and 3.18), and from a biquadratic VDOS after Eq. (3.41), ($\tilde{\Theta}_D$), for different iQCs systems. For the sake of comparison the mean value of the Debye temperatures of the composing elements, and the QCs melting temperature are also listed.

compound	Type	Θ_D^S (K)	Θ_D^α (K)	$\tilde{\Theta}_D$ (K)	$\langle\Theta_D\rangle$ (K)	T_m (K)	Ref
TiZrNi	B	$317 - 325$	$270 - 411$		325	1080	[416]
ZnMgY	B		$319 - 348$	201	259	934	[978]
AlLiCu	B	$490 - 501$	$346 - 465$		370	895	[194]
ZnMgGa	B		243		264	680	[660]
AlPdRe	M		$235 - 450$	$206 - 217$	315	1300	[176]
AlPdMn	M	$475 - 504$	$420 - 492$	$240 - 316$	356	1140	[365]
AlCuFe	M	$495 - 514$	$370 - 560$	$171 - 185$	376	1170	[295]
AlCuRu	M		$496 - 600$	415	375		

which reduces to Eq. (3.34) when $b_v \to 0$, and where we have introduced a *generalized* Debye temperature $\tilde{\Theta}_D$ (see Eq. (3.111)). This temperature can be experimentally determined by fitting the low-temperature specific heat curve c_k/T^3 to obtain the α' and δ parameters (see Fig. 3.5), and then solving the quintic equation (**Exercise 3.6**)

$$\tilde{\Theta}_D^5 + \frac{100\pi^2\alpha'}{21\delta}\tilde{\Theta}_D^3 - \frac{80\pi^6 R}{7\delta} = 0. \tag{3.41}$$

By inspecting Tables 3.6 and 3.18 we see that the obtained $\tilde{\Theta}_D$ values are significantly smaller than those obtained by assuming the applicability of the standard Debye model, Θ_D^α.[18] Therefore, the discrepancy between the Θ_D^S and Θ_D^α data set values cannot be accounted for in terms of non-linear effects in the frequency dispersion relation, and it should be related to the presence of non-propagating lattice excitations that do not manifest themselves in acoustic experiments [941]. To this end, one may consider additional contributions to the VDOS stemming from the motion of groups of atoms in a collective manner [92], which may be signaling the presence of cluster related oscillation modes.

---

[18]It is interesting to note that the derived smaller $\tilde{\Theta}_D$ values compare well with the reported Debye temperature ($\Theta_D = 298 \pm 7$ K) of the pentagonal surface of a single grain i-AlPdMn QC [521].

Along with the presence of a pseudogap close to the Fermi energy the theoretical considerations presented in Chapter 2 predict the existence of a singular continuous component in the electron spectrum and, consequently, one may expect the presence of some additional contributions in the specific heat curves naturally arising from this additional component. The possible presence of very narrow spectral features in the electronic DOS, described in terms of a series of δ-Dirac functions in the form [533, 534]

$$N_\delta(E) = \sum_{i=1}^{\nu} N_i \delta(E - E_i), \tag{3.42}$$

where E_i indicates the location of the peaks, would lead to an additional internal energy contribution to the electron system given by

$$U_e^s = \int_{-\infty}^{+\infty} E N_\delta(E) f(E, T) dE, \tag{3.43}$$

in Eq. (3.15). By plugging Eqs.(3.8) and (3.42) into Eq. (3.43) and making use of the relation $\int_{-\infty}^{+\infty} f(x)\delta(x - x_0) = f(x_0)$ involving δ-Dirac function integrals, we obtain

$$
\begin{aligned}
U_e^s &= \frac{1}{2} \sum_{i=1}^{\nu} N_i \int_{-\infty}^{+\infty} E \left(1 - \tanh\left(\frac{E - E_F}{2 k_B T}\right)\right) \delta(E - E_i) dE \\
&= \cdot \frac{1}{2} \sum_{i=1}^{\nu} N_i E_i \left[1 - \tanh\left(\frac{\Delta E_i}{2 k_B T}\right)\right],
\end{aligned}
\tag{3.44}
$$

where $\Delta E_i \equiv |E_i - E_F|$ measures the position of the i-th DOS peak. Therefore, the singular electronic contribution to the specific heat is given by (we will assume E_F does not depend on temperature)

$$c_e^s = \left(\frac{\partial U_e^s}{\partial T}\right)_V = k_B \sum_{i=1}^{\nu} N_i \left(\frac{\Delta E_i}{2 k_B T}\right)^2 \operatorname{sech}^2\left(\frac{\Delta E_i}{2 k_B T}\right), \tag{3.45}$$

where we adopted $E_F \simeq 0$ for convenience. We note that the functional form of each term in Eq. (3.45) is completely equivalent to a Schottky type curve given by

$$C_i(T) = N_i \left(\frac{\Delta E_i}{k_B T}\right)^2 \frac{e^{\frac{\Delta E_i}{k_B T}}}{\left(1 + e^{\frac{\Delta E_i}{k_B T}}\right)^2}. \tag{3.46}$$

Quite interestingly, experimental evidence for the presence of two Schottky type peaks in the specific heat curves of several iQCs has been reported in the literature [711], as it is shown in Fig. 3.8. The proper assignation of the physical mechanisms originating these peaks is still open. For one thing, the presence of Schottky contributions stemming from crystal field splitting effects has been reported in several rare-earth bearing QCs [384]. This contribution leads to a characteristic upturn in the $C_P(T)$ curve at temperatures below 2 K. This temperature range is much lower than that shown in Fig. 3.8, where the lower temperature peak is located at about 90 K. In order to account for this peak, instead of invoking to a two-level electronic system one could alternatively consider the motion of groups of atoms in a collective manner (localized cluster modes), which can interact with the extended sound waves related to phonon states, as it is illustrated in Fig. 3.9. Evidence

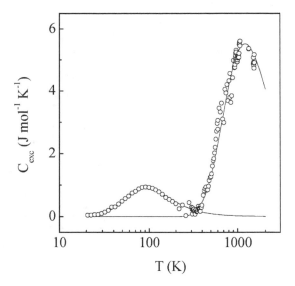

Figure 3.8   Temperature dependence of the excess specific heat $C_{exc}(T) = C_V(T) - \gamma T - c_l(T)$, where the lattice contribution c_l has been derived making use of the Grüneisen law, for an i-$Al_{63}Cu_{25}Fe_{12}$ QC. Solid lines show the fit to two Schottky type curves with excitation energies $\Delta E_1 = 0.02$ eV and $\Delta E_2 = 0.25$ eV, respectively. (Reprinted from Prekul A F and Shchegolikhina N, Two-level electron excitations and distinctive physical properties of Al-Cu-Fe quasicrystals, *Crystals* 2016 6, 119; doi:10.3390/cryst6090119, Creative Commons Atribution License CC BY 4.0).

for these collective modes was provided by inelastic neutron scattering experiments of i-AlCuFe QCs showing that the partial VDOS differ greatly between Cu atoms and Fe ones [92]. The contribution of these collective vibration modes can be described in terms of an Einstein-like model VDOS. To this end, we include in the VDOS a Dirac-comb term [530]

$$D_\delta(\omega) = \sum_{i=1}^{\nu} N_i \delta(\omega - \omega_i), \qquad (3.47)$$

where $N_i \equiv 3\eta_i N_A$ measures the δ-functions heights expressed in terms of the fractional contribution η_i of each mode to the total VDOS, and the frequencies ω_i describe localized vibration modes. Plugging Eq. (3.47) into (3.29), and taking into account the δ-Dirac property $\delta(\lambda x) = |\lambda|^{-1}\delta(x)$, we get

$$c_k^\delta(T) = \frac{3R}{4} \sum_{i=1}^{\nu} \eta_i \int_0^{\Theta_D/T} y^2 \operatorname{csch}^2\left(\frac{y}{2}\right) \delta(y - y_i)dy = 3R \sum_{i=1}^{\nu} \eta_i \int_0^{\frac{\Theta_D}{2T}} G(z)\delta(z - z_i)dz, \qquad (3.48)$$

For temperatures low enough as compared to Θ_D we can express Eq. (3.48) as

$$c_k^\delta(T) \simeq \frac{3R}{2} \sum_{i=1}^{\nu} \eta_i \int_{-\infty}^{+\infty} G(z)\delta(z - z_i)dz = \frac{3R}{2} \sum_{i=1}^{\nu} \eta_i z_i^2 \operatorname{csch}^2 z_i, \qquad (3.49)$$

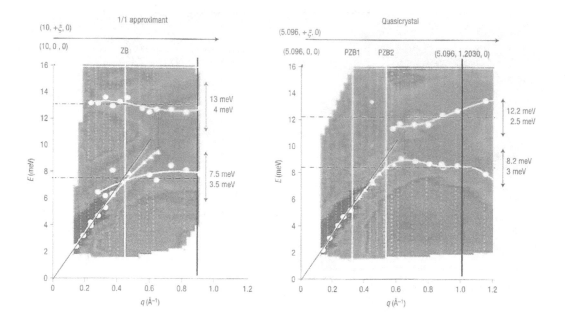

Figure 3.9 Comparison between the measured dispersion relation (symbols) and the simulated response function $S(Q, E)$ in the 1/1 approximant (left panel) and an i-Al$_{63}$Cu$_{25}$Fe$_{12}$ QC (right panel). Vertical lines indicate the position of the (pseudo) Brillouin zone boundaries. (Reprinted from de Boissieu et al, Two-level electron excitations and distinctive physical properties of Al-Cu-Fe quasicrystals, *Crystals* 2016 6 119; doi:10.3390/cryst6090119, Creative Commons Atribution License CC BY 4.0).

where we have exploited the even symmetry of the $G(z)$ function. By introducing the effective Einstein temperatures $\hbar\omega_i = k_B\Theta_i$, Eq. (3.49) can be rewritten in the form [530]

$$c_k^\delta(T) = \frac{3R}{8} \sum_{i=1}^\nu \eta_i \frac{\Theta_i^2}{T^2} \operatorname{csch}^2\left(\frac{\Theta_i}{2T}\right) \simeq \frac{3R}{2} \sum_{i=1}^\nu \eta_i \frac{\Theta_i^2}{T^2} e^{-\Theta_i/T}, \qquad (3.50)$$

at low enough temperatures. Remarkably enough, in the limit $\beta\Delta E_i \gg 1$, Eq. (3.46) coincides with Eq. (3.50) upon substitution $\Delta E_i = k_B\Theta_i$. Accordingly, the excess specific heat peaks shown in Fig. 3.8 could be attributed to either two-level electronic features in the electronic DOS or to localized collective oscillation modes in the VDOS. In fact, the contribution due to each term in Eq. (3.50) is characterized by a peaked curve whose maximum is located at $T_i = \Theta_i/2$. As we previously mentioned a relatively low-lying optical mode (almost merging with the higher frequency acoustic modes) has been identified at about 8 meV ($\Theta_1 \simeq 93$ K) in the phonon spectrum of several iQCs, whereas another optical mode is located at about 12 meV ($\Theta_2 \simeq 139$ K), as shown in Fig. 3.9 [164, 551]. In the same vein, higher energy Einstein-like vibrational modes could also make some contribution to the high-temperature lattice specific heat in QCs (**Exercise 3.7**).

3.4 ELECTRICAL TRANSPORT PROPERTIES

"The outstanding characteristic of metals is good electrical and thermal conductivity". (C. M. Hurd, 1981) [356]

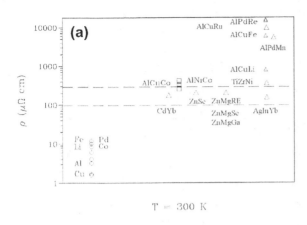

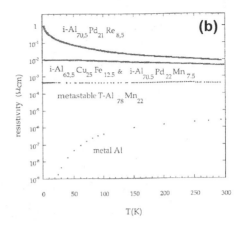

Figure 3.10 (a) Room temperature electrical resistivity values of icosahedral (triangles) and decagonal (squares) QCs belonging to different alloy systems. The room temperature resistivity values of their respective constituent elements (circles) is indicated for the sake of comparison. The dashed lines indicate the resistivity range of usual ternary alloys. (b) The temperature dependence of electrical resistivity of representative Al-based iQCs is compared with that corresponding to elemental aluminium, its main alloying element (Courtesy of Claire Berger).

3.4.1 Electrical conductivity

The electrical conductivity of both icosahedral and decagonal QCs has been extensively studied during the last three decades. Measurements comprise different stoichiometric compositions for samples belonging to the same alloy system, and cover different temperature windows within the broad interval 0.1–1000 K. In the case of iQCs the collected data disclosed several *general trends* in the electrical resistivity $\rho(T)$ curves, namely:

- Electrical resistivities take unusually high values for alloys made of good metals (Fig. 3.10a and Tables 3.7–3.8).

- In most samples $\rho(T)$ steadily decreases as the temperature increases up to the melting point (Figs.3.10b and 3.13).

- The $\rho(T)$ curve is extremely sensitive to minor variations in the sample stoichiometry (Table 3.8 and Fig. 3.11b).

- The electrical resistivity increases when the structural order of the sample is improved by annealing (Fig. 3.12), in contrast to the decrease in resistivity that usually accompanies removal of defects in common alloys.

Fig. 3.10a displays the room temperature electrical resistivities for different QCs as well as for their constituent pure elements, which are good metals. As we see, iQCs belonging to Tsai-type have ρ_{300K} values comparable to those reported for conventional metallic alloys (both crystalline and amorphous), whose representative values fall in the region between the horizontal dashed lines in Fig. 3.10a. This is also the case for Bergman-type i-ZnMgGa compounds. dQCs show somewhat higher values (250–450 $\mu\Omega$cm) along their QP symmetry axis. These ρ_{300K} figures are similar to those measured for Bergman-type TiZrNi

TABLE 3.7   Room temperature resistivity and resistivity ratio of iQCs belonging to different alloy systems and cluster-types arranged by decreasing R_{4K} ratio values.

Alloy system	type	ρ_{300K} ($\mu\Omega$cm)	R_{4K}	Reference
AlPdRe	M	$(1.1 - 7.1) \times 10^3$	$2 - 300$	[728]
AlCuRu	M	$(0.2 - 4.9) \times 10^4$	$1.2 - 5.2$	[371] [866][756]
AlCuOs	M	$(2 - 3.3) \times 10^4$	$4.0 - 4.5$	[341]
AlCuFe	M	$(2.2 - 4.4) \times 10^3$	$1.2 - 2.5$	[866]
AlPdMn	M	$\sim 5 \times 10^3$	$1.1 - 2.5$	[725]
AlCuLi	B	$670 - 880$	1.1	[417] [938]
TiZrNi	B	$150 - 520$	$1.02 - 1.08$	[461]
ZnMgRE	B	$150 - 190$	$1.05 - 1.09$	[249]
ZnMgGa	B	$100 - 150$	$0.9 - 1.1$	[939, 604]
ZnSc	T	~ 200	~ 1.05	[655]
ZnMgSc	T	$160 - 330$	1.07	[324]
CdMgYb	T	$100 - 180$	$0.85 - 1.1$	[869]
CdYb	T	$140 - 180$	$0.2 - 0.85$	[867] [324]

iQCs, whereas Bergman-type AlLiCu iQCs electrical resistivity values jump to $\rho_{300K} \sim 800$ $\mu\Omega$cm. Finally, Mackay-type iQCs in the AlCu(Fe,Ru) and AlPd(Mn,Re) alloy systems show electrical resistivities which are higher than those of usual ternary alloys by 2–3 orders of magnitude.

For typical metals the electrical resistivity decreases as the temperature is decreased and it can even completely vanish at low enough temperatures for those materials reaching the superconducting state. Such a behavior is illustrated for the case of aluminium (the main constituent of an important class of iQCs) in Fig. 3.10b. Conversely, the electrical resistivity of most QCs progressively increases as the temperature is decreased, exhibiting a negative temperature coefficient of resistivity, defined as $\alpha_\rho \equiv \rho^{-1} d\rho/dT$. Thus, the very possibility of reaching a metal-insulator transition at low temperatures in high-quality iQCs belonging to the AlPdRe alloy system was scrutinized [175, 478, 698, 755]. Indeed, in Fig. 3.10b, we see that below 10 K the resistivity of i-$Al_{70.5}Pd_{21}Re_{8.5}$ sample reduces to about 1 Ωcm, a value more than ten orders of magnitude smaller than those corresponding to usual metallic materials. In order to properly quantify the resistivity variation rate with temperature it is convenient to introduce the so-called resistivity ratio $R_{4K} \equiv \rho(4 \text{ K})/\rho(300 \text{ K})$.[19] By inspecting Table 3.7, we see that the more resistive samples (i-AlPdRe, i-AlCuOs, and i-AlCuRu) exhibit the largest R_{4K} values as well [728]. We also see that the binary icosahedral phases exhibit fairly small $R_{4K} \lesssim 1$ values, indicative of low-temperature positive α_ρ coefficients, which become slightly negative at temperatures above Θ_D^α for this Tsai-type iQCs [869].

The use of a Mooij-like plot allows one to classify different iQCs by attending to their electrical resistivity values at different temperatures.[20] To this end, in Fig. 3.11a the difference $\Delta\rho \equiv \rho_{4K} - \rho_{295K} = (R_{4K} - 1)\rho_{295K}$ is plotted versus ρ_{295K}, thereby disclosing a

---

[19]The choice of the low, and high-temperature values is a matter of experimental convenience. Lower temperature values, approaching the residual resistivity value $\rho_0 = \lim_{T \to 0} \rho(T)$, can also be found in the literature (see Table 3.8).

[20]The original Mooij's plot disclosed a correlation between the resistivity temperature coefficient and the resistivity value, both measured at *room temperature*, of the form $\alpha_\rho(300K) = \alpha_\rho^0 - \zeta\rho_{300K}$, in many *disordered* metallic systems, where $\alpha_\rho^0 \sim 2 \times 10^{-4}$ K^{-1} and $\zeta > 0$. Accordingly, α_ρ changes its sign in a relatively narrow range of resistivity, within 100–150 $\mu\Omega$cm [621]. Later on, Tsuei included a larger number of samples, reporting that such a correlation is not universal due to a competition between the quantum mechanical effects of incipient localization and the classical Boltzmann electron transport. Notwithstanding

TABLE 3.8   Room temperature resistivity and temperature coefficient ratio as a function of the e/a ratio for i-$Al_{65}Cu_{35-x}Ru_x$ QCs.

compound	e/a	ρ_{300K} ($\mu\Omega$cm)	$R_{1.4K}$	Ref.
$Al_{65}Cu_{19}Ru_{16}$	2.3129	15468	3.3	[466]
$Al_{65}Cu_{19.5}Ru_{15.5}$	2.3127	10500 ± 213	5.20	[371]
$Al_{65}Cu_{20}Ru_{15}$	2.3125	6900 ± 140	2.78	[371]
$Al_{65}Cu_{20}Ru_{15}$	2.3125	10030	3.5	[466]
$Al_{65}Cu_{20}Ru_{15}$	2.3125	3983	3.7	[695]
$Al_{65}Cu_{20.5}Ru_{14.5}$	2.3123	5360 ± 109	2.81	[371]
$Al_{65}Cu_{21}Ru_{14}$	2.3121	4850 ± 98	2.57	[371]
$Al_{65}Cu_{21}Ru_{14}$	2.3121	5983	2.2	[466]
$Al_{65}Cu_{21.5}Ru_{13.5}$	2.3119	1850 ± 38	1.26	[371]
$Al_{65}Cu_{22}Ru_{13}$	2.3117	2595	1.4	[466]
$Al_{63}Cu_{25}Ru_{12}$	2.2711	$\sim 2 - 4 \times 10^3$	$\sim 1.1 - 2.4$	[865]
$Al_{68}Cu_{17}Ru_{15}$	2.3728	1600	1.25	[605]

power-law relationship of the form $\Delta\rho \sim \rho_{295K}^{r(R_{4K})}$ over seven orders of magnitude, where the exponent $r(R_{4K})$ significantly increases for the more resistive iQCs ongoing a metal-insulator transition [728].

Another characteristic feature of QCs, which is not exhibited by usual metallic alloys, is the extreme sensitivity of most of their transport properties to minute changes in the sample stoichiometry. This is illustrated in Table 3.8 and Fig. 3.11b for the room temperature electrical resistivity of a series of samples belonging to the i-AlCuRu system. By inspecting this figure, we readily appreciate that the ρ_{300K} value systematically increases as we increase the number of electrons per atom. This is a quite unexpected behavior if we think of it in terms of the free electron model, but it can be better understood if instead we consider that most electrons are mainly employed in chemical bonding, hence becoming essentially localized, rather than extended in character. We also note that the $R_{1.4K}$ values generally increase with increasing ρ_{300K}, which in turn increases with the Ru content.

An important open question in the field regards whether the reported anomalies in the transport properties of high-quality QCs can be satisfactorily accounted for by merely invoking band structure effects or, conversely, they must be traced back to the critical nature of the electronic states. At this stage, it seems quite reasonable that the proper answer should likely require a proper combination of both kinds of effects. In fact, on the one hand, certain experimental facts, such as the relative insensitivity of the specific heat electronic term γ_e to thermal annealing as compared to the strong dependence of the electrical conductivity, suggests that the low values of residual conductivity $\sigma(0)$ cannot be satisfactorily explained by solely invoking the existence of the pseudogap. This conclusion is further stressed by the unrelated variations of σ_{4K} and γ_e among different AlPdRe samples [176]. On the other hand, it has been suggested that when the energy spacing between the electronic bands in the vicinity of Fermi level becomes very small, as it occurs in the case of quasicrystalline approximants, the transport may turn out to be anomalous because tunneling occurs between different bands, causing the instability of the wave packet coherence [748].

The high resistivity values and the large R ratios have been attributed to a remarkably low DOS value close to the Fermi level (see Table 3.2), along with the critical nature of

---

this, a clear trend of a negative $d\rho/dT$ at larger ρ values could be still appreciated in most considered samples, with a critical resistivity that can vary from 30 to 400 $\mu\Omega$cm [911].

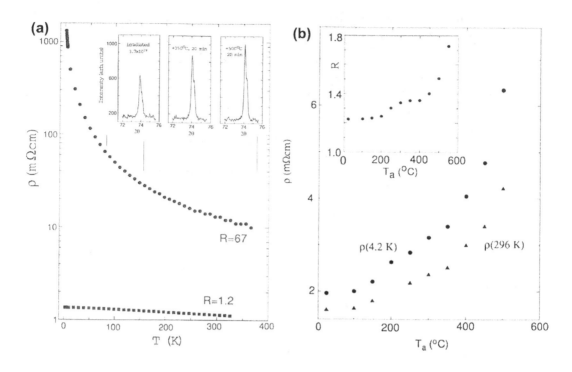

Figure 3.11 (a) Generalized Mooij-like plot for iQCs covering a broad resistivity ratios range. The dashed lines are estimates of the metal-insulator transition in i-AlPdRe samples [728]. (Rapp Ö, Generalization of the Mooij correlation to quasicrystals, J. Phys. Condens. Matt. 25, 065701 (7pp) 2013. © IOP Publishing. Reproduced with permission. All rights reserved). (b) Electronic resistivity values as a function of the e/a ratio for the i-Al$_{65}$Cu$_{35-x}$Ru$_x$ QCs listed in Table 3.8.

these states (see Sec. 2.7). Indeed, according to the Einstein relation

$$\sigma(T \to 0) = \frac{e^2}{V} D(E_F) N(E_F), \qquad (3.51)$$

the low temperature electrical conductivity is determined by both the number of available electronic states close to the Fermi energy and their diffusivity value, $D(E_F)$.[21] Therefore, a material is definitively an insulator if $N(E_F)$ is zero, though the possible existence of localized states in the energy spectrum with $D(E_F) \simeq 0$ in the Fermi energy neighborhood would lead to $\sigma(0) \to 0$, even in the presence of a finite $N(E_F)$ value. After Eqs.(3.22) and (3.51), we get

$$D(E_F) = L_0 \frac{\sigma(0)}{\gamma_e} \simeq 2.842 \times 10^{-4} \frac{\sigma(4 \text{ K}) \ [\Omega^{-1}\text{cm}^{-1}]}{\gamma_e \ [\text{mJ mol}^{-1}\text{K}^{-2}]}, \qquad (3.52)$$

which provides a rough estimation of charge carriers diffusivity. By inspecting the values listed in Tables 3.2–3.4 several conclusions can be drawn: 1) ZnMgGa sample exhibits the largest $D(E_F)$ value among all the considered iQCs,[22] 2) AlPdRe iQCs exhibit the lowest

---

[21] Note that Eq. (3.51) reduces to the spectral conductivity function given by Eq. (3.5).
[22] For the sake of comparison typical metallic compounds have $D(E_F) \simeq 50 \text{ cm}^2\text{s}^{-1}$.

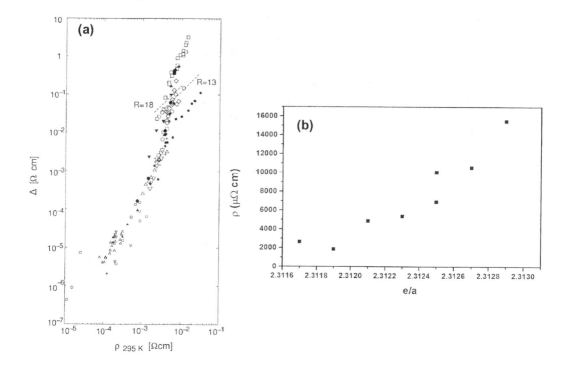

Figure 3.12   (a) Temperature dependence of an i-AlPdRe sample before ($R = 67$) and after a 6.8×10^{19} cm^{-2} neutron irradiation dose ($R = 1.2$) (inset: variation of the (4,6,0,0,0,0) diffraction peak intensity of i-AlPdRe after irradiaton with 3.7×10^{19} neutrons cm^{-2} followed by subsequent annealings for 20 min. at 350 and 500°C). (b) Evolution of the low and room temperature resistivity values of an i-AlPdRe sample irradiatiated with 3.7×10^{19} neutrons cm^{-2} upon increasing annealing temperatures T_a. (inset: the variation of the resistivity ratio R_{4K} with T_a). (Rapp Ö, Karkin A A, Goshchitskii B N, Voronin V I, Srinivas V, Poon S J, Electronic and atomic disorder in icosahedral AlPdRe, J. Phys. Condens. Matt. **20** 114120 (6pp) 2008. © IOP Publishing. Reproduced with permission. All rights reserved).

$D(E_F)$ values, which are about four orders of magnitude smaller than those of typical metallic compounds, 3) the electrical diffusivity values of different QC systems cluster around different, relatively narrow intervals, namely, 1–2 cm^{2}s^{-1} for i-AlLiCu, 1.3–2.3 cm^{2}s^{-1} for i-ZnMgY, and 0.14–0.24 cm^{2}s^{-1} for i-AlCu(Fe,Ru). Thus, Bergman-type QCs display similar figures, which are about one order of magnitude higher than those reported for Mackay-type QCs belonging to the i-AlCu(Fe,Ru) system, 4) Tsai-type QCs display a comparatively broader palette of $D(E_F)$ values, ranging from 0.2 cm^{2}s^{-1} (i-Cd$_{85}$Yb$_{16}$, i-Au$_{51}$Al$_{34}$Yb$_{15}$) up to 1.7 cm^{2}s^{-1} (i-Cd$_{85.1}$Yb$_{14.9}$). It is worthy to note that we can find iQCs exhibiting remarkably low diffusivity values ($\sim$0.2 cm^{2}s^{-1}) among both i-AlCu(Fe,Ru) and i-CdYb representatives, which display low temperature electrical conductivity values differing by more than one order of magnitude.

Fig. 3.12a shows the $\rho(T)$ curves of an i-Al$_{70.5}$Pd$_{21}$Re$_{8.5}$ sample before neutron irradiation ($R_{4K} = 67$) and after a maximum irradiation dose with 6.8×10^{19} neutrons cm^{-2} ($R_{4K} = 1.2$). After irradiation the sample was annealed for 20 min. at a series of low temperatures T_a, which were increased from 100°C in steps of 50°C and its $\rho(T)$ curve determined from 4.2 K to about 330 K (Fig. 3.12b). X-ray diffraction patterns showed that the heights of the peaks decreased with increasing irradiation dose, whereas their heights increased after subsequent annealing (see the inset to Fig. 3.12a), hence signaling a recovery of the icosahedral phase, which directly correlated with an increase of the electrical resistivity and the R_{4K} ratio (see the inset to Fig. 3.12b) as a function of T_a at both 4.2 K and room temperature. This experiment clearly demonstrates the standard QCs property of an increase of both ρ and R_{4K} values with improved structural perfection in a given sample under controlled conditions [727]. This counterintuitive behavior makes it convenient to distinguish between electronic localization and atomic disorder effects in QCs transport properties.

3.4.2 Inverse Matthiessen rule

In 1993, it was reported that the $\sigma(T)$ conductivity curves of a series of AlCuFe and AlPdMn iQCs, as well as some related approximant phases, were nearly parallel up to about 1000 K (Fig. 3.13a), so that one can write

$$\sigma(T) = \sigma(0) + \Delta\sigma(T), \tag{3.53}$$

where $\sigma(0)$ measures the (strongly sample dependent) residual conductivity at $T \to 0$, and $\Delta\sigma(T)$ is proposed to be a general function satisfying the condition $\lim_{T\to 0} \Delta\sigma(T) = 0$ [593]. According to Eq. (3.53), the contribution to the sample conductivity due to different sources of scattering seems to be additive, which is just the opposite to what happens to usual metallic alloys.[23] This remarkable behavior, referred to as *inverse Matthiessen rule*, is further illustrated for a broader collection of QCs in Fig. 3.13b. It has been also observed in most quasicrystalline approximants [716], and even in amorphous phases prior to their thermally driven transition to the QC phase (see Fig. 3.29). These findings indicate that the inverse Matthiessen rule may be a quite general property of structurally complex alloy phases closely related to quasicrystalline compounds.

The near parallelism of $\rho(T)$ curves over the temperature range $10 - 300$ K has been reported in i-AlCuFe thin films and ribbons irradiated at fluences of 10^{11} to a few 10^{13} ions cm^{-2}. Both the resistivity and R_{4K} values of thin films decrease after irradiation, though in ribbons a slight increase of the resistivity ratio is observed at high enough ion fluences [124], thereby suggesting that the energy deposited by incident ions may be used to improve the long-range QPO of iQCs.

Attending to the overall variation of their $\sigma(T)$ curves, Al-based QCs can be separated into two main sets. For one thing, we have AlCuFe and AlPdMn representatives, which are characterized by the presence of *broad minima* at about 10–20 K and 80–150 K, respectively. On the other hand, we have the relatively more resistive AlCuRu and AlPdRe alloys (see Table 3.7), exhibiting a *monotonous growth* over the entire considered temperature range. The $\sigma(T) - \sigma(0)$ curves of samples belonging to AlPdRe, AlCuRu, and AlCuFe samples can be fitted in terms of power law functions of the form $\Delta\sigma(T) \sim AT^\upsilon$, $(1/2 \leq \upsilon \leq 5/3)$ over a broad temperature range [287, 466, 537]. Notwithstanding this, at very low temperatures, the $\sigma(T) - \sigma(0)$ curves of the highly resistive AlPdRe samples are best fitted in terms of stretched exponential functions of the form $\Delta\sigma(T) \sim \sigma_0 \exp(-(T_0/T)^m)$, with $m = 1/2$, or $m = 1/4$ [174].

---

[23]The Matthiessen rule states that the electrical resistivities due to different sources of disorder are additive in dilute alloys, that is, $\rho(T) = \rho_r + \rho_L(T)$, where ρ_r is temperature independent and depends on the nature and concentration of impurities, while ρ_L describes the resistivity of the ideal metal. This relationship results from the additivity of the inverse scattering time τ_s, since $\rho \sim 1/\tau_s$ in conventional metals, as far as the scattering of electrons by impurities and lattice vibrations can be independently treated [1009].

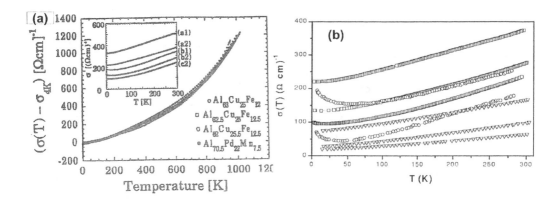

Figure 3.13 (a) Temperature dependence of the electrical conductivity for four different iQCs up to 1000 K. The inset illustrates the sensitivity of the residual conductivity value to minor variations in the samples composition. (Reprinted figure with permission from Mayou D, Berger C, Cyrot-Lackmann F, Klein T, Lanco P, Evidence for unconventional electronic transport in quasicrystals, Phys. Rev. Lett. 70, 3915-3918, 1993. Copyright (1993) by the American Physical Society). (b) Plot comparing the electrical conductivity temperature dependence for iQCs belonging to the AlCuFe ($\square$), AlCuRu ($\triangledown$), and AlPdMn ($\bigcirc$) alloy systems. From top to bottom their chemical compositions read as follows: $Al_{63}Cu_{24.5}Fe_{12.5}$, $Al_{62.8}Cu_{24.8}Fe_{12.4}$, $Al_{70}Pd_{20}Mn_{10}$, $Al_{62.5}Cu_{25}Fe_{12.5}$, $Al_{70}Pd_{20}Mn_{10}$, $Al_{65}Cu_{21}Ru_{14}$, $Al_{65}Cu_{20}Ru_{15}$, and $Al_{65}Cu_{19}Ru_{16}$. Data for AlCuFe samples were kindly provided by Claire Berger. Data for AlCuRu and AlPdMn samples after Refs. [466], [704], and [903], respectively. (Reprinted figure with permission from Maciá E, Universal features in the electrical conductivity of icosahedral Al-transition-metal quasicrystals, Phys. Rev. B 66, 174203, 2002 Copyright (2002) by the American Physical Society).

Then, the question arises concerning the possible existence of a suitable physical mechanism supporting the presumed universality of the $\Delta\sigma(T)$ function. In fact, the parallelism of the $\sigma(T)$ curves is difficult to understand in terms of a classical thermally activated mechanism, since the temperature dependence of $\sigma(T)$ does not follow an exponential law of the form $\exp(-E_g/k_BT)$. The inadequacy of this fitting implies the absence of a conventional semiconducting-like gap in QCs [697]. Additional evidence comes from the fact that, whereas for heavily doped semiconductors the $\sigma(T)$ curve decreases at high enough temperatures (as soon as all the impurity levels have become ionized), no evidence of such a limiting threshold has been observed in QCs. On the other hand, the existence of similar fitting curves for different systems does not guarantee the presence of similar physical mechanisms at work in those systems. For instance, curves of the form given by Eq. (3.53) have been used to describe the temperature dependence of the electrical conductivity in conducting polymers, where $\Delta\sigma(T)$ is interpreted within the model of weak localization and electron–electron interaction effects [197]. In the same vein, signatures of electron–electron scattering, spin-orbit interaction and chemical disorder effects have been inferred from the temperature dependence of $\sigma(T)$ curves in iQCs, although their relative role is still awaiting for a precise experimental and theoretical clarification [174].

Making use of the DOS model given by Eqs.(3.2) and (3.4) into Eqs.(3.5) and (3.11) one can derive the following approximate analytical expression for the electrical conductivity

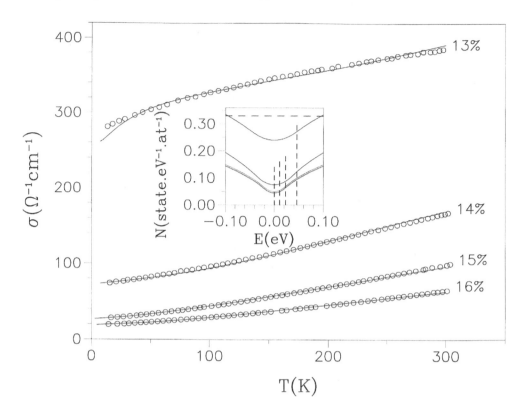

Figure 3.14 Comparison between analytical (solid lines) and experimental (circles) $\sigma(T)$ curves for i-AlCuRu samples, arranged according to their increasing content of Ru from top to bottom. The inset shows the DOS structure around the Fermi energies (dashed vertical lines) for the different samples. (Reprinted figure with permission from Maciá E, Modeling the electrical conductivity of icosahedral quasicrystals, Phys. Rev. B **61** 8771-8777 (2000). Copyright (2000) by the American Physical Society).

curve,

$$\sigma(T) = \sigma(0)\left[1 + \frac{\Omega b_W^2}{N(E_F^0)}F(T)\right],\tag{3.54}$$

where we have assumed that the electrons diffusivity is a constant, $D(E) \approx D_0$, so that $\sigma(0) = \frac{e^2 D_0}{V}N(E_F^0)$, and $F(T)$ is a dimensionless function including information regarding the electronic structure close to the Fermi level and satisfying the low temperature limit $\lim_{T\to 0}F(T) = 0$ [531]. According to Eq. (3.54), the $\sigma(T)$ curve can be expressed as a product involving two different contributions.[24] The first one is given by the $\sigma(0)$ factor which describes the residual conductivity of the sample. This factor will ultimately be the responsible for the overall low conductivity values observed in these materials. The second factor in Eq. (3.54) describes the temperature dependence of the electrical conductivity as the temperature is increased. It is worth noting that Eq. (3.54) essentially reduces to the

---

[24]Note that, if one reasonably approximates $\sigma_{4K} \simeq \sigma(0)$, this factor reduces to the resistivity ratio value R_{4K}.

empirically proposed inverse Matthiessen rule given by Eq. (3.53) by just identifying

$$\Delta\sigma(T) \equiv \sigma(0) \frac{\Omega b_W^2}{N(E_F^0)} F(T) = \frac{e^2 D_0}{V} \Omega b_W^2 F(T), \qquad (3.55)$$

so that, this contribution does not depend on the DOS value at the Fermi level, whereas $\sigma(0)$ does, as required. In addition, the $\Delta\sigma(T)$ contribution increases as both the curvature and the width of the pseudogap are increased, hence indicating that the positive temperature coefficient and resistivity ratio values are enhanced for more resistive iQCs.[25] Albeit the assumption of a constant diffusivity is too simplistic, expression (3.54) can reproduce experimental $\sigma(T)$ curves to a high degree of accuracy (Fig. 3.14), providing a unified description of the electrical conductivity for different i-AlCuRu samples over a broad temperature range. By inspecting the inset, we realize the importance of the precise location of the Fermi energy relative to the DOS minimum, which in turn is shifted by minor changes in the stoichiometry of the considered QC [531].

In order to further analyze the role of the states diffusivity in the electronic transport properties it is convenient to use the spectral conductivity function. Making use of Eqs.(3.6), (3.7), and (3.11), keeping terms up to $O(\beta^{-6})$ in the Taylor expansion of the $\sigma(x)$ spectral conductivity function around E_F^0, one obtains

$$\sigma(T) = \frac{J_0}{4} = \sigma(0)[1 + bT^2(\xi_2 + \xi_4 bT^2 + \xi_6 b^2 T^4)], \qquad (3.56)$$

which holds up to ~400 K, where $b \equiv e^2 L_0$ [475, 537]. The coefficients ξ_n contain detailed information about the electronic structure of the sample and they can be explicitly expressed in the form,

$$\xi_1 = \frac{1}{2} \left(\frac{d \ln \sigma(E)}{dE} \right)_{E_F}, \quad \xi_2 = 2\xi_1^2 + \frac{1}{2} \left(\frac{d^2 \ln \sigma(E)}{dE^2} \right)_{E_F}, \qquad (3.57)$$

which depends on the topology of the spectral conductivity function $\sigma(E)$. Thus, from the knowledge of the phenomenological coefficients ξ_1 and ξ_2 we can obtain suitable information concerning the slope and curvature of the DOS close to E_F. Eq. (3.56) is completely analogous to Eq. (3.54), hence accounting for the inverse Matthiessen rule given by Eq. (3.53) by identifying $\Delta\sigma(T) \equiv \sigma(0)bT^2(\xi_2 + \xi_4 bT^2 + \xi_6 b^2 T^4)$. However, although Eq. (3.56) renders a power law increase of the electrical conductivity with the temperature, the exponents are all of them integer, instead of the rational values obtained from fitting the experimental curves.

Broadly speaking, one would expect that different fits to the experimental data may be more or less adequate depending on the temperature ranges considered, since the relative importance of different physical mechanisms at work will depend on their own temperature scales. Thus, the electrical conductivity curve, when considered over a wide temperature range, can be expressed as a sum of two main contributions, namely $\sigma(T) = \sigma_l(T) + \sigma_h(T)$, where $\sigma_{l,(h)}(T)$ describes the low (high) temperature behavior. A systematic fitting analysis studied indicated that the following mathematical expression,

$$\sigma(T) = \sigma_0 + A_0 e^{-\gamma T} + A_1 T^v, \qquad (3.58)$$

where σ_0, A_1, and A_0 are measured in $(\Omega \text{ cm})^{-1}$ units and v and γ take on real values, yield

---

[25]Quite interestingly, a fit of the form $\Delta\sigma(T) \approx 1.3\sqrt{T}$ (i.e., almost independent of $\sigma(0)$) was reported for a series of highly resistive i-Al$_{70}$Pd$_{22.5}$(Re$_{1-x}$Mn$_x$)$_{7.5}$ QCs [497].

good enough correlation coefficients, though best fits usually were obtained by considering power law trial functions [537]. Temperature dependences of the form $\sigma(T) \sim T^v$ are predicted in different physical scenarios. An illustrative instance is provided by metal-insulator transition referred to as *Anderson transition*, after Philip W. Anderson (1923–2020). This transition is related to the wave functions localization due to the presence of disorder in the sample and is described in terms of a spectral conductivity given by the step function

$$\sigma(E) = \begin{cases} 0, & \text{if } E < E_C \\ \sigma_A \left| E - E_C \right|^v, & \text{if } E \geq E_C \end{cases} \tag{3.59}$$

where σ_A is a constant measured in $(\Omega \text{ cm})^{-1}(\text{eV})^{-v}$ units, E_C gives the mobility edge location separating localized (delocalized) states on the left (right), respectively, and $v > 0$ is a universal critical exponent [932]. We see that for $v \neq 0$ this step function is continuous at $E = E_C$, whereas for $v = 0$ Eq. (3.59) reduces to the Mott's spectral conductivity model describing the so-called Mott insulators. Making use of the spectral conductivity function one deduces that the electrical conductivity obeys a power law temperature dependence of the form $\sigma(T) = AT^v$ (**Exercise 3.8**), so that $\sigma(T) \to 0$ in the $T \to 0$ limit. Interestingly enough, a good adjustment to a spectral conductivity function of the form $\sigma(E) = \sigma_0 + \sigma_A \left| E - E_F^0 \right|^v$, was reported for i-$Al_{62.5}Cu_{25}Fe_{12.5}$ and i-$Al_{65}Cu_{20}Ru_{15}$ samples within the temperature ranges $\sim 100-400$ K and $\sim 4-500$ K, respectively [696]. These results indicated that these iQCs still exhibited a metallic-like behavior (with $\sigma(T) \neq 0$ in the $T \to 0$ limit), and the Fermi energy was playing a role similar to that of the mobility edge in full-fledged Anderson's systems. In fact, as we discussed in Sec. 2.7 the transport properties of different critical states can vary over a wide range (in terms of their related transmission coefficient value), though most of them exhibit rather low $T_N(E)$ figures in the QP $N \to \infty$ limit, thereby behaving in a way analogous to that typical of localized eigenstates in disordered systems. These analytical results are additionally supported by detailed numerical studies on the nature of electronic states in 1/1 and 2/1 QC approximants showing that nearly insulating behavior can coexist (in a very narrow energy interval) with a non vanishing, metallic-like DOS [447]. Analogous results were reported for a 5/3 approximant of a 3D Penrose lattice [931]. We should note that, in this case, localization is caused by constructive interference of electronic states stemming from the characteristic symmetry of QCs, at variance with the Anderson localization arising from disorder [928–930].

One can also explain the emergence of a power law temperature dependence for the electrical conductivity over a broad temperature interval by considering the very nature of electronic states in QCs. As we saw in Sec. 2.6.2, critical electronic states typically decay in space following a power law given by $\psi \propto r^{-\alpha}$, with $\alpha = \frac{1}{2} \frac{\ln(n(\lambda))}{\ln \lambda}$, where λ is a scaling factor describing the QC self-similar geometry and $n(\lambda)$ gives the number of clusters of size L in a given volume of size λL (Conway's theorem, see Sec. 1.4.3). Within the so-called variable range hopping framework it is assumed that the electron transport is mediated via thermally activated hoppings involving localized states inside atomic clusters nested within a self-similar hierarchical arrangement [376, 377, 379]. Thus, the resulting conductivity is proportional to the jump frequency between neighboring clusters, which is given by

$$f(L) \simeq |\psi(L)|^2 \exp\left(-\frac{\Delta E_L}{k_B T}\right), \tag{3.60}$$

where ΔE_L measures the separation between successive energy levels in the clusters. Assuming that these clusters can be modeled in terms of square potential wells, we have $\Delta E_L \sim L^{-2}$.[26] Thus, $\sigma(T) \sim f(L) \simeq L^{-2\alpha} e^{-\frac{g}{L^2 T}}$, where g is a suitable constant. If we

---

[26] For a square-well potential we have $\Delta E_n \equiv E_{n+1} - E_n = \left(\frac{hc}{2e}\right)^2 \frac{2n+1}{2mL^2}$, where m is the electron's mass.

further assume that at a given temperature value the more probable hopping distance, L_o, optimizes the jumping frequency, we get $L_o^2(T) = \frac{g}{\alpha T}$, so that $\sigma_{L_o}(T) = A_\alpha T^\alpha$, with $A_\alpha \equiv (\alpha/g)^\alpha e^{-\alpha}$ [376, 377]. Accordingly, the spatial structure of critical states, parametrized in terms of the exponent α, ultimately determines the power law exponent of the electrical conductivity. Alternative approaches, exploiting the multiple-valley fractional Fermi surface model for the electronic structure of QCs, have been considered to explain the $\sigma(T)$ curve as well [928–930].This approach appears as the natural extension of the fractal energy spectrum obtained for 1D models to the 3D case, where Fermi sphere is replaced by a fractal surface in 3D [703].

3.4.3 Optical conductivity

The study of optical properties, performed over a very broad spectral range, is a powerful experimental tool for identifying the spectrum of excitations in a solid. In this way, several intrinsic parameters, such as the plasma frequency, relevant excitations due to phonons, or the strength of interband transitions can be evaluated. To this end, one experimentally obtains the reflectivity curve, $R(\omega)$, as a function of the incoming electromagnetic radiation frequency, and derives from it the optical conductivity curve $\sigma(\omega)$ by means of the so-called Kramers-Krönig transformation of the reflectivity spectrum. This transformation requires the knowledge of the optical responses at very low and very high frequencies, which are generally obtained from suitable extrapolations, since the measurements of optical properties is generally carried out at the infrared to UV regime (0.1−50 eV).

The $\sigma(\omega)$ curve of conventional metallic alloys is determined by several contributions. Firstly, we have intraband transitions involving conduction electrons which can be analyzed using the complex conductivity function derived from the Drude model for free electrons [608],

$$\sigma(\omega, T) = \frac{\sigma(0, T)}{1 - i\omega\tau} = \frac{\sigma(0, T)}{1 + (\omega\tau)^2}(1 + i\omega\tau) \equiv \sigma_1(\omega, T) + i\sigma_2(\omega, T), \qquad (3.61)$$

where $\sigma(0, T) = e^2 n\tau/m_o^*$ is the dc conductivity, τ is the relaxation time, n is the number of electrons per unit volume, and m_o^* is the optical effective mass of the conduction electrons. The $\sigma_1(\omega)$ contribution dominates the optical response at low frequencies and results in a characteristic Lorentzian function centered at the zero frequency, known as the Drude peak, followed by a rapid decay of the optical conductivity at large frequencies. A second contribution (mainly affecting the far-infrared region of the spectrum) is related to the presence of optical phonon modes, which are activated when the incoming radiation frequency is equal to or exceeds the necessary excitation energy. Additional contributions come from transitions involving both the valence and conduction bands (interband transitions), which are usually located in the visible spectral range. Thus, good conductors show a reflectance close to 100% at frequencies below the onset of absorption due to interband transitions and a characteristic sudden decay (known as the plasma edge) as the frequency increases approaching the value

$$\omega_p^2 \equiv \frac{ne^2}{m_o^*\varepsilon_0} = \frac{4\pi ne^2}{m_o^*} \text{ (CGS)}, \qquad (3.62)$$

referred to as the plasma frequency, where ε_0 is the vacuum dielectric constant (**Exercise 3.9**). This frequency defines a threshold value. At low frequencies (i.e., $\omega < \omega_p$), the free

---

The relationship becomes more involved in the case of a spherical square-well, whose energy eigenvalues read $E_{n,\ell} = \frac{\hbar^2}{2mL^2}\chi_{n,\ell}$, where $\chi_{n,\ell}$ are the zeros of the Bessel functions $J_{\ell+\frac{1}{2}}(kr)$ [375], so that the L^{-2} dependence is still preserved.

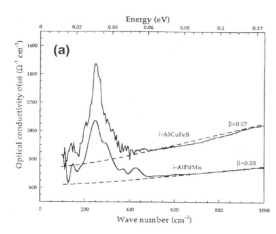
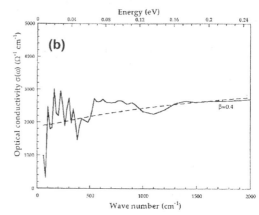

Figure 3.15 (a) Optical conductivity in i-AlCuFeB and i-AlPdMn samples (solid lines). The fit curves (dashed lines) are obtained after Eq. (3.63). The peaks around 0.03 eV are associated to phonons. (b) Same as in (a) for an AlCrFe approximant sample. (Reprinted figure with permission from Demange V, Milandri A, de Weerd M C, Machizaud F, Jeandel G, and Dubois J M, Optical conductivity of Al-Cr-Fe approximant compounds, Phys. Rev. B **65** 144205, 2002 Copyright (2002) by the American Physical Society).

electrons can couple to the oscillating electromagnetic field of incoming photons giving rise to a collective motion referred to as plasma oscillation. Accordingly, no radiation can propagate and the radiation field falls exponentially inside the solid. On the contrary, at large enough frequencies (i.e., $\omega > \omega_p$), the electromagnetic wave can propagate and the medium becomes transparent. Thus, a metal is basically transparent to light for wavelengths smaller than the plasmon cut-off and absorbing and reflecting above. On the other hand, in semiconducting materials the absorption of a photon of energy $\hbar\omega_g$ is possible as soon as it equals the gap width E_g (direct transitions) or if the top of the valence band and the minimum of the conduction band in reciprocal space are separated by a wave vector belonging to the lattice (indirect transitions).

Attending to the experimental data the optical conductivity of QCs studied so far is quite different from that of either a metal or a semiconductor. For instance, reflectance of high quality samples was found to be significantly small (about 60%) over a wide wavelength region from about 300 nm (UV region) to 20 μm (IR region). In addition, the following unusual features were observed in the optical conductivity curves:

1. The far infrared signal is very weak and Drude peak is either highly suppressed or completely absent (Fig. 3.15). Notwithstanding this, extrapolation of the $\sigma(\omega)$ curves to the zero frequency yield conductivity values in good agreement with the measured dc conductivity ones [22, 58, 173, 178, 340, 964]. Furthermore, using spectrometers of sufficient resolution and samples of excellent lattice perfection, signatures of the expected hierarchical nature of the Brillouin zone, showing up in the IR light absorption coefficient as a series of van Hove singularities in the DOS, were reported [935].

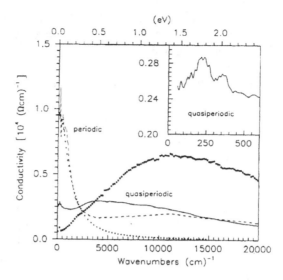

Figure 3.16 The optical conductivity of the decagonal AlCoCuSi quasicrystal for the periodic (short-dashed line) and the quasiperiodic (solid line) directions is compared with the conductivity of the icosahedral AlCuFe studied in Ref. [340] (solid dots). In the inset the quasiperiodic conductivity in the far-infrared part of the spectrum is shown. (Reprinted figure with permission from Basov D N, Timusk T, Barakat F, Greedan J, and Grushko B, Anisotropic optical conductivity of decagonal quasicrystals, *Phys. Rev. Lett.* **72** 1937–1940 1994 Copyright (1994) by the American Physical Society).

2. All the studied QCs exhibit a typical absorption feature overlapping the low frequency tail of the far-infrared region (Fig. 3.15). This relatively broad feature usually splits into several separate contributions (at about 25 and 35 meV in most cases) which may be related to the presence of fine structure in the electronic DOS and/or the phonon VDOS.

3. At higher energies ($\sim$0.4 eV) the optical conductivity progressively rises exhibiting a broad intraband transition within the intervals around 0.7 eV (i-ZnMgY, i-ZnMgTb, i-ZnMgHo), 1.2–1.5 eV (i-AlCuFe, i-AlPdMn), or 2.6–2.9 eV (i-AlPdRe), with optical conductivities between 1600 and 13000 $(\Omega\text{cm})^{-1}$, after which the conductivity decreases. This absorption feature is commonly ascribed to excitations across the characteristic DOS pseudogap (Sec. 3.3.1), since the location of the pseudogap absorption feature correlates with the width and depth of the related pseudogap for the different QC alloy systems.

Two complementary explanations have been proposed to account for the unusual absence of a Drude peak. For one thing, the small $\sigma(\omega)$ value at low frequencies can be assigned to a low density of states at the Fermi level due to the presence of a pseudogap in the band structure of QCs, hence leading to a substantially small value of $\sigma(0)$ in Eq. (3.61) [97]. Since, according to Eq. (3.62), $\sigma(0) = \omega_p^2 \varepsilon_0 \tau$, one should expect relatively small plasma frequency values in this case. In fact, ω_p values within the ranges 8–9 eV and 11.5–12 eV have been reported for i-AlPd(Mn,Re) and i-ZnMgRE, respectively [58, 173, 406], which are smaller than that of aluminum metal (14.9 eV). In the same vein, it has been reported that i-ZnMgRE QCs, exhibiting higher dc conductivity values than those of Al-based QCs, also have a larger $\sigma(\omega)$ signal in the low frequency region [406]. On the other hand, the absence of a Drude peak could also be explained as stemming from a low electronic mobility arising from very small τ values, related to the localization of charge carriers due to the quasiperiodicity of the structure, which leads to an anomalous diffusion mechanism. In that case, Drude's formula for the optical conductivity adopts the form [594],

$$\sigma(\omega) = Ae^2 N(E_F)\Gamma(2\beta + 1)\left(\frac{\tau}{1 - i\omega\tau}\right)^{2\beta-1}, \tag{3.63}$$

where A is a constant, Γ is the Gamma function, and β is a diffusion exponent which depends on the energy ($0 \leq \beta \leq 1$). This expression reduces to Eq. (3.61) in the case $\beta = 1$, whereas when $\beta < 1$, $\sigma(\omega)$ displays a dip at $\omega \to 0$ that increases monotonically with photon energy. Values as low as $\beta = 0.07$ and $\beta = 0.03$ were found from fitting analysis of the $\sigma(\omega)$ curves of AlCuFeB and AlPdMn QCs, respectively (Fig. 3.15a) [178]. Significantly larger values were obtained from a fitting analysis comparing i-GdYb$_{7.8}$ QCs ($0.2 \leq \beta \leq 0.4$) with their related approximant phases GdCd$_6$ ($0.6 \leq \beta \leq 0.8$) and YCd$_6$ ($0.5 \leq \beta \leq 0.75$) via Eq. (3.63) [22]. Accordingly, the diffusion of charges in QCs is subdiffusive ($\beta < 1/2$), while it is superdiffusive ($\beta > 1/2$) in approximant crystals

In summary, unlike disordered metals (where a Drude model is still applicable) or semi-conductors (with a well developed conductivity gap), the reflectivity spectra of icosahedral phases display low optical conductivity on the far-infrared energy range and a marked absorption in the visible. These characteristic features are also observed in typical approximant phases, such as 1/1 AlMnSi, although the optical conductivities of both 1/1-GdCd$_6$ and 1/1-YCd$_6$ approximants have distinct Drude peaks in the low frequency regime [22]. Therefore, the question regarding as to whether these unusual optical properties are (or not) related to long-range QPO effects cannot be definitively answered to date. In this sense, we note that different behaviors of the $\sigma(\omega)$ curve can be clearly established in dQCs between the quasicrystalline and the periodic directions (Fig. 3.16) [40]. Thus, a Drude peak is observed when light is irradiated within a narrow area parallel to the periodic axis, whereas no peak is detected in a plane perpendicular to it. The analysis of the optical data shows that contrarily to the case of iQCs, there is no clear evidence for the presence of a marked pseudo-gap at the Fermi level.

The optical behavior of both synthetic and natural icosahedrite and decagonite QCs belonging to the AlCuFe alloy system (see Sec. 1.2) was recently explored by measuring their reflectance in the range $400-700$ nm. In this way, significant differences were observed by comparing the obtained $R(\lambda)$ curves, which have been tentatively ascribed to the higher degree of structural perfection of mineral versus synthetic QCs (quantified in terms of the absence of phason strain in their electron diffraction patterns, see **Exercise 1.5**) [76].

3.4.4 Current-voltage curves

In contrast to the non-metallic behavior of the $\sigma(T)$ curves the characteristic current–voltage (I–V) curves of i-AlCuFe QCs exhibit a perfect Ohmic behavior for bias voltages which vary by seven (five) orders of magnitude in thin films (sintered bulk samples), respectively, suggesting that a linear I–V behavior may be a common property of QCs [425]. Such a linear behavior still holds when the sample temperature is progressively increased (inset to Fig. 3.17), indicating that the Kubo's linear response theory of electrical conductivity can be safely applied to QCs. Nevertheless, in the light of the electronic structure discussed in Sec. 3.2.2, one would have expected to observe some non-linearity in the I–V curves as soon

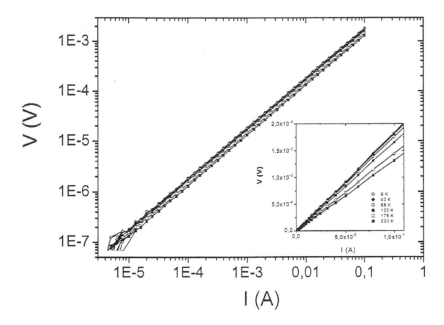

Figure 3.17 Double logarithmic I-V plots of an i-Al$_{63}$Cu$_{25}$Fe$_{12}$ QC (kindly provided by Jean Marie Dubois) at $T = 9, 45, 65, 100, 175$ and 230 K. The inset shows the linear representation of the same data. (Courtesy of Javier García-Barriocanal).

as the energy change of the charge carriers involved in the measurement process are in the range 0.01–1 eV due to existence of fine structure features in the DOS on the scales of about 0.1 eV (dip pseudogap) to 0.01 eV (resonant effects due to the quasiperiodic distribution of transition metal clusters) [901]. Now, the highest electric field applied in these experiments is about $\mathcal{E} = 50$ V/cm, so that we get the electron energy $E \simeq e\mathcal{E}\ell_0 \simeq 10^{-5}$ eV, where $\ell_0 \simeq 2$ nm, is a roughly estimate of the electronic mean free path in these materials [425]. Certainly, this figure is small enough to play a subsidiary role in the considered I–V measurements. In other words, stronger electric fields should be applied in order to observe any possible effect related to finer electronic structure features in these materials.

In fact, according to a detailed study of Fibonacci chain described in terms of a diagonal Hamiltonian model (see Sec. 2.3) with a uniform electric field $\mathcal{E}$ applied along the chain direction,[27] it is observed that the singular spectrum related to the QP potential tends to be smoothed in the case of a weak applied field, so that the gaps widths decrease linearly with the field strength. When the electric field and the QP potential have the same order of magnitude it is possible to observe a delocalization effect due to local resonances. Only in the limit of a strong field, a ladder spectrum with a structure that has the shape of the QP potential is observed, and the electronic states tend to be localized in this case [767], albeit due to quantum tunnelling effect it is possible to observe delocalization signatures in some states as well.

The presence of an external electric field tilts the energy level profile leading to the presence of an oscillating motion of some electrons, which are consequently removed from

---

[27]The system Hamiltonian is then obtained by adding the term $H_F = \mathcal{E}\sum_k |k\rangle k \langle k|$ to Eq. (2.7), with $t_{m,k} \equiv t$.

the dc current flow. In periodic crystals the period of these so-called Bloch oscillations depends on the electron's lattice momentum k as $\tau_B = \frac{\hbar k}{e\mathcal{E}}$, which amounts to $\tau_B \simeq 10^{-6}$ s in typical conditions. This figure is much larger than the typical relaxation time value $\tau_R \simeq 10^{-14}$ s in metallic systems, which explains the experimental difficulty in observing Bloch oscillation effects in periodic crystals. In QCs k is not bounded below, so that the condition $\tau_B \simeq \tau_R$ could be attained in principle for small enough k values. In that case, a significant number of charge carriers might be sequestered from dc conductivity in the presence of a strong enough electrical field [378]. In this regard, a detailed study of the dynamics of electronic wave packets in Fibonacci semiconductor superlattices subjected to homogeneous electric fields perpendicular to the layers, demonstrated that periodic Bloch oscillations observed in periodic superlattices,[28] are replaced in Fibonacci superlattices by more complex oscillations displaying QP signatures instead [184].

3.4.5 Superconductivity?

A large number of metallic elements and alloys become superconductors, that is they lost all trace of electrical resistivity below a well-defined critical temperature T_c, ranging from about 0.3 K for $ZrZn_2$ up to 23.2 K for Nb_3Ge [198]. In order to present convincing evidence for the emergence of superconductivity, not only zero resistance must be shown, but also the presence of the Meissner effect, showing that the superconducting state is a thermodynamically equilibrium state and that superconductors are perfect diamagnets which expel all magnetic flux from their interiors.[29] In addition, to provide evidence that the superconductivity is of bulk origin, one must observe a heat capacity jump at the transition temperature T_c.

Attending to these criteria the $Mg_{37.5}Zn_{37.5}Al_{25}$ alloy was earlier tentatively reported to become a superconducting QC [304], but it was later shown to be an approximant crystal instead [861]. Therefore, it is important to examine whether the studied alloy is an approximant phase or a full-fledged QC, since many approximant crystals show bulk superconductivity at low temperatures but the related QCs exhibit non-zero residual resistivity values. By all indications, the presence of bulk superconductivity in an i-QC was first reported in 2018, in samples belonging to the Al-Zn-Mg alloy system, with a very low transition temperature of $T_c \simeq 0.05$ K [405].[30] The precise geometric structure of the alloy samples depends on the ratio of the three constituent elements. When reducing the Al content, while keeping the Mg content almost constant, the alloy is an approximant crystal, which transforms to the iQC phase at 15% Al content [405, 861]. All of the approximant crystals show bulk superconductivity at low temperatures; the transition temperature T_c gradually decreasing from ~0.8 to ~0.2 K as the Al content is reduced, and T_c goes down to ~0.05 K when the critical Al concentration is reached to obtain the i-$Mg_{44.1}Zn_{41.0}Al_{14.9}$ QC ($e/a = 2.171$).

In Fig. 3.18, the relation between the superconducting transition temperature T_c and the inverse of the electronic specific heat coefficient γ_e (see Sec. 3.3.1) is shown for Al-Zn-Mg samples of different compositions (including seven 1/1 approximants, one 2/1 approximant,

---

[28]The period and amplitude of the Bloch oscillations in periodic superlattices in real as well as in k space are given by $\tau_B = \frac{h}{e\mathcal{E}L}$ and $x_B = \frac{\Delta W}{e\mathcal{E}}$, where L is the superlattice constant and ΔW is the bandwidth.

[29]It should be noted that in the so-called Type-II superconductors the Meissner effect is not complete, and their magnetic properties can be described in terms of a vortex lattice model.

[30]For the sake of comparison, this T_c value is about six times lower than that of $ZrZn_2$ alloy, one order of magnitude lower than that of the heavy electron intermetallic compound UPt_3, and only comparable to some representatives of the $SiTiO_{3-x}$ perovskites. Among metallic elements only W ($T_c = 0.01$ K) and Rh ($T_c = 0.0003$ K) exhibit lower figures [198].

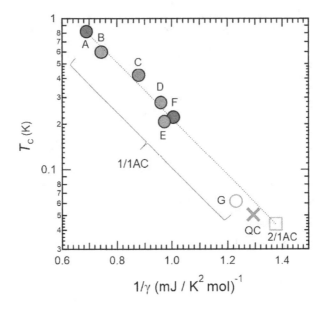

Figure 3.18 Correlation between the superconducting transition temperature T_c and the inverse of the electronic specific coefficient γ_c for the Mg-Zn-Al QC and its approximant crystals. A to G denote the 1/1 approximants. (Kamiya K, Takeuchi T, Kabeya N, Wada N, Ishimasa T, Ochiai A, Deguchi K, Imura K, Sato N K, Discovery of superconductivity in quasicrystal, *Nat. Commun.* **9,** (2018) 154; doi: 10.1038/s41467-017-02667-x. Creative Commons Attribution 4.0 International License (CC BY)).

and one iQC). A linear relation of the form $\ln T_c \sim \gamma_e^{-1}$ can be clearly appreciated. According to the BCS theory of superconductivity, $T_c = 1.14\Theta_D e^{-1/\Lambda N(E_F)}$, where Θ_D is the Debye temperature, $N(E_F)$ is the DOS at the Fermi energy, and Λ measures the electron–electron interaction within the weak coupling condition $\Lambda N(E_F) \ll 1$. As $\Theta_D \simeq 300$ K is almost independent of the Al content in the considered samples, the linear relationship plotted in Fig. 3.18 indicates that: (1) T_c is fully determined by $N(E_F)$, and (2) the effective pairing interaction between electrons is kept attractive and its magnitude remains the same when the long-range order present in the system is varied from periodic to QP in nature.

The observation of superconductivity in QCs raises an interesting question of whether the emerging superconductivity shows a weak-coupling, spatially extended Cooper pair or a strong-coupling, local pair (reflecting the possible role of electronic critical states). According to the experimental observations the temperature dependences of the thermodynamic properties and the upper critical filed can be understood within the weak-coupling framework BCS theory, suggesting the formation of spatially extended pairs. However, this does not necessarily mean that the superconducting state of this QC is the same as that generally observed in periodic crystals. Indeed, according to a recent theoretical study [766], the pairing state of quasicrystal is expected to differ from the conventional Cooper pairing of the BCS theory.

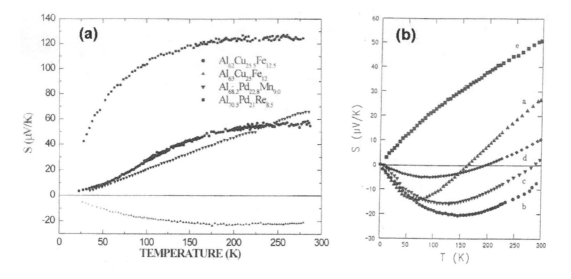

Figure 3.19 (a) Temperature dependence of the Seebeck coefficient for different Al-based, TM bearing iQCs. (Courtesy of Roberto Escudero). (b) Seebeck coefficient curves for i-Al$_{65}$Cu$_{20}$Ru$_{15}$ (a), i-Al$_{68}$Cu$_{17}$Ru$_{15}$ (b), i-Al$_{70}$Cu$_{15}$Ru$_{15}$ (c), i-Al$_{64.5}$Cu$_{20}$Ru$_{15}$Si$_{0.5}$ (d), i-Al$_{64}$Cu$_{20}$Ru$_{15}$Si$_1$ (e). (Reprinted figure with permission from Pierce F S, Bancel P A, Biggs B D, Guo Q, and Poon S J, Composition dependence of the electronic properties of Al-Cu-Fe and Al-Cu-Ru-Si semimetallic quasicrystals, Phys. Rev. B **47** 5670-5676 (1993) Copyright (1993) by the American Physical Society).

3.5 SEEBECK COEFFICIENT

The Seebeck coefficient, $S(T)$, describes the conversion of thermal energy into electrical energy in the form of an electrical current. The magnitude of this effect can be described in terms of the Seebeck voltage, $\Delta V = S(T)\Delta T$, which relates the electric response of a sample due to the application of an external temperature gradient ΔT. For materials whose electronic states exhibit a nearly constant diffusivity D_0, one has $\sigma(E) = e^2 D_0 N(E)/V$, and the Seebeck coefficient takes the celebrated Mott's expression [26, 27, 608],

$$S(T) = -\frac{\pi^2 k_B^2}{3|e|}\left(\frac{d\ln N(E)}{dE}\right)_{E_F} T = -|e|L_0 \frac{1}{N(E_F)}\left(\frac{dN(E)}{dE}\right)_{E_F} T. \quad (3.64)$$

By inspecting this expression one realizes that, in order to exhibit large S values, two main features should be present in the considered material DOS, namely, (1) a sharp peak close to the Fermi level, resulting in a large DOS slope, and (2) a substantial depletion of states, leading to a small value of the DOS close to the Fermi level. In this way, from very general considerations, one concludes that the *simultaneous* presence of a small DOS background (i.e., a pseudogap) along with a narrow peak near the Fermi level, will generally contribute to enhance the thermoelectric performance of the considered material.

During the last decade the $S(T)$ curves of samples belonging to different QCs have been measured. Reported data include a broad range of stoichiometric compositions and cover different temperature ranges in the interval from 1 K to 900 K. From the collected data the following general conclusions can be drawn for high-quality iQCs containing transition metals [63, 317, 695, 696].

- The temperature dependence of the Seebeck coefficient usually deviates from the linear behavior prescribed by Eq. (3.64) as the temperature is increased in Mackay-type iQCs, exhibiting pronounced curvatures (either positive or negative) at temperatures above ~50–100 K (Fig. 3.19a). Conversely, both Bergman-type and Tsai-type iQCs exhibit markedly linear temperature dependences, with slopes ranging from approximately 0.015 μV K^{-2} to 0.030 μV K^{-2} [285, 705].

- Room temperature Seebeck coefficients ranging from -30 μVK^{-1} up to $+120$ μVK^{-1} have been reported for Mackay-type iQCs. These values are relatively large when compared to those observed in both crystalline and disordered metallic systems ($|S| \simeq 1-10$ μVK^{-1}).

- The $S(T)$ curves can exhibit well-defined extrema in the temperature range $80-250$ in several iQCs. Both the magnitude and position of the extrema observed in these curves are extremely sensitive to minor variations in the chemical stoichiometry of the sample. As a consequence, small variations in the chemical composition (of just a few atomic percent) can give rise to sign reversals in the $S(T)$ curve within the temperature interval $160-350$ K (Fig. 3.19b).

Making use of the spectral conductivity function given by Eq. (3.6) into Eq. (3.9) the Seebeck coefficient curve can be written in the form[537, 539],

$$S(T) = -2|e|L_0 T \frac{\xi_1 + \xi_3 b T^2}{1 + \xi_2 b T^2 + \xi_4 b^2 T^4},$$ (3.65)

where $b \equiv eL_0$. Thus, in the low temperature limit Eq. (3.65) takes the linear form

$$S(T \to 0) = -2|e|L_0 \xi_1 T.$$ (3.66)

The sign of the slope is determined by the sign of the parameter ξ_1 which, in turn, depends on the electronic structure of the sample according to Eq. (3.57). Therefore, Eq. (3.66) reduces to Eq. (3.64) in the low temperature limit. It then follows that Mott's formula will properly describe the Seebeck coefficient of QCs as far as the terms $\xi_2 b T^2$, $\xi_3 b T^2$, and $\xi_4 b^2 T^4$ in Eq. (3.65) remain negligible as compared to ξ_1 as the temperature increases. Therefore the range of validity of Mott's formula is strongly dependent on the electronic structure of the sample. According to Eq. (3.65) the $S(T)$ curve will change its sign at a temperature given by $T_0 = \sqrt{-\frac{\xi_1}{\xi_3 b}}$, which requires that coefficients ξ_1 and ξ_3 take on opposite signs, thereby justifying why not all QCs exhibit such a signs reversal on the basis of possible differences in the finer details of their electronic structure close to the Fermi energy. Furthermore, by imposing to Eq. (3.65) the extreme condition $dS(T)/dT = 0$, we get

$$\xi_3 \xi_4 y^3 + (3\xi_1 \xi_4 - \xi_2 \xi_3) y^2 + (\xi_1 \xi_2 - 3\xi_3) y - \xi_1 = 0,$$ (3.67)

where $y \equiv bT^2$. By means of Eq. (3.67) the possible existence of maxima or minima in the $S(T)$ curve can be related to the electronic structure in a precise, although somewhat involved way, by means of the phenomenological coefficients ξ_n. This fact helps us to understand the physical reasons motivating the strong dependence of these extrema on minor stoichiometric changes, since by changing the electronic structure of the sample, we are substantially modifying the values of the ξ_n coefficients which determine the solutions of Eq. (3.67).

3.6 HALL COEFFICIENT AND MAGNETORESISTANCE

Let us consider that an external electric field, $\mathcal{E}_x$ is applied to a solid along the X direction, so that a current density $j_x = nqv_x$ flows, where n is the charge carriers density per unit volume and q is the carrier's charge. If a transverse magnetic field B_z pointing in the positive Z direction is simultaneously applied the resulting Lorentz force $\mathbf{v} \times \mathbf{B}$ will deflect the charge carriers in the Y direction. This gives rise to an electrostatic field $\mathcal{E}_y$ (the so called Hall field) in that direction, that opposes further charge motion along Y axis. In the equilibrium the Hall field will balance the Lorentz force, and current will flow only in the X direction. In this setup there are two magnitudes of interest. For one thing, we have the magnetoresistance, which is defined by the ratio $\rho(B_z) = \mathcal{E}_x/j_x$, and the other is the Hall coefficient, defined as the ratio of the Hall field to the Lorentz force [608],

$$R_H \equiv -\frac{\mathcal{E}_y}{j_x B_z} = \frac{1}{nq} = \frac{1}{nqc} \text{ [CGS]}. \tag{3.68}$$

Therefore, the value of the Hall coefficient measures the density of carriers and its sign indicates their nature (i.e., electrons $q = -e$, or holes $q = +e$). According to Eq. (3.68) the Hall coefficient is temperature independent, albeit R_H actually depends of temperature in real solids, mainly due to a temperature dependence of the relaxation time in metals and also due to an increase of charge carriers with temperature in semimetals and semiconductors.

An important relationship between electrical conductivity and the Hall coefficient allows one to determine the Hall mobility of charge carriers (measured in $cm^2 V^{-1} s^{-1}$)

$$\mu_H = \sigma R_H = c\sigma R_H \text{ [CGS]}, \tag{3.69}$$

so that Hall coefficient measurements also provide information regarding the charge carriers diffusivity. In fact, making use of Eqs.(3.51) and (3.68) into Eq. (3.69), we get $\mu_H(T \to 0) = R_H(T \to 0)\sigma(T \to 0) \sim |q|D(E_F)$, so that we can estimate the charge carriers diffusivity by extrapolating the $\sigma(0)$ and $R_H(0)$ values from the experimental $\sigma(T)$ and $R_H(T)$ curves. As expected from Eq. (3.69), the $R_H(T)$ curve increases as the electrical conductivity $\sigma(T)$ decreases (and vice versa), and eventually diverges when the residual conductivity $\sigma(T \to 0)$ goes to zero (insulating case), as it is shown in Fig. 3.20a.

Experimental measurements of the Hall coefficient of samples covering a broad collection of different QC alloy systems have been reported during the last decades [60, 61, 63, 131, 289, 317, 318, 423, 498, 504, 695, 697, 709, 865]. Both positive and negative values have been observed at low temperatures: $R_H(0) > 0$ for AlPd(Mn,Re) QCs, $R_H(0) < 0$ for AlCuRu and ZnMgRE QCs, and $R_H(0)$ can take on both positive and negative values for AlCuFe QCs depending on the sample's stoichiometry (see Table 3.9). The corresponding $R_H(0)$ values indicate small carriers densities, ranging from $n \simeq 10^{21}$ cm^{-3} for ZnMgY QCs, to $n \simeq 10^{20} - 10^{21}$ cm^{-3} for AlCu(Fe,Ru) and AlPdMn QCs, or even $n \simeq 10^{19}$ cm^{-3} for AlPdRe QCs, which are unusually low figures for alloys made of good metals. In Table 3.9, we also list the low temperature electrical conductivity and related Hall mobility values obtained after Eq. (3.69) for several QCs.

Several anomalous properties have been reported in the temperature dependence of the Hall coefficient within the interval 4 K–300 K:

- The temperature dependence of R_H in QCs and their approximants is generally stronger than in amorphous alloys.

- The $R_H(T)$ curves of QCs are strongly sensitive to minor variations in the sample stoichiometry and annealing conditions.

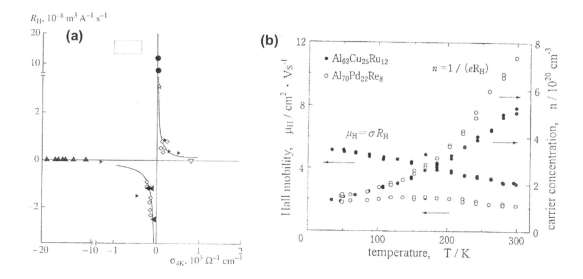

Figure 3.20 (a) Dependence of the low temperature Hall coefficient on the residual conductivity for AlCuMg (▲),[703] AlPdRe (●)[498], AlCuFe (◇),[504] AlCuRu (◄),[60] AlCuFe (►),[708] AlMn (▼),[464] and AlPdRe (★) iQCs. We clearly observe the trend of R_H to diverge for the most resistive QCs [318, 504, 708, 709]. (Reprinted by permission from Springer Nature Customer Service Centre GmbH, Prekul A F and Shchegolikhina N, Correlations between electric, magnetic, and galvanomagnetic quantities in stable icosahedral phases based on aluminium, *Crystallogr. Rep.* **52** 996-1005, 2007. Copyright 2007). (b) Temperature dependence of charge carrier density and Hall mobility of i-$Al_{63}Cu_{25}Ru_{12}$ and i-$Al_{70}Pd_{22}Re_8$ QCs [419]. (Courtesy of Kaoru Kimura).

- The $R_H(T)$ curve of ZnMgY QCs remains relatively constant, taking on the values between -1×10^{-9} and -5×10^{-9} m^{3}C^{-1} in the 1.8−250 K interval [131].

The $R_H(T)$ curves of Al-based QCs can be interpreted as indicating that the number of available charge carriers increases as temperature is increased [318, 709, 710]. A temperature dependence of the form $n(T) = n_0 + AT^2$ was proposed from the study of AlCuRu QCs [865], and a similar expression may also apply to AlPdRe iQCs with relatively small resistivity ratios below 200 K (Fig. 3.20b). On the other hand, the Hall coefficient follows a curve of the form $R_H(T) = A_H \exp(T_c/T)^{1/4}$, where T_c and A_H depend on the considered sample, for AlPdRe iQCs with resistivity ratios above a critical value, so that they can be regarded as located in the insulating side of a metal-insulator transition [126].

Previous results have been discussed within a single band description, which is valid for metallic systems but may become unreliable for high resistivity QCs resembling semimetals instead. Along this line of though two-band models have been sometimes used to explain the experiments [63, 289, 317, 318]. In that case, where both electrons and holes are present, the charge carriers concentration is described in terms of a function of the form $n_{e,h}(T) = n_{e,h}^0 + A \exp(-E_g/k_B T)$, where E_g plays the role of an effective gap energy, and the Hall coefficient is given by [608],

$$R_H = \frac{\sigma_e^2 R_e + \sigma_h^2 R_h}{(\sigma_e + \sigma_h)^2} = \frac{n_e \mu_e^2 - n_h \mu_h^2}{|e|(n_e \mu_e + n_h \mu_h)^2}. \tag{3.70}$$

TABLE 3.9   Low temperature Hall coefficient, charge carrier concentration, $n(0)$, and mobility $\mu(0)$ values for several QCs determined after Eqs.(3.68) and (3.69), respectively. The samples are arranged by increasing n value in each alloy system. To avoid possible inaccuracies we have sampled among those works where measurements of electrical conductivity and Hall coefficient are simultaneously reported for a given QC.

Sample	$R_H(0)$	$n(0)$	$\sigma(0)$	$\mu_H(0)$	Reference
	$\times 10^{-8}$ m^{3}C^{-1}	10^{20} cm^{-3}	$(\Omega$ cm$)^{-1}$	cm^{2}V^{-1}s^{-1}	
Zn$_{57}$Mg$_{34}$Y$_9$	-0.4	15.6	4910	19.6	[131]
Al$_{62.5}$Cu$_{25.05}$Fe$_{12.45}$	-2.25	2.8	138	3.10	[504]
Al$_{63.2}$Cu$_{24.8}$Fe$_{12}$	-1.80	3.5	240	4.32	[317]
Al$_{62.8}$Cu$_{24.8}$Fe$_{12.4}$	-1.70	3.7	135	2.30	[504]
Al$_{63}$Cu$_{25}$Fe$_{12}$	-1.20	5.2	205	2.46	[63]
Al$_{63.5}$Cu$_{24.5}$Fe$_{12}$	-1.00	6.3	230	2.30	[61]
Al$_{62.5}$Cu$_{25}$Fe$_{12.5}$	-0.71	8.8	110	0.78	[317]
Al$_{62.5}$Cu$_{26.5}$Fe$_{11}$	-0.66	9.5	270	1.78	[695]
Al$_{62.5}$Cu$_{25}$Fe$_{12.5}$	$+0.18$	34.7	100	0.18	[504]
Al$_{62.3}$Cu$_{24.9}$Fe$_{12.8}$	$+0.60$	10.4	165	0.99	[504]
Al$_{62.5}$Cu$_{24.5}$Fe$_{13}$	$+0.74$	8.4	163	1.21	[695]
Al$_{62}$Cu$_{25.5}$Fe$_{12.5}$	$+0.90$	6.9	160	1.44	[708]
Al$_{62}$Cu$_{25.5}$Fe$_{12.5}$	$+0.98$	6.4	150	1.47	[63]
Al$_{65}$Cu$_{20}$Ru$_{15}$	-2.43	2.6	75	1.82	[60]
Al$_{68}$Cu$_{17}$Ru$_{15}$	-1.26	5.0	186	2.34	[60]
Al$_{70}$Cu$_{15}$Ru$_{15}$	-1.26	5.0	100	1.26	[60]
Al$_{70.4}$Pd$_{20.8}$Mn$_{8.8}$	$+0.83$	7.5	143	1.19	[701]
Al$_{70.5}$Pd$_{21}$Re$_{8.5}$	$+16.4$	0.4	0.07	0.01	[698]
Al$_{70}$Pd$_{22.5}$Re$_{7.5}$	$+8.0$	0.8	2.3	0.18	[498]

Hence, the sign of R_H is determined by the balance between the electrons and holes concentrations and mobilities, respectively. In such two-carrier systems $R_H(T)$ curves usually exhibit strong temperature dependences, since the Hall coefficient reflects the anisotropy of the electronic structure, generally leading to electrons and holes having different effective masses, $m^*_{e,h}$, and relaxation times $\tau_{e,h}$, naturally yielding different mobilities $\mu_{e,h} = \tau_{e,h}/m^*_{e,h}$.

To conclude, let us briefly comment on the magnetoresistance, defined in terms of the expression $[\rho(B,T) - \rho(0,T)]/\rho(0,T) \equiv \Delta\rho(B,T)/\rho(0,T)$. Earlier results, for bad quality QCs, reported on $\Delta\rho(B,T)/\rho(0,T)$ values up to $+10^{-3}$, which are comparable to values obtained for amorphous metals. Subsequently, much larger magnetoresistance values, within $10-20\%$ for i-AlCuFe samples and in excess of 100% for AlPdRe QCs, were reported [726]. For QCs with resistivity values $\rho(4$ K$) \leq 0.1$ $\mu\Omega$ cm, the obtained magnetoresistance curves can be properly interpreted within the framework of quantum interference effects, so that $\Delta\rho(B,T)/\rho(0,T) = A\sqrt{T}$, where A is a constant at low temperatures.

3.7   MAGNETIC PROPERTIES

The response of a material in the presence of an external magnetic field **H** is determined by the value of the magnetization vector **M** through the relationship $\mathbf{M} = \chi\mathbf{H}$, where $\chi(T)$ measures the magnetic susceptibility of the sample. Diamagnetic materials are characterized by negative values of χ, indicating that the magnetic particles in the material act against the applied magnetic field. On the contrary, paramagnetic materials are characterized by positive values of χ. In a metal composed of non-magnetic atoms (i.e., atoms with no intrinsic magnetic moments) the magnetic susceptibility is determined by two main contributions: the Lenz response of ion core electron orbitals to the external field (Larmor's diamagnetism, χ_L) and the conduction electrons contribution χ_e, which in turn, can be split into the effect of spin electrons aligning in a direction parallel to **H** (Pauli paramagnetism, χ_p) and the Lenz response of free electrons (Landau diamagnetism, χ_l). In the free electron approximation the Pauli's contribution is proportional to the density of states at the Fermi level, $\chi_p = \mu_B^2 N(E_F)$, where μ_B is the Bohr magneton,[31] and the Landau's contribution amounts to $\chi_l = -\chi_p/3$, so that $\chi_e = \chi_p + \chi_l = 2\chi_p/3 > 0$, and the conduction electrons contribution becomes paramagnetic. The core electrons contribution is more difficult to determine but in most metals and metallic alloys the magnetic susceptibility at low temperatures, $\chi_0 = \chi_L + \chi_e$, takes on positive values typical of a paramagnetic response. For instance, one gets $\chi_0 = +0.6 \times 10^{-6}$ emu/g for both aluminium metal and the β−AlCuFe alloy phase.

At variance with this typically metallic behavior, iQCs exhibit a weak diamagnetism over a wide temperature range. For instance, i-AlCuFe and i-AlPdMn QCs, as well as their related approximant phases, are diamagnetic over a broad temperature range, with magnetic susceptibility values comprised within the interval $[-0.6, -0.4] \times 10^{-6}$ emu/g and $[-0.7, -2.1] \times 10^{-3}$ emu/mol, respectively [372]. A similar behavior has been reported for i-AlPdRe and i-GaMgZn representatives as well[336], whereas $\chi_0 \simeq -3 \times 10^{-7}$ emu/g for i-CdY, which is about two orders of magnitude smaller than the value reported for other i-CdRE members (RE = Gd − Tm) [444], but it is close to the magnetic susceptibility of i-ZnMgY QCs. It is interesting to note that the more diamagnetic samples are generally the more resistive ones as well. The emergence of a diamagnetic behavior in QCs is attributed to a weak Pauli term contribution due to the existence of a pseudogap close to the Fermi level and to an anomalously strong Landau term $\chi_l = -(m/m_*)^2 \chi_p/3$, resulting from a peculiar band structure, characterized by flat bands with large effective masses m_* [472].

Quite remarkably a diamagnetic behavior is also observed in QCs containing magnetic atoms such as i-AlPdMn and i-ZnMg(Ho,Yb,Tb,Er). In the low temperature regime the magnetic susceptibility of these phases obeys a Curie-Weiss law

$$\chi(T) = \chi_0 + \frac{C}{3k_B(T - \Theta_p)}, \tag{3.71}$$

where the second term describes the contribution due to the presence of ions with incomplete orbitals giving rise to a net angular momentum J (Curie paramagnetism), where Θ_p is referred to as the paramagnetic Curie temperature and it can be regarded as describing the net interaction among the spins, so that positive Θ_p values indicate a ferromagnetic net interaction on each spin. The Curie constant is usually expressed as $C = g^2 \mu_B^2 J(J+1)N_m = N_A \mu_{eff}^2 \mu_B^2$, where g is the Landé factor, N_m measures the number of magnetic atoms and μ_{eff} is the so-called effective magnetic moment. In the case of the i-AlPdMn phases the analysis of the obtained measurements indicate that only a minor fraction (i.e., 0.04%–4%) of the Mn atoms ($g = 2$) present in the QC carry a magnetic moment [337]. This fraction increases rapidly with the Mn concentration in the quasicrystalline alloy. Detailed studies of the evolution of magnetic properties of a single-grain i-$Al_{70.1}Pd_{21.4}Mn_{8.5}$ QC upon increased structural equilibration show that different thermal treatments of the same sample can induce reversible changes of its paramagnetic magnetization up to a factor 26, with more equilibrated samples being systematically less magnetic [372]. Thus the magnetic momentum formation on Mn atoms is very sensitive to environmental effects determined by atomic distances, coordination number, and the kinds of atoms around Mn ones.

---

[31]$\mu_B = \frac{\hbar e \mu_0}{2m} = 1.165 \times 10^{-29}$ Wb m $= 9.273 \times 10^{-24}$ JT^{-1}.

For instance, Mn sites with a low Al coordination experience weaker sp-d hybridization effects, which favour the appearance of a magnetic moment [900]. ZnMgRE (Bergman type) and CdMgRE (Tsai type) (RE=Gd, Tb, Dy, Ho, Er, and Tm) iQCs closely follow the Curie-Weiss law at high enough temperatures, with μ_{eff} values that are consistent with +3 valence values for the RE atom, and the Θ_p values obtained from $\chi(T)$ fits are negative (within the interval $-55 \leq \Theta_p \leq -2$ K), indicating primarily antiferromagnetic interactions between the rare-earth ions [298, 463].[32] Thus, whereas the magnetic properties of QCs containing Mn atoms are mainly determined by the number of Mn atoms carrying a magnetic moment, the magnetic behavior of i-ZnMgRE QCs containing rare-earth atoms with particularly strong magnetic moments such as Ho,Yb,Tb, and Er, is strongly influenced by the nature of the magnetic ordering of these atoms. The presence of relatively large, negative values of Θ_p (from -5 K to -26 K) in Eq. (3.71) indicate the existence of dominant antiferromagnetic exchange interactions between magnetic atoms. In fact, the direct confirmation of the presence of short-range spin antiferromagnetic correlations in i-ZnMgHo [783], spurred the interest in revisiting previous theoretical works which had predicted that long-range ferromagnetic order is possible in QCs [492].[33] In this regard, the i-$Zn_{77}Sc_{16}Fe_7$ QC is of special interest since the μ_{eff} value is intermediate between Fe^{+2} and Fe^{+3} free ion values, with a positive $\Theta_p \simeq +6.5$ K, indicating primarily ferromagnetic interactions [407]. On the contrary, albeit the effective moments for i-YbAuAl QCs are also intermediate between Yb^{+2} and Yb^{+3} (mixed valence compound), their Θ_p are negative and significantly higher than previously studied magnetic QCs, indicating strong antiferromagnetic interactions (for the parent i-YbCd QC $\Theta_p = -153$ K). In addition i-YbAuAl QC exhibits strong quantum critical effects (i.e., $\rho \to 0$ and $\chi \to \infty$ as $T \to 0$).

On the other hand, evidences of the so-called spin glass transition (rather than long-range magnetic order) have been observed in i-AlPdMn (Mackay type), i-CdMgRE (Tsai type), and i-ZnMgRE (Bergman type) QCs, at relatively low freezing temperatures ($T_f \simeq 5$ K), as compared to those usually observed in conventional alloys which are one order of magnitude larger [336]. In fact, due to the icosahedral symmetry, atoms in a QC find themselves in a variety of different local environments, which means that magnetic interactions often become "frustrated". In other words, there is no possible configuration that allows magnetic moments to align in their preferred directions. A well-known example is that of antiferromagnetic spins on a triangular lattice: the three spins cannot be arranged so that all neighboring spins are antiparallel. The low value of the transition temperature is explained by the relatively small fraction of magnetic atoms present in QC phases. The very existence of such a transition indicates that QPO is sufficient to effectively coupling the magnetic atoms (via delocalized d electrons through the so-called Ruderman-Kittel-Kasuya-Yoshida mechanism) in order to induce a transition leading to the spin glass state. Magnetic properties of decagonal phases in AlCuCo and AlPdMn systems also indicate the presence of anisotropy effects. Thus the value of the local magnetic moments for the i-AlPdMn are about twice as large as those for the d-AlPdMn [326].

At elevated temperatures the dependence of the magnetic susceptibility of most Al-based QCs can be fitted to the form $\chi(T) \sim AT^2$ [591], where the parabolic term is ascribed to a temperature dependent Pauli susceptibility. In fact, the Pauli paramagnetic susceptibility

---

[32]In ferromagnetism all the magnetic moments point in the same direction, whereas in antiferromagnetism neighboring atoms point in alternate directions.

[33]By all indications, previous claims of both antiferromagnetic [123], and ferromagnetic [735], QC phases may be probably related to the presence of secondary magnetic crystalline phases in the considered samples [370].

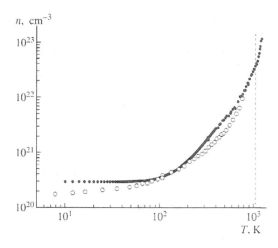

Figure 3.21 Temperature dependence of the charge carrier concentration for $Al_{63}Cu_{25}Fe_{12}$ (filled dots) and AlPdRe (open dots) QCs. The vertical dashed line indicates the boundary of the region of existence of the homogenous phase. (Reprinted by permission from Springer Nature Customer Service Centre GmbH, Prekul A F and Shchegolikhina N, Correlations between electric, magnetic, and galvanomagnetic quantities in stable icosahedral phases based on aluminium, *Crystallogr. Rep.* **52** 996-1005, 2007. Copyright 2007).

can be expressed as

$$\chi_p(T) = \mu_B^2 N(E_F) \left\{ 1 + \frac{\pi^2}{6}(k_B T)^2 \left[\frac{1}{N(E)} \frac{d^2 N}{dE^2} - \left(\frac{1}{N(E)} \frac{dN}{dE} \right)^2 \right]_{E_F} \right\}$$
$$= \mu_B^2 N(E_F) \left[1 + bT^2 \left(\xi_2 - 2\xi_1^2 \right) \right], \tag{3.72}$$

where we have used Eq. (3.57).

Nevertheless, this expression can not account for the linear dependence $\chi(T) = \chi_0 + AT$ reported for several i-AlPdRe samples. The temperature dependence of the magnetic susceptibility over the broader temperature interval ranging from 4 K to 1200 K can be nicely accounted for (after substracting of the Curie-Weiss type contribution) in terms of the Langevin equation

$$\chi(T) = \frac{n(T)\mu_{eff}^2}{3k_B T}, \tag{3.73}$$

where $n(T)$ measures the thermally activated charge carriers concentration. This is shown in Fig. 3.21 when we compare the charge carrier density curve derived from Hall coefficient measurements after Eq. (3.68) with the $n(T)$ curve obtained from Eq. (3.73) for two Al-based iQCs. In the temperature range below the respective Debye temperatures the $n(T)$ curves grow relatively smoothly, but the curve slopes dramatically increase for temperatures above Θ_D [709, 711].

No ferromagnetic QC has been reported since their discovery in 1984. In contrast, ferromagnetic order has been observed in the lowest order Tsai-type approximants 1/1-AuSiRE, 1/1-AuAlRe, and 1/1-AuSiTb during the last decade. The magnetic ground state of these approximant phases is tunable in terms of the e/a ratio parameter, which has opened up an efficient route to synthesize a variety of magnetic 1/1 approximant crystals. By following this approach higher order 2/1 approximants in the (Au,Cu)-(Al,In)-(Gd,Tb) system have been reported, exhibiting a ferromagnetic transition at significantly high Curie temperature within the range $15-30$ K [362].

3.8 THERMAL TRANSPORT PROPERTIES

3.8.1 General considerations

According to the so-called Fourier's law, the presence of a temperature gradient ∇T (measured in Km^{-1}) induces in the material a heat current density **h** (measured in Wm^{-2} units), which is given by $\mathbf{h} = -\kappa\,\nabla T$, where κ is a characteristic property of the considered material, referred to as its thermal conductivity (measured in $Wm^{-1}K^{-1}$ units). In general, the thermal conductivity depends on the material's temperature, that is, $\kappa(T)$, and it always takes on positive values ($\kappa > 0$), so that the minus sign properly describes the thermal current propagation sense. In electrically insulating materials the thermal conductivity $\kappa_l(T)$ is entirely determined by the vibration of atoms around their equilibrium positions in the crystal lattice. In non-insulating materials thermal energy is transported by both lattice oscillations and the motion of charge carriers, so that $\kappa(T) = \kappa_e(T) + \kappa_l(T)$, though in metallic systems the thermal conductivity is mainly governed by the motion of electrons at any temperature (i.e., $\kappa_e(T) \gg \kappa_l(T)$). Since charge carriers motion also determines their electrical conductivity, one should expect that the transport coefficients $\kappa_e(T)$ and $\sigma(T)$ will be tied up in these materials. Indeed, experiments disclosed that the thermal and electrical conductivities of most metallic materials are mutually related through the relationship

$$\kappa_e(T) = L_0 T \sigma(T), \tag{3.74}$$

referred to as the Wiedemann-Franz's law (WFL). Eq. (3.74) also holds for semiconducting materials, provided that the L_0 number is replaced in Eq. (3.74) by smaller values ranging from $L_s = 2.0 \times 10^{-8}$ V^2K^{-2} to $L_s = 2(k_B/e)^2 \simeq 1.48 \times 10^{-8}$ V^2K^{-2} [608].

Physically, the WFL expresses a transport symmetry arising from the fact that the motion of the carriers determines both the electrical and thermal currents in both metallic and semiconducting materials at relatively low temperatures. As the temperature of the sample is progressively increased, the validity of WFL will depend on the nature of the interaction between the charge carriers and the different scattering sources present in the solid. In general, the WFL applies as far as elastic processes dominate the transport coefficients, and usually holds for arbitrary band structures provided that the change in energy due to collisions is small compared with k_BT [27]. Accordingly, one expects some appreciable deviation from WFL when electron–phonon interactions, affecting in a dissimilar way to electrical and heat currents, start to play a significant role. On the other hand, at high enough temperatures the heat transfer is dominated by the charge carriers again, due to Umklapp phonon scattering processes, and the WFL is expected to hold as well.

From a practical viewpoint, the importance of the WFL can be seen by considering that only the total thermal conductivity $\kappa(T)$ can be experimentally measured in a straightforward way, and the contributions $\kappa_e(T)$ and $\kappa_l(T)$ must be somehow separated from empirical data. This is usually done by explicitly assuming the applicability of the WFL to the considered sample, so that the lattice contribution to the thermal conductivity is obtained from the expression

$$\kappa_l(T) = \kappa(T) - LT\sigma(T), \tag{3.75}$$

where $L = L_0$ for metallic systems and $L = L_s$ for semiconducting ones. Actually, this estimation of the lattice contribution should be regarded as a mere approximation, since one generally lacks a precise knowledge of the L value in real applications. In fact, for a given material the L value usually varies with the temperature according to the expression given by Eq. (3.10). A direct measurement of the Lorenz function value can be experimentally done by utilizing a transverse magnetic field in order to suppress the charge carrier contribution to the thermal conductivity. In so doing, the reported values in single crystal Al, Cu, and Zn metals deviate form the Sommerfeld's value $L_0 = 2.44 \times 10^{-8}$ V^{2}K^{-2} in the intermediate temperature range 5−60 K, exhibiting systematically decreasing $L(T)/L_0$ curves as the temperature is increased, reaching $L(T)/L_0 \simeq 0.2$−0.4 at about 50 K [977]. On the other hand, at 100 K values relatively close to L_0, namely $L = 2.16 \times 10^{-8}$ V^{2}K^{-2} and $L = 2.33 \times 10^{-8}$ V^{2}K^{-2}, have been reported for BiSb and BiSeTe alloys [519]. One of the limitations of this technique, however, is the requirement of high mobility values for the considered sample charge carriers, along with the necessity of applying intense magnetic fields. Another way of experimentally determining the Lorenz number, not subjected to these requirements, is through the controlled introduction of impurities in the sample, thereby inducing a correlated change in the electrical conductivity and the κ_e contribution, from which one can determine κ_l [102].

In terms of the vibrational DOS $D(\omega)$ and the Planck distribution function $p(\omega, T)$ introduced in Sec. 3.3.1, the lattice thermal conductivity can be expressed as [27],

$$\kappa_l(T) = \frac{v^2 \hbar}{3V} \int_0^{\omega_D} \left(\frac{\partial p}{\partial T} \right) D(\omega)\tau(\omega, T)\omega d\omega, \tag{3.76}$$

where v is the sound velocity of the considered material, V is the sample's volume, and $\tau(\omega, T)$ is the average time between heat current degrading collisions involving phonons at a given temperature (the so-called phonon relaxation-time). In the simplest approach, the relaxation-time may be regarded as independent of both the phonon frequency and the temperature. In that case, Eq. (3.76) can be rewritten in the form

$$\kappa_l(T) = \frac{v^2 \tau}{3V} \int_0^{\omega_D} \hbar\omega D(\omega) \left(\frac{\partial p}{\partial T} \right) d\omega, \tag{3.77}$$

where the expression in the integral can be readily identified as the phonon contribution to the specific heat at constant volume c_k given by Eq. (3.28), so that Eq. (3.77) reduces to the well-known formula $\kappa_l = c_k v\ell/3$, where $\ell \equiv v\tau$ is the phonon mean-free-path [27, 608].

In the general case, Eq. (3.76) can be rewritten in the form

$$\kappa_l(T) = \frac{v^2 k_B^2 T}{12 V \hbar} \int_0^{\Theta_D/T} y^2 \operatorname{csch}^2 \left(\frac{y}{2} \right) D(y)\tau(y, T)dy, \tag{3.78}$$

in terms of the dimensionless scaled energy variable $y \equiv \beta\hbar\omega$. Thus, within the Debye model approximation given by Eq. (3.24), Eq. (3.78) adopts the form

$$\kappa_l(T) = \frac{3}{4} v^2 k_B n_a \left(\frac{T}{\Theta_D} \right)^3 \int_0^{\Theta_D/T} y^4 \operatorname{csch}^2 \left(\frac{y}{2} \right) \tau(y, T)dy, \tag{3.79}$$

where we have made use of Eq. (3.31) to introduce the atomic density n_a.

The mean relaxation time of heat-carrying phonons is determined by the various scattering mechanisms phonons may encounter when propagating through the solid, such as grain boundaries, point defects (i.e, atomic isotopes, impurity atoms, or vacancies), phonon-phonon interactions, or resonant dynamical effects (e.g., rattling atoms in interstitial voids or molecular cages). Thus, the overall phonon relaxation time can be expressed in the general form

$$\tau^{-1}(\omega, T) = \frac{v}{L} + A_1\omega^4 + A_2\omega^2 T \exp\left(-\frac{\Theta_D}{3T}\right) + \frac{A_3\omega^2}{(\omega_0^2 - \omega^2)^2}, \tag{3.80}$$

where L is the crystal size in a single-grained sample or measures the average size of grains in a poly-grained sample, A_1 (measured in s³), A_2 (measured in sK⁻¹), and A_3 (measured in s⁻³), are suitable constants and ω_0 is a resonance frequency. The first term on the right side of Eq. (3.80) describes the grain-boundary scattering, the second term describes scattering due to point defects, the third term describes anharmonic phonon-phonon Umklapp processes, and the last term describes the possible coupling of phonons to localized modes present in the lattice via mechanical resonance. The ω^4 dependence of the second term in Eq. (3.80) indicates that point defects are very effective in scattering short-wavelength phonons, and they have a lesser effect on longer wavelength phonons. Remarkably enough, short-wavelength phonons make the most important contribution to the thermal current. In the absence of dynamical resonance effects, Eq. (3.80) can be expressed in the form

$$\tau^{-1}(y, T) = \frac{v}{L} + c_0^2 y^2 \left[A_1 c_0^2 y^2 T + A_2 \exp\left(-\frac{\Theta_D}{3T}\right)\right] T^3, \tag{3.81}$$

where $c_0 = k_B/\hbar$. In the low temperature regime the average phonon frequency is small and only long wavelength phonons will be available for heat transport, which are mostly unaffected by both point defects and phonon-phonon interactions. These long wavelength phonons are chiefly scattered by grain-boundaries (polycrystalline samples) and crystal dimensions (single crystals). Accordingly, $\tau \simeq L/v$ and Eq. (3.79) reads

$$\kappa_l^L(T) = \frac{4\pi^4}{5} L k_B v n_a \left(\frac{T}{\Theta_D}\right)^3 \tag{3.82}$$

where we have used the Mellini's transform given by Eq. (3.33). Thus, in the low temperature regime the thermal conductivity will show a cubic dependence with the temperature. Eq. (3.82) also shows that at low enough temperatures the thermal conductivity can be expressed in the well-known form $\kappa_l = c_k v^2 \tau / 3$, where the specific heat is given by Eq. (3.34).

On the other hand, in the high temperature limit (i.e., $T > \Theta_D$), $\exp\left(-\frac{\Theta_D}{3T}\right) \to 1$ in Eq. (3.81), and the phonons wavelength is significantly shorter than the sample dimensions, so that it can be regarded as effectively infinite in size ($L \to \infty$). Thus, $v/L \to 0$ and Eq. (3.81) can be written

$$\tau^{-1}(y, T) = c_0^2 y^2 (A_1 c_0^2 y^2 T + A_2) T^3. \tag{3.83}$$

Plugging this relaxation time expression into Eq. (3.79) and making use of Eq. (3.31), we obtain

$$\kappa_l(T) = \frac{\hbar}{8\pi^2 v A_1 T} \int_0^{\Theta_D/T} \frac{y^2}{y^2 + A_4} \operatorname{csch}^2\left(\frac{y}{2}\right) dy, \tag{3.84}$$

where $A_4 \equiv (\hbar/k_B)^2 A_2 (A_1 T)^{-1}$ is a dimensionless constant. This expression can be further simplified by taking into account that at high enough temperatures ($y \ll 1$), we can approximate $\sinh(y/2) \simeq y/2$ in Eq. (3.84), which can then be explicitly integrated to get

$$\kappa_l(T) = \frac{k_B}{2\pi^2 v \sqrt{A_1 A_2 T}} \tan^{-1}\left(\frac{\Theta_D}{\sqrt{A_4} T}\right). \tag{3.85}$$

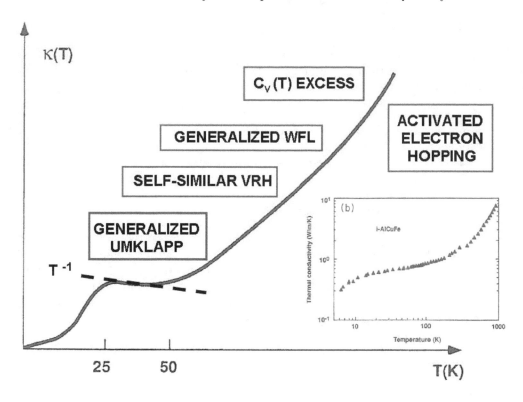

Figure 3.22 (Main frame) Sketch depicting the main features of the thermal conductivity in iQCs over a broad temperature interval. Diferent physical processes taking place at diferent temperature ranges are indicated. (Inset) Measured thermal coefficient curve for an i-AlCuFe QC [688]. (Reprinted figure with permission from Janot C, Conductivity in quasicrystals via hierarchically variable-range hopping, Phys. Rev. B **53** 181–191 (1996) Copyright (1996) by the American Physical Society).

Finally, we must take into account that, at the high temperature regime we are now considering, the phonon-phonon Umklapp processes generally overshadow the scattering due to impurities as a major mechanism degrading the thermal current, so that $A_1/A_2 \ll 1$. Therefore, one can make the approximation $\tan^{-1} \alpha \simeq \alpha$, and Eq. (3.85) can be rewritten in the form

$$\kappa_l(T) = \frac{(6\pi^2 n_a)^{1/3}}{2\pi^2 A_2} k_B T^{-1},\qquad(3.86)$$

in agreement with experimental data obtained at high temperatures [26].

In periodic crystals the transition from low to high temperature regimes is characterized by the presence of a relatively shallow peak in the $\kappa_l(T)$ curve due to an exponential increase in the number of occupied high frequency phonon states, which allow for the occurrence of phonon-phonon Umklapp process.

3.8.2 Thermal conductivity of quasicrystals

The temperature dependence of the thermal conductivity of QCs belonging to different alloy systems has been measured, over different temperature ranges, and the following general conclusions can be drawn from the collected data:

- The thermal conductivity of QCs is unusually low, even lower than that observed for thermal insulators of extensive use in aeronautical industry, such as titanium carbides or nitrides, doped zirconia, or alumina. For example, in i-AlPdMn QCs the thermal conductivity at room temperature is comparable to that of zirconia ($1\,\mathrm{Wm^{-1}K^{-1}}$), and this value decreases to about $10^{-4}\,\mathrm{Wm^{-1}K^{-1}}$ below $0.1\,\mathrm{K}$ [210]. The thermal diffusivity of these alloys is extraordinarily low as well, being even lower than that of zirconium oxide [206].

- The overall behavior of the $\kappa(T)$ curve is quite sensitive to the microstructure of the sample. Thus, for poly-grained samples, the lattice thermal conductivity monotonically increases with T, showing a marked tendency to saturation for temperatures above $10\text{--}20\,\mathrm{K}$, and exhibiting a characteristic plateau extending from about 25 to $55\,\mathrm{K}$. On the contrary, the lattice thermal conductivity of single-grained samples first increases with increasing T, it reaches a shallow maximum at about $20\,\mathrm{K}$, and then smoothly decreases with further increasing T [131].

- Above room temperature the $\kappa(T)$ curve steeply increases with the T up to the melting point.

The low thermal conductivity of QCs can be understood in terms of two main facts. In the first place, as we have learnt in previous sections, the charge carrier concentration is low, so that heat must mainly propagate by means of atomic vibrations (phonons). In turn, within the energy window where lattice thermal transport is expected to be most efficient, the frequency spectra is highly fragmented. As a consequence, the corresponding eigenstates become more localized and thermal transport is further reduced. Physically, this effect stems from the fact that QCs have a fractal reciprocal space, lacking a well defined lower bond as that provided by the lattice parameter in the case of periodic crystals. Consequently, the transfer of momentum to the lattice is not bounded below, which gives rise to a significant degradation of thermal current through the sample [640]. These processes are expected to be dominant in the temperature range $20 \lesssim T \lesssim 100$. Thus, instead of the usual exponential term, in the case of QCs the expression for the Umklapp processes must be modified to properly account for their characteristic self-similar symmetry, and the corresponding relaxation-time expression adopts a power law dependence with the temperature of the form $\tau^{-1} = A_U \omega^2 T^n$, where ω is the phonon frequency and A_U is measured in sK^{-n} units [131]. Making use of this scattering rate value in Eq. (3.79) one obtains (**Exercise 3.10**)

$$\kappa_l^U = \frac{\pi^2 k_B n_a^{1/3}}{(6\pi^2)^{2/3} A_U} \frac{T^{1-n}}{\Theta_D}. \tag{3.87}$$

Fitting analyses to experimental κ_l curves lead to a temperature dependence of the form $\kappa_l \sim T^{-1}$ in all considered cases [65, 189, 286], thereby suggesting that the generalized Umklapp processes relaxation-time takes the form $\tau^{-1} = A_U \omega^2 T^2$ in this temperature interval,[34] in agreement with experimental data reported for i-AlCuFe and i-AlPdMn

---

[34] Alternatively, one may consider the relationship $\tau^{-1} = A_U \omega^3 T$ to this end, as reported in the case of i-ZnMgY and i-AlCuFe QCs [189, 286].

TABLE 3.10 Reported values of the power law exponent describing the lattice thermal conductivity temperature dependence $\kappa_l(T) \sim T^\alpha$, for different QCs. The α values were obtained from a fitting analysis of the experimental curves in the indicated temperature ranges.

SAMPLE	T range (K)	α	REF
i-CdYb	$110 - 300$	1.2	[631]
i-AlPdRe	$150 - 300$	1.4	[459]
i-AgInYb	$150 - 300$	1.5	[460]
i-AlPdMn	$200 - 300$	1.7	[64]
i-ZnMgY	$140 - 300$	1.7	[286]

QCs [65, 64]. Accordingly, in QCs the rate of Umklapp processes becomes weaker, since the momentum of vibrational excitations can be transferred to the lattice in inelastic scattering events by arbitrarily small amounts, not limited from below. This feature is directly related to the underlying long-range QPO of the structure in reciprocal space.

On the other hand, starting at $T \gtrsim 100$ K one expects the variable range hopping mechanism will start to play an increasingly significant role due to electron–phonon interactions. By adopting the jump frequency is given by Eq. (3.60), one gets $\tau_{\mathrm{VRH}}^{-1} = A_V T^\alpha$, where α essentially coincides with the value obtained from the power law fitting of the electrical conductivity $\sigma(T)$ curves [376]. According to this model the thermal conductivity curve should rise following a power law of the form $\kappa_l \sim T^\alpha$ within the temperature interval $100 \lesssim T \lesssim 400$. The fitting analysis to experimental κ_l curves shown in Table 3.10 indicates that most considered samples reasonably agree with the expected behavior in this temperature range. At $T \simeq \Theta_D$ the $\kappa_l(T)$ curve increases due to an increase of the specific heat (Fig. 3.7), which may be related to the contribution of optical-like modes in the VDOS (Fig. 3.9) according to Eq. (3.50). Finally, as the temperature is further increased, the total thermal conductivity remarkably increases in a marked non-linear way (Fig. 3.22). By all indications, this rapid enhancement arises from a parallel increase of the charge carriers concentration (Fig. 3.21), ultimately leading to the breakdown of the WFL.

3.8.3 Assessing Wiedemann-Franz law in quasicrystals

The WFL has been routinely applied in order to estimate the phonon contribution to the thermal conductivity of QCs. In doing so, it was concluded that the contribution of electrons to the thermal transport is, at least, one order of magnitude lower than that due to phonons over a wide temperature range ($0.1\,K \leq T \leq 200\,K$) [130, 400]. Indeed, room temperature κ_e/κ_l ratios within $1-5$, ~ 1, or within $0.5-0.01$, have been determined for i-ZnMgY, [286] i-AgInYb,[459], and i-AlPd(Mn,Re) QCs [64, 460], respectively. Keeping in mind that the κ_e/κ_l ratio takes on values within the range $10 - 100$ for conventional alloys, one realizes that the thermal transport of QCs is largely dominated by phonons at room temperature. However, since transport properties of most QCs are quite unusual by the standard of common metallic alloys, it seems convenient to check up on the validity of the WFL for these materials [475, 538, 545, 595]. A suitable experimental measure to this end is provided by the magnitude $\kappa(T)/\sigma(T) = (\kappa_e + \kappa_l)/\sigma \equiv TL(T) + \varphi(T)$, where

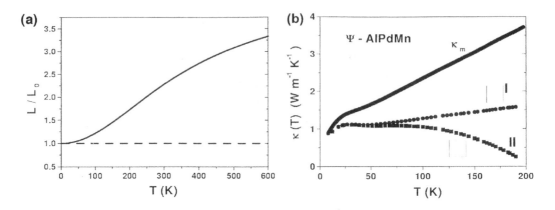

Figure 3.23   (a) Temperature variation of the normalized Lorenz function analytically derived for iQCs described in terms of the spectral conductivity function given by Eq. (3.6). The WFL is obeyed at very low temperatures. At high temperatures $L(T)$ aproaches the asymptotic limit value 21/5. A significant enhancement of the Lorenz number with respect to the Sommerfeld's value takes place over a wide temperature range [475, 545]. (b) The phonon contribution to the thermal conductivity is derived by substracting to the experimentally measured thermal conductivity (κ_m, experimental data by courtesy of J. Dolinšek) the charge carrier contribution $\sigma(T)$ was reported in Ref. [188]), by assuming: I) the validity of the WFL according to Eq. (3.74) (circles), or II) by considering a temperature dependent Lorenz function (squares). (Maciá E, Theoretical aspects of thermal transport in complex metallic alloys: A generalization of the Wiedemann-Franz law, Croat. Chem. Acta 83 65-68, 2010 (Creative Commons Attribution 4.0 International License CC BY).

the so-called Lorenz function is defined by Eq. (3.10) and $\varphi(T)$ accounts for the phonon contribution to the heat transport. A study of the temperature variation of the κ/σ ratio in several intermetallic compounds showed that the experimental data may be fitted to a linear temperature dependence of the form $\kappa/\sigma = LT + B$ over the wide temperature range 350-800 K [200, 688]. By comparing the slopes obtained at high temperatures for pure aluminium and i-AlCuFe samples the ratio $L_{QC}/L_{Al} \simeq 1.21$ was obtained, hence indicating an enhanced Lorenz number for quasicrystalline alloys. In a similar way, room temperature $L(T)$ values larger than the Sommerfeld's value L_0 were experimentally reported for other complex metallic alloys (ranging from $L_{300}/L_0 = 1.15$, Ref. [62] to $L_{300}/L_0 = 1.43$, Ref. [286]), hence suggesting the convenience of introducing a slightly modified WFL of the form

$$\kappa_e(T) = (1 + \varepsilon)L_0 T \sigma(T). \qquad (3.88)$$

with $\varepsilon = 0.43$ and $\varepsilon = 1.1$ for i-$Zn_{57}Mg_{34}Y_9$ and i-$Al_{64}Cu_{23}Fe_{13}$, respectively [189, 286]. In this vein, a generalized WFL of the form $\kappa_e(T) = L(T)T\sigma(T)$, which is characterized by a *non-linearly* temperature dependent Lorenz number (Fig. 3.23a), has been proposed on theoretical basis [545, 550].

The impact of the Lorenz's function temperature dependence in a proper analysis of the phonon contribution to the thermal conductivity is illustrated in Fig. 3.23b, where we compare the measured thermal conductivity (including contributions from both charge carriers and phonons) with the phonon contribution derived from the application of the WFL by either assuming a constant value for the Lorenz number (Eq. (3.74), circles)

TABLE 3.11   Elastic constants and related moduli for AlPdMn QCs. The data for elemental Al are given at the bottom row for the sake of comparison.

Compound	C_{11}	C_{44}	C_{12}	B	E	v	$\frac{G}{B}$	Ref.
	(GPa)	(GPa)	(GPa)	(GPa)	(GPa)			
$Al_{72}Pd_{19.5}Mn_{8.5}$	211	69.3	72.4	118	174	0.255	0.586	[879]
	218	72.5	73.0	122	181	0.252	0.594	[879]
$Al_{68.7}Pd_{21.7}Mn_{9.6}$	215 ± 1	65 ± 0.2	85.0	128	166	0.283	0.507	[12]
	216 ± 1	65 ± 0.2	86.0	129	166	0.285	0.502	[12]
$Al_{70.4}Pd_{21.2}Mn_{8.4}$	195 ± 15	64 ± 1	67.0	110	161	0.256	0.583	[185]
Al	120.3	28.7	62.8	82.0	77.2	0.343	0.351	[879]

or explicitly taking into account its temperature dependence as given by the $L(T)$ curve plotted in Fig. 3.23a. As we see, the usual procedures significantly overestimate the phonon contribution at room temperature and above.

3.9   MECHANICAL AND TRIBOLOGICAL PROPERTIES

The relation between stress and strain (both magnitudes described in terms of a second rank tensor) in a solid is provided by a fourth rank symmetric tensor, whose elements are referred to as elastic constants. The number of independent elastic constants is 21 in the general case, but it reduces to 2 for isotropic solids, namely, $C_{12} \equiv \lambda$ and $C_{44} \equiv \mu$, which are called the Lamé coefficients, satisfying $C_{11} = \lambda + 2\mu$. Mechanical properties of isotropic solids are characterized by the so-called elastic moduli, namely, the bulk modulus

$$B \equiv \frac{C_{11} + 2C_{12}}{3} = \lambda + \frac{2}{3}\mu, \qquad (3.89)$$

the Young modulus

$$E \equiv \frac{C_{11}^2 + C_{11}C_{12} - 2C_{12}^2}{C_{11} + C_{12}} = \mu\frac{3\lambda + 2\mu}{\lambda + \mu} = 3B\frac{\mu}{\lambda + \mu} = 2(1 + v)\mu, \qquad (3.90)$$

where

$$v \equiv \frac{C_{12}}{C_{11} + C_{12}} = \frac{\lambda}{2(\lambda + \mu)}, \qquad (3.91)$$

is the Poisson's ratio, and the shear modulus

$$G \equiv \mu = \frac{E}{2(1 + v)} = \frac{3}{2}\frac{1 - 2v}{1 + v}B, \qquad (3.92)$$

where we have used Eqs.(3.90) and (3.91). According to Eq. (3.92) the normalized shear modulus G/B only depends on the Poisson ratio. In Table 3.11 we list the elastic constants and their related moduli for some representatives of the i-AlPdMn system. As we see, i-AlPdMn QCs exhibit a relatively large value of the G/B ratio (0.5−0.6), while values close to $G/B \simeq 0.7$ were reported for i-AlCuFe representatives. These figures are comparable to those observed in semiconductors (0.5 − 0.75), rather than the values of typical metals (i.e., $G/B = 0.27$ for Pd, $G/B = 0.38$ for Cu, or $G/B = 0.52$ for Fe) [875]. This is an indication that interatomic bonding in QCs is rather directional, and not as isotropic as in metals [210].

The so-called Grüneisen parameter provides a direct measure of the anharmonicity of the bonds in a solid. For materials exhibiting an isotropic thermal expansion this dimensionless parameter can be determined in terms of accessible experimental magnitudes

$$\gamma_G = \frac{BV_m\alpha_L}{C_V}, \tag{3.93}$$

where B is the bulk modulus, V_m is the molar volume, C_V is the constant volume specific heat, and

$$\alpha_L = \frac{1}{V}\left(\frac{\partial V}{\partial T}\right)_P,$$

is the linear thermal expansion coefficient. Thermal expansion coefficient values ranging from $\alpha_L = 1.3 \times 10^{-5}$ K^{-1} for i-AlPdMn[979], $\alpha_L = 1.55 \times 10^{-5}$ K^{-1} for both synthetic and natural i-AlCuFe QCs [852, 832], up to $\alpha_L = 2.765 \times 10^{-5}$ K^{-1} for i-CdCa QCs [389], have been reported in the literature.[35] The Grüneisen parameter is related to the pressure and volume constant specific heats through the relationship

$$\frac{C_P - C_V}{C_V} = 3\alpha_L\gamma_G T.$$

Values of γ_G typically range from ± 1 to ± 4. For most non-metallic solids, the Grüneisen parameter is generally of the order of unity. For fullerite simple cubic crystals $\gamma_G \simeq 3$ below $T = 90$ K, and $\gamma_G \simeq 1$ in the interval $90-200$ K [957]. The low temperature high γ_G value quantifies the high degree of anharmonicity in fullerite. This comes from the essentially incompressible hard core in a C_{60} molecule, which makes the fractional change of the intermolecular distance on compression substantially smaller that it would be in the absence of rigid C_{60} building blocks in this molecular crystal. It is interesting to note that the γ_G values reported for fullerite compare well with those obtained for graphite along the c-axis, which decreases gently from $\gamma_G = 3$ at $T = 40$ K to $\gamma_G = 1$ at $T = 200$ K [36]. From a study of the dependence of the elastic constants with the temperature the value $\gamma_G = 1.9-2$ was reported for i-Al$_{72}$Pd$_{19.5}$Mn$_{8.5}$ [879], and i-Al$_{71}$Pd$_{21}$Mn$_8$ [852], and $\gamma_G \simeq 1.6$ for i-Al$_{71.7}$Pd$_{19.4}$Mn$_{8.9}$ and d-Al$_{72}$Co$_{16}$Ni$_{12}$ [361], single grain QCs. Broadly speaking, these figures support the possible consideration of QCs as clusters aggregates.

QCs are noteworthy for their hardness, low surface energy, and low friction [578]. A remarkable characteristic property of QCs is that they are brittle at room temperature and ductility sets in at temperatures of the order of 70% and higher of their respective melting points. This feature is at variance with the behavior observed in most periodic intermetallic compounds, which typically show ductile behavior at room temperature and even below. On the other hand, numerous experiments using spherical indenters have demonstrated that friction is lower on QCs than on more conventional materials of similar hardness.[36] Thus, the value $\mu = 0.046 \pm 0.008$ has been reported from scratch tests using a diamond spherical indenter of 0.8 mm diameter on i-AlCuFeB QC based coating, and the steady state friction coefficient $\mu = 0.055 \pm 0.008$ was reported from pin-on-disk tests (Fig. 3.24), whereas for hard Cr-steel it comes to $\mu = 0.07 \pm 0.01$ [312]. With more conventional riders, such as mild steel or tungsten carbide (WC–Co), the comparison is less favorable due to the transfer of material from the indenter to the surface. During such friction measurements, a major drawback arises from the brittleness of the contacting bodies and hence formation of wear particles. However, scratch tracks indicate that QCs

---

[35]For the sake of comparison we have $\alpha_L \simeq 2.6 \times 10^{-5}$ K^{-1} for conventional Al-based alloys.

[36]However, the friction coefficient observed in air is low in a transient regime only. It increases with time during the test and many artifacts due to oxidation effects generally appear [829]. Indeed, investigations performed in vacuum have shown that the friction coefficient strongly varies with the oxidation state of the surface [674]. In this case, the friction of QC against steel is typically as high as $\mu = 0.6-0.8$ [312].

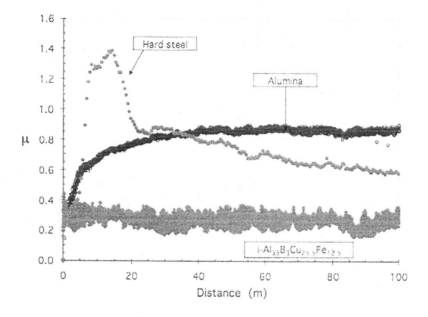

Figure 3.24 Friction coefficient recorded as a function of sliding distance during pin-on-disk tests managed under vacuum conditions (10^{-7} mbar). The materials under test are: hard Cr-steel, sintered alumina, and sintered i-AlCuFeB QC. The indenter was a Cr-steel sphere of 6 mm diameter under 2 N load. (Reprinted from Guedes de Lima B A S, Medeiros Gomes R, Guedes de Lima S J, Dragoe D, Barhtes-Labrousse M G, Kouitat-Njiwa R, Dubois J M, Self-lubricating, low-friction, wear-resistant Al-based quasicrystalline coatings, *Sci. Tech. Adv. Mats.* **17** 71-79 (2016); doi:10.1080/14686996.2016.1152563, Creative Commons Atribution License CC BY 4.0).

subjected to repetitive and severe shear develop some ductility during this process, and consequently they acquire an ability to self-repair. This brittle-to-ductile behavior might be associated with the nucleation of crystalline nanometer-sized grains within the quasicrystalline matrix [962].

The influence of atomic commensurability on friction has been examined by a number of experimental and theoretical studies [328, 722]. In the ideal case, when two workpieces with incommensurate lattices are brought in contact, the minimal force required to achieve sliding (known as the static frictional force) vanishes, provided the two substrates are stiff enough [94]. Thus, it has been observed that friction becomes negligible for incommensurate surfaces sliding under conditions of elastic contact. In real situations, however, physical contact between two (uncontaminated) surfaces is generally mediated by third bodies acting like a lubricant film. In that case the sliding interface should be properly described in terms of three characteristic lengths, corresponding to the periods of both substrates and the lubricant layer.

Scanning tunneling microscopy studies have revealed the presence of atomic structures closely related to the golden mean in quasicrystalline surfaces. For example, on the surface of i-AlPdMn QCs atomic terraces are separated by steps of three different heights whose values are related, not by a simple integer, but by the irrational number $\tau \simeq 1.618$ [800, 919]. This means that the spacing between similar planes of the bulk structure has not one but two dominant spacings, and this is reflected in the step heights observed on the surface. In this way, the bulk QPO of the sample naturally emerges to its surface, hence suggesting that QC surfaces can act as templates for the growth of thin films having quasicrystalline order as well [482, 105]. Theoretical studies indicated that the best low-friction regime is achieved for incommensurabilities related to cubic irrational numbers rather than to quadratic irrationals, like the golden mean [95]. A suitable example of cubic irrational number is provided by the so-called spiral mean, which satisfies the equation $\omega^3 - \omega - 1 = 0$. Its rational approximants are generated by the recursion relation $G_{n+1} = G_{n-1} + G_{n-2}$ with $G_{-2} = G_0 = 1$ and $G_{-1} = 0$, leading to the sequence $G_n = \{1, 0, 1, 1, 1, 2, 2, 3, 4, 5, 7, 9, 12, 16, 21, 28, ...\}$ whose terms satisfy the asymptotic limit, $\lim(G_{n+1}/G_n) = \omega \simeq 1.3247...$ According to these results, the low friction observed in QCs cannot be simply explained in terms of their characteristic Fibonacci-based surface ordering.

Quite interestingly, experimental studies on friction anisotropy of a clean d-AlNiCo QC, whose surface terminations exhibit periodic as well as Fibonacci-like atomic ordering along different directions, reveal a strong connection between interface atomic structure and the mechanisms by which energy is dissipated [675]. This result suggests that electronic and phononic contributions probably play a significant role in the tribological properties of QCs. In fact, experiments carried out on various materials (including i-AlCuFe and i-AlPdMn QCs, several approximants of the decagonal phase, Al-based intermetallic compounds, pure Al and pure Cu) have shown that the friction coefficient scales with the inverse of the hardness of the material, and the density of d-states at the Fermi energy able to interact with the d-states of the iron-containing rider. Consequently, a low DOS at E_F naturally leads to low interaction, hence low adhesion and friction [208, 209].

Other appealing properties of QC surfaces, which are currently intensively explored, include oxidation resistance [385, 883], low surface energy [46], and catalytic activity [334, 386, 764, 694]. These properties will be discussed, within the context of possible applications, in Secs.5.2.1 and 5.4.1.

3.10 CORRELATIONS AMONG TRANSPORT COEFFICIENTS

So far we have reviewed the main features of different transport coefficients in QCs, comparing them to their respective behavior in periodic metallic alloys. In this Section we will compare several transport coefficients among them, searching for possible relationships of physical interest. For the sake of illustration in Fig. 3.25 the temperature dependences of the electrical conductivity, magnetic susceptibility and reciprocal Hall coefficient of two Al-based iQCs samples are shown. As we see, the $\sigma(T)$, $\chi(T)$, and $R_H^{-1}(T)$ curves exhibit a remarkably similar behavior over a broad temperature range.

By substracting the residual $T \to 0$ contribution (including the Curie-Weiss term in the case of the magnetic susceptibility) these curves can be fitted to power law functions of the form $f_\zeta(T) = A_\zeta T^{v_\zeta}$ in the $4 - 300$ K interval, where f_ζ stands for the considered transport coefficient, with $v_\sigma = 1.5$, $v_\chi = 1.8$, and $v_{R_H^{-1}} = 2.9$. Accordingly, this fitting is compatible with a variable range hopping mechanism (similar to that described to account for the temperature dependence of the $\kappa_l(T)$ curve in Sec. 3.8.2, see the inset to Fig. 3.22). Furthermore, since R_H^{-1} is proportional to the charge carrier density, Fig. 3.25c essentially coincides with Fig. 3.21 in the considered temperature interval, hence suggesting that the behavior of the $\sigma(T)$ and $\chi(T)$ curves is similarly controlled by the $n(T)$ curve, particularly in the high temperature regime.

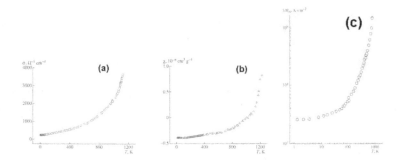

Figure 3.25   Temperature dependence of the (a) electrical conductivity, (b) magnetic susceptibility, for an i-Al$_{63}$Cu$_{25}$Fe$_{12}$ QC; and (c) reciprocal Hall coefficient for an i-AlPdRe QC. (Reprinted by permission from Springer Nature Customer Service Centre GmbH, Prekul A F and Shchegolikhina N, Correlations between electric, magnetic, and galvanomagnetic quantities in stable icosahedral phases based on aluminium, *Crystallogr. Rep.* **52** 996-1005, 2007. Copyright 2007).

From a theoretical viewpoint we can combine the analytical expressions given by Eqs.(3.11) and (3.13) to disclose the noteworthy relationships

$$\frac{S(T)}{R_H(T)\sigma(T)T} = \frac{L_0}{D_0} \equiv \lambda_0 \tag{3.94}$$

$$\frac{S(T)L(T)}{R_H(T)k_e(T)} = \frac{L_0}{D_0} \equiv \lambda_0, \tag{3.95}$$

which naturally lead to the WFL upon equating to each other. After Eq. (3.94), we can estimate the electronic states diffusivity value making use of the transport coefficient values at a given temperature.[37] In Table 3.12, we list the λ_0 values obtained from the knowledge of experimental $R_H(300 \text{ K})$, $\sigma(300 \text{ K})$, and $S(300 \text{ K})$ data for several Al-based iQCs. The samples are arranged according to their average valence in order to discriminate for possible stoichiometric effects. From the data listed in Table 3.12, we get $\bar{\lambda}_0 = 0.002$ and $\bar{\lambda}_0 = 0.005 \text{ V}^2\text{K}^{-2}\text{m}^{-2}\text{s}$ for the AlCuRu and AlCuFe QCs, respectively, so that $\bar{\lambda}_0 \simeq 0.0035$ $\text{V}^2\text{K}^{-2}\text{m}^{-2}\text{s}$ on average, yielding $\bar{D}_0 = 0.07 \text{ cm}^2\text{s}^{-1}$. This figure is 2−4 times smaller than the diffusivity values listed in Table 3.3 for similar compounds. In this regard, it is instructive to consider the ratio

$$\frac{\gamma_e}{\sigma(0)} = \frac{L_0}{D(E_F^0)} \equiv \lambda_0 \frac{D_0}{D(E_F^0)}, \tag{3.96}$$

where we have used Eqs. (3.5) and (3.22). In Table 3.13, we list the values for this ratio as derived from the low temperature experimental results reported in Tables 3.2–3.4 for γ_e and $\sigma(0)$. Two main conclusions can be drawn from these data: (1) the considered iQCs can be grouped in four different classes according to their $\gamma_e/\sigma(0)$ values, which differ by about one order of magnitude from one another, (2) within a given class the $\gamma_e/\sigma(0)$ value is remarkable constant, particularly if one takes into account that each $\gamma_e/\sigma(0)$ ratio class includes samples belonging to different cluster-type classes.

---

[37]We recall that $k_e(T)$ is not directly accesible experimentally, so that Eq. (3.95) cannot be used to this end.

Table 3.12 Room temperature transport coefficients values for several Al-based iQCs, along with the related λ_0 value determined from Eq. (3.94). (*) Value measured at 250 K. (**) Value measured at 200 K. The slopes of $R_H(T)$ and $S(T)$ curves exhibit the same sign in the low temperature regime as expected from Eqs.(3.64) and (3.69).

Sample	$\frac{S(300\ K)}{300\ K}$ (μVK^{-2})	$R_H(300\ K)$ $\times 10^{-8}$ (m^3C^{-1})	$\sigma(300\ K)$ $(\Omega\ cm)^{-1}$	$\lambda_0(300\ K)$ $(V^2K^{-2}m^{-2}s)$	Reference
$Al_{70}Cu_{15}Ru_{15}$	+0.008	−0.04	227	−0.001	[60]
$Al_{68}Cu_{17}Ru_{15}$	−0.03	−0.04	357	+0.002	[60]
$Al_{65}Cu_{20}Ru_{15}$	+0.09	+0.15	154	+0.004	[60]
$Al_{62.5}Cu_{24.5}Fe_{13}$	+0.15	+0.02*	311	+0.024	[695]
$Al_{62}Cu_{25.5}Fe_{12.5}$	+0.13	+0.19	284	+0.002	[63]
$Al_{63}Cu_{25}Fe_{12}$	−0.03	−0.19	355	+0.0004	[63]
$Al_{63.5}Cu_{24.5}Fe_{12}$	−0.06	−0.12**	357	+0.001	[61]
$Al_{63.5}Cu_{24.5}Fe_{12}$	−0.06	−0.11*	352	+0.002	[695]
$Al_{62.5}Cu_{26.5}Fe_{11}$	−0.08	−0.06	441	+0.003	[695]

Quite interestingly, the empirical relationship

$$R_H(T) = R_H(0)\frac{\sigma(0) - \Delta\sigma(T)}{\sigma(0) + \Delta\sigma(T)}, \tag{3.97}$$

was reported to hold for a series of i-AlCuFe QCs obeying the inverse Matthiessen rule $\sigma(T) = \sigma(0) + \Delta\sigma(T)$ [289]. This noteworthy correlation between the $R_H(T)$ and $\sigma(T)$ transport coefficients curves can be accounted for in terms of the two-band model given by Eq. (3.70) by assuming that these QCs behave as compensated two-band systems, so that $R_h = -R_e \equiv R_H(0)$, and adopting $\sigma_h \equiv \sigma(0)$ and $\sigma_e \equiv \Delta\sigma(T)$.

3.11 ON THE NATURE OF CHEMICAL BOND IN QUASICRYSTALS

As it is well-known, metallic substances exhibit a number of characteristic physical properties which are directly related to the presence of a specific kind of chemical bond among their atomic constituents: the so-called metallic bond [608]. Now, in Table 3.1 we listed a number of representative physical properties of both metals and QCs, clearly indicating that quasicrystalline alloys significantly depart from metallic behavior, resembling either ionic or semiconducting materials, hence providing an intriguing example of solids made of typical metallic atoms which do not exhibit any of the physical properties usually signaling the presence of metallic bonding. In this way, the fundamental question arises concerning whether the unusual properties observed in QCs should be mainly attributed (or not) to the characteristic QPO of their underlying structure.

3.11.1 Comparison among different alloy systems

By carefully comparing the transport properties of high structural quality iQCs belonging to different alloy systems and cluster-types, we observe that they can be distinguished attending to the degree their properties depart from an ideal metallic behavior. To illustrate this point the electronic specific heat coefficient and the Debye temperatures of several iQCs are plotted in Fig. 3.26. As it can be seen the more semiconductor-like representatives are mainly observed in Mackay type Al-based iQCs containing transition metals (see also

TABLE 3.13 Electronic specific heat coefficient γ_e (in mJmol^{-1}K^{-2}), low temperature electrical conductivity σ (in $(\Omega\ \text{cm})^{-1}$), γ_e/σ ratio (in JΩm^{-2}K^{-2} units) and electronic difussivity (in cm^{2}s^{-1}) for iQCs arranged according to their diffusion coefficient.

Sample	Cluster	e/a	γ_e	$\sigma(4.2\ \text{K})$	$\frac{\gamma_e}{\sigma(4.2\ \text{K})}$	$D(E_F)$	Ref.
$Zn_{52}Mg_{32}Ga_{16}$	B	2.184	0.18	10200	0.000018	16.1	[939]
$Zn_{56.8}Mg_{34.6}Y_{8.6}$	B	2.125	0.623	4910	0.00013	2.24	[360] [131]
$Al_{54.8}Li_{33.9}Cu_{11.3}$	B	2.108	0.17	1250	0.00014	2.09	[938]
$Cd_{85.1}Yb_{14.9}$	T	2.026	1.1	6757	0.00016	1.74	[705]
$Zn_{59}Mg_{29}Y_{12}$	B	2.165	0.6	3571	0.00017	1.69	[705]
$Zn_{81}Sc_{13}Mg_6$	T	2.155	0.5	2670	0.00019	1.52	[324]
$Zn_{62}Mg_{30}Y_8$	B	2.120	0.796	3570	0.00022	1.27	[441]
$Al_{56.1}Li_{33.7}Cu_{10.2}$	B	2.134	0.35	1500	0.00024	1.22	[938]
$Ag_{42}In_{42}Yb_{16}$	T	2.017	1.27	4878	0.00026	1.09	[79]
$Al_{55.0}Li_{35.8}Cu_{9.2}$	B	2.113	0.318	1140	0.00028	1.02	[417] [441]
$Al_{64}Cu_{20}Ru_{15}Si_1$	M	2.322	0.21	280	0.00075	0.38	[695]
$Al_{62.5}Cu_{24.5}Fe_{11}$	M	2.262	0.32	270	0.0012	0.24	[695]
$Al_{63}Cu_{25}Fe_{12}$	M	2.272	0.26	220	0.0012	0.24	[424]
$Al_{62.5}Cu_{24.5}Fe_{11}$	M	2.262	0.32	270	0.0012	0.24	[695]
$Al_{63}Cu_{25}Fe_{12}$	M	2.272	0.26	220	0.0012	0.24	[424]
$Al_{68}Cu_{17}Ru_{15}$	M	2.377	0.23	180	0.0013	0.22	[695]
$Au_{51}Al_{34}Yb_{15}$	T	1.933	6.5	5050	0.0013	0.22	[383]
$Cd_{85}Yb_{15}$	T	2.026	7.5	5714	0.0013	0.22	[148]
$Al_{63.5}Cu_{24.5}Fe_{12}$	M	2.282	0.31	230	0.0013	0.21	[695] [765]
$Cd_{84.6}Yb_{15.4}$	T	2.025	2.87	2029	0.0014	0.20	[867, 631]
$Al_{65}Cu_{20}Ru_{15}$	M	2.313	0.11	75	0.0015	0.19	[695]
$Al_{64.5}Cu_{20}Ru_{15}Si_{0.5}$	M	2.317	0.27	180	0.0015	0.19	[695]
$Al_{70.5}Pd_{21}Re_{8.5}$	M	2.443	0.10	63	0.0016	0.18	[701]
$Al_{70}Cu_{15}Ru_{15}$	M	2.413	0.20	100	0.0020	0.14	[695]
$Al_{70}Pd_{20}Re_{10}$	M	2.439	0.22	100	0.0022	0.13	[697]
$Al_{70.5}Pd_{21}Re_{8.5}$	M	2.443	0.11	2	0.055	0.005	[707]
$Al_{70.5}Pd_{21}Re_{8.5}$	M	2.443	0.11	1.5	0.074	0.004	[698]
$Al_{70.5}Pd_{21}Re_{6.5}Mn_2$	M	2.436	0.38	5	0.076	0.004	[314]

Tables 3.2–3.4 and Fig. 3.10a). In a similar way, i-AlCu(Fe,Ru,Os) and i-AlPd(Mn,Re) QCs exhibit higher room temperature resistivity and residual resistivity ratio R_{4K} values than i-AlCuLi ones, which, in turn, have larger values than those reported for TiZrNi, ZnSc, ZnMgRE, Cd(Ca,Yb), CdRE, and ZnMgGa iQCs (see Table 3.7). Accordingly, we realize that QCs belonging to the Mackay-type exhibit significantly higher resistivity and R_{4K} values than those observed in QCs containing either Bergman- or Tsai-type clusters, which display room temperature resistivity values comprised within the range 100−300 $\mu\Omega$cm, typical of conventional ternary alloys (Fig. 3.10a). A similar trend is observed for other transport coefficients as well. For instance, iQCs belonging to the systems AlCuLi, TiZrNi, or ZnMgRE exhibit almost linear $S(T)$ curves over a wide temperature range, at variance with the non-linear $S(T)$ curves reported for i-AlCu(Fe,Ru) and i-AlPd(Mn,Re) representatives (Fig. 3.19). In comparison, the non-metallic feature which is more generally shared by all the studied QCs to date refers to thermal conductivity, since remarkably low κ values have been observed in all of them, even for QCs exhibiting relatively large electrical conductivity values and a linear temperature dependence of the Seebeck coefficient.

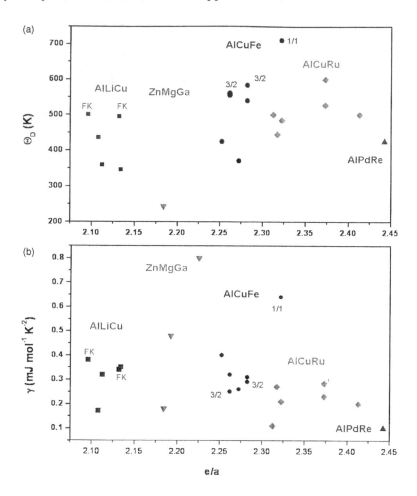

Figure 3.26 Debye temperature (a) and specific heat electronic coefficient (b) values for different thermodynamically stable QCs representatives and some related approximants as a function of their e/a ratio.

By inspecting Table 3.7, we clearly see a systematic increase of both the room temperature resistivity and the resistivity ratio maximum values as the Z number of the group 7 (group 8) third element is increased in i-AlCu(Fe,Ru,Os) and i-AlPd(Mn,Re) alloy systems, respectively. Thus, one would expect the highest electrical resistivity to be observed in QCs belonging to the systems i-AlPdOs and i-AlPdOsRe. These expectations were partially fulfilled: whereas the room temperature resistivities of several i-AlPdOs QCs of different compositions range within the interval $\rho = 2 - 8 \times 10^3$ $\mu\Omega$cm, with $R_{4K} = 1.2–2.0$ [25],[38] a somewhat higher room temperature resistivity $\rho = 1.2 \times 10^4$ $\mu\Omega$cm with $R_{4K} = 5$ was obtained for i-Al$_{71.5}$Pd$_{19}$Os$_{5.5}$Re$_{4.5}$, hence confirming the peculiar role of rhenium atoms in these iQCs [868]. Indeed, Re-bearing iQCs exhibit the highest low-temperature resistivity record, along with huge $R_{4K} \sim 300$ ratios [25]. The correlation between large Z number and

---

[38]These results may be related to the metastable nature of these phases, which transform into a 1/0-cubic approximant upon further annealing.

TABLE 3.14 Melting temperatures of iQCs belonging to different alloy systems and cluster-types arranged by decreasing T_m value. The Debye temperatures are taken from Tables 3.2, 3.4 and 3.6. (a) Dodecahedral polygrain of 30μm edge, (b) T_m is at least 500 K higher at $P = 5$ GPa.

Compound	Type	T_m (K)	Reference			Θ_D^α (K)
$Al_{70.5}Pd_{21}Re_{8.5}$	M	1298	[176]			425
$Al_{63}Cu_{25}Fe_{12}^{(a)}$	M	1169	[295]			$370 - 509$
$Al_{63}Cu_{24}Fe_{13}^{(b)}$	M	1145	[832]			~ 540
$Al_{65}Cu_{20}Fe_{15}^{(a)}$	M	1163	[295]			
$Al_{65}Cu_{20}Fe_{15}$	M	1138	[904]	[127]	[305]	
$Al_{72}Pd_{20}Mn_8$	M	1140	[365]			~ 504
$Ti_{40}Zr_{40}Ni_{20}$	B	1083	[416]			$270 - 410$
$Zn_{60}Mg_{30}Y_{10}$	B	934	[476]			$320 - 350$
$Cd_{85.1}Yb_{14.9}$	T	909	[908]			140
$Al_{60}Li_{30}Cu_{10}$	B	895	[194]			$350 - 465$
$Au_{44.2}In_{41.7}Ca_{14.1}$	T	848	[501]			
$Zn_{88}Sc_{12}$	T	778	[108]			
$Zn_{43}Mg_{37}Ga_{20}$	B	680	[660]			~ 240

high resistivity value may be due to the relativistic contraction of the s and p states relative to the d and f states, which lowers the orbital energies of s and p states and screens the nucleus effective charge, thereby causing the outer d electrons to experience lesser binding, and therefore to display a larger spatial extent. This relativistic effect brings the s, p, and d bands closer to each other, hence enhancing s-p-d hybridization, and ultimately leading to an increase of cohesive energy and a localization of electrons involved in the formation of covalent bonds [308].

The cohesive energy of a solid refers to the energy necessary to separate the constituent atoms from each other and bring them to an assembly of neutral free atoms. Therefore, in forming solids cohesive energy is gained by lowering the total energy relative to that of the free atoms through the formation of different bonding types, namely, ionic, covalent, and metallic. Since ionic bonding originates from electrostatic interaction among charged atoms it is not strictly necessary to consider the overlap of wave function between neighboring atoms to evaluate the cohesive energy of these compounds. Conversely, in both covalent and metallic bonding types one must consider the formation of a valence band stemming from the quantum resonance effects involving the atomic orbitals of condensing atoms. Thus, as atoms get closer and closer to each other the atomic orbitals in the original assembly of neutral atoms are lifted and split leading to the formation of a valence band upon hybridization.

Accordingly, an abrupt change in physical properties related to the cohesion energy (e.g., melting temperature) in a series of compounds may indicate an abrupt change in the bond type, say from a mainly metallic to a mainly covalent one, for instance. In Table 3.14, we list the melting temperatures of several QCs based on different cluster types. Whereas the characteristic melting point of metals and ionic solids is $\sim$1200 K, most molecular solids melt well below $\sim$600 K. Thus, the relatively high melting points observed in Al-based QCs may be indicating stronger than expected bonds among neighboring clusters. In this regard, it is interesting to note that the melting point of the $Al_{65}Cu_{20}Fe_{15}$ cubic approximant alloy is $T_m = 1281$ K, which is about 12% higher than that of the related icosahedral phase [127]. Alternatively, one may also think of the hierarchical spatial arrangement of clusters

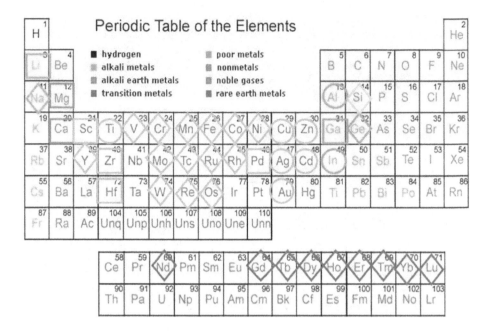

Figure 3.27 Chemical elements found in thermodynamically stable QC alloys. Main forming elements (Al, Ga, In, Zn, Cd, Cu, Ag, Au, Ti) are circled. The second major constituents are squared. Minor constituents are framed within a diamond.

precluding an easy separation of each other even if they interact weakly among them. This may be indicative of long-range QPO effects in QCs.

3.11.2 The chemical synthesis route to new quasicrystals

Most atomic elements composing thermodynamically stable QCs observed to date belong to the broad chemical family of metals, including representatives from alkaline, earth-alkaline, transition metals, or rare-earth groups. Thus, most metallic atoms are able to participate in the formation of quasicrystalline compounds (Fig. 3.27). As a matter of fact all the stable QCs discovered to date have very sharp composition ranges in their respective phase diagrams, suggesting that the alloy chemistry strongly affects the stability of these materials. For instance, the minor atoms constituent in the systems

$$Al_{70}Pd_{20}\begin{pmatrix} Mn \\ Tc \\ Re \end{pmatrix}_{10}, \quad Al_{63}Cu_{25}\begin{pmatrix} Fe \\ Ru \\ Os \end{pmatrix}_{12}, \quad Zn_{60}Mg_{30}\begin{pmatrix} Gd \\ Tb \\ Dy \\ Ho \\ Er \end{pmatrix}_{10}$$

belong to the same group of the periodic table, hence indicating the importance of their electronic structure for the stability of the compound. A similar argument holds for the main forming elements as well. For instance, the Mg-substituted alloys of $Cd_{84-x}Mg_xYb_{16}$ form stable iQCs in a wide composition range of $x = 0-61$ at.%. On the other hand, since Cd is located in the same column as Zn in the periodic table, the family of stable iQCs $Cd_{65}Mg_{20}(Y,Gd-Lu)_{15}$ was obtained by replacing Zn with Cd in the $Zn_{62}Mg_{30}RE_8$

compound. Note, however, that the relative content of second major constituent Mg decreases (and the rare-earth elements content increases) in the Cd-Mg-RE compounds as compared to the Zn-Mg-RE ones. It is worth noticing that RE elements with atomic radii larger than 1.8 Å apparently do not form stable structures in the Cd-Mg-RE and Zn-Mg-RE alloy systems, hence suggesting that there exists a restriction on the atomic size in this case [910]. To this end, a parameter, referred to as the effective atomic size ratio, was introduced in order to analyze the role of size effects in QCs stability [145]. This parameter takes into account both atomic radius r_α and element concentration C_α in the form $R_{r,e} = \frac{r_M C_M}{r_A C_A}$, where the subscript A indicates the major constituent, and M stands for secondary or minor elements in the alloy.

Substitution of minor constituents by next neighbor elements along a given row in the periodic table has been exploited in order to obtain the family of stable quaternary QCs given by the formula

$$\text{Al}_{70}\text{Pd}_{20} \begin{pmatrix} \text{Mn} \\ \text{Tc} \\ \text{Re} \end{pmatrix}_{10} \longmapsto \text{Al}_{70}\text{Pd}_{20} \begin{pmatrix} \text{V} \\ \text{Cr} \\ \text{Mo} \\ \text{W} \end{pmatrix}_5 \begin{pmatrix} \text{Co} \\ \text{Fe} \\ \text{Ru} \\ \text{Os} \end{pmatrix}_5 .$$

The ternary alloy systems

$$\text{Ag}_{42}\text{In}_{42} \begin{pmatrix} \text{Yb} \\ \text{Ca} \end{pmatrix}_{16}, \quad \text{Au}_{65-70} \begin{pmatrix} \text{Si} \\ \text{Ge} \end{pmatrix}_{16-19} \begin{pmatrix} \text{Yb} \\ \text{Gd} \end{pmatrix}_{14-16},$$

have been obtained in a similar vein, the former from the parent binary $\text{Cd}_{84}(\text{Ca},\text{Yb})_{16}$ iQCs by replacing Cd atoms by equal amounts of Ag and In ones, which flank Cd element in the periodic table. Finally, it is interesting to note the existence of a great variety of stable iQCs (all of them belonging to the Tsai-type cluster class) based on elements located in different groups of the periodic table, but all of them bearing Sc atoms as the minor element, as it is shown in Table 3.15. Taking into account that Sc atoms have two possible effective valences depending on whether they are alloyed with non-TM or TM elements, respectively (see Fig. 3.1), we can identify two different possible broad e/a intervals yielding stable QCs, namely, $1.8 \leq e/a \leq 1.9$ and $2.0 \leq e/a \leq 2.2$.

The important role played by the presence of a deep DOS minimum close to the Fermi energy in the stabilization of QCs (Sec. 3.2.1), inspired a chemical synthesis exploration project aimed at obtaining new quasicrystalline compounds via pseudogap electronic tuning, which has rendered successful results in the ZnScCu, CaAuIn, and MgCuGa systems [499, 500, 501]. Indeed, for a given alloy composition the Fermi surface-Brillouin zone interaction mechanism can give rise to the formation of pseudogaps at energy values which are not close enough to the Fermi energy to significantly contribute to reduce its electronic energy. Now, if that pseudogap occurs above (below) the Fermi energy then one can change the alloy's composition by properly alloying the original compound with electron rich (poor) atoms in order to change its e/a ratio, so that to shift the Fermi level to properly match it inside the deeper region of the pseudogap. This strategy was applied to $\text{Zn}_{17}\text{Sc}_3$ alloy ($e/a = 2.175$) whose structure appeared to be isotypic with that of the prototypical Tsai-type 1/1 approximant Cd_6Yb. The $\text{Zn}_{17}\text{Sc}_3$ alloy has an electron concentration ratio somewhat larger than that observed in $\text{Cd}_{85}(\text{Ca},\text{Yb})_{15}$ iQCs ($e/a = 2.026$), but substitution of some Zn by Cu yields a novel ternary icosahedral phase with the stoichiometry $\text{Zn}_{71.5}\text{Sc}_{16.2}\text{Cu}_{12.3}$ ($e/a = 2.058$). The electronic tuning via compositional change route was subsequently applied to $\text{Zn}_{11}\text{Mg}_2$ alloy precursors to get the quaternary i-$\text{Cu}_{48}\text{Ga}_{34}\text{Sc}_{15}\text{Mg}_3$ ($e/a = 2.001$) phase [500], and the ternary $\text{Zn}_{82.1}\text{Sc}_{14.6}\text{Mg}_{3.3}$

TABLE 3.15 Scandium bearing QCs belonging to different alloy systems grouped by their e/a ratio value. [499].

Compound	e/a
$Zn_{71.5}Cu_{12.3}Sc_{16.2}$	1.797
$Cu_{46}Al_{38}Sc_{16}$	1.817
$Zn_{74}Sc_{16}\left(\begin{array}{c}Ag\\Au\end{array}\right)_{10}$	1.823
$Zn_{75}Sc_{16}Pd_9$	1.829
$Zn_{74}Sc_{16}Pd_{10}$	1.838
$Zn_{77}Sc_{16}Fe_7$	1.857
$Zn_{78}Sc_{16}Co_6$	1.866
$Zn_{74}Sc_{16}Pt_{10}$	1.885
$Zn_{77}Sc_8\left(\begin{array}{c}Ho\\Er\\Tm\end{array}\right)_8 Fe_7$	1.991
$Al_{54}Pd_{30}Sc_{16}$	2.126
$Zn_{88}Sc_{12}$	2.148
$Zn_{81}Sc_{15}Mg_4$	2.174

($e/a = 2.170$) and $Au_{44.2}In_{41.7}Ca_{14.1}$ ($e/a = 2.15$) compounds [501, 502]. Remarkably enough, the systematic exploration of the gold-post-transition-metal Na-Au-Ga system led to the discovery of the first Na-containing icosahedral phase, i-$Ga_{37.5}Au_{30.0}Na_{32.5}$, which has a comparatively low $e/a \simeq 1.753$ value[39] [820]. However, the existence of the compound i-$Au_{62.7}Al_{23.0}Ca_{14.3}$, with the even lower $e/a \simeq 1.605$ ratio, has been reported recently [690]. It is also interesting to note that in the i-$Mg_{44.1}Zn_{41.0}Al_{14.9}$ QC ($e/a = 2.171$) Al is the minor constituent [405].

3.11.3   Assessing the quasiperiodic order role

Are the unusual properties observed in QCs unique to the QPO present in their underlying structure? In the earlier times it was tacitly assumed that these specific properties should be somehow related to the new kind of order present in QCs, so that they could be regarded as the fingerprints of QPO in matter. A clear indication on the significant role played by the kind of order (i.e., periodic or QP) in the transport properties of a given compound, was provided by a series of studies on the transport properties of dQCs belonging to the AlCo(Cu,Ni) system. The atomic arrangement in these thermodynamically stable QCs is periodic in the 10-fold growth direction, and QP in the plane perpendicular to it. Therefore, the study of their physical properties allows for the comparison between the transport properties through QP planes and those along the periodic direction in the *same sample*. Quite interestingly, a remarkable anisotropy in the electrical, thermal and optical conductivities, as well as in Seebeck and Hall coefficients was reported for high-quality, single-grained dQCs (see Table 3.16) [40, 221, 805, 993]. Thus, for example, when measured along the decagonal axis (periodic order) the electrical resistivity increases with temperature, as usually occurs in metals. On the contrary, when the electrical resistivity is measured along a QP plane it decreases with temperature (Fig. 3.28). In a similar way, the electronic contribution to the thermal conductivity appears to be almost completely suppressed in the quasicrystalline

---

[39]Note that Ga is the main constituent in this iQC.

TABLE 3.16   Room temperature values of the transport coefficients for dQCs belonging to different alloy systems.

Compound	σ $(\Omega^{-1}\mathrm{cm}^{-1})$	S $(\mu\mathrm{VK}^{-1})$	κ $(\mathrm{Wm}^{-1}\mathrm{K}^{-1})$	Ref.
$Al_{70}Ni_{15}Co_{15}$				
periodic	37037	+3.5	15.0	[806, 585]
quasiperiodic	6098	−2.5	2.5	[806, 585]
$Al_{69.7}Ni_{20.3}Co_{10}$				
periodic	25000	+1.9	22.0	[191]
quasiperiodic	3300	+4.0	5	[191]
$Al_{65}Cu_{20}Co_{15}$				
periodic	25641	+4.5		[805]
quasiperiodic	2469	−2.0		[805]
$Al_{64}Cu_{20}Co_{15}Si_{1}$				
periodic	20000	+7.5	4.0	[805, 182]
quasiperiodic	2330	−4.5	1.8	[805, 182]

plane, whereas the heat transport along the periodic direction behaves like that observed in usual periodic crystalline metals [805]. These transport anisotropy measurements provide compelling evidence on the existence of physical effects intrinsically related to the kind of order present in the underlying lattice (**Exercise 3.11**).

Notwithstanding this, there are several hints signaling the importance of short range effects in the emergence of some unusual physical properties of QCs as well. For instance, electrical conductivity measurements show that (1) amorphous precursors of AlCuFe icosahedral phase already exhibit an anomalous increase of conductivity with temperature (Fig. 3.29a), and (2) the structural evolution from the amorphous to the quasicrystalline state is accompanied by a progressive enhancement of the electronic transport anomalies [318, 758]. It is worthy to note that the $\sigma(T)$ curves shown in Fig. 3.29a for both amorphous and QC phases are almost parallel to each other, hence displaying the inverse Mathiessen rule.

In an analogous way, crystalline approximants, which exhibit a local atomic environment very similar to their related QC compounds, appear as natural candidates to investigate the relative importance of short-range versus long-range order effects on the transport properties. For one thing, certain typical transport properties of QCs, such as a high electrical resistivity value or a negative temperature coefficient of electrical resistivity, can also be observed in some crystalline alloys consisting of metallic elements as, for instance, the Heusler-type Fe_2VAl alloy, whose structure is unrelated to the structure of QCs, although this compound also exhibits a narrow and relatively deep pseudogap near the Fermi energy [652]. On the other hand, many of the characteristic physical properties of QCs are also found in approximant phases, albeit to a lesser extent. Thus, experimental measurements indicate that 3/2, 2/1, or 1/1 approximants, with lattice parameters exceeding about 2 nm, display transport properties similar to those observed in full-fledged QCs, whereas lowest-order 1/0 approximants systematically show a typically metallic behavior (Fig. 3.29b).

For instance, by inspecting Table 3.17, we see that approximant samples share similar electronic specific heat γ_e values with their QC counterparts, and the Debye temperature of approximant phases are systematically larger (smaller) than those of Mackay (Tsai) type QCs. On the other hand, albeit both QCs and approximant phases exhibit similar γ_e values, the low temperature electrical conductivity values of approximant phases are significantly higher than those measured in QCs, as a consequence of the larger diffusivity values of their

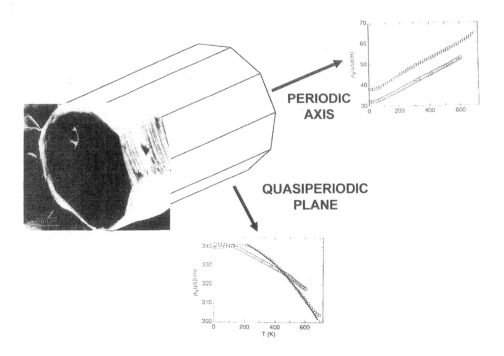

Figure 3.28 The electrical resistivity temperature dependence in the d-$Al_{65}Cu_{15}Co_{15}$ and d-$Al_{70}Ni_{15}Co_{15}$ QCs, measured along the periodic and QP directions, exhibits quite differerent behaviors (dQC picture by courtesy of A. P. Tsai.). Note that, at any given temperature, the resistivity values differ by about an order of magnitude depending on the kind of order present in the underlying substrates. ($\rho(T)$ curves insets reprinted with permission from Martin S, Hebard A F, Kortan A R, and Thiel F A, Transport properties of $Al_{65}Cu_{20}Co_{15}$ and $Al_{70}Ni_{15}Co_{15}$ decagonal quasicrystals, Phys. Rev. Lett. 67, 719 (1991). Copyright 1991, American Physical Society).

electronic states. These results indicate that the electronic properties of QCs and their related approximant phases are quite similar, provided that the order of the considered approximants is high enough, thereby guaranteeing the local atomic arrangements are essentially the same. Notwithstanding this, we must keep in mind that significant differences between physical properties of QCs and their related approximants have been disclosed as well. For instance: (1) only one QC has been reported to be superconductor to date, whereas most approximant phases may attain the superconductor state at low enough temperatures (see Sec. 3.4.5), (2) the melting temperature of approximants is higher than that of their related QC compounds, (3) the question regarding as to whether the unusual optical properties described in Sec. 3.4.3 are (or not) related to long-range QPO effects cannot be definitively answered to date [76], and (4) ferrimagnetic behavior has only been observed in approximant compounds to date (see Sec. 3.7). Magnetic properties of Tsai-type 1/1 approximants have been intensively investigated in several systems during the last decade. The binary Cd_6RE and ternary $Au(Si,Ge)Gd$ compounds are of particular interest since they exhibit long-range antiferromagnetic and ferromagnetic orderings, respectively [463]. In addition, 2/1 ferromagnetic approximants obtained by simultaneous substitution of Ca and In for Au and Al, respectively, into the AuAl(GdYb) 1/1 approximant have been recently reported [362].

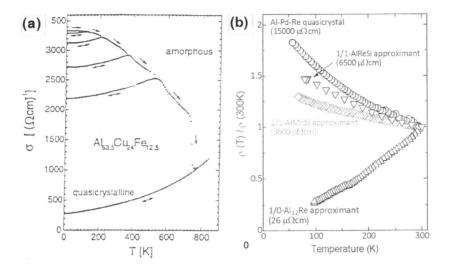

Figure 3.29 (a) Temperature dependence of the electrical conductivity of an i-Al$_{63.5}$Cu$_{24}$Fe$_{12.5}$ thin film for different annealing states for the amorphous and for the iQC phase. The conductivity progressively decreases (spanning over more than one order of magnitude at low temperatures) when the long-range order quality is improved, and the fingerprints of the inverse Mathiessen rule are clearly observed in the $\sigma(T)$ curves. (Reprinted from Haberken R, Khedhri K, Madel C, Häussler P, Electronic transport properties of quasicrystalline thin films, Mater. Sci. Eng. A 2000, 294-296, 475-480. Copyright (2000), with permission from Elsevier). (b) Temperature dependence of the electrical resistivity of an i-AlPdRe QC (circles) is compared to that of 1/1-AlReSi, 1/1-AlMnSi, and 1/0-Al$_{12}$Re approximants. The low-temperature electrical resistivity values are indicated in the graph. (Reprinted from Takagiwa Y, Kirihara K, Metallic-covalent bonding conversion and thermoelectric properties of Al-based icosahedral quasicrystals and approximants, Sci. Technol. Adv. Mater. 2014, 15, 044802; doi:10.1088/1468-6996/15/4/044802, Creative Commons Atribution-NonCommercial-ShareAlike 3.0 License).

On the contrary, transport properties of metallic alloys with complex unit cells, with a similar number of atoms in the unit cell as that of approximant phases, but lacking the local icosahedral symmetry characteristic of both QCs and related approximants, are typically metallic instead [823]. Therefore, mere structural complexity is not a sufficient condition to give rise to the emergence of non-metallic transport properties in QCs. In this regard, the study of the physico-chemical and transport properties of cubic QCs endowed with classical 2-, 3-, and 6-fold rotational symmetries (see Sec. 1.1.5), will be quite clarifying to this end.

Taken together all these observations clearly indicate the importance of short-range effects on the emergence of several transport anomalies, which are appreciably intensified as soon as long-range QPO progressively pervades the overall structure. It is then tempting to assign the origin of most non-metallic properties of ideal QCs to the synergic action of two main mechanisms:

- Local effects involving chemical bonding among neighboring atoms, giving rise to some characteristic features in the electronic structure close to the Fermi level (such

TABLE 3.17 Electronic specific heat coefficient γ_e (in mJmol^{-1}K^{-2}), DOS at the Fermi level (in state eV^{-1} atom^{-1}), Debye temperature Θ_D^α (in K), low temperature electrical conductivity σ (in $(\Omega$ cm$)^{-1}$), and electronic difussivity (in cm^{2}s^{-1}) for several iQCs and their related approximants.

Sample	γ_e	$N(E_F)$	Θ_D^α	$\sigma(4.2$ K$)$	$D(E_F)$	R_{4K}	Ref.
1/1$-$Al$_{72.6}$Re$_{17.4}$Si$_{10}$	0.26	0.11	482	113	0.12	2.7	[862]
3/2$-$Al$_{63.5}$Cu$_{24.5}$Fe$_{12}$	0.29	0.12	583	222	0.22	1.8	[61]
i-Al$_{63.5}$Cu$_{24.5}$Fe$_{12}$	0.31	0.13	539	230	0.21	1.5	[61]
1/1$-$Cd$_{85.7}$Yb$_{14.3}$	7.60	3.22	144	11000	0.41	0.6	[872]
i-Cd$_{84.6}$Yb$_{15.4}$	2.87	1.22	142	2040	0.20	0.9	[872]
1/1$-$Zn$_{85}$Sc$_{15}$	0.71	0.30	226	6667	2.67	0.6	[324]
2/1$-$Zn$_{82}$Sc$_{13}$Mg$_5$	0.42	0.18	235	3333	2.66	1.05	[324]
i-Zn$_{81}$Sc$_{13}$Mg$_6$	0.50	0.21	240	2857	1.62	1.07	[324]
1/1-Mg$_{44.1}$Zn$_{41.0}$Al$_{14.9}$	0.81	0.35	305	20000	6.99	0.89	[405]
2/1-Mg$_{43.0}$Zn$_{42.1}$Al$_{14.9}$	0.73	0.31	310	6667	2.61	1.06	[405]
i-Mg$_{44.1}$Zn$_{41.0}$Al$_{14.9}$	0.72	0.30	310	6667	2.66	1.09	[405]

as the presence of a narrow pseudogap), which are generic (though not specific) of QCs [610]. By all indications, these local effects are closely related to the icosahedral symmetry of atomic clusters. This is nicely illustrated by the observation that non-metallic behavior in 1/1-AlReSi approximants is enhanced when the transition metal sites in the icosahedral clusters are exclusively occupied by Re atoms, so that by just removing a Re atom from the corresponding icosahedron shell significantly destroys the localization tendency [868].

• The establishment of a long-range chemical network extending cluster related resonance effects throughout the overall structure as a consequence of the isomorphism property [560], which guarantees the existence of suitable interference conditions at multiple scales (see Sec. 1.3.3). In fact, Conway's theorem, along with the introduction of QP generalized Bloch functions (see Secs.1.3.2 and 2.6.3), suggest a possible route towards a more general notion of metallic state, naturally encompassing some specific features observed in QCs.

In this picture, the nature of chemical bonding would play the primary role in the very onset of most intrinsic properties of QCs, whose magnitude will be subsequently enhanced as a consequence of their characteristic long-range QPO. Within this scenario, rather than trying to explain the specific properties of QCs in terms of the conceptual schemes previously introduced to describe classical periodic solids, one should properly exploit specific features of QCs atomic structure. Accordingly, since the main building blocks of QCs are assumed to be hierarchically arranged atomic clusters, one should consider bonds among atoms inside clusters along with cluster-cluster interactions giving rise to the extended chemical network. Following this line of thought, the bonding nature of icosahedral clusters of the group 13 elements was earlier investigated to understand the relation between properties of clusters and their own aggregates. By means of molecular orbital quantum calculations it was found that Al$_{12}$ and Al$_{13}$ atomic icosahedra (the latter with one atom located at the center) exhibited covalent and metallic-type bonding, respectively (see Fig. 3.30a, top panel). Thus, atomic occupation of the icosahedron center induces a metallic-covalent bonding conversion phenomenon in these monoatomic clusters [265]. When going from a metallic to a covalent bonding scenario one may expect to observe some general features in the atomic arrangement of the system, such as a smaller coordination number around most atoms, a

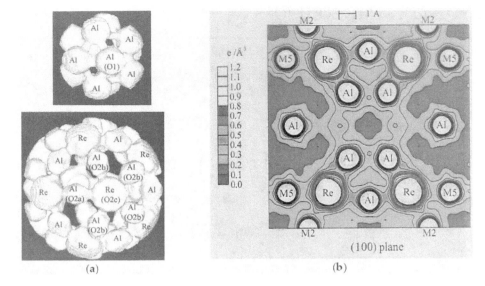

Figure 3.30 (a) Electronic charge equidensity surfaces (0.35 eÅ^{-3}) of the Mackay icosahedral cluster first (top panel) and second (bottom panel) shells in α-AlReSi approximant crystal. (b) Section contour map of charge density of α-AlReSi in the range 0.00–1.20 eÅ^{-3} with a step of 0.10 eÅ^{-3}. (Reprinted figure with permission from Kirihara K, Nakata T, Kimura K, Kato K, Takata M, Kubota Y, Nishibori E, Sakata M, Covalent bonds and their crucial effects on pseudogap formation in α-Al(Mn,Re)Si icosahedral quasicrystalline approximant, Phys. Rev. B 2003, 68, 014205 Copyright (2003) by the American Physical Society).

sharper directionality of the electronic charge distribution among them, as well as shorter interatomic distances. Indeed, it is well known that different bonding styles (i.e., ionic, covalent, metallic, Van der Waals) are characterized by different values of their corresponding effective radii. Accordingly, it was found that the distance between neighboring Al atoms in Al-based iQCs is 0.24 nm, approximately 10% shorter than that of fcc-type Al elemental crystals (0.286 nm). On the other hand, detailed X-ray diffraction studies indicated that the quasilattice constant of i-AlPdRe QCs slightly increases when the TM concentration is increased in the sample, whereas their average atomic radius decreases, so that the atomic density (measured in atoms per nm^3) decreases due to the related quasilattice expansion. However, the atomic density of the samples, determined from their bulk density (measured in gcm^{-3}) divided by the average atomic weight, decreases more rapidly than what is estimated from the quasilattice expansion. This result was interpreted as resulting from the presence of a covalent bonding network involving Al and TM atoms in AlPdRe iQCs [421]. Indeed, studies of the electronic charge density in the approximant crystals α-AlMnSi and α-AlReSi (both containing Mackay-type clusters) disclosed a strong directional Al-Mn (Al-Re) covalent bond in the second shell of the Mackay cluster (see Fig. 3.30a, bottom panel), with similar bonds existing between the Mn atom in the second shell and the so-called Al glue atoms connecting neighbor clusters (labeled M2 and M5 in Fig. 3.30b). In this way, the Al-TM bonds in the second shell are considered to play an important role in the stabilization of the corresponding clusters [420, 422].

3.12   EXERCISES WITH SOLUTIONS

3.1 *Making use of Eq. (3.1) determine the threshold e/a values for the following Hume-Rothery alloys: (a) α-brass (fcc, N = 4, {1, 1, 1} diffracting planes), (b) β-brass (bcc, N = 2, {1, 1, 0} diffracting planes), and (c) γ-brass (bcc, N = 52, {3, 3, 0} and {4, 1, 1} diffracting planes) [610].*

The Fermi surface radius can be expressed as

$$k_F = \frac{1}{a_0} \sqrt[3]{3\pi^2 \left(\frac{e}{a}\right) N},$$

and for cubic lattices we have

$$K_{hkl} = \frac{2\pi}{a_0} \sqrt{h^2 + k^2 + l^2},$$

where (h, k, l) are the Miller's indices of the diffraction planes. Thus, Eq. (3.1) can be written in the form

$$\frac{e}{a} = \frac{\pi}{3N} \left(h^2 + k^2 + l^2\right)^{3/2}. \tag{3.98}$$

Therefore, plugging into Eq. (3.98) the diffraction data given above, we obtain

(a)

$$\frac{e}{a} = \frac{\pi}{3 \times 4} \left(1^2 + 1^2 + 1^2\right)^{3/2} = \frac{\pi\sqrt{3}}{4} \simeq 1.360,$$

for the α-brass.

(b)

$$\frac{e}{a} = \frac{\pi}{3 \times 2} \left(1^2 + 1^2\right)^{3/2} = \frac{\pi\sqrt{2}}{3} \simeq 1.481,$$

for the β-brass.

(c)

$$\frac{e}{a} = \frac{\pi}{3 \times 52} \left(3^2 + 3^2\right)^{3/2} = \frac{9\pi\sqrt{2}}{26} \simeq 1.538,$$

$$\frac{e}{a} = \frac{\pi}{3 \times 52} \left(4^2 + 1^2 + 1^2\right)^{3/2} = \frac{9\pi\sqrt{2}}{26} \simeq 1.538,$$

for the γ-brass.

3.2 *Making use of the data given in Fig. 3.31, determine the energy gain due to the formation of a rectangular pseudogap in the DOS with a depth N_p and width ΔE close to the Fermi level.*

The electronic energy at $T = 0$ in the absence of a pseudogap is given by[610],

$$U_0 = \int_0^{E_F} EN(E)dE = N_0 \int_0^{E_F} EdE = \frac{N_0}{2} E_F^2. \tag{3.99}$$

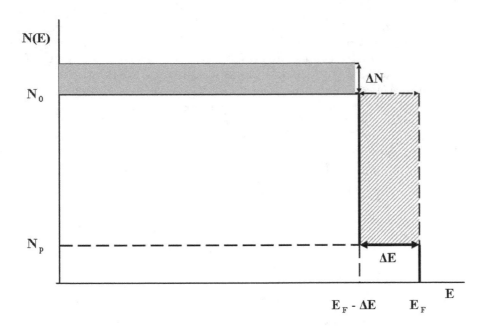

Figure 3.31 Diagram depicting the formation of a rectangular pseudogap of depth N_p and width ΔE in the DOS close to the Fermi level E_F. The original DOS was given by a rectangle of high N_0 and length E_F. As a consequence of the pseudogap formation the states closest to E_F (dashed area) are shifted to lower energies (shadowed upper rectangle of sides ΔN and $E_F - \Delta E$).

The electronic energy in the presence of a rectangular pseudogap is given by

$$
\begin{aligned}
U_p &= \int_0^{E_F - \Delta E} (N_0 + \Delta N)E\,dE + \int_{E_F - \Delta E}^{E_F} N_p E\,dE \\
&= \frac{1}{2}(N_0 - N_p + \Delta N)(E_F - \Delta E)^2 + \frac{N_p}{2}E_F^2.
\end{aligned}
\tag{3.100}
$$

Due to the charge conservation the dashed and shadowed areas in Fig. 3.31 must be equal. Thus, we get the relationship

$$
(E_F - \Delta E)\Delta N = (N_0 - N_p)\Delta E.
\tag{3.101}
$$

After Eqs. (3.99) and (3.100) then we have

$$
\begin{aligned}
\Delta U &= U_p - U_0 = \frac{1}{2}(N_0 - N_p + \Delta N)(E_F - \Delta E)^2 + \frac{N_p - N_0}{2}E_F^2, \\
&= \frac{1}{2}\left[(N_0 - N_p)\Delta E^2 - 2E_F\Delta E(N_0 - N_p) + \Delta N(E_F - \Delta E)^2\right],
\end{aligned}
\tag{3.102}
$$

and making use of Eq. (3.101) in Eq. (3.102), we obtain

$$
\Delta U = -\frac{\Delta E}{2}N_0(1 - r)E_F,
\tag{3.103}
$$

where $r \equiv N_p/N_0$. For the sake of illustration we can take $N_0 = 0.21$ states $(eV)^{-1}at^{-1}$,

$E_F = 7$ eV, $r = 0.2$, and $\Delta E = 1$ eV in Eq. (3.103), to obtain $\Delta U = -56.7$ kJ mol^{-1}. We should highlight that only when the total energy difference between two structurally competing phases is proved to originate solely from the difference in kinetic energy of valence electrons, one may confidently take it as a theoretical proof for the Hume-Rothery electron concentration rule being the ultimate reason for that phase stabilization. Current band structure calculations generally prevent us from keeping the accuracy in determining a total energy difference between two competing phases within the level of 0.1% [611].

3.3 *Making use of Eq. (3.2), determine the energy gain due to the formation of a parabolic pseudogap with a depth a, and width ΔE close to the Fermi level (see Fig. 3.4).*

The electronic energy in the presence of a pseudogap described by Eq. (3.2) is given by

$$U_p = \int_0^{b/2} (d + c\sqrt{|E|})E dE + \int_{b/2}^{E_F} (a + \alpha E^2)E dE$$
$$= \frac{E_F^2}{2}\left(a + \alpha\frac{E_F^2}{2}\right) - \frac{3\alpha b^4}{320}. \tag{3.104}$$

The electronic energy in the absence of a pseudogap is given by (see **Exercise 3.1**)

$$U_0 = \int_0^{E_F} EN(E)dE = N_0\int_0^{E_F} E dE = \frac{N_0}{2}E_F^2. \tag{3.105}$$

Due to the conservation of the DOS area we have the relationship

$$N_0 E_F = \int_0^{b/2} (d + c\sqrt{|E|})dE + \int_{b/2}^{E_F} (a + \alpha E^2)dE = E_F\left(a + \alpha\frac{E_F^2}{3}\right) - \frac{\alpha b^3}{12}. \tag{3.106}$$

Then, making use of Eq. (3.106), Eq. (3.105) can be written as

$$U_0 = \frac{E_F^2}{2}\left(a + \alpha\frac{E_F^2}{3} - \frac{\alpha b^3}{12 E_F}\right). \tag{3.107}$$

Thus, substracting Eqs.(3.104) and (3.107), we get[40]

$$\Delta U = U_0 - U_p = -\frac{\alpha}{4}\left[\frac{E_F^4}{3} + \frac{b^3}{2}\left(\frac{E_F}{3} - \frac{3b}{40}\right)\right]. \tag{3.108}$$

Therefore, the electronic energy gain is proportional to the DOS curvature value at the Fermi energy. For the sake of illustration, we can take $\alpha = 40$ states (eV)$^{-3}$at^{-1}, $E_F = 1$ eV, and $b = 0.06$ eV [531], in Eq. (3.108) to obtain $\Delta U = -3.33$ eVat$^{-1} = -321.6$ kJ mol^{-1}. This energy gain is about 5.5 times larger than that obtained in **Exercise 3.2**.

3.4 *From low temperature specific heat measurements of a centimeter sized single grain of i-$Al_{55.0}Li_{35.8}Cu_{9.2}$ the following fitting parameters to Eq. (3.14) were reported: $\gamma_e = 0.318$ mJmol^{-1}K^{-2} and $\alpha = 4.19 \times 10^{-2}$ mJmol^{-1}K^{-4} [417]. Determine the DOS at the Fermi level and the Debye temperature.*

In order to determine the DOS at the Fermi level it is convenient to express Eq. (3.22) in the form [608],

$$N(E_F)[\text{states (eV)}^{-1}\text{atom}^{-1}] = \frac{\gamma_e[\text{mJmol}^{-1}\text{K}^{-2}]}{2.358} = \frac{0.318}{2.358} = 0.135.$$

[40]Please, do note that the occupied DOS is defined over a negative energy interval in Fig. 3.4.

Making use of Eq. (3.34), we get

$$\Theta_D^\alpha = \sqrt[3]{\frac{12\pi^4 R}{5\alpha}} = \sqrt[3]{\frac{12\pi^4 \times 8.3144}{5 \times 4.19 \times 10^{-2}} \frac{\text{Jmol}^{-1}\text{K}^{-1}}{10^{-3}\ \text{Jmol}^{-1}\text{K}^{-4}}} \simeq 359\ \text{K},$$

where $R = k_B N_A \simeq 8.3144\ \text{Jmol}^{-1}\text{K}^{-1}$ is the perfect gas constant.

3.5 *Derive Eq. (3.40).*

Plugging Eq. (3.26) into Eq. (3.29) in terms of the variable $z = y/2$ we get

$$c_k(T) = a_v k_B q^3 2^3 \int_0^{\frac{\Theta_D}{2T}} z^4 \operatorname{csch}^2 z\, dz + b_v k_B q^5 2^5 \int_0^{\frac{\Theta_D}{2T}} z^6 \operatorname{csch}^2 z\, dz, \qquad (3.109)$$

where $q(T) \equiv k_B T/\hbar$. In the low-temperature limit, $\Theta_D/T \to \infty$, the integrals appearing in Eq. (3.109) can be evaluated in terms of Mellini's transformation given by Eq. (3.33) to obtain

$$c_k(T) = \frac{4\pi^4}{3} k_B q^3 \left(\frac{a_v}{5} + \frac{4\pi^2}{7} b_v q^2 \right), \qquad (3.110)$$

where we have used $\Gamma(n) = (n-1)!$, $\zeta(4) = \pi^4/90$, and $\zeta(6) = \pi^6/945$. Now, we introduce the generalized Debye temperature $\hbar\omega_D \equiv k_B \tilde{\Theta}_D$, satisfying the normalization condition

$$3N_A \equiv \int_0^{\omega_D} D(\omega)d\omega = \frac{a_v}{3}\omega_D^3 + \frac{b_v}{5}\omega_D^5 \equiv q^3 \left(\frac{\tilde{\Theta}_D}{T} \right)^3 \left(\frac{a_v}{3} + \frac{b_v}{5} q^2 \left(\frac{\tilde{\Theta}_D}{T} \right)^2 \right), \qquad (3.111)$$

so that, we have

$$a_v = 9N_A q^{-3} \left(\frac{T}{\tilde{\Theta}_D} \right)^3 - \frac{3b_v}{5} q^2 \left(\frac{\tilde{\Theta}_D}{T} \right)^2. \qquad (3.112)$$

Hence, substituting Eq. (3.112) into Eq. (3.110) and rearranging the coefficients we finally obtain Eq. (3.40).

3.6 *Making use of the low-temperature specific heat data listed in Table 3.18 derive the values for the Debye Θ_D^α, and generalized Debye $\tilde{\Theta}_D$, temperatures of the corresponding iQCs.*

TABLE 3.18　Low-temperature specific heat fit parameters for iQCs belonging to different alloy systems, along with their Debye and generalized Debye temperatures.

Compound	α	δ	Θ_D^α	$\tilde{\Theta}_D$	$\Delta\Theta$	Ref
	$(\text{Jmol}^{-1}\text{K}^{-4})$	$(\text{Jmol}^{-1}\text{K}^{-6})$	(K)	(K)	(%)	
$\text{Zn}_{50}\text{Mg}_{42}\text{Y}_8$	5.59×10^{-5}	2.16×10^{-7}	326.4	200.6	38.5	[325]
$\text{Al}_{70}\text{Pd}_{21.4}\text{Re}_{8.6}$	3.60×10^{-5}	1.52×10^{-7}	378.0	217.5	42.5	[941]
$\text{Al}_{70.5}\text{Pd}_{21}\text{Re}_{8.5}$	2.20×10^{-5}	2.20×10^{-7}	445.4	206.3	53.7	[698]
$\text{Al}_{70.5}\text{Pd}_{21}\text{Re}_{6.5}\text{Mn}_2$	2.30×10^{-5}	2.10×10^{-7}	438.8	207.9	52.6	[314]
$\text{Al}_{68.2}\text{Pd}_{22.8}\text{Mn}_9$	2.63×10^{-5}	9.21×10^{-8}	419.6	240.5	42.7	[941]
$\text{Al}_{70}\text{Pd}_{21}\text{Mn}_9$	1.91×10^{-5}	2.00×10^{-8}	467.0	316.0	32.3	[129]
$\text{Al}_{68}\text{Cu}_{17}\text{Ru}_{15}$	0.90×10^{-5}	0.50×10^{-8}	600.0	424.6	29.2	[605]

Plugging the α values listed in the second column of Table 3.18 into Eq. (3.34) we get the Θ_D^α values listed in the 4th column. On the other hand, making use of both α and δ values into Eq. (3.41), and solving it numerically, we obtain the generalized Debye temperature values listed in the 5th column of Table 3.18. As we see, both Debye temperatures significantly differ (the $\tilde{\Theta}_D$ values being systematically lower), and their relative difference $\Delta\Theta \equiv (\Theta_D^\alpha - \tilde{\Theta}_D)/\Theta_D^\alpha$, amounts to about 30–50%.

3.7 *Estimate the contribution of the VDOS given by Eq. (3.47) to the lattice specific heat at high temperatures.*

In the high-temperature limit $G(z) = 4z^2(e^z - e^{-z})^{-2} \simeq 1$, so that Eq. (3.48) reduces to

$$
\begin{aligned}
c_k^\infty(T) &= 3R\sum_{i=1}^{\nu}\eta_i\int_0^{\frac{\Theta_D}{2T}}\delta(z - z_i)dz = 3R\sum_{i=1}^{\nu}\eta_i\int_0^{\frac{\Theta_D}{2T}}\lim_{\varepsilon\to 0}\left(\frac{1}{\pi}\frac{\varepsilon}{\varepsilon^2 + (z - z_i)^2}\right)dz \\
&= \frac{3R}{\pi}\sum_{i=1}^{\nu}\eta_i\lim_{\varepsilon\to 0}\left(\int_{-\frac{z_i}{\varepsilon}}^{u}\frac{du'}{1 + u'^2}\right) \\
&= \frac{3R}{\pi}\sum_{i=1}^{\nu}\eta_i\left[\arctan\left(\lim_{\varepsilon\to 0}\frac{2\varepsilon T\Theta_D}{4T^2\varepsilon^2 - \Theta_i(\Theta_D - \Theta_i)}\right)\right] = 3R\sum_{i=1}^{\nu}\eta_i,
\end{aligned}
\tag{3.113}
$$

where we have introduced the auxiliary variable $u \equiv (z - z_i)/\varepsilon$, and made use of the relation $\arctan A + \arctan B = \arctan\frac{A+B}{1-AB}$. Therefore, at high enough temperatures the lattice contribution of the specific heat attains an asymptotic limit value larger than the Dulong-Petit figure by a factor $\sum_{i=1}^{\nu}\eta_i$.

3.8 *Obtain the electrical conductivity temperature dependence for a material undergoing a metal-insulator Anderson transition.*

Expressing Eq. (3.59) in terms of the scaled variable x, and plugging it into Eq. (3.12) we obtain

$$
J_n(\beta) = \sigma_A\beta^{-\upsilon}\int_{x_C}^{\infty}x^n\,|x - x_C|^\upsilon\,\text{sech}^2(x/2)dx.
\tag{3.114}
$$

Experimentally one usually approaches the Anderson transition from the metallic side.[41] Thus, we will consider the limit $E_F \to E_C$, with $E_F > E_C$ (i. e., $x_C \to 0^-$) in Eq. (3.114), which takes the form $J_n(\beta) = \sigma_A\beta^{-\upsilon}\tilde{J}_m$, where we have introduced the integrals

$$
\tilde{J}_m \equiv \int_0^{\infty}x^m\,\text{sech}^2(x/2)dx,
\tag{3.115}
$$

with $m \equiv n + \upsilon > 0$, $n = 0, 1, 2$. These integrals can be calculated analytically. To this end, we express them in the form

$$
\tilde{J}_m \equiv \int_0^{\infty}x^m\,\text{sech}^2(x/2)dx = 2\int_0^{\infty}x^m\frac{d}{dx}\left(\tanh\frac{x}{2}\right)dx,
\tag{3.116}
$$

and integrate by parts to obtain

$$
\tilde{J}_m = 2\lim_{x\to\infty}\left(x^m\tanh\frac{x}{2}\right) - 2m\int_0^{\infty}x^{m-1}\tanh\frac{x}{2}dx.
\tag{3.117}
$$

---

[41] For instance, by systematically increasing the disorder amount in the sample.

Making use of Eq. (3.8) one can express Eq. (3.117) as

$$\tilde{J}_m = 2 \lim_{x \to \infty} (x^m - x^m) + 4m \int_0^\infty \frac{x^{m-1}dx}{1 + e^x} = \begin{cases} 4\ln 2, & \text{if } m = 1 \\ 4mI_m, & \text{if } m \neq 1 \end{cases}, \tag{3.118}$$

where $I_m \equiv (1 - 2^{1-m})\Gamma(m)\zeta(m)$, $\Gamma(m)$ is the Gamma function and $\zeta(m)$ is the Riemann zeta function. By plugging Eq. (3.118) into the expression $J_n(\beta) = \sigma_A \beta^{-\upsilon} \tilde{J}_m$, the reduced kinetic coefficients can be explicitly obtained in terms of the model parameters σ_A and υ, and substituting the obtained values in Eq. (3.11) one finally gets,

$$\sigma(T) = \begin{cases} \ln 2\, \sigma_A k_B T, & \text{if } \upsilon = 1 \\ (1 - 2^{1-\upsilon})\upsilon\Gamma(\upsilon)\zeta(\upsilon)\sigma_A k_B^\upsilon T^\upsilon, & \text{if } \upsilon \neq 1 \end{cases},$$

so that $\sigma(T) = A_\upsilon T^\upsilon, \forall \upsilon \geq 1$.

3.9 *Obtain the value of the integral $\int_0^\infty \sigma_1(\omega)d\omega$ (the so-called sum rule) for materials obeying Eq. (3.61).*

Making use of Eq.(3.61), we have

$$\int_0^\infty \sigma_1(\omega)d\omega = \frac{ne^2}{m_0^*} \int_0^\infty \frac{\tau d\omega}{1 + (\omega\tau)^2} = \frac{ne^2}{m_0^*} \int_0^\infty \frac{du}{1 + u^2} = \frac{ne^2}{m_0^*}\frac{\pi}{2} = \frac{\omega_p^2}{8},$$

where we have introduced the auxiliary variable $u \equiv \omega\tau$, and made use of Eq. (3.62). This expression allows us to experimentally derive the plasma frequency value from the knowledge of the $\sigma_1(\omega)$ curve.

3.10 *Derive Eq. (3.87).*

Making use of the scattering rate value $\tau^{-1}(\omega) = A_U \omega^2 T^n \Rightarrow \tau(y) = A_U^{-1}\left(\frac{\hbar}{k_B}\right)^2 T^{-(2+n)}y^{-2}$ in Eq. (3.79), we have

$$\begin{aligned} \kappa_l^U(T) &= \frac{3}{4}v^2 k_B n_a \left(\frac{T}{\Theta_D}\right)^3 A_U^{-1}\left(\frac{\hbar}{k_B}\right)^2 T^{-(2+n)} \int_0^{\Theta_D/T} y^2 \operatorname{csch}^2\left(\frac{y}{2}\right) dy \\ &= \frac{6 k_B n_a^{1/3} A_U^{-1}}{(6\pi^2)^{2/3}} \frac{T^{1-n}}{\Theta_D} \int_0^{\Theta_D/T} y^2 \operatorname{csch}^2\left(\frac{y}{2}\right) dy, \end{aligned} \tag{3.119}$$

where we have made use of Eq. (3.31). In the low temperature regime the integral appearing in Eq. (3.119) can be explicitly evaluated by means of Eq. (3.33) to obtain Eq. (3.87).

3.11 *The mean value of the Hall coefficient for a number of decagonal quasicrystals is about $R_H \simeq 3 \times 10^{-4}$ cm^{3}C^{-1} (with the magnetic field perpendicular to the periodic stacking direction) and the volume of their quasi-Brillouin zone is $V \simeq 17.7$ Å^{-3} [993]. Estimate: a) the charge carriers density, and b) the carriers mobility. Comment on the obtained results.*

a) The Hall coefficient directly measures the charge carrier density according to the expression $R_H = (en)^{-1}$, so that

$$n = \frac{1}{eR_H} = \frac{1}{1.602 \times 10^{-19}\ \text{C}\ 3 \times 10^{-4}\text{cm}^3\text{C}^{-1}} = 2.08 \times 10^{22}\ \text{cm}^{-3}$$

On the other hand, if one assumes that the Fermi surface completely fills the available Brillouin zone volume one would have

$$V = \frac{4}{3}\pi k_F^3 = \frac{4}{3}\pi(3\pi^2 n),$$

and then

$$n = \frac{V}{4\pi^3} = \frac{17.7 \text{ Å}^{-3} \times 10^{24} \text{ Å}^3 \text{cm}^{-3}}{4\pi^3} = 1.43 \times 10^{23} \text{ cm}^{-3}.$$

As we see, the charge carrier density values obtained from both procedures differs by a factor of about seven, and it is reasonable to conclude that the carriers concentration is not enough to allow the Fermi sphere to completely fill the Brillouin zone in this case.

b) Making use of Eq. (3.69), we have

$$\mu_H = \sigma R_H = 3.5 \times 10^3 \text{ } \Omega^{-1}\text{cm}^{-1} \times 3 \times 10^{-4} \text{ cm}^{-3}\text{C}^{-1} = 1.05 \text{ cm}^2\text{V}^{-1}\text{s}^{-1}$$

where the electrical conductivity value along the QP plane has been estimated by averaging the values listed in table 3.16, and we have made use of the dimensional relationship [V] = [ΩA]= [ΩCs^{-1}]. By comparing with the results listed in Table 3.9, we conclude that the carriers mobility is significantly low in the QP plane of these dQCs, and comparable to the values reported for certain iQCs.

CHAPTER 4

Photonic and Phononic Quasicrystals

4.1 QUASIPERIODIC ORDER BEYOND ATOMIC DOMAIN

"Quasi-periodic structures produced in accordance with the present invention are produced simply and inexpensively without need for very stringent manufacturing tolerances and controls". (Roberto Merlin and Roy Clarke, 1990) [599]

Shortly after the pioneering paper by Steinhardt and Levine was published on 24 December 1984 (see Sec. 1.1.1), the notion of QPO in condensed matter was extended from the atomic domain observed in metallic alloys by Shechtman and co-workers to structures based on larger building blocks. Thus, on 21 October 1985, Roberto Merlin and co-workers reported in *Physical Review Letters*, the first realization of a semiconductor-based QP superlattice [598]. The sample, grown by molecular beam epitaxy, consisted of two types of building blocks, each one consisting of AlAs/GaAs bilayers of different thickness arranged according to the Fibonacci sequence. The AlAs layers (acting as electronic barriers) have all the same nominal thickness (1.7 nm), whereas the GaAs layers (serving as electronic wells) take on two different values: 4.2 nm (A layers) and 2.0 nm (B layers). The resulting bilayers satisfied the thickness ratio $d_A/d_B = 59/37 \simeq 1.595$ (a value relatively close to the golden mean value $\tau = 1.618...$) and the whole sample consisted of $F_{13} = 377$ bilayers with a total length of ~ 1.85 μm (nominal value $L = d_A F_{12} + d_B F_{11} = 1.9075$ μm).

The room-temperature X-ray diffraction pattern of this Fibonacci superlattice (FSL) showed a significant number of very narrow peaks superimposed to the main satellite reflections of the GaAs layers (Fig. 4.1). These finer peaks can be labeled as a series of golden mean powers given by the expression

$$k_{m_1,m_2} = \frac{2\pi}{\tau d_A + d_B} m_1 \tau^{m_2}, \tag{4.1}$$

where m_1 and m_2 are integers (**Exercise 4.1**) [886]. Eq. (4.1) properly highlights the self-similar arrangement of the diffraction peaks stemming from the bilayers QP arrangement (**Exercise 4.2**), whereas the distance between successive peaks arising from the GaAs periodic crystal atomic planes is approximately constant (Fig. 4.1 inset), clearly demonstrating the presence of two kinds of order coexisting in the *same* sample at *different length scales* [544]. In fact, at the atomic level we have the usual crystalline order determined by the periodic arrangement of atoms in each layer, whereas at longer scales we have the QPO determined by the sequential deposition of the different layers.

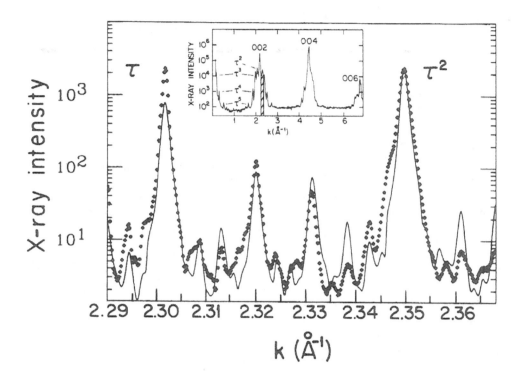

Figure 4.1 High-resolution diffraction profile of the first grown AlAs/GaAs Fibonacci superlattice. The dots represent synchrotron X-ray data, and the line is the calculated Fourier intensity. A low-resolution scan showing the overall appearance of the X-ray scattering is plotted in the inset. The shaded region indicates the range of the high-resolution scan shown in the main figure. For reciprocal space vectors within the range $1 - 8$ Å^{-1} one observes a series of evenly spaced reflections, corresponding to the (002), (004), and (006) GaAs planes, while for reciprocal space vectors within the shadowed range one obtains a dense set of peaks which can be properly labeled by successive powers of τ. (Reprinted figure with permission from Todd J, Merlin R, Clarke R, Mohanty K M, and Axe J D, 1986 Phys. Rev. Lett. 57 1157. Copyright (1986) by the American Physical Society).

Although the first physical realization of a FSL was based on layers of semiconducting materials, it is clear that one may think of using the same QP design employing other kinds of materials instead. Indeed, in the original patent file four broad classes of different materials were included in the claims as possible constituents for the layers, namely, metallic, semiconductor, dielectric, and transparent oxides [599]. In the intervening years, a lot of diverse materials, including piezoelectrics, multiferroics [878],[1] high-temperature yttrium barium copper oxide superconductors,[721] alkali-metal doped C_{60} fullerene thin films [719], or dielectric layers separated by graphene, to name just a few [142], have been added to that list as possible components of aperiodic multilayers, with a view in different potential applications.

As it was stated in the Merlin and Clarke's patent file quoted in the opening of this Section, aperiodic order can be precisely controlled during the growth process in these man-made structures. This novel degree of freedom distinguishes QP superlattices and multilayers from usual periodic ones, opening promising avenues in new materials design. In fact, since different physical phenomena have their own relevant physical scales, by properly matching the characteristic length scales of elementary excitations propagating through the multilayer, we can efficiently exploit the aperiodic order we have introduced in the system, hence promoting technological innovation. Quite interestingly, detailed structural characterization studies have shown that structural imperfections which are inevitably introduced during the growth process do not seriously disrupt the overall coherence of QP multilayers, so that most physical properties related to QPO are robust enough. This feature certainly prompted the interest in exploiting the potential applications of multilayered structures composed of different materials arranged according to different kinds of aperiodic sequences.[2] [599]

In a similar vein a remarkable two-dimensional QP set up was designed in 1988 in order to study the nature of the frequency spectrum of a *macroscopic* acoustic system consisting of 150 tuning forks (natural frequency 440 Hz) glued into a heavy aluminum plate patterned according to a standard Penrose tile (Fig. 4.2a). The tuning forks are mounted at the centers of the rhombuses, with the two tines oriented in line with the shorter diagonal. Using the four sides of each rhombus as a reference, four nearest neighbors are identified, and each tine of a tuning fork is coupled to the two nearest tines of the adjacent tuning forks by means of 1-mm-diameter steel wire arcs, forming a QP network of coupled resonators. An electromagnet is positioned near one tine of the array, and an AC current is passed through it in order to drive the system, and its response is monitored with four electric guitar pickups positioned next to random tines in the array. The resulting resonant frequencies correspond to the eigenvalues of the 2D QP system. The obtained frequency spectrum is shown in Fig. 4.2, along with its related DOS. The frequency spectrum of this Penrose QC exhibits four characteristic bands (labeled A, B, C, and D) separated by the corresponding gaps α, β, and γ. In addition, a series of subsidiary gaps appear at different locations within the four main bands, giving the overall spectrum the typical appearance of a codebar pattern. Quite remarkably, the widths of these bands and gaps are in ratios involving the golden mean, namely, $C/B = \tau$, $A/B = \tau^2$, $\beta/\alpha = \gamma/\beta = \tau$ [330, 592]. Accordingly, most characteristic features of energy spectra in atomic based 1D QP systems, namely,

---

[1]Materials having *several* different ferroelectric properties in the same phase, such as ferroelectricity, ferromagnetism, or ferroelasticity, for instance.

[2]In this regard, it is worth noticing that in the original patent file 6th claim the authors extend their design principle to other possible aperiodic sequences other than the Fibonacci one in terms of their respective substitution matrices by stating that they should have *"a non-zero determinant and a first matrix element equal unity"*. Note, however, that these mathematical conditions exclude Thue-Morse, Rudin-Shapiro, or precious means based multilayers from patent scope (cf. Sec. 2.2).

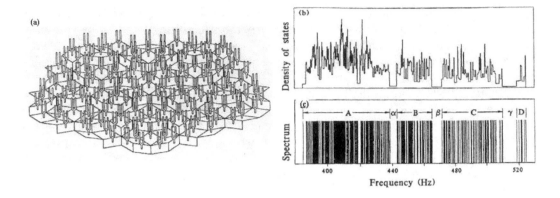

Figure 4.2 (a) Schematic drawing of a macroscopic acoustic QC based on the Penrose tile geometry (for the sake of drafting simplicity the tuning forks are not drawn with their actual orientation in the experimental setup). The DOS (b), and the frequency spectrum (c) of the tuning fork QC were determined as a composite of the resonant spectra obtained from twenty different positions in the Penrose lattice. The DOS is obtained as the inverse of the difference in frequency for neighboring eigenvalues in the frequency spectrum. (Reprinted figure with permission from He S and Maynard J D , Phys. Rev. Lett. 62, 1888, 1989. Copyright (1989) by the American Physical Society).

hierarchical fragmentation and scalability, are similarly present in the frequency spectra of 2D macroscopic ones.

4.2 LAYERED OPTICAL STRUCTURES

Light transmission through aperiodic media has deserved a progressively growing attention in order to understand the interplay between optical properties and the underlying aperiodic order of the substrate. To this end, the mathematical analogy between Schrödinger and Helmholtz equations:

$$\frac{d^2\psi}{dz^2} + \frac{2m}{\hbar^2}[E - V(z)]\psi = 0, \tag{4.2a}$$

$$\frac{d^2\mathcal{E}}{dz^2} - \frac{\omega^2}{c^2}\left[\left(\frac{ck_\parallel}{\omega}\right)^2 - n^2(z)\right]\mathcal{E} = 0, \tag{4.2b}$$

provides a powerful tool to relate previous knowledge about electron motion in superlattices to electromagnetic waves propagating in multilayers. Eq. (4.2a) describes the motion of an electron with effective mass m, energy E, and a wave function ψ, under the action of a 1D potential $V(z)$, whereas Eq. (4.2b) describes a monochromatic electromagnetic wave of frequency ω propagating in a lossless, dispersionless medium with a variable refractive index profile $n(z)$, where $\mathcal{E}$ is the transversal component of the electric field, $k_\parallel$ is the wave vector in the XY plane (perpendicular to the propagation direction Z), and c is the vacuum speed of light. In the multilayer case the refractive index profile straightforwardly connects a physical magnitude to the aperiodic sequence describing the stacking order along the multilayer (Fig. 4.3a) in terms of a suitable aperiodic function (Fig. 4.3b,c). Thus, the isomorphism of Shrödinger and Helmholtz equations allows us to borrow some basic

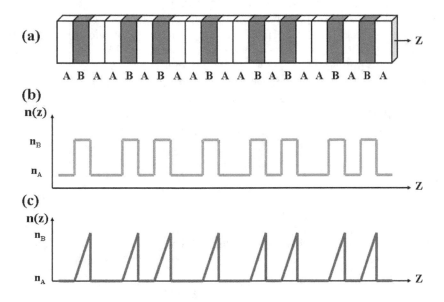

Figure 4.3 (a) Sketch of a Fibonacci dielectric multilayer grown along the Z direction, (b) refractive index profile n(z) for an arrangement of layers with constant refraction index, (c) refractive index profile n(z) for an arrangement of layers with a linearly graded refraction index. (Reprinted from Maciá E, Thermal emission control via bandgap engineering in aperiodically designed nanophotonic devices, Nanomaterials 5 814-825 2015, doi:10.3390/nano5020814, Creative Commons Atribution License CC BY 4.0).

notions from solid state physics, such as Bloch waves, Brillouin zones or band gaps in order to describe several optical properties of many devices of common use in optoelectronics and optical communication applications nowadays [274]. In this way, the so-called photonic crystal concept was introduced to describe optical systems which exhibit wide frequency stop bands due to interference effects photonic band gaps (PBGs), in close analogy with the presence of band structure in crystalline lattices, or the formation of energy minibands in superlattices. Certainly, the very notion of photonic crystal can be extended to describe the properties of QP photonic structures as well. To this end, one simply considers that the optical properties of the medium are given by a QP refraction index function, such as those shown in Fig. 4.3b,c, instead of a periodic one. The resulting structure can then be properly referred to as a *photonic quasicrystal* (PQC).

Long-range QPO, by its own, endows PQCs with certain characteristic properties which are not exhibited by their periodic counterparts. This feature stems from the richer structural complexity of aperiodic sequences, which arises from the presence of QP and self-similar order related fingerprints, and naturally leads to the presence of more resonant frequencies stemming from multiple interference effects throughout the structure. For instance, due to their highly fragmented frequency spectrum, aperiodic multilayers offer more full transmission peaks (alternatively, absorption dips) than periodic ones in a given frequency range for a given system length, and the inflation symmetry gives rise to a denser Fourier spectrum structure in reciprocal space. Another important feature stemming from the underlying structural order in PQCs is the existence of the so-called critical modes, which are spatially more complex than those related to both extended Bloch and exponentially localized defect modes in periodic and random photonic crystals, respectively [592]. The presence of these critical states allows for the possible existence of either relatively localized or significantly extended electromagnetic intensity patterns depending on the considered frequency. The occurrence of highly localized optical modes in the absence of disorder, that is, in defect-free PQCs, is of interest for the fabrication of high quality factor resonators as well as coupled-resonator waveguides (see Sec. 5.5). Therefore, in contrast to periodic systems, spatial QPO naturally introduces a full set of non-equivalent sites in aperiodic arrangements. In this way, a great diversity of electromagnetic spatial patterns can be obtained for different frequencies. Thus, by properly exploiting the additional degrees of freedom inherent to the aperiodic arrangements, one can design optical devices offering a broader and more flexible design capability than their periodic counterparts for certain specific applications [87, 154, 548, 552, 554].

In order to properly compare the optical responses of periodic and QP multilayered systems the transfer matrix technique is particularly well suited, since it describes the light behavior in contiguous layers sequentially. The simplest approach focuses on systems composed of linear, homogeneous, lossless materials with no optical activity, embedded in a medium of refractive index n_M. The unimodular local transfer matrices describing the light propagation from the i-th layer to the $(i+1)$-th layer are

$$\mathbf{K}_{i,i+1} = \begin{pmatrix} \cos \delta_i & -q_{i,i+1} \sin \delta_i \\ q_{i,i+1}^{-1} \sin \delta_i & \cos \delta_i \end{pmatrix}, \qquad (4.3)$$

where $\delta_i \equiv n_i d_i k \cos \theta_i$ is the layer phase thickness, n_i is the layer refractive index, d_i is the layer width, $k = 2\pi/\lambda$ is the wave vector of the incident radiation in vacuum, and θ_i is the refraction angle determined by Snell's law. The parameter $q_{i,i+1}$ describes the propagation across the interface separating two neighboring layers and can be written as

$$q_{i,i+1}(s) = \frac{n_i \cos \theta_i}{n_{i+1} \cos \theta_{i+1}}, \qquad q_{i,i+1}(p) = \frac{n_{i+1} \cos \theta_i}{n_i \cos \theta_{i+1}}, \qquad (4.4)$$

for each basic polarization s (TE) or p (TM), respectively. We note that $q_{i+1,i} = q_{i,i+1}^{-1}$. The transmission of light through a binary multilayer composed of N layers of two different materials, labelled A and B, can be properly described in terms of a product involving the matrices $\mathbf{K}_{AA}$, $\mathbf{K}_{BB}$, $\mathbf{K}_{AB}$, and $\mathbf{K}_{BA}$, along with suitable boundary matrices $\mathbf{K}_{LA}$, $\mathbf{K}_{AL}$, $\mathbf{K}_{LB}$, and $\mathbf{K}_{BL}$, where $L = \{0, N+1\}$ [552]. The transmittance spectrum can be obtained from the standard expression $T_N(\omega) = 4(\|\mathcal{M}_N\| + 2)^{-1}$, where $\mathcal{M}_N \equiv \prod_{i=N}^{0} \mathbf{K}_{i,i+1}$ is the multilayer global transfer matrix (we note that $\det \mathcal{M}_N = 1$), and the norm $\|\mathcal{M}_N\|$ denotes the sum of the squares of the four elements of $\mathcal{M}_N$. The interested reader is addressed to recent literature for a detailed presentation of transfer matrix techniques in the study of aperiodic multilayered systems [78, 142, 548, 552].

4.2.1 A broad palette of possible designs

The rapid progress achieved in growth technologies, such as molecular beam epitaxy, magnetron sputtering, or vacuum deposition, has made possible to grow artificial structures with different aperiodic modulations of chemical composition along the growth direction. In the earlier times Fibonacci dielectric multilayers (FDMs) were intensively studied, since they provide a canonical example of QP structures for optical applications, but the interest progressively moved on towards other classes of aperiodic structures based on substitution rules (see Sec. 2.2), which may also exhibit interesting photonic properties. Thus, different generalizations of the Fibonacci sequence, preserving both the inflation self-similarity and the long-range QPO characteristic of the original one, have been considered in the literature for optical applications [93]. Beyond strictly QP structures the optical properties of multi-layers based on the Thue-Morse sequence (whose Fourier spectrum is singular continuous, see Sec. 2.2) have received a considerable attention as well [214, 510, 923, 924].

Certainly, there exist many possible ways of arranging a series of different layers in an ordered aperiodic way without recurring to substitution sequences at all. For instance, one may consider a binary multilayer where each layer is labelled by a natural number in such a way that, say, A layers correspond to prime numbers [762]. Since the number of primes progressively decreases as one goes to larger integers the relative frequency of A layers approaches zero as the system grows longer. Therefore, A layers can be regarded as some sort of diluted impurity layers embedded in a B type substrate. Quite remarkably, the Fourier spectrum of 1D lattices where atoms are arranged according to the prime numbers sequence exhibit a sharp distribution of Bragg-peaks, hence indicating the presence of well-defined long-range order (see Sec. 6.3.3).[3] [890, 891, 994] To the best of my knowledge, no rigorous mathematical results regarding the related energy (or frequency) spectrum have been reported yet (as July 2020) Nevertheless, an interesting property of primes is that they do not come infrequently in pairs (called twin primes) such as the couples (11,13) or (101,103), for instance. In this way, twin primes introduce a short-range correlation among the A layers, and one expects that their presence will give rise to the emergence of resonance effects leading to the presence of delocalized states minibands around the resonance energies in this sort of multilayers [691, 770].

As another instance of aperiodic multilayer not depending on substitution rules we may consider a structure in which the refractive index of the layers follows a self-similar arithmetical series named the 1s-counting sequence. This sequence is defined by taking into account the number of "1"s in the binary representation of natural numbers, namely, $0 = \mathbf{0}$, $1 = \mathbf{1}$, $2 = \mathbf{10}$, $3 = \mathbf{11}$, $4 = \mathbf{100}$, $5 = \mathbf{101}$, $6 = \mathbf{110}$, $7 = \mathbf{111}$, etc. Thus, by counting the number of "1"s in each $n \in \mathbb{N}$ binary representation we obtain the sequence $A_j = \{0, 1, 1, 2, 1, 2, 2, 3...\}$, $j = 0, 1, 2 \ldots$. Quite conveniently, this series can also be generated from the recursive formula $A_j = \{\{A_{j-1}\}, 1 + \{A_{j-1}\}\}$, with $A_0 = 0$ as the initial value. The self-similarity property of the sequence can be checked by decimating every second term in a sequence of order j, which leads to the corresponding $j - 1$ order sequence, as it is illustrated below

$$A_4 = \{0, 1, 1, 2, 1, 2, 2, 3, 1, 2, 2, 3, 2, 3, 3, 4\} \to \{0, , 1, , 1, , 2, , 1, , 2, , 2, , 3\} = A_3. \tag{4.5}$$

On the basis of this numerical sequence one can construct a dielectric multilayer composed of $N = 2^j$ layers, each one having a refractive index given by $n_l = n_* + \epsilon\{A_l\}$, $l = 1, ..., N$, where n_* is a suitable initial reference value and ϵ is a small increment [768]. Reflectivity spectra of porous silicon multilayers based on this sequence show that the number of null transmission gaps is reduced in the aperiodic multilayer, as compared to the periodic one. This is due to the *absence* of long-range QPO in the considered A_j sequence. Nevertheless, the transmittance of the aperiodic multilayer exhibits exact scalability, since the transmittance spectra of smaller structures is contained in the spectrum of the larger one upon proper rescaling [240].

Alternative generalization routes may consider *modular designs* constructed by concatenating different sorts of aperiodic lattices. For instance, the optical properties of structures

---

[3]According to the terms of reference introduced by the Commission on Aperiodic Crystals of the ICrU, this system should be properly regarded as a 1D aperiodic crystal (see Sec. 1.1.5).

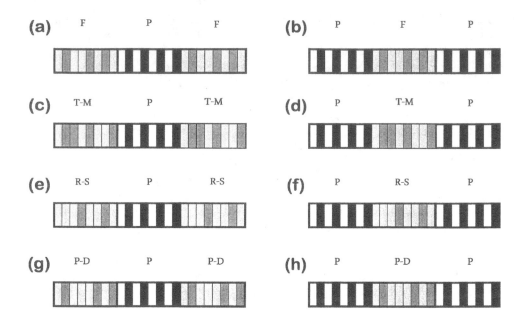

Figure 4.4 Diagram showing different modular designs of hybrid systems obtained by combining periodic (P) and Fibonacci (F), Thue-Morse (T-M), Rudin-Shapiro (R-S), or period-doubling (P-D) aperiodic multilayers blocks. Most of the sketched structures have been studied by different groups: (a) [896, 237, 720], (b) [685, 237, 889, 720], (c) [896], (e) [301], (f) [301]. Modular designs containing different types of aperiodic blocks (e. g., F-(T-M)-F or (T-M)-F-(T-M)) have been considered as well [896]. (Updated with permission from Maciá Barber E, *Aperiodic Structures in Condensed Matter: Fundamentals and Applications*, Taylor & Francis CRC Press, FL, 2009).

composed of a series of concatenated Fibonacci multilayers containing an increasing number of layers as their basic building blocks were studied, and it was concluded that their transmission spectra closely resemble those corresponding to periodic Bragg mirrors [354]. Following this modular approach one may take a step further and think of blending periodic and aperiodic multilayers in order to fabricate devices endowed with new capabilities. In these *hybrid order devices*, we have two different kinds of subunits, exhibiting a different stacking sequence of layers, thereby providing an additional design parameter related to the kind of topological order in each subunit [544, 535]. Certainly, there exist many ways of constructing different hybrid order structures by concatenating different kinds of aperiodic building blocks among them, hence opening a broad avenue to design novel multilayered systems. A graphical account of some possible designs is shown in Fig. 4.4. One may reasonably expect that the competition between highly fragmented spectra supporting critical eigenstates (a characteristic feature of QP systems) and absolutely continuous spectra possessing Bloch eigenfunctions (typical of periodic systems) should give rise to some peculiar eigenstates in these hybrid structures. In fact, the existence of selectively localized optical modes in Fibonacci/periodic/Fibonacci porous silicon based multilayers has been reported [236].

We may also consider modular designs based on QP building blocks whose layer arrangements only differ by the role played by the A and B layers, which are respectively interchanged. In that case, the long-range QPO is preserved under the conjugation

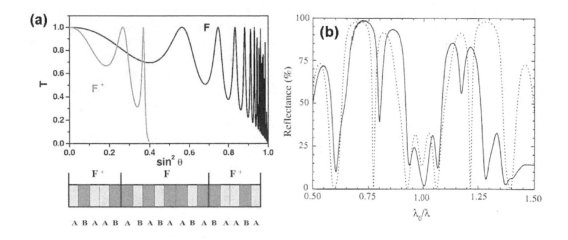

Figure 4.5 (a) Dependence of the transmission coefficient with the incidence angle for the conjugated order device sketched in the bottom panel, with $n_A = 1.46$ (SiO$_2$), $n_B = 2.35$ (TiO$_2$) (type F subunit), and $n_A = 2.35$ (TiO$_2$), $n_B = 1.46$ (SiO$_2$) (type F^+ subunit). Bottom panel: Scheme of a conjugated order optical resonating microcavity based on three QP subunits, where a FDM (with $N = 8$), showing high transmission, is encased between two FDMs (with $N = 5$), acting as optical mirrors. Dark (light) layers correspond to low (high) refractive index materials, respectively. (Reprinted figure with permission from Maciá E, Physical Review B 63, 205421, 2001. Copyright (2001) by the American Physical Society). (b) Experimental (solid line), and theoretical (dashed line) reflectance spectra at an incidence angle of $5°$ of a symmetric Fibonacci structure S_6/S_6^+ made of porous silicon ($\lambda_0 = 600$ nm). (Courtesy of Gerardo G. Naumis).

operation $n_A \leftrightarrow n_B$. In Fig. 4.5a bottom panel, we sketch a multilayer containing two kinds of FDM modules. In the first one (labelled F), the A (B) layers are composed of low (high) refractive index materials. In the second one (labelled F^+), the values of the refractive indices assigned to the layers A and B are reversed, so that the total internal reflection angle condition is achieved when $\theta_0 \simeq 40°$. Consequently, the F^+ module (on the left) and its mirror reversed F^+ module (on the right) behave as perfect mirrors for incidence angles larger than θ_0. Accordingly, these conjugated modules could be used to construct efficient optical microcavities by properly selecting the refraction indices of the layers materials. Thus, we may achieve broad multidirectional reflection devices based on QP structures, as it is shown in the top panel of Fig. 4.5a. For the sake of illustration, the experimental reflectance curve of a Fibonacci/Fibonacci$^+$ multilayer based on porous silicon is displayed in Fig. 4.5b) [647]. A good agreement between the experimental and theoretical reflectance spectrum is observed, except by a shift around $\lambda_0/\lambda = 1$, which increases as the wavelength decreases. This result is a consequence of deviations from the quarter wavelength condition due to the refractive index dispersion of porous silicon layers.

4.2.2 Exploiting local symmetries

Along with their characteristic long-range order aperiodic structures are generally endowed with different sorts of local symmetry features, which can be described in terms of mirror operations or palindromic reversals. For instance, depending on the parity of the generation order, the Thue-Morse sequence naturally exhibits either conjugation or mirror symmetries involving different spatial scales. Although other substitutional sequences considered so far, such as Fibonacci, period-doubling or Rudin-Shapiro ones, lack these symmetries on a global scale, one can intentionally design related systems exhibiting overall mirror symmetry by (1) reversing the order of the letters in the original sequence (i.e., $ABAAB \rightarrow BAABA$), and (2) concatenating the original sequence to the reversed one to obtain the string of letters $ABAAB|BAABA$ [352]. Alternatively, aperiodic sequences with mirror symmetry can be obtained making use of their respective substitution rules in terms of the iterative formula $S_{j+1} = S_{j-1}S_jS_jS_{j-1}$, which guarantees the resulting sequence has mirror symmetry by construction. In so doing, the structure inherits a nested series of mirror symmetry planes (i.e., $A\{B[A\{A[B/B]A\}A]B\}A$). This structural correlation enhances resonant transmission effects leading to the presence of additional PBGs in a given frequency interval (as compared to those present in an aperiodic multilayer of similar length without mirror symmetry) [137]. This interesting feature was used to manipulate the resonant transmission properties at specific wavelengths by designing and growing a symmetric Fibonacci/Fibonacci^{-1} multilayer (the symbol $^{-1}$ here indicates the letters reversal operation) based on TiO$_2$ and SiO$_2$ dielectric layers. The measured transmission spectra supported the potential applications of this kind of aperiodic structures in multiwavelength narrow band filters and wavelength division multiplexing devices [681, 682]. On the other hand, the propagation of light waves in period-doubling multilayers with mirror symmetry has been numerically studied by considering sequences based on the iteration formula $S_j = S_j S_j^{-1}$. The obtained results suggest that the main effect of introducing the mirror symmetry is to enhance the number of self-similar features in the transmission spectra [137].

In summary, making use of certain formal operations (letter conjugation, letter reversal), followed by concatenation of the obtained sequences (regarded as structural building blocks), one can usefully exploit additional symmetries in the resulting aperiodic structures. Indeed, several works have clearly illustrated that the very concept of local symmetry provides a powerful tool in describing transport phenomena in complex aperiodic structures [396, 397, 398, 399].

4.2.3 Stacking metamaterials

Metamaterials are artificial compounds exhibiting a negative electrical permittivity ϵ together with a negative magnetic permeability μ in the same frequency range, which leads to a negative refractive index: $n = \sqrt{-|\epsilon|}\sqrt{-|\mu|} = \sqrt{i^2|\epsilon|}\sqrt{i^2|\mu|} = i^2\sqrt{|\epsilon|}\sqrt{|\mu|} = -\sqrt{\epsilon\mu}$.[4] The main effect of having $n < 0$ is that the electric field **E**, the magnetic field **H**, and the wave vector **k** of the EM waves propagating throughout such a material form a left-handed triplet instead of the usual right-hand one.[5] As a consequence, metamaterials can support EM waves with the phase velocity opposite to the direction of the energy flow (Poynting vector). Moreover, their phase and group velocity are antiparallel, since the group velocity usually has the same direction as the energy flow [821].

The study of photonic systems consisting of alternating layers of left-handed (say A, with $n_A < 0$) and right-handed materials (say B, with $n_B > 0$) has recently received a great deal of attention since they can exhibit new types of PBGs which are not based on

---

[4]There are mainly two kinds of metamaterials: double-negative metamaterials whose permitivity and permeability are simulteanously negative, and single-negative metamaterials, which include the μ-negative materials (with $\varepsilon > 0$) and the ε-negative materials (with $\mu > 0$).

[5]Due to this feature these materials are also referred to as left-handed materials.

interference effects. The first kind of the so-called non-Bragg gaps is given by the condition that the multilayer *volume average* refractive index equals zero, so that the structure cannot support propagating waves [489]. For a binary aperiodic multilayer containing N_A and N_B layers, respectively, the average refractive index reads

$$\bar{n} = \frac{n_A d_A N_A + n_B d_B N_B}{d_A N_A + d_B N_B}. \tag{4.6}$$

Therefore, the effective refractive index will generally depend on the filling factor (i.e., the relative proportions of the different types of constituent layers) but not on the specific order of the layers themselves [140].[6] For aperiodic multilayers based on substitution sequences the N_A and N_B values will in general depend on the generation order of the considered substitution rule, so that for most aperiodic multilayers we have an additional design parameter given by the N_A/N_B ratio (which is lacking in periodic ones), albeit some notable exceptions exist. This ratio converges towards certain limiting value as the generation order is progressively increased. In fact, several studies indicate that the zero-$\bar{n}$ gaps are more significantly affected by layer widths and material choice than by the considered generation order of the aperiodic multilayer [179, 343]. For example, the average refraction indices for n-th order Fibonacci and Thue-Morse multilayers are respectively given by [614, 925],

$$\bar{n}_F = \frac{n_A d_A F_{n-1} + n_B d_B F_{n-2}}{d_A F_{n-1} + d_B F_{n-2}}, \qquad \bar{n}_{TM} = \frac{n_A d_A + n_B d_B}{d_A + d_B} = \bar{n}. \tag{4.7}$$

Thus, making use of the well-known relationship $\lim_{k \to \infty}(F_k/F_{k-1}) = \tau$ in (4.7) one gets the zero-$\bar{n}$ gap condition $n_A = -n_B \eta \tau^{-1}$ for Fibonacci multilayers, where $\eta \equiv d_B/d_A$ (**Exercise 4.3**). Conversely, $\bar{n}_{TM}$ does not depend on the system size, since the number of A layers and B layers is just the same in a Thue-Morse sequence of any order (i.e., $N_A = N_B = 2^n$). As a consequence, the location of the zero-$\bar{n}$ gap in Thue-Morse multilayers coincides with that present in periodic multilayers made of the same materials.[7] In fact, from (4.6), we see that the zero-$\bar{n}$ gap condition for a periodic multilayer with $N_A = N$ and $N_B = \ell N$, $\ell \in \mathbb{N}$, is determined by the relationship $n_A = -\ell n_B \eta$, so that, it is independent of the system size. Similar results have been obtained for photonic multilayers arranged according to generalized Thue-Morse sequences given by the substitution rule $g(A) = A^n B^m$ and $g(B) = B^n A^m$, for which $N_A/N_B = n/m$ [809]. Furthermore, albeit both n_A and n_B will generally depend on the EM wave frequency, the zero-$\bar{n}$ gap is generally rather robust and insensitive to both the light incidence angle and polarization effects, as compared to the ordinary Bragg gaps [388, 995].

On the contrary, the second kind of non-Bragg gaps appear in dispersive substances at frequencies where either $\mu(\omega)$ or $\epsilon(\omega)$ vanishes in the metamaterial. Unlike zero-$\bar{n}$ gap, these gaps occur at frequencies where only a single constituent material of the multilayered structure shows zero refractive index. In addition, zero-μ and zero-ϵ gaps are polarization dependent and they can interact with zero-$\bar{n}$ gaps giving rise to new behaviors [180, 613]. For instance, omnidirectional PBGs have been reported in Fibonacci multilayers containing both $\epsilon < 0$ (say, A layers) and $\mu < 0$ (say, B layers) so that this structure is entirely composed of metamaterials [179]. In this case, the PBG width in the QP limit is given by $\Delta\omega = \omega_+ - \omega_-$, where the ω_+ (ω_-) band-edges are determined from the condition $\langle \epsilon \rangle = 0$

---

[6]This property is analogous to that observed in the algebraic invariants related to substitution matrices (see Sec. 2.2).

[7]We recall that the substitution matrices of periodic and Thue-Morse substitution rules exactly coincide (see Sec. 2.2).

and $\langle \mu \rangle = 0$, respectively, with

$$\langle \epsilon \rangle = \frac{\tau \epsilon_A(\omega) + \eta \epsilon_B}{\tau + \eta}, \quad \langle \mu \rangle = \frac{\tau \mu_A + \eta \mu_B(\omega)}{\tau + \eta}, \tag{4.8}$$

where $\epsilon_A(\omega) = \epsilon_a - (\omega_e/\omega)^2$, $\mu_B(\omega) = \mu_b - (\omega_p/\omega)^2$ and $\omega_{e(p)}$ are the electronic (magnetic) plasma frequencies, respectively [179]. Making use of (4.8) one checks that the omnidirectional PBG is very sensitive to the design parameter η adopted value (**Exercise 4.4**).

The optical properties of Fibonacci structures composed of alternating *chiral* and achiral isotropic layers have also been studied. Chiral objects are 3D bodies that cannot be brought into congruence with their mirror image by translation and rotation operations. Isotropic chiral metamaterials are macroscopically continuous media composed of equivalent chiral objects that are uniformly distributed and randomly oriented [41]. Self-similarity, scalability and sequential splitting of the spectra of QP chiral photonic structures has been investigated [913], along with the influence of chirality on the critical modes localization [914].

4.3   PHOTONIC QUASICRYSTAL TILINGS

In the previous section, we have studied the role of aperiodic order on the physical properties of multilayered systems along the stacking direction, so that they can be treated as effectively one-dimensional. Now, conclusions drawn for 1D structures can not straightforwardly be extended to higher dimensions. For instance, PBGs appear even for arbitrarily small refractive index contrasts in 1D multilayers, whereas a relatively large contrast is required in 2D and 3D photonic crystals. In this section we will study aperiodically ordered 2D structures, generally designed by following two basic steps:

- Firstly, one defines a suitable lattice of aperiodically arranged points in 2D, such as those related to the Vogel's spirals (Fig.1.14), the Fibonacci square (Fig.1.15, see also Sec. 2.2), or those based on a grid of points located at the vertices or centers of rhombuses in the Penrose tile, to name just a few.

- The previously obtained aperiodic lattice is decorated with appropriate optical nanoelements, such as metallic particles or dielectric rods. For instance, planar arrays of Au nanoparticles arranged according to Vogel's spirals with divergence angles $\phi_d = 2\pi/\tau^2$, $\phi_d = 137.3°$, and $\phi_d = 137.6°$ were fabricated by electron beam lithography on quartz substrates [902]. In this way, polarization-insensitive planar light diffraction in the visible spectral range was demonstrated as stemming from their almost circularly symmetric Fourier space [155]. Alternatively, one can decorate the lattice points by etching a pattern of air holes in a homogenous dielectric slab. This relatively simple procedure enables one to fabricate planar optical devices with varying light transport properties, including different realizations of PQCs.

Two main routes have been exploited to date in order to design 2D aperiodic patterns. For one thing, we can use suitable generalizations of 1D substitution sequences introduced in section 2.2 to the 2D case. On the other hand, we can consider structures based on aperiodic tilings of a plane from the start. Let us begin by introducing the first route. A simple method to achieve our goal is based on alternating the iterations of 1D substitution sequences along two *orthogonal* directions. For instance, a 2D Fibonacci lattice can be obtained from a seed letter A (or B) by applying two complementary Fibonacci sequence substitution rules $g_A : A \to AB$, $B \to A$ and $g_B : A \to B$, $B \to BA$ (depending on whether the *first* element encountered in the 2D letter expansion is A or B) along the horizontal (from left to right) and the vertical (up-down) directions, alternately [151, 158]. For the sake of illustration, the first generations of this Fibonacci structure are shown below in matrix form

$$A \rightarrow AB$$
$$\begin{pmatrix} A & B \\ B & A \end{pmatrix} \rightarrow \begin{pmatrix} A & B & A \\ B & A & B \end{pmatrix}$$
$$\begin{pmatrix} A & B & A \\ B & A & B \\ A & B & A \end{pmatrix} \rightarrow \begin{pmatrix} A & B & A & A & B \\ B & A & B & B & A \\ A & B & A & A & B \end{pmatrix} \cdots . \tag{4.9}$$

A generalization of the Thue-Morse substitution rule is somewhat simpler, since the same standard substitution rule $g : A \rightarrow AB$, $B \rightarrow BA$ can be applied irrespective of the character of the first letter encountered. In this way, starting from the seed letter A one obtains [151],

$$A \rightarrow AB$$
$$\begin{pmatrix} A & B \\ B & A \end{pmatrix} \rightarrow \begin{pmatrix} A & B & B & A \\ B & A & A & B \end{pmatrix} \downarrow$$
$$\begin{pmatrix} A & B & B & A \\ B & A & A & B \\ B & A & A & B \\ A & B & B & A \end{pmatrix} \cdots . \tag{4.10}$$

The two dimensional extension of the Rudin-Shapiro lattice is obtained similarly, with the proviso that the inflation rule for this sequence now acts upon a two-letter combination instead of a one-letter one. In fact, in a two-letter alphabet based on the assignment $A \rightarrow AA$, $B \rightarrow AB$, $C \rightarrow BA$, and $D \rightarrow BB$, the Rudin-Shapiro sequence can be obtained by the iteration of the following substitution rule [452],

$$AA \rightarrow AAAB, \quad AB \rightarrow AABA,$$
$$BA \rightarrow BBAB, \quad BB \rightarrow BBBA.$$

As a result the inflation seed is given by the 2×2 matrix [158],

$$\begin{pmatrix} B & A \\ A & B \end{pmatrix} \rightarrow \begin{pmatrix} B & B & A & B \\ A & A & B & A \end{pmatrix} \downarrow$$
$$\begin{pmatrix} B & B & A & B \\ B & B & A & B \\ A & A & B & A \\ B & B & A & B \end{pmatrix} \cdots . \tag{4.11}$$

An alternative way of obtaining the Thue-Morse lattice is based on the recursive rule given in the matrix form [1010],

$$\mathbf{M}_{k+1} = \begin{pmatrix} \mathbf{J}_k - \mathbf{M}_k & \mathbf{M}_k \\ \mathbf{M}_k & \mathbf{J}_k - \mathbf{M}_k \end{pmatrix}, \tag{4.12}$$

where $\mathbf{J}_k$ is a $2^k \times 2^k$ matrix in which each element is equal to one, and the initial matrices to start iteration are

$$\mathbf{M}_1 = \begin{pmatrix} 1 & 0 \\ 0 & 1 \end{pmatrix}, \quad \mathbf{J}_1 = \begin{pmatrix} 1 & 1 \\ 1 & 1 \end{pmatrix}. \tag{4.13}$$

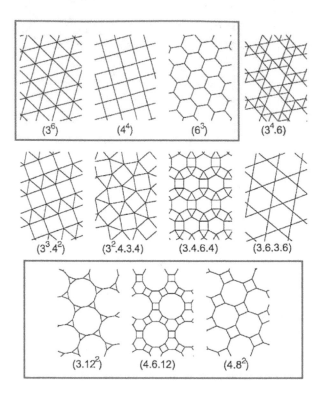

Figure 4.6 Archimedean tilings. A set of integers $(n_1.n_2.n_3.\cdots)$ denotes the vertex type in the way that n_1-gon, n_2-gon, n_3-gon,$\cdots$ different polygons consecutively meet on each vertex. For instance, the symbol $(3^2.4.3.4)$ represents a tiling in which two equilateral triangles, a square, an equilateral triangle, and a square successively gather edge-to-edge around any given vertex. The three first tilings shown at the top row, corresponding to the notation (3^6), (4^4), and (6^3), are composed of just one type of polygon tile and are usually referred to as Platonic tiles. The last three tilings at the bottom are of interest for structure analysis of QC approximants exhibiting octagonal and dodecagonal symmetries.

It can be readily checked that the pattern of letters given by Eq. (4.10) is completely equivalent to the number pattern appearing in the matrix $\mathbf{M}_2$ given by Eq. (4.12) by simply identifying $A \to$ "0" and $B \to$ "1". It is also easily verified that any row and any column of the symmetric $\mathbf{M}_k$ matrix containing 2^k elements is a 1D Thue-Morse sequence of order k.

The experimental realization of 2D aperiodic structures is a hard fabrication challenge. Notwithstanding this, several large area (up to 800 μm^2) 2D Thue-Morse lattices (up to $k = 10$ order) made of air rods ($n = 1$, $r = 100$ nm) embedded into polymethyl-methacrylate (PMMA, $n \sim 1.5$) have been fabricated with nanometric resolution (10 nm) by means of electron beam lithography techniques [588]. The obtained samples were characterized by successfully comparing experimental far-field diffraction patterns with theoretical Fourier spectra corresponding to Thue-Morse patterns with lattice constants ranging from $a = 140 - 1080$ nm.

Let us now consider aperiodic patterns generated from tilings of the plane. The word tiling is generally used to describe a pattern or structure made of one or more polygonal shapes (tiles) that pave a plane exactly, leaving no spaces between them. Squares, equilateral triangles, and hexagons are particularly easy to tile with in order to achieve a periodic pattern that repeats itself at regular intervals. The obtained patterns are endowed with the characteristic symmetries – 3-fold, 4-fold, 6-fold – of the tiles they are respectively made up from. Regular pentagon tiles, however, cannot be used to fill the entire plane to form a periodic tiling pattern: unfilled gaps will always remain. Certainly, one may relax the tiling rules by allowing for *different* types of polygons to be simultaneously used to construct the tile. If one imposes all the vertices to be of the same type at every joining point, only eleven types of tiling of the plane by regular polygons can be found, the so-called *Archimedean tiling* patterns shown in Fig. 4.6 (we note that the so-called regular or *Platonic tilings* are a subset of Archimedean ones) [263].

A significant step forward towards more general tiling patterns consists in relaxing the periodicity condition as well. Thus, 2D *aperiodic tilings* are collections of polygons capable of covering a plane with neither gaps nor overlaps in such a way that the resulting overall pattern lacks any translational symmetry. The *simpler* aperiodic tiling was discovered by Roger Penrose in 1974 [683], and consists of just two different tiles: a skinny (acute angle $\pi/5$) and a fat (acute angle $2\pi/5$) rhombi with equal edge length a (see Figs. 1.11 and 4.7b). Their frequency of appearance in the infinite tiling is $1 : \tau$ and their areas ratio is τ. A close inspection to the Penrose tiling reveals the existence of eight different vertex configurations in the pattern and five relevant distances, $\sqrt{1 + \tau^2}a > \tau a > a > \sqrt{3 - \tau}a > \sqrt{2 - \tau}a$, between different close vertices. In addition to the regular Penrose lattice other related tilings, such as the decagonal and dodecagonal Penrose lattices have been also considered in the design of 2D PQCs, along with the *Ammann-Beenker* octagonal lattice (Fig. 4.7a) and the *Stampfli* dodecagonal lattice (Fig. 4.7c). The Ammann-Beenker lattice is formed by squares and $45°$ rhombi of equal edges a, leading to the presence of three relevant distances, $\sqrt{2}a > a > \sqrt{2 - \sqrt{2}}a$, between different close vertices throughout the pattern. In turn, each vertex can be classified as belonging to six different possible environments according to their next-neighbors number (coordination index). The ratio of the frequencies of the square to the rhombus tiles in the infinite tiling is $1 : \sqrt{2}$, since the ratio of the areas of a square to a rhombus tile is $\sqrt{2}$, the total area of the tiling covered by squares equals that covered by rhombi. The Stampfli tile is formed by equilateral triangles and squares of length a. The ratio of squares to triangles is $\sqrt{3}/4$. There exist three different vertex configurations and two relevant distances, $\sqrt{2}a$ and a between close vertices [834].

4.3.1 Photonic band structures

The optical response of 1D aperiodic multilayers is generally analyzed in terms of two main physical magnitudes: the reflectance/transmittance curves and the electrical field intensity profiles along the multilayer growth direction. Whereas the physical interpretation of the electrical field intensity spatial pattern can be readily extended from one to two dimensions, the transmission coefficient $T_N(\omega, \theta)$ is not so useful in 2D, and it must be complemented by other physical magnitudes, such as the optical band structure diagram, the optical DOS, and the power radiation spectrum of a line source embedded in the structure. To this end, one integrates the output energy flux through a closed contour L, according to the expression

$$\oint_L \mathbf{S}(\mathbf{r}) \cdot \mathbf{n}dr, \tag{4.14}$$

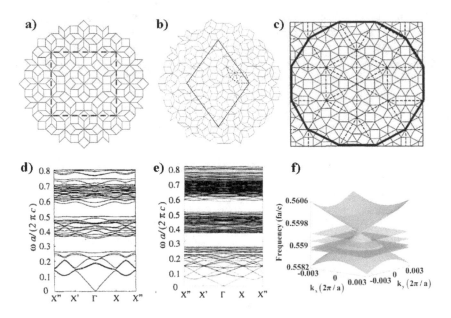

Figure 4.7 Portions of the (a) Ammann-Beenker, (b) regular Penrose, and (c) Stampfli tilings along with their optical band diagrams obtained for the approximant unit cells highlighted by thick line polygons in the tiles (a), (b) and (c), which are shown in panels (d) and (e), respectively. Two main PBGs with midgap normalized frequencies $(\omega a/2\pi c)$ at 0.32 and 0.55, and relative widhts of about 30% and 15%, respectively, characterize the band diagram of both PQCs. (Adapted from [946, 948, 947]). In (f) a conical band dispersion near $\mathbf{k} = \mathbf{0}$ for a 12-fold Stampfli lattice based PQC is shown. A Dirac-like point can be readily seen at the point where the two cones touch, along with a flat band intersecting the Dirac frequency $\omega = 0.5595c/a$. The lower cone is intersected by additional flat sheets corresponding to localized ring-like optical modes. (Reprinted figure with permission from Dong J W, Chang M L, Huang X Q, Hang Z H, Zhong Z C, Chen W J, Huang Z Y, Chan C T, Physics Review Letters. 114, 163901 (5pp), 2015. Copyright (2015) by the American Physical Society).

where $\mathbf{S}$ is the Poynting vector, $\mathbf{r}$ is a two-dimensional vector, and $\mathbf{n}$ is a unit vector normal to the contour, and divide it by the total flux without the sample. A band gap then corresponds to a dip in the radiation power spectrum, which deepens as the sample size is systematically increased. In this way, it is observed that the spectral positions of low-frequency PBGs in nano-element based photonic crystals largely depend on the resonant properties of the individual optical elements (Mie resonances), so that they approximately coincide with the PBGs positions in their periodic counterparts with similar geometries and material compositions [87, 623, 625]. However, aperiodic structures also feature a number of additional optical modes originating from diffraction effects in the quasi-lattice high density planes [946], as well as multiple scattering resonances in self-similar local environments [177]. Both the number and the precise frequency values of these modes depend on the aperiodic lattice filling factor (i.e., the ratio of surface area covered by the optical elements to the total area of the photonic structure), which is in turn determined by the adopted geometrical parameters in the considered structure. In addition, the existence of higher order rotational axes, including 10-fold, 12-fold, and even larger symmetries, approaching the "Bragg-ring" limit in the case of spiral lattices, leads to more isotropic complete photonic band gaps in PQCs, generally requiring lower refractive index contrast values than those observed for periodic photonic crystals [87, 552].

Indeed, to open a complete gap in 2D photonic crystals the refractive index needs to be larger that 2, which rules out the possibility of using polymer materials to fabricate periodic photonic crystals because their refractive indices are typically lower than 1.7. Compared with periodic photonic crystals, PQCs have a smaller refractive index threshold value to open a complete gap. For example, the threshold value is $n = 1.26$ for an octagonal quasilattice, which indicates that optoelectronic components based on PQCs may be obtained in low refractive index materials such as silica, glasses, and polymers. Experimental evidence for complete bandgaps for both TE and TM polarizations has been reported for different setups, including octagonal arrangements of dielectric Ammann-Beenker cylinders in air or dodecagonal Stampfli tiles. In contrast to periodic lattices, the TM bandgap lies exactly in the middle of the TE bandgap, leading to an efficient overlapping for both polarization states in the resulting complete PBG. According to numerical simulations this bandgap remains open for very low refraction index materials such as glass ($n = 1.45$) with similar filling fractions [1013].

Some time ago it was suggested that complete PBGs would be more readily achievable by considering deterministic aperiodic arrangements beyond quasicrystalline structures, hence showing neither translation *nor discrete rotational* symmetries [152]. This feature originates from the near circular symmetry of such structures in reciprocal space. For instance, if one compares the diffraction patterns of deterministic array structures based on the Archimedean lattices, a pinwheel tiling and Vogel spiral lattices one can observe an increasing degree of circular symmetry in the reciprocal space ranging from discrete spectra with pure rotational axes, characteristic of both periodic and QP crystals, to more diffuse spectra with continuous components (see Fig.2.2) [155, 477].

During the last decade a number of systematic numerical studies regarding the optical band diagram of periodic approximant structures in relation to their parent QP lattices have been reported [946, 947, 948, 949]. It has been found that: (1) even the lowest order approximants can support quite *isotropic* PBGs, hence allowing for a more uniform light reflection (Fig. 4.7d–e). This property stems from the fact that the first Brillouin zone has more symmetries in QP lattices as compared to conventional photonic crystals based on periodic square or hexagonal lattices, hence favoring the possible appearance of a complete gap in these systems. Though quite isotropic PBGs are also obtained for periodic structures displaying high order local rotational symmetries, such as Archimedean tilings, the resulting PBGs are significantly narrower (gap-midgap ratios ~0.05) and they appear at relatively higher frequency values ($a/\lambda \geq 0.6$) in this case [730, 917]. (2) Most bands are very flat indicating that a significant reduction of the group velocity (within the range $0.005c–0.02c$) occurs in all directions [625]. Indeed, recent experiments indicate that transverse light transport throughout 2D Fibonacci structures is significantly hampered in comparison with that observed in a periodic lattice [80]. (3) PGBs in PQCs are remarkably almost independent of the angle of incidence of the incoming light [43]. (4) It was both theoretically and experimentally shown that some 2D PQCs based on the Stampfli dodecagonal design can exhibit conical band structure at $\mathbf{k} = \mathbf{0}$ (Fig. 4.7f), and their finite related approximants can behave like a zero-refractive-index medium [193].

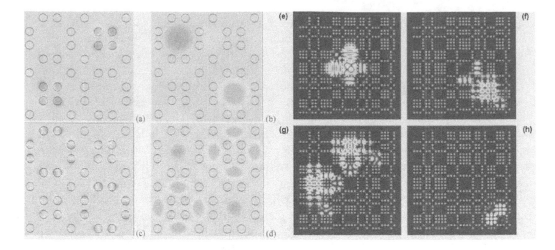

Figure 4.8 (Left panels) Comparison between the electric field patterns at the Γ point of a M_3 generation order Thue-Morse lattice (see Eq. (4.10)) containing $N = 64$ dielectric rods (located at the 0's of the $\mathbf{M}_3$ matrix) for two states belonging to the band edges of the frequency spectrum (a,c), and two states located near the center of the band gaps (b,d). The modes shown in (a) and (c) are mainly concentrated on the high-dielectric regions, and can be classified as monopole and dipole modes. On the other hand, the states shown in (b) and (d), which can be regarded as defect modes, are concentrated in the low-dielectric region between the rods [623]. (Courtesy of Luigi Moretti). (Right panels) Electric field patterns of selected critical modes in and around the first TM band gap of a Rudin-Shapiro optical structure with $N = 512$ dielectric cylinders [87]. (Courtesy of Luca Dal Negro).

4.3.2   Electric field patterns

Most states in 2D aperiodic systems are characterized by light wave patterns localized in high symmetry square, hexagonal and octagonal symmetry local environments of the underlying tile. By comparing the electric field patterns shown in Fig. 4.8, we can see that the optical modes supported by the Rudin-Shapiro lattice are generally more localized than those in the Thue-Morse lattices, which agrees with the absolutely continuous nature of the Rudin-Shapiro's Fourier spectrum. Consistently, an increase in the structure size does not have a significant effect on the localization properties of the Rudin-Shapiro modes [87]. Unlike classical Anderson localization phenomena, however, in these aperiodic systems the localization is not triggered by destructive interference phenomena, but rather determined by resonant effects between local nearest-neighbor scatterers via Mie resonance modes [947], thereby suggesting a local interaction mechanism. The nature of the band-edge states in 2D aperiodic lattices, however, cannot be properly addressed by considering relatively small systems containing about 100–200 optical elements only, and it is reasonably expected that the transport properties of very large systems may be more complicated as compared to approximants based on the finite-size unit cells.

For instance, ring-ring interactions mediated by significantly extended critical modes, exhibiting self-similar patterns, may introduce novel optical phenomena. In fact, experimental studies on the emission in a Penrose-type QC laser provided clear indication that the lasing action stems from well-defined extended modes that are coherently spread (with a coherent length $\geq 100a$) through the sample due to the long-range QPO of the tile [657, 658]. Quite remarkably numerical studies analyzing how the properties of band-edge states evolve with sample size in Stampfli lattices containing up to 10^4 dielectric cylinders show that these states are characterized by the presence of a hierarchy of local resonances which progressively appear as soon as the size of the considered sample allows for it. In this way, the self-similarity of the structure properly manifests itself in the QP limit [465].

On the other hand, point defects (missing cylinders or rods) or line defects (missing rows of cylinders or rods) can be respectively used to create highly localized defect modes or to form waveguides in PBG systems in both periodic and aperiodic systems alike. The existence of such localized modes can be fruitfully exploited in order to design single-cell resonators with high quality factors [415], or coupled-resonator optical waveguides [954]. Numerical evidence of guided resonances in Ammann-Beenker octagonal tilings obtained by placing air holes in a silicon slab at the tiling vertices positions has been reported [741], hence providing new degrees of freedom in the engineering of guided resonances. In this regard it should be noted that the choice of orientational symmetry, quasiperiodicity and the fundamental repeating units does not uniquely specify a QC, since the local isomorphism property guarantees the existence of many space-filling arrangements, involving the same repeating units, with the same overall symmetry, which can be grouped into local isomorphism classes. These different representatives, though defect free, possess different kinds of critical states, exhibiting different light localization features [496].

4.3.3 Optical lattices

In the so-called optical lattices several monochromatic travelling light waves are combined in vacuum space to create an standing interference pattern which exhibits long-range order. Then, such a light field is used to trap small particles or atoms inside it. In the resulting structure the atoms adopt a regular arrangement in space, but they are weakly coupled among them, at variance with the usual situation in conventional solid matter. In fact, arrays of atoms trapped in optical lattices do not present either defects or phonons, offering a powerful tool for investigating the quantum behavior of condensed matter systems under unique control possibilities. The symmetry of the optical potential created by the interference of lasers is completely determined by the geometric arrangement of their beams. Therefore, one can in principle design both periodic and QP optical potentials at will, allowing for a precise tuning of QPO parameters in the latter case [814]. In this way, optical lattices technology allow for the arrangement of atoms of the same kind to be quasiperiodically distributed throughout the space, thereby leading to the physical realization of a *mono-component* QC.[8] Indeed, the formation of a QP optical lattice of Cs atoms by the interference of five or six laser beams was reported some time ago [313]. Furthermore, by properly combining different optical lattices one can obtain ordered systems displaying interesting phenomena. For instance, a localization-delocalization wave packed transition was disclosed in *Pythagorean aperiodic* potentials created by two mutually rotated square patterns which allow for a continuous transformation between incommensurate and periodic structures upon variation of the rotation angle θ [349].[9]

Quite interestingly, a physical realization of the geometrical cut-and-project method (see Sec. 1.4.1) was illustrated by creating a Fibonacci optical lattice. To this end, four laser

---

[8]The existence of monocomponent metallic QCs has not been reported to date (July, 2020).

[9]Such lattices become periodic only for rotation angles $\cos\theta = a/c$, $\sin\theta = b/c$, where a, b, $c \in \mathbb{N}$ are Pythagorean triples satisfying the relationship $a^2 + b^2 = c^2$.

beams were used to build a two-dimensional square lattice by their interference. A fifth laser beam is then used to drive Rb atoms along the projection direction at an appropriate irrational angle. By properly selecting the on-site energies of these atoms one can then obtain a physical realization of a Fibonacci lattice of atoms [228]. This class of optical lattice based QC offer dramatic possibilities for designing a wide range of geometrical arrays able to confine Bose-Einstein condensates of both fundamental and practical interest [780]. Thus, a Bose-Einstein condensate of K atoms has been recently studied to probe a QC optical lattice in a matter-wave diffraction experiment which allows one to observe a self-similar diffraction pattern similar to those obtained by using electron diffraction in QC alloys [934]. In addition, these experiments have demonstrated the ability to simulate tight-binding models in one to four dimensions by observing the light-cone-like spreading of particles in reciprocal space [782]. Thus, by properly comparing the physical properties of optical lattice based QCs with those observed in both usual metallic QCs and soft matter QCs (see Sec. 6.1.3), we can gain additional understanding on the role of chemical bonding in both the stability and physical properties of different sorts of QC phases in condensed matter. In this regard, a scheme for realizing a novel quantum matter state where both superfluidity and long-range QPO coexist has recently been proposed [342].

4.4 PHOTONIC QUASICRYSTALS IN THREE DIMENSIONS

Although currently available nanotechnology offers several ways to construct aperiodic structures in 1D and 2D, the fabrication of 3D aperiodic structures is a very challenging task. Notwithstanding this, three methods have successfully been used to this end, namely, stereolithography, optical interference holography and direct laser writing [481, 965, 990]. In particular, technological advances in the latter technique have enabled aperiodic structures of increasing thickness to be grown in order to study their optical properties [736, 737].

4.4.1 Axial photonic quasicrystals

An obvious extension of the 2D aperiodic patterns considered in Sec. 4.3 to three dimensions consists in to simply pile up 2D QP planes along the third, orthogonal direction. In so doing, one can stack the QP layers either periodically or aperiodically. In the former case, one obtains a hybrid order structure which is periodic along the stacking direction and QP in the perpendicular planes, which we will refer to as *axial* PQCs. These structures then mimic at a micron scale the overall structure observed in dQCs at the atomic scale. Following this approach a PQC exhibiting the Penrose tile pattern in the XY plane was fabricated by using ten-beam laser (488 nm Ar) holographic lithography in a polymer resin with $n = 1.62$. Normal incidence reflection and transmission spectra (along the Z axis) indicated the presence of band gaps which approximately follow Bragg's diffraction relation $\lambda_{gap} \simeq 2d_z n_{eff}$, where d_z is the z axis periodicity and n_{eff} measures the effective refraction index as a function of the sample's filling factor [953]. In a similar way, making use of a reconfigurable, optically induced nonlinear photonic lattice approach, several axial PQCs exhibiting 8-, 19-, 27-, and 32-fold symmetry axes have been obtained as well [965].

In the previously considered structures the distance between successive 2D QP planes was kept constant, generally determined by the QP layer thickness itself. One may then think of relaxing this design constraint in order to assemble a 3D QP structure by stacking a series of QP planes in such a way that the spacing among them is given by a QP sequence too. In this way, one obtains a QP tiling in 3D where three identical but orthogonal sets of parallel QP planes are superimposed. This goal was recently accomplished by

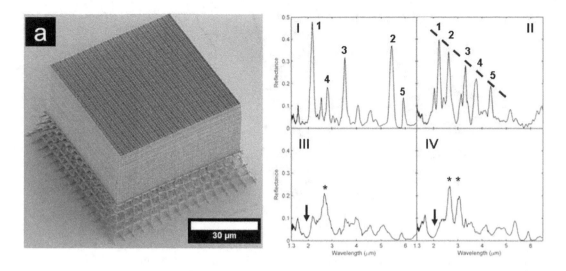

Figure 4.9 (a) Scanning electron micrograph of a 30 μm thick 3D Fibonacci structure fabricated on top of a base to release shrinkage induced strain. (Right panels) Reflectance spectra of Fibonacci (I), Thue-Morse (II), Rudin-Shapiro (III), and a random (IV) 3D structures with about 72 μm in height (containing about 60 (80) elements along the parallel (perpendicular) direction to the substrate). (Reprinted from Renner M and von Freymann G, Spatial correlations and optical properties in three-dimensional deterministic aperiodic structures, Scientific Reports 5, 13129; doi:10.1038/srep13129 (2015), Creative Commons Atribution International License CC BY 4.0).

piling up square Fibonacci grids (see Fig. 1.11). To this end, each Fibonacci grid is decorated with cross-shaped nano-elements that enable to build up successive Fibonacci grids on top to each other along the perpendicular stacking direction. We note that this 3D QP tile provides an illustrative instance of a microscopic *orthorhombic* QC without forbidden rotational symmetries (see Sec. 1.1.5). This approach can be readily extended to design 3D aperiodic structures based on other 2D aperiodic tiles, such as the Thue-Morse or Rudin-Shapiro ones, aiming at exploring the optical properties of different representative structures located at different boxes in the spectral chart displayed in Fig. 2.2 [736, 737]. In Fig. 4.9a, we show a micrograph of the grown 3D *Fibonacci prism* and in the right panels we compare the reflectance spectra obtained for Fibonacci, Thue-Morse, and Rudin-Shapiro 3D aperiodic prisms, as well as a random 3D structure. As expected from the results summarized in the spectral chart introduced in Fig. 2.2, the reflectance spectra of both Fibonacci and Thue-Morse prisms display a series of well defined, narrow reflection peaks (stemming from PBGs in the frequency spectra), whereas the Rudin-Shapiro reflectance spectrum only exhibits a few, quite broad peaks of low intensity. Two main conclusions can be drawn from a careful inspection of these reflection spectra. In the first place, we see that the more intense peaks (labeled by decreasing order intensity in Fig. 4.9I–II) are inversely correlated with the radiation wavelength (dashed line) in the Thue-Morse 3D structure (whose Fourier transform is singular continuous), while such a dependence is not observed in the Fibonacci 3D structure (with a discrete point Fourier spectrum), where the reflection peaks are hierarchically arranged in a characteristic self-similar pattern previously known from studies of 1D multilayer systems [276]. In the second place, we observe that the reflectance spec-

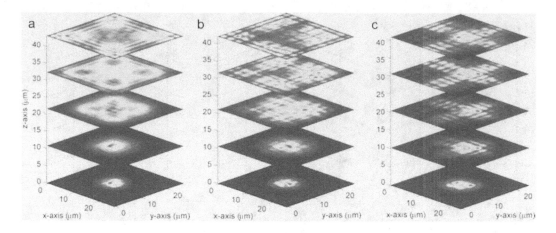

Figure 4.10 Calculated transmission patterns for (a) a periodic 3D structure with equal elements, (b) a weakly modulated Rudin-Shapiro 3D structure, and (c) a strongly modulated one. Patterns are obtained from 20 individual simulation runs for a propagating wave with $\lambda = 1.7$ μm. (Reprinted from Renner M and von Freymann G, Spatial correlations and optical properties in three-dimensional deterministic aperiodic structures, Scientific Reports 5, 13129; doi:10.1038/srep13129 (2015), Creative Commons Atribution International License CC BY 4.0).

trum of Rudin-Shapiro and random 3D structures are very similar and relatively flat (as expected, since both share an absolutely continuous Fourier spectrum). Nonetheless, we can still appreciate some distinctive features between them: (1) the broad dip close to $\lambda = 2$ μm (arrowhead in Fig. 4.9III–IV) exhibits inside a weak spectral peak in Rudin-Shapiro's structure that the random one completely lacks, (2) the main reflection peak in Rudin-Shapiro structure splits in the random one (asterisks in Fig. 4.9III–IV). Therefore, the different 3D aperiodic structures can be properly distinguished in terms of their respective optical properties, in complete agreement with theoretical expectations based on previous studies of 1D aperiodic systems [544]. Indeed, since the 3D structure Fourier spectrum can be naturally decomposed in terms of the corresponding 1D spectra, the construction scheme used to design these 3D aperiodic structures preserves all spectral features related to their underlying 1D aperiodic sequences.

In order to study the transport properties of the above described 3D aperiodic structures the spatial broadening of locally excited optical mode patterns was measured at different z coordinates throughout the samples. To this end, a set of structures with increasing thickness (from $z = 13$ up to $z = 170$ μm) was fabricated and the light distribution pattern $I(x, y)$ was measured on the top surface in a systematic way in order to estimate the effective width, $W_{\text{eff}} = P_z^{-1/2}$, of a given optical mode in terms of the inverse participation ratio

$$P_z = \frac{\int I^2(x, y)dxdy}{\left[\int I(x, y)dxdy\right]^2},$$

evaluated at different z values. Assuming a power-law broadening of the effective width of the form $W_{\text{eff}} = z^v$, the values $v = 0.28$, $v = 0.12$, and $v = 0.01$ were extracted for Fibonacci, Thue-Morse, and Rudin-Shapiro structures, respectively [737]. The obtained exponents indicate a subdiffusive light transport in both Fibonacci and Thue-Morse prisms, whereas an almost complete localization along the transversal direction is observed in the Rudin-Shapiro aperiodic prism (though no largely enhanced photon dwell times inside the structure were observed in time-resolved transmission experiments) [737]. These experimental results were supported by numerical calculations of the mode light distribution upon modification of the structural parameters, as it is illustrated in Fig. 4.10.

4.4.2 Icosahedral photonic quasicrystals

Complete PBGs[10] appear due to the overlapping of band gaps in all the possible directions. For a 3D periodic distribution of the dielectric function, different directions generally correspond to different periodicities and, consequently, to different frequency values for the stop band centers. The required overlapping of the different stop bands can only be ensured if these bands are wide enough. Motivated by this basic principle, several theoretical studies concluded that lattices with diamond symmetry have the minimum permittivity modulation contrast necessary for the appearance of a complete PBG among periodic crystals. Now, it is well known that QCs have higher order point group rotational symmetries, so that the Brillouin zone is close to being spherical and the existence of a complete band gap in icosahedral PQCs was theoretically inferred from band structure calculations of 3D cubic approximants of a 6D bcc cubic hyperlattice. The approximants of successive 1/0 and 1/1 orders exhibit quite isotropic and relatively large PBGs ($\Delta\omega/\omega = 17.6\%$ and $\Delta\omega/\omega = 10.3\%$ respectively) for a wide range of relative permittivity values with a threshold at $\epsilon \simeq 6$ [217].

Spurred by these results a *centimeter* scale PQC based on a generalized version of 3D Penrose tiling was fabricated by ultraviolet laser photopolymerization stereolithography and microwave transmission measurements within the range $8-42$ GHz were performed through the structures. The sample exhibited large stop bands in certain directions closely related to the main symmetry axes of the icosahedral group [577], albeit the transmission spectrum is less structured than originally expected. Indeed, subsequent numerical calculations provided evidences for intrinsic photonic wave localization in a 3D QP structure based on the hierarchical arrangement of several building blocks exhibiting icosahedral symmetry [387]. Later on, similar icosahedral PQCs were obtained for the infrared [481], and visible ranges [972]. Despite their relative low dielectric contrast, transmission spectra showed angular-dependent band gaps at 560 nm, 500 nm, and 400 nm in the visible range [972]. Furthermore, observation of multidirectional lasing exhibiting the icosahedral symmetry at wavelengths in the range 580–600 nm was reported for Rhodamine dye-doped PQCs fabricated in dichromate gelatin emulsions by the seven-beam optical interference holographic method [438].

4.4.3 Spherical and cylindrical geometries

Since a sphere has the highest possible symmetry, one may reasonably expect that devices based on a concentric arrangement of spherical shells would be ideal candidates for achieving complete PBGs. In addition, such a design would naturally lead to the formation of PBGs in any direction pointing towards the center of the structure, not requiring high dielectric contrast materials to this end. Certainly, to fabricate such a device is not an easy task using common microfabrication techniques, though holographic lithography methods appear as a promising approach, since one can obtain a spherical light pattern from a point source located at the center of a spherical mirror which interferes with the own mirror's reflected light. Based on this approach, spherical layered structures with different curvatures (ranging

---

[10]That is, for both TE and TM polarizations.

from 1.0 to 9.5 mm) were fabricated in dichromate gelatin emulsions [355]. As expected, the obtained structures behave like curved mirrors and exhibit complete PBGs in the visible, within the 500–600 nm spectral range (depending on the curvature of the sample) and with typical bandwidths of about 50 nm, despite the low dielectric contrast of the employed material.

Following the wake of these results the potential use of aperiodic structures based on a dielectric microsphere which is coated by a series of alternate materials arranged according to the Fibonacci sequence was numerically analyzed. The mathematical treatment of wave propagation through spherical multilayers is similar to that used for planar multilayers by expressing Helmholtz equation (4.2b) in terms of a scalar function $\Pi(r)$ as

$$\frac{d^2\Pi}{dr^2} + \left[\frac{\omega^2}{c^2}n^2(r) - \frac{l(l+1)}{r^2}\right]\Pi = 0, \qquad (4.15)$$

where l is the angular momentum [132]. This expression can be solved in terms of the spherical Hankel functions and the propagation of the wave described in terms of a series of transfer matrices containing these functions. Nevertheless, in the spherical multilayer case these transfer matrices are no longer unimodular, since their determinant is given by the ratio $(r_{i+1}/r_i)^2 > 1$ involving consecutive shells radii [98]. Another consequence of the spherical geometry is that the transfer matrix depends not only on the thickness of the layer, but also on the distance of the considered block to the center. As a result, in the study of QP stacking of shells, simple concatenation rules giving the Fibonacci stacking sequence in 1D layered systems must be replaced by the following one for the spherical Fibonacci stack, $\mathbf{K}_{n+1}(n+1) = \mathbf{K}_n(n)\mathbf{K}_{n-1}(F_{n+2})$, where $\mathbf{K}_i(n)$ denotes the transfer matrix for the nth Fibonacci generation [99, 100]. Different realizations of this spherical arrangement have been considered based on both conventional materials and left-handed materials. It was found that (1) as the number of layers increases, the frequency spectrum becomes more and more fragmented, exhibiting extremely narrow transmission peaks (say ∼0.04 nm in width for a F_9 Fibonacci generation containing 34 layers), and (2) there exists a particular choice for the shells thickness ratio η_S, such that the band gap width is considerably wider in spherical structures containing left-handed materials (see Sec. 4.2.3) than those composed of conventional materials. In addition, the precise location of the narrow transmission peaks can be changed by properly tuning the structural parameter η_S value, allowing for the design of optical filters with extremely narrow passbands [98–100].

On the other hand, cylindrical geometries have also been considered in the design of coaxial optical waveguides and it has been experimentally shown that light waves with different frequencies in the visible spectral range can be selectively guided and spatially separated along the radial direction in a self-similar dielectric waveguide, where a hollow cylindrical core is surrounded by a concentric series of cylindrical layers of two alternating materials radially arranged according to the Thue-Morse sequence [346]. In this way, a "rainbow" where the light waves with different frequencies separate spatially can be trapped in Thue-Morse coaxial multilayered waveguides. The physical origin of this feature can be traced back to the self-similar nature of the refractive index profile in the radial direction, which allows for the appearance of a progressive multifurcation of PBGs by increasing the generation of the Thue-Morse sequence. An extension of this design using dielectric (A layers) and liquid crystal (B layers) materials was subsequently fabricated, so that both photonic bands and transmission modes can be tuned by changing temperature, thereby the rainbow pattern change the colors in the waveguide. This effect can be applied to get temperature-dependent integrated photonic devices of interest [347].

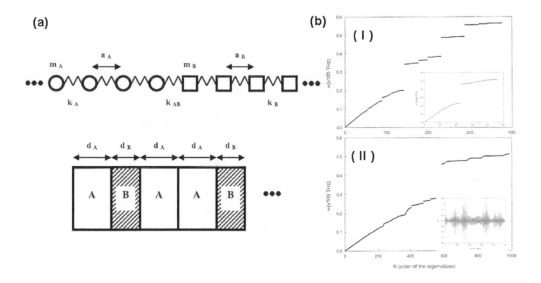

Figure 4.11 (a) Sketch illustrating the hierarchical arrangement of a FSL. At the atomic scale (top frame) the system can be modelled as a lattice chain composed of two kinds of atoms m_A and m_B coupled via force constants K_A (layer A), K_B (layer B), and K_{AB} (interfaces). At a larger scale (bottom frame) the system is described in terms of a sequence of layers of different width $d_A = n_A a_A$ and $d_B = n_B a_B$, respectively, where n_υ is the number of atoms composing the layer, and a_υ is its lattice constant [542]. (Reprinted figure with permission from Maciá E, Hierarchical description of phonon dynamics on finite Fibonacci superlattices, 2006 Physical. Review B **73** 184303. Copyright (2006) by the American Physical Society). (b) Frequency spectra for (I) a binary periodic chain with $N = 244$ atoms (inset), and a FSL with $N = 377$ atoms (main frame); (II) a hybrid Al/Ag metallic multilayer formed by a periodic chain (including 116 atoms) sandwiched between two FSLs including 377 atoms each. The model parameters are $m_A = 4.805 \times 10^{-23}$ g, $m_B = 1.791 \times 10^{-22}$ g, $K_A = 4.416 \times 10^4$ dyn/cm, $K_B = 5.032 \times 10^4$ dyn/cm, and $K_{AB} \equiv (K_A + K_B)/2$. In the inset the normalized atomic displacements distribution corresponding to the eigenvalue $\omega = 15.68$ GHz is shown (Courtesy of Victor R.Velasco).

4.5 PHONONIC QUASICRYSTALS

1D phononic QCs consist of arrangements of layers whose thickness and/or acoustic impedance $Z = \rho_m v$, where ρ_m is the density and v the sound velocity, are quasiperiodically arranged. For instance, the fabrication of a FSL based on porous silicon with alternating high and low porosity layers was performed using electrochemical etching, where the thickness layers (in the μm range) was controlled by the etching duration [11].

4.5.1 Frequency spectrum of Fibonacci superlattices

In Fig. 4.11a, we illustrate the characteristic two-level structure of a FSL. Within the transfer matrix formalism the dynamic response of a FSL composed of N layers can then be expressed in terms of the global transfer matrix $\mathcal{M}_N = \prod_{j=N}^{1} \mathbf{L}_j$, where the *layer matrices*

$$\mathbf{L}_A \equiv \mathbf{T}_A^{n_A} = \begin{pmatrix} 2 - \lambda & -1 \\ 1 & 0 \end{pmatrix}^{n_A}, \qquad \mathbf{L}_B \equiv \mathbf{T}_B^{n_B} = \begin{pmatrix} 2 - \alpha\lambda & -1 \\ 1 & 0 \end{pmatrix}^{n_B} \qquad (4.16)$$

describe the phonon propagation through layers A and B as a product of the local *atomic matrices* $\mathbf{T}_A \equiv \mathbf{T}_{AAA}$ and $\mathbf{T}_B \equiv \mathbf{T}_{BBB}$, respectively (see Sec. 2.5). As a first approximation, the physical description has been substantially simplified by assuming all the force constants to be equal, that is, $K_A = K_B = K_{AB} \equiv K$. The atomic matrices are characterized by the normalized frequency $\lambda \equiv m_A \omega^2 / K$, and the number of atoms in each layer $n_{A(B)}$, respectively. In a FSL the layers are arranged according to the Fibonacci sequence, which determines their order of appearance in the global transfer matrix product $\mathcal{M}_N$. Making use of the Cayley-Hamilton theorem the power matrices given by Eq. (4.16) can be readily expressed as [542],

$$\mathbf{L}_r = \begin{pmatrix} U_{n_r}(x_r) & -U_{n_r-1}(x_r) \\ U_{n_r-1}(x_r) & -U_{n_r-2}(x_r) \end{pmatrix}, \qquad r = \{A, B\} \qquad (4.17)$$

where $U_{n_r}(x_r) \equiv \sin((n_r + 1)\varphi_r)/\sin\varphi_r$, with $\varphi_r = \cos^{-1} x_r$, are Chebyshev polynomials of the second kind, $x_r \equiv tr\mathbf{T}_r/2$, and we have explicitly used Eq. (2.34). Since $\mathbf{L}_r$ is a product of SL(2,$\mathbb{R}$) group elements, the layer matrices are unimodular themselves, and one can exploit this property in order to extend Kohmoto's theorem,[11] originally introduced to describe a Fibonacci lattice of atoms, to the FSL case. In fact, since $\det(\mathbf{L}_r) = 1$, and the set of $\mathbf{L}_r$ matrices are arranged according to the Fibonacci sequence in the FSL transfer matrix $\mathcal{M}_N$, we realize that $\mathbf{L}_r$ matrices satisfy the conditions of Kohmoto's theorem. Accordingly, the dynamical map given by Eq. (4.19) can be properly applied to FSLs, provided the initial conditions

$$z_{-1} = \frac{1}{2} tr\mathbf{L}_B = T_{n_B}(x_B), \qquad z_0 = \frac{1}{2} tr\mathbf{L}_A = T_{n_A}(x_A), \qquad (4.21)$$

$$z_1 = \frac{1}{2} tr(\mathbf{L}_B \mathbf{L}_A) = T_{n_A}(x_A)T_{n_B}(x_B) + (x_A x_B - 1)U_{n_A-1}(x_A)U_{n_B-1}(x_B), \qquad (4.22)$$

where $T_{n_r}(x_r) = \cos[n_r \cos^{-1}(x_r)]$ are Chebyshev polynomials of the first kind, and we have used the relationship $U_n - U_{n-2} = 2T_n$. Making use of Eqs. (4.21)–(4.22), the invariant of the dynamical map given by Eq. (4.20) reads

$$I = (x_A - x_B)^2 U_{n_A-1}^2(x_A)U_{n_B-1}^2(x_B). \qquad (4.23)$$

---

[11]Consider a set of matrices $\mathbf{M}_n$ belonging to the SL(2,$\mathbb{R}$) group and satisfying the concatenation rule $\mathbf{M}_{n+1} = \mathbf{M}_{n-1}\mathbf{M}_n$, then [431]

$$tr\mathbf{M}_{n+1} = tr\mathbf{M}_n tr\mathbf{M}_{n-1} - tr\mathbf{M}_{n-2}, \quad n \geq 2. \qquad (4.18)$$

By defining $z_n \equiv tr\mathbf{M}_n/2$, Eq. (4.18) is rewritten as the dynamical map

$$z_{n+1} = 2z_n z_{n-1} - z_{n-2}, \quad n \geq 2, \qquad (4.19)$$

usually referred to as the trace map. This map has the constant of motion [431]

$$I = z_{-1}^2 + z_0^2 + z_1^2 - 2z_{-1}z_0 z_1 - 1, \qquad (4.20)$$

determined by the initial conditions $z_{-1} = tr M_0/2$, $z_0 = tr M_1/2$, and $z_1 = tr M_2/2$.

In this way, the trace map formalism is extended to discuss the phonon propagation through FSLs characterized by the presence of two relevant physical scales. In fact, by equating $z_{-1} = T_{n_B}(x_B) \equiv \cos(qd_B)$ and $z_0 = T_{n_A}(x_A) \equiv \cos(qd_A)$, where q is the wave vector, we readily obtain the dispersion relation corresponding to the A and B layers

$$\omega = 2\sqrt{\frac{K}{m_{A(B)}}} \left| \sin\left(\frac{qa_{A(B)}}{2}\right)\right|, \tag{4.24}$$

respectively. Analogously, the equation $z_1 \equiv \cos[q(d_A + d_B)]$, leads to the dispersion relation corresponding to the binary periodic superlattice with unit cell AB [810, 927]. Accordingly, the initial conditions implementing the generalized trace map are directly related to the phonon dispersion relations corresponding to the constituent layers (z_{-1} and z_0) and the lowest order periodic approximant to the FSL (z_1). Consequently, the expression $z_n \equiv \cos(qD)$, ($n \geq 2$), with $D = F_n d_A + F_{n-1} d_B$, can be properly regarded as the dispersion relation corresponding to successive FSL approximants obtained from a continued iteration of the trace map [339]. The phonon spectrum of the FSL can then be obtained as the asymptotic limit of a series of approximants whose dispersion relations are determined by the successive application of the trace map recursion relation given by Eq. (4.19). In doing so, the dispersion relation of a given FSL approximant can be generally split into two complementary contributions. The first one describes a periodic binary lattice, where the layers alternate in the form $ABABABA$.... The other one includes the progressive emergence of QPO effects in the system [542, 638, 639].

In order to properly describe the phonon dynamics in FSLs where two kinds of order (periodic at the atomic scale and QP beyond the layer scale) are present in the same sample at different scale lengths, it is convenient to express the trace map in terms of nested Chebyshev polynomials of the form $T_{F_\nu}[T_{n_r}(x_r)]$ and $U_{F_\nu - 1}[T_{n_r}(x_r)]$, where the variable x_r describes the atomic scale physics and the function $T_{n_r}(x_r)$ describes the dynamics at the layer scale. This representation can be regarded as describing a *scale transformation,* formally expressed as $T_{n_r}(x_r) \to X_r$. By applying this transformation to Eqs. (4.21)–(4.22) the trace map initial conditions now read, $z_{-1} = X_B$, $z_0 = X_A$, and $z_1 = X_A X_B + \sqrt{I + Y_A^2 Y_B^2}$, where $Y_r = \sqrt{1 - X_r^2}$, and we have made use of the constant of motion I given by Eq. (4.23) [542]. Since the trace map itself can be interpreted as giving the dispersion relation of a given FSL realization in terms of the dispersion relations corresponding to lower order approximants, this nested structure provides a suitable unified description of the phonon dynamics in FSLs, able to encompass their characteristic hierarchical structure in a natural way. The adoption of different values for the elastic constants K_A, K_B, and K_{AB} renders a much more involved mathematical description, although it does not significantly affect the underlying physics of the approach [542]. This is illustrated in Fig. 4.11b (panel I) where the frequency spectra of a periodic and a FSL are shown. In the FSL spectrum one can readily see the characteristic fragmentation scheme, as well as the presence of three allowed bands in the region corresponding to the gap in the binary chain [616]. The high fragmentation of the phonon spectra and the critical nature of their related eigenstates significantly affects the dependence of the thermal conductivity with the temperature over a wide temperature range [530, 302, 779], and at variance with periodic structures it monotonously decrease with the SL length [971].

The nature of frequency spectra in hybrid order structures is an interesting open problem. A representative example of the obtained numerical results is shown in the panel II of Fig. 4.11b for the frequency spectrum of a Fibonacci/periodic/Fibonacci hybrid structure (see Fig. 4.4a). By inspecting this plot one realizes that the spectrum of the hybrid structure shares some characteristic features with both the periodic and the QP frequency spectra shown in panel I above. Thus, the hybrid spectrum preserves certain fragmentation degree over all the considered frequency range, but the two energy bands located at the intermediate frequency region in the isolated FSL are shifted towards lower (higher) frequency values respectively, nearly closing the gaps located at about 20 and 30 GHz in the FSL spectrum. Along with these modifications in the overall spectrum's structure, one may expect that the competition between highly fragmented spectra supporting critical eigenstates (a characteristic feature of QP systems) and continuous spectra possessing Bloch eigenfunctions (typical of periodic systems) should give rise to some peculiar eigenstates in these hybrid structures. An illustrative example is shown in the inset to Fig. 4.11b (panel II) where one can appreciate an extended vibration pattern, corresponding to a frequency which simultaneously belongs to the energy spectra of both Fibonacci and periodic SLs in the Fibonacci/periodic/Fibonacci hybrid structure. The vibration pattern looks very regular in the periodic central portion of the structure (a reminiscent feature of a typical Bloch-like function), whereas it exhibits fluctuations at several scales (showing typical self-similar fingerprints) in the QP blocks flanking it [616]. Similar results were obtained for hybrid structures where the periodic unit is sandwiched between Thue-Morse or Rudin-Shapiro SLs, so that by properly choosing the frequency value one can obtain a selective confinement of the atomic motions in one of the subunits forming the global structure, in close analogy with the case of photonic cavities [617, 618]. In this way, the presence of two different kinds of order in the underlying substrate of the hybrid order structure naturally affects the energy spectrum and the spatial distribution of its eigenstates. Quite interestingly, these confining features can also be observed when the full 3D geometry of the hybrid order SLs is explicitly taken into account [586, 619].

Certainly, a more realistic description of aperiodic SL requires a full 3D treatment. A straightforward way to do that is by considering that the aperiodic multilayers are made of 3D slabs rather than to describe them in terms of 1D harmonic chains. The dynamical equation describing elastic waves propagating throughout a 3D multilayered structure can be written as

$$\frac{\partial^2 u_j^i}{\partial t^2} - \frac{1}{\rho_j} \left\{ \frac{\partial}{\partial x_i} \left(\lambda_j \frac{\partial u_j^i}{\partial x_l} \right) + \frac{\partial}{\partial x_l} \left[\mu_j \left(\frac{\partial u_j^i}{\partial x_l} + \frac{\partial u_j^l}{\partial x_i} \right) \right] \right\} = 0, \qquad (4.25)$$

where u_j^i is the i-th component of the displacement vector $\mathbf{u}(\mathbf{r})$ in the j-th layer, $\rho_j(\mathbf{r})$ is the layer mass density, and $\lambda(\mathbf{r})$ and $\mu(\mathbf{r})$ are the so-called Lamé coefficients related to the elastic constants (see Sec. 3.9) [893]. In this way, one can properly consider the propagation of acoustic waves in solid/liquid aperiodic structures as well [10, 717].

An alternative approach to design fully 3D phononic QCs is by assembling arrangements of cylinders, spheres or rods of different materials (steel, polymer) embedded in air or some material substrate (either solid or liquid) according to certain aperiodic pattern. In this way, both 3D icosahedral Penrose tilings of steel spheres embedded in a polyester matrix [851], and 2D QP steel/water heterostructures with 8-, 10-, 12-, and 14-fold axial symmetries have been grown, using steel rods with different regular-polygonal cross sections (including 3, 4, 5, and 8 sides) [849, 850]. The existence of gaps in such materials provides an opportunity to confine and control the propagation of acoustic waves, leading to the possibility of designing flat lenses [997]. In this case, the motivation for using the high-symmetry QC is to maintain an efficient interference of waves (long-range order) while reducing the orientational order of the system (crystallographical restriction theorem is relaxed) to get a more isotropic propagation. On the other hand, thermal transmission in 3D Fibonacci waveguides is lower than the corresponding value for periodic systems [612, 970], in agreement with the results reported from the study of 1D Fibonacci lattices [530].

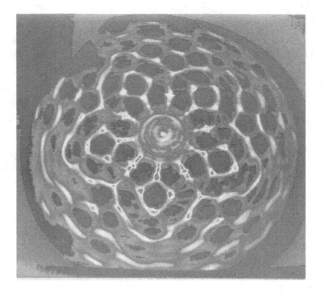

Figure 4.12   Spontaneous nonlinear quasicrystalline structure with five-fold symmetry obtained in the Faraday wave experiment (courtesy of Manuel Torres).

4.5.2   Surface acoustic waves

Interesting examples of pattern formation can be observed in the study of Faraday oscillations in which surface waves on a liquid are amplified by forcing a vertical oscillatory motion of the container. In this way, the effective gravity is periodically modulated according to the expression $g(t) = -g - A\cos\omega t$, where A and ω are the driven amplitude and frequency, respectively. When the driving amplitude exceeds a critical threshold A_c a standing-wave instability occurs with temporal frequency $\omega_0 = \omega/2$. The characteristic spatial wavelength, $\lambda_0 = c/\omega_0$, of the standing-wave pattern is determined by the gravity-capillary dispersion relation of the fluid

$$\omega_0^2 = gk_0 + \frac{\sigma k_0^3}{\rho} \tanh k_0 h, \qquad (4.26)$$

where $k_0 = 2\pi/\lambda_0$, σ is the surface tension, ρ is the fluid density, and h is the liquid depth [38]. Accordingly, patterns of various symmetries can be observed depending on the driven frequency and fluid properties, the most usual ones being parallel stripes (single standing waves) and square patterns (two waves at a 90° angle). These results raised the question of whether other types of regular patterns can be formed that are not spatially periodic but do have rotational symmetry. Indeed, more unusual patterns, including octagonal, dodecagonal, and pentagonal symmetries have been reported [300, 892] and theoretically discussed [493]. An example of five-fold symmetry pattern is illustrated in Fig. 4.12. This pattern was obtained in a circular vessel under free boundary conditions. Thus, the symmetry breaking consists of the infinity symmetry reduction of the synchronous circular wave at the forcing frequency, and originates in the meniscus. Quite remarkably, some of the five-fold patterns observed   in these experiments resemble the evolutionary sequence of some sea urchins as purported by paleontological fossil records, hence suggesting a possible mechanism for inducing the pentagonal symmetry observed in the early development of these animals [18, 38]. An interesting relationship between QCs and the morphological structure of some extant sea urchins was further explored by using the mathematical concept of eutectic star [894], a notion which had been extensively used to describe transformations from QP to periodic tilings. In this way, mathematical tools introduced to deal with strictly condensed matter notions can prove extremely useful for studying biological problems as well.

The dynamics of Rayleigh acoustic waves on a QP corrugated solid substrate formed by a series of thousands of grooves engraved at the surface of a piezoelectric LiNbO$_3$ substrate according to a Fibonacci sequence was early used in order to visualize the shape of surface waves propagating on QP structures [181, 572]. It was reported that the structure of the observed waves could be described in terms of a product of sinusoids with periods related to the different quasi-periods of the system. This resulted in a broad modulated wave containing a rich spiky structure at different spatial scales, displaying an amplitude profile similar to that shown in Fig. 2.16b. Around its maximum the intensity profile of the fundamental mode corresponding to consecutive terms in the Fibonacci sequence reasonably agreed with the Gaussian function

$$I(x) = I_0 \exp\left[-\frac{F_j T}{2\sqrt{\tau}}\left(\frac{2\pi x}{L}\right)^2\right], \tag{4.27}$$

where F_j is related to the system length by $L = F_j F_{j+1} d$, with d measuring the grooves width (so that in the large size limit $F_j \sim \sqrt{L}$), T is the transmission coefficient through a given groove, and τ is the golden mean. It is interesting to note that the norm of this function diverges as L increases (**Exercise 4.5**), so that this state can not be regarded as localized in the usual sense. Subsequently, an experiment consisting of pulse propagation on a shallow fluid covering a 2D Octonacci pattern drilled in the bottom of a transparent vessel, confirmed the existence of plane transverse waves of the Bloch-like form prescribed by Eq. (2.77), with $u_{\mathbf{k}}(\mathbf{r})$ given by a suitable QP function in such systems [895].

4.6   EXERCISES WITH SOLUTIONS

4.1 *Show that Eq. (4.1) satisfies the relationship* $k_{m_1,m_2+1} = k_{m_1,m_2} + k_{m_1,m_2-1}$.

From explicit substitution, we have

$$k_{m_1,m_2} + k_{m_1,m_2-1} = \frac{2\pi}{\tau d_A + d_B} m_1 \tau^{m_2}(1 + \tau^{-1}) = \frac{2\pi}{\tau d_A + d_B} m_1 \tau^{m_2+1} \equiv k_{m_1,m_2+1},$$

where we have used the basic relation $1 + \tau = \tau^2$.

4.2 *Show that Eq. (4.1) reduces to the Fibonacci lattice Fourier transform.*

The Fourier transform of the ideal Fibonacci lattice consists of a dense set of diffraction peaks given by the wave vectors [486],

$$k_{n_1,n_2} = \frac{2\pi}{\tau d_A + d_B}(n_1 + n_2\tau), \tag{4.28}$$

where n_i are integers. When n_1 and n_2 are successive Fibonacci numbers, say F_{n-2} and F_{n-1}, $n_1 + n_2\tau = F_{n-2} + \tau F_{n-1} = \tau^n$, where we have used Eq. (1.3). In that case, Eq. (4.28) reads

$$k_{n_1,n_2} = \frac{2\pi}{\tau d_A + d_B}\tau^n. \tag{4.29}$$

In actual experiments the entire diffractogram profile is dominated by peaks which can be properly labeled in terms on integers belonging to the set $\{1, 2, 3\}$, which are Fibonacci numbers themselves. This fact indicates that Eq. (4.1) and (4.29) are essentially equivalent in practice.

4.3 *Obtain the zero-$\bar{n}$ gap condition for aperiodic multilayers based on the (a) period-doubling, (b) copper mean, (c) silver mean sequences in the asymptotic $N \to \infty$ limit.*

According to Eq. (4.6), the zero-$\bar{n}$ gap condition for a given aperiodic multilayer containing N_A (N_B) layers of type A (B) can be written in terms of the layers relative frequency $v_i \equiv N_i/N$, with $N = N_A + N_B$, as

$$n_A = -n_B \eta \frac{N_B}{N_A} = -n_B \eta \frac{v_B}{v_A}, \tag{4.30}$$

where $\eta = d_B/d_A$. Now, in Sec. 2.2, we learnt that the components of the eigenvector related to the leading eigenvalue, λ_+, of the substitution matrix, once normalized, respectively indicate the frequencies of letters A and B of a given aperiodic sequence in the infinite sequence $N \to \infty$ limit. Thus, making use of Eq. (2.2), we obtain:

(a) The leading eigenvalue of the period-doubling sequence is $\lambda_+ = 2$ and the components of the related eigenvector are (see Table 2.3):

$$v_A = \frac{\lambda_+ - n_B[g(B)]}{\lambda_+ + n_B[g(A)] - n_B[g(B)]} = \frac{2}{3} \qquad v_B = \frac{n_B[g(A)]}{\lambda_+ + n_B[g(A)] - n_B[g(B)]} = \frac{1}{3},$$

so that Eq. (4.30) reads $n_A = -n_B \eta/2$.

(b) The leading eigenvalue of the copper mean sequence is $\lambda_+ = 2$ also, and the components of the related eigenvector are (see Table 2.3):

$$v_A = \frac{\lambda_+ - n_B[g(B)]}{\lambda_+ + n_B[g(A)] - n_B[g(B)]} = \frac{1}{2} \qquad v_B = \frac{n_B[g(A)]}{\lambda_+ + n_B[g(A)] - n_B[g(B)]} = \frac{1}{2},$$

so that Eq. (4.30) reads $n_A = -n_B \eta$.

(c) The leading eigenvalue of the silver mean sequence is $\lambda_+ = 1 + \sqrt{2}$, and the components of the related eigenvector are (see Table 2.3):

$$v_A = \frac{\lambda_+ - n_B[g(B)]}{\lambda_+ + n_B[g(A)] - n_B[g(B)]} = \frac{\sqrt{2}}{2} \qquad v_B = \frac{n_B[g(A)]}{\lambda_+ + n_B[g(A)] - n_B[g(B)]} = 1 - \frac{\sqrt{2}}{2},$$

so that Eq. (4.30) reads $n_A = -n_B \eta(\sqrt{2} - 1)$.

4.4 *Determine the photonic band gap width dependence on the layer thickness ratio $\eta = d_B/d_A$ for a Fibonacci multilayer composed of metamaterials with $\varepsilon < 0$ (A layers) and $\mu < 0$ (B layers).*

Making use of the band-edge conditions $\langle \epsilon \rangle = 0$ (for ω_+) and $\langle \mu \rangle = 0$ (for ω_-) into Eq. (4.8) we get $\epsilon_A(\omega_+) = -\eta\tau^{-1}\epsilon_B$ and $\mu_B(\omega_-) = -\eta^{-1}\tau\mu_A$, so that by plugging these expressions in the constitutive relations: $\epsilon_A(\omega) = \epsilon_a - (\omega_e/\omega)^2$, $\mu_B(\omega) = \mu_b - (\omega_p/\omega)^2$, where $\omega_{e(p)}$ are the electronic (magnetic) plasma frequencies, and solving for the respective band-edge frequencies, we have

$$\Delta\omega = \omega_+ - \omega_- = \frac{\omega_e}{\sqrt{\epsilon_a + \eta\tau^{-1}\epsilon_B}} - \frac{\omega_p}{\sqrt{\mu_b + \eta^{-1}\tau\mu_A}}. \tag{4.31}$$

According to Eq. (4.31) in the $d_B \ll d_A$ limit, $\Delta\omega \to \omega_e/\sqrt{\epsilon_a}$ and the gap width $\Delta\omega$ progressively decreases as η increases, vanishing when

$$\eta_* = \frac{\tau}{2\epsilon_B}\left(\Omega^2\mu_b - \epsilon_a + \sqrt{(\epsilon_a - \Omega^2\mu_b)^2 + 4\Omega^2\epsilon_B\mu_A}\right), \qquad (4.32)$$

with $\Omega \equiv \omega_e/\omega_p$ (the $\eta < 0$ solution has no physical meaning), and hence progressively increasing (in absolute value, since the roles of the $\omega_\pm$ frequencies are interchanged when μ becomes negative) as η is further increased ($\lim_{\eta\to\infty}\Delta\omega = -\omega_p/\sqrt{\mu_b}$).

In the particular case given by the choice $\epsilon_a = \mu_b = 1$, $\epsilon_B = \mu_A = 3$ and $\omega_e = \omega_p$, Eq. (4.32) reduces to $\eta_* = \tau$, so that the omnidirectional band gap closes when the thickness ratio equals the golden ratio value [179].

4.5 *Determine de norm of the fundamental acoustic wave given by Eq. (4.27).*

The norm is given by

$$\|\Psi\| \equiv \int_{-L/2}^{+L/2} I(x)dx = 2I_0\int_0^{L/2} e^{-Ax^2}dx = \frac{\tau^{1/4}I_0 L}{\sqrt{2\pi F_j T}}\,\mathrm{erf}\left(\frac{\pi\sqrt{2F_j T}}{2\tau^{1/4}}\right), \qquad (4.33)$$

where $A \equiv \frac{2F_j T}{\tau^{1/2}}\left(\frac{\pi}{L}\right)^2$, and we have used the integral

$$\int e^{-Ax^2}dx = \frac{1}{2}\sqrt{\frac{\pi}{A}}\,\mathrm{erf}\left(\sqrt{A}x\right),$$

along with the relation $\mathrm{erf}(0) = 0$. Now, in the large size limit we can make the approximation $F_j \sim \sqrt{L}$, so that taking into account $\lim_{x\to\infty}\mathrm{erf}(x) = 1$ in Eq. (4.33), we get $\|\Psi\| \sim L^{3/4}$ [572]. Therefore, $\|\Psi\|$ diverges as L increases and this wave can not be regarded as localized in the standard sense.

CHAPTER 5

Actual and Prospective Applications of Quasicrystals

5.1   INTERMETALLIC QUASICRYSTALS: AN OVERVIEW

> *"Although the quasicrystal concept is mathematically sound, it was not clear at first whether it had physical relevance. A patent disclosure filed in 1983 was rejected on the grounds that it was unlikely to find real materials with this symmetry".* (Paul J. Steinhardt, 2013) [836][1]

> *"We had with this experiment all reasons to believe that quasicrystalline alloys could be produced at industrial scale and low cost, using all the tricks and well-known methods of classical metallurgy".* (Jean-Marie Dubois, 2005) [210]

In order to be of technological relevance a material must be easy to produce in the desired shape (say, bulk, powder or coating), stable in working conditions, environmentally friendly, of low cost and non-toxic.[2] In addition, a novel material must benchmark already existing competitive materials for any given specific application. It is certainly difficult to fulfil all these criteria at once and, consequently, only a few new materials ultimately find a successful route to the market. To this end, a major drawback of macroscopically sized QCs is their brittleness at low and intermediate temperatures. For instance, the toughness of single iQC grains is less than 0.5 MPa m$^{\frac{1}{2}}$, whereas metallic alloys of common use are characterized by a toughness value about two orders of magnitude higher. This property is likely related to the nature of chemical bonding is these compounds. Indeed, as far as QCs can be regarded as hierarchical cluster aggregates (see Sec. 1.5.2) one expects relatively weak inter-cluster bonds along with relatively strong intra-cluster ones [419, 857, 562]. As a result, QCs have little resistance to crack propagation below 450°C, and plastic deformation has been found to only be possible at relatively high temperatures or under hydrostatic pressures. Quite

---

[1]Patent disclosure UPENN-9-23-83. D. Levine and P. J. Steinhardt, *Crystalloids* (1983).

[2]Being made of non-toxic elements does not necessarily make a compound non-toxic. For instance, Zn_3P_2 is made of non-toxic elements but itself is highly toxic (rat poisson). Conversely, being made of toxic elements does not necessarily make a compound toxic. Rather, it is the actual chemical form and bioavailability that makes it so. Thus, the lead in the PbS compound is extremely strongly bound, thereby not bioavailable and therefore non-toxic [183].

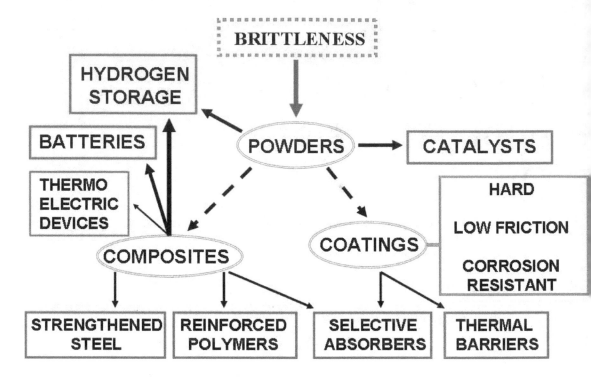

Figure 5.1 Diagram showing the possible niches of application (boxes) of materials based on quasicrystalline powders, composites and coatings (ovals). The physical properties of QC powders exploited in the different applications are listed in Table 5.1.

interestingly, a brittle-to-ductile transition, driven by a sample size reduction, was recently reported for i-AlPdMn QCs at room temperature, the critical size being located around $500 - 350$ nm[1012]. This finding certainly opens up the possible application of nano-sized QCs in the design of small scale devices.

Conversely, QCs are not very useful as bulk materials at macroscopic scale due to their characteristic brittleness, which restricts their use as structural materials. Instead, they can be easily crashed into powders in order to fruitfully be employed as surface coatings or as composites. In such a case, the mechanical integrity is supplied by the substrate or the matrix, respectively, while the QC can be used to improve required functions by exploiting their specific properties, hence promoting their applications as functional materials. The brittle nature of QCs can also be exploited to get fine-grained particles of interest for surface catalysis or hydrogen storage. The related technological niches are sketched in Fig. 5.1 and listed in Table 5.1, and they will be thoroughly described in Secs. 5.2–5.4.2. These applications, which include hard-low-friction-corrosion-resistant coatings, thermal barriers, solar selective absorbers, composites, catalysts, and hydrogen storage materials, have been systematically demonstrated in laboratories, and some of them have even reached marketplace. In Secs.5.4.3 and 5.4.4, we will describe the potential application of QCs for their use in rechargeable batteries and thermoelectric devices, respectively. Finally, in Sec. 5.5, we will consider recent progresses in the growing field of optical applications of PQCs, including photonic bad gap engineering, photovoltaic solar cells, thermophotovoltaic devices, thermal emission control, daytime radiative coolers, Fibonacci lenses, and high harmonic generation.

TABLE 5.1 Commercial applications of different intermetallic QCs grouped according to their shape as coatings (top), composites (middle), or powders (bottom). The exploited physical properties and proposed applications are listed in the second and third columns, respectively. SGCS stands for St. Gobain Coating Solutions, Avignon, France.

Compound	Exploited property	Application	Company
i-AlCuFe	Low adhesion	Surgical scalpels	PHILIPS
	Low friction	Razor blades	TYCO Int. Ltd.
	Hardness	Cookware	SITRAM
	Low $\kappa(T)$	Frying pans	AT&T
d-AlCuFeCr	Corrosion resistance	Frying pans	AT&T
d-AlCoFeCr	Low $\kappa(T)$	Thermal barriers	Aeronautical
	High heat diffusivity	Turbines, engines	Automotive
i-AlCuFe	Low reflectance $\sigma(\omega)$	Selective absorbers	Green
	High IR absorption	Bolometers, filters	energy
i-AlMgZn	Low adhesion	health industry	SANDVIK STEEL
	Low friction	Shavers	PHILIPS
	Hardness		
i-AlCuFe	Strengthen	Prosthetics	NIPPON STEEL
i-TiNiZr	Reinforcing	Mechanical industry	YKK Corp.
i-AlCuFeB	High IR absorption	Coating films	SGCS
i-AlCuFe	Interstitials	Batteries	
i-AlPdMn	Clustered structure	H$_2$ Production	
i-AlCuFe	Hardness		Catalysts
i-AlCuRu		Methanol dissociation	industry
i-AlNiCu			
i-AlPdFe		CH$_3$OH	
i-AlCuFeCr		Steam reforming	
i-AlCuFeB		Antimicrobial	
i-TiZrNi	Interstitials	H storage	Green energy
i-AlCuFe	Hardness		
i-TiZrNi		Batteries	Rechargable
i-TiNiV			batteries
i-AlCuFe			

5.2 APPLICATIONS BASED ON TRIBOLOGICAL PROPERTIES

"The very first usefulness I found of quasicrystals (...) was by not means a high-tech product. Rather, it was prosaically a frying pan that could combine a metallic substrate for fast cooking and a coating made of a specifically designed quasicrystalline alloy offering surface hardness, chemical resistance and substantially low adhesion to food". (Jean-Marie Dubois, 2005) [210]

5.2.1 Hard, low-friction, and corrosion-resistant coatings

Albeit brittle, at room temperature QCs are harder than related periodic alloys in the phase diagram, and quasicrystalline alloys resist corrosion better as well.[3] Furthermore,

---

[3]Corrosion is a kind of gradual destruction of materials by chemical and/or electrochemical reaction with their environment, such as atmospheric or immersion in a specific solution.

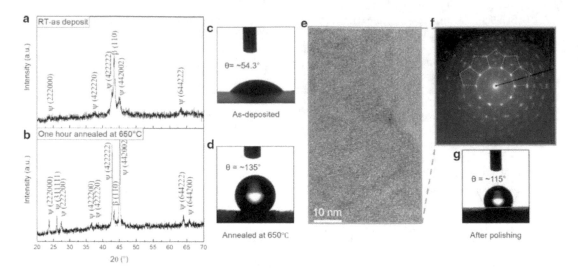

Figure 5.2 X-ray diffraction spectra for (a) as-deposited material and (b) after annealing for 1h at 650°C. Contact angles measured against water for the sample (c) before and (d) after annealing, respectively. (e) HRTEM image for the annealed sample, whose diffraction pattern is shown in (f). (g) contact angle of the annealed sample after polishing. (Reprinted from Parsamehr H, Chen T S, Wang D S, Leu M S, Han I, Xi Z, Tsai A P, Shahani A J, Lai C H, Thermal spray coating of Al-Cu-Fe quasicrystals: Dynamic observations and surface properties, Materalia 8 100432, 2019 Copyright (2019), with permission from Elsevier).

it was recognized soon after the discovery of stable QCs that the friction coefficient of these materials is significantly reduced compared with the friction coefficient measured in comparable conditions on conventional metallic alloys and ceramic materials of similar mechanical characteristics (see Sec. 3.9) [39]. In good metals like aluminium, the adhesion energy of the clean-surface is typically 1 Jm^{-2} or greater. Conversely, lower values, in the range of 0.40−0.80 Jm^{-2} (that is, lower than the surface energy of their constitutive elements), have been reported for Al-based QCs [46, 208].

Additional, although more indirect evidence, that the surface energy of QCs is indeed small, can be gained from a comparison of contact angles measured by depositing small liquid droplets on the surface of a quasicrystalline sample, and on reference samples such as Teflon®, metals, or a bulk oxide such as alumina (Al_2O_3). For nonreactive metal liquids on solids, the intrinsic contact angles result from two types of competing forces: adhesion forces between the liquid and the solid phases and cohesion forces of the liquid. Then, in the absence of barriers to wetting such as oxide films, good wetting, that is, contact angles of a few degrees or tens of degrees, is observed if the interactions occurring at the interface are significant. This is fulfilled for liquid metals on metallic substrates because the interfacial bond is strong (metallic). By making certain assumptions, the contact angle can be fairly simply related to the reversible adhesion energy W of a given liquid onto the surface, larger angles and poorer wetting corresponding to smaller values of W (see Fig. 5.2c,d,g, and **Exercise 5.1**). The fundamental origin of this behavior is not yet understood, naturally rising the question concerning as to whether it is controlled by the properties of the surface, or those of the bulk in QCs [46, 208]. In this regard, one must keep in mind that, when in contact with pure O_2 or air, the QC surface is soon covered by a thin layer of pure alumina [883]. Indeed, the oxidation resistance of QCs does not significantly depart from that of aluminium alloys or compounds. However, bulk alumina is completely wet by water, indicating a high value of W. Thus, the outstanding values of the contact angles on QC surfaces were attributed to a combination of intrinsic and extrinsic factors, namely, the specific electronic density of states within the bulk material, underneath the oxide layer, and thickness of the oxide layer.[4] In a recent study, using liquid lead as a probe, notable intrinsic low wetting behaviors (i.e., contact angles close to or larger than 90°) were measured for two intermetallic substrates considered as approximants to dQCs, namely, Al_5Co_2 and $Al_{13}Co_4$. Electronic effects were found to have a significant influence on the wetting properties of the considered QC-related substrates, suggesting a possible correlation of the contact angle values with the density of states at the Fermi energy [13].

In the light of the results discussed in the paragraphs above, most of the proposed commercial applications of intermetallic QCs to date take advantage of their specific tribological properties, which combine wear resistance with low friction and low adhesion [201, 202]. In this way, the low adhesion of biological tissues to coatings based on i-AlCuFe QC led to their use in electrosurgical scalpels or razor blades [853], as well as in cookware [199]. Thus, in the late 1990s, the company Sitram developed and sold a line of aluminium frying pans spray-coated with partly-quasicrystalline Al-Cu-Fe alloy under the trademark Cybernox™. The resulting QC-based coating is not directly competitive with Teflon® in terms of its non-stick property, but adhesion for many foods is reduced relative to a metal pan, and of course the coating is much harder than Teflon®. The hardness means that normal cutlery can be used with the QC coating, an attractive feature relative to Teflon®. Furthermore, the characteristic low thermal conductivity of QCs provides an additional advantage in the cookware, since it leads to even surface heating. Unfortunately, the cookware was not commercially successful and Sitram withdrew the QC coated product after a few years. Detailed analysis of the defective pans revealed a mixture of iQC phase and either a decagonal phase or a difficult-to-distinguish approximant. A large fraction of unmelted starting material was also observed along with microcracks [828]. By all indications, these defects arose due to the manufacturer's change of the originally patented process, skipping an essential annealing step at 700°C, in order to use a bulk metallic core with a lower melting temperature. This modification ultimately resulted in the presence of phases with low corrosion resistance in the coating, which ruined the pan lifetime expectations [211]. Indeed, the relative amounts of QC versus periodic crystal phases in the coating significantly affect the surface properties of interest, namely, wettability and hardness, so that the lower the presence of secondary phases the better the resulting surface properties. By using in-situ X-ray diffraction and TEM measurements of high-velocity oxygen-fuel thermal spraying i-AlCuFe based coatings it has been recently confirmed that, whereas the as-deposited coating contains more than 68% of secondary phases, the volume fraction of QC phase significantly increases from about 32% to 93% after an annealing treatment for 1h at 650°C [676]. The crucial importance of the annealing process in the wettability of the coating is readily appreciated by comparing panels a-c versus b-d in Fig. 5.2. In order to rule out oxidation as well as surface roughness effects, the coating was polished after annealing, which resulted in a slightly decreasing contact angle from 135° to 115°, hence still yielding a good enough performance (**Exercise 5.1**).

During the last decade the company Technology Assessment and Transfer (AT&T) has conducted extensive research on i-AlCuFe, d-AlCuFeCr and d-AlCoFeCr QCs for

---

[4]Note that this discussion is only demonstrably valid for water; common organic solvents wet QCs well, and this indicates that the wetting may depend on the polarity of the liquid.

applications related to friction-, wear-, and corrosion-resistant coating, including frying pans market, by using magnetron sputtering instead of plasma spray coatings, thereby avoiding both cracks and porosity related effects [244]. In this regard, a patent secured the manufacture of articles comprising a quasicrystalline AlCuFe alloy film less than 300 nm thick, formed by depositing in sequence on a substrate through radio frequency sputtering a stoichiometric amount of each respective alloy material, and then annealing those layers to form the film through solid state diffusion [853].

5.2.2 Thermal barriers

Thermal barriers based on quasicrystalline alloys can also be manufactured as relatively thick coatings by thermal spraying or magnetron sputtering techniques. This application relies upon two bulk properties of QCs, namely, their low thermal conductivity and high heat diffusivity,[5] along with plasticity at high enough temperature. In fact, the characteristic brittleness of QCs is overcome at above about 450°C, when they become plastic and progressively deform with applied stress. This deformation is facilitated by dislocations, as it occurs in periodic crystals. However, in contrast to classical crystals, there is no work hardening in QCs. This proves the absence of pinning centers and indicates a viscous flow regime that might be due to some internal friction between relatively rigid atomic entities [205]. In this way, QCs may be used as thermal insulating barriers sandwiched between two materials in order to dampen the stress generated at the interface with the substrate due to the difference in thermal expansion coefficients. Indeed, in contrast to zirconia, such a difference is small, since expansion coefficients for QCs, in the range of $(13-16) \times 10^{-6}$ K^{-1}, are close to those of metallic alloys. Furthermore, QCs also resist oxidation and corrosion by sulfur at high temperature, a property of interest for aeronautical applications. In particular, coatings consisting of an i-AlCoFeCr approximant have been extensively studied for components of power generation turbines. The main drawback in using QCs to this end comes from their comparatively low melting temperature and appreciable atomic mobility, leading to significant material diffusion at the interfaces. Thus, a specific composition with nearly congruent melting at 1170°C was developed to enhance the temperature range in which quasicrystalline thermal barriers may be useful. Although not competitive with doped zirconia above 1050−1100°C, such barriers were successfully tested at 950°C in a real-time ground test of an aircraft engine. Excellent results were obtained in terms of thermal and corrosion properties [20, 771]. However, a strong interdiffusion between the coating and the substrate was observed at temperatures above 950°C, which drastically hinders potential applications to aeronautical devices, even though the use of suitable diffusion barriers proved reasonably efficient to slow the interdiffusion effect. Notwithstanding this, it is likely that the excellent thermal and mechanical properties exhibited by QC based coatings may be useful in less demanding fields in terms of temperature range, such as heat insulation of fast-moving mechanical parts [210, 536]. For instance, it has been reported that cylinder liners and piston QC coatings in motor-car engines would increase engine lifetimes and reduce air pollution due to the thermal barrier properties of such coatings [974].

---

[5]The value of the thermal conductivity of typical QCs at room temperature (about 1 Wm^{-1}K^{-1}) is close to that of pure zirconia and two orders of magnitude lower than that of simple metals, such as Al and Cu (see Sec. 3.8).

5.2.3 Solar selective absorbers

There is an increasing demand for solar thermal energy at high temperature levels above 100°C up to 500°C, where electrical power is generated with heat produced in parabolic collectors concentrating solar light onto suitable selectively absorbing surfaces. Solar selective absorbing materials must have a high absorptance α, and small thermal emittance ϵ, in order to reduce radiative losses. In addition, thermal and chemical stability to prevent degradation over long periods of utilization are also convenient. Within the Drude theoretical framework the emissivity of a metal can be expressed in the form $\epsilon = a_1\sqrt{\rho T} + a_2\rho T + \ldots$, where a_i are numerical constants, ρ is the electrical resistivity and T is the temperature [290]. Since iQCs exhibit significantly higher resistivity values than periodic metallic alloys of similar composition one may expect their emissivity values to be higher as well. However, the reported absence of Drude peak in iQCs clearly indicates that the above expression for $\epsilon(\rho, T)$ does not hold for these compounds. Indeed, iQCs show optical properties which are quite different compared to those of usual insulators, semiconductors or metals (see Sec. 3.4.3). In particular, iQCs exhibit a significantly low reflectance ($\sim$60%) in a wide wavelength region from about 300 nm to 20 μm, combined with a significant thermal stability and corrosion resistance, a set of properties which can be exploited in order to design solar selective absorbers for heat production.

Preliminary numerical studies indicated two possible designs to this end: a sandwich dielectric/QC/dielectric multilayer or a cermet film where small quasicrystalline particles are surrounded by a dielectric or ceramic matrix [224, 225]. In the multilayer films the QC layer must be very thin ($\sim$10−12 nm), since it must be transparent in the infrared region, where a substrate of a good conductor metal (e.g., copper) reflects back the incident radiation. Such thin QC layers were grown by sputtering at hot ($\sim$480°C) substrate in order to get copper/50 nm Al_2O_3/10 nm i-$Al_{12}Cu_{25}Fe_{13}$/70 nm Al_2O_3 multilayers exhibiting $\alpha \simeq 0.9$ and $\epsilon \simeq 0.05$ values [226]. On the other hand, a cermet film containing i-$Al_{64.5}Cu_{22.5}Fe_{13}$ particles embedded in an Al_2O_3 matrix, with a filling factor of about 30%, were produced exhibiting $\alpha \simeq 0.8$ and $\epsilon \simeq 0.05$ values [227], which are comparable to those reported for the multilayer structure.[6] In addition, it was confirmed that these absorptance and emissivity values are very robust, and degradation in air at 400°C remained small over a time period of the order of 100h in both kinds of QC based solar selective absorbers [227]. At higher temperatures, however, the practical use of QCs must take into account important modifications of their optical properties due to oxidation [759]. On the other hand, the reported high absorption of QC in the infrared region of the spectrum can be exploited in applications such as bolometers, IR-filters or thermocouples [207].

5.3 QUASICRYSTAL BASED COMPOSITES

The so-called composites are structures made of two or more constituent materials with significantly different physical or chemical properties which remain separate on a macroscopic level within the whole structure. In this way, one can design novel materials with improved properties for specific applications by properly combining the main features of the different materials in a given composite. Aluminum matrix composites are widely used in automotive and aerospace applications due to their excellent mechanical properties. The reinforcements in these composites are mostly ceramics, such as TiB_2, SiC, or BN. By adding these reinforcement particles, the alloys are strengthened due to the dispersion hardening effect.

---

[6]For the sake of comparison the experimental values of absorptivity and emissivity at $T \simeq 290$ K are $\alpha = 0.14$ and $\epsilon = 0.018$ for aluminium, and $\alpha = 0.022$ and $\epsilon = 0.016$ for copper [290].

Among QC based composites two main designs have been explored to date. The first type of considered composites are alloys where small sized Al-based QCs are formed by precipitation after rapid solidification and powder processing, yielding nanoscale quasicrystalline particles surrounded by an Al matrix [203, 210]. Following this approach, light Al-Mg-Zn alloys with outstanding mechanical properties were produced by self-precipitation of small QC particles. In the same vein, a QC based precipitation-strengthened steel relatively ductile, corrosion-resistant, and resistant to overaging was marketed by Sandvik Steel in Sweden. A close collaboration between shaver manufacturers at Philips and Sandvik specialists led to the development of a unique QC based stainless steel, which was named Sandvik Nanoflex^{TM} [781]. Its non-adhesive, low surface energy makes this steel a potentially useful material for the manufacture of engines and all sorts of moving mechanical parts, including key shaver components. The steel is also attractive for tools in the health industry, such as surgical staples, retractable syringes, or needles used for vaccination and acupuncture.

The second type of promising QC based composites involves the use of i-AlCuFe or i-TiNiZr small particles (say, about $0.5-2$ μm in size, on average) as fillers to reinforce polymers. In this way, one can significantly improve the mechanical, tribological, and thermal properties of composites with polymer matrices, including epoxy resin, polyethylene, polyamide, ethylene-vinyl acetate, polysulphone, or polyphenylene sulfide among others [918]. Tests with certain high-performance thermoplastic resins have shown that i-AlCuFe particles significantly enhance the wear resistance of the polymers, probably due to a combination of their intrinsic hardness, low friction, and low thermal conductivity. For instance, composite materials based on thermoplastic matrices of linear low-density polyethylene and i-$Al_{65}Cu_{22}Fe_{13}$ particles were obtained by mixing the components in the polymer melt. The dynamic modulus of elasticity of the composites exceeds the modulus of the unfilled polymer matrix over a wide temperature range [918]. At the same time, key thermochemical characteristics of the polymer, that is, its glass and melting transitions, are not degraded, and this indicates that the QC does not catalyze cross-linking or other disadvantageous reactions in the resin, so that it may be potentially used as a prosthetic biomaterial [14].

Another niche of application of quasicrystalline powders in polymer based composites has been demonstrated in the direct prototyping of reinforced mechanical parts [411, 412]. The whole process is divided in two basic steps: first, gas atomization from molten alloy yields quasicrystalline $Al_{59}Cu_{25}Fe_{13}B_3$ powder grains,[7] which are then mixed with a suitable polymer component and the resulting composite is finally manufactured by selective laser sintering, which is one of the current most powerful additive manufacturing technologies in use in the mechanical industry. One important feature of the obtained materials due to the usage of QC powders is the absence of porosity, likely as a consequence of their specific electronic properties. Indeed, the absence of the Drude peak in QCs results in the absorption of infrared light being much higher in QCs as compared to that of conventional metals, such as steel or aluminium alloys. This property, combined with the intrinsic low thermal conductivity of QCs, ultimately results in the powder grains to store a significant amount of heat, and during a long enough time as to completely melt the surrounding polymer, thereby fully wetting it, so that no pores are left around. This result is an important technological advantage since it eliminates the post-processing impregnation treatment that is usually mandatory to obtain water or gas tight components in other composites. Thus, this application niche ultimately relies on the specific electronic properties of QCs leading to their high absorption in the IR spectral window.

---

[7]Batches of these atomized powders can be bought from the international materials producers Saint Gobain Coating Solutions (France). Other chemical solids distributors including QC powders in their commercial catalogue are Sigma-Aldrich (Germany) or READE® International Corporation (USA).

It is instructive to estimate the temperature increase of a powder grain of mass density ρ_m, diameter D, specific heat at constant pressure C_P, and absorption (emission) coefficient α (ϵ), during the time interval it is irradiated by an IR laser of power P_L and beam diameter D_L (where $D_L \gg D$), making use of the expression,[411]

$$\frac{\pi}{6}\rho_m D^3 C_P \frac{dT}{dt} = \alpha P_L \left(\frac{D}{D_L}\right)^2 - h\pi D^2(T - T_b) - \sigma\epsilon\pi D^2 T^4, \qquad (5.1)$$

where h is the convection exchange coefficient at the QC-polymer interface, σ is the Stefan-Boltzmann constant, and T_b is the embedding polymer temperature. The first term on the right hand of Eq. (5.1) gives the energy brought into the particle by the laser beam per unit time, whereas the second and third terms measure the convective and radiative losses, respectively. By assuming that these losses are negligible as a first approximation, one gets a grain temperature increase given by

$$\Delta T = \frac{6\alpha P_L}{\pi \rho_m D D_L C_p v}, \qquad (5.2)$$

where the time interval of irradiation of the grain by the laser beam is assumed to be given by $\Delta t = D_L/v$, where v is the drift velocity of the laser beam on the composite surface (**Exercise 5.2a**). For the sake of illustration we can plug the following i-AlCuFe QC representative values $\alpha = 0.9$ (see Sec. 5.2.3), $\rho_m = 4370$ kg m^{-3} (see Table 3.5), $C_p = 23$ JK^{-1} mol$^{-1} = 581$ JK^{-1}kg^{-1} (see Fig. 3.6), along with $P_L = 5 \times 10^4$ W, $D = 75$ μm, $D_L = 1$ mm, and $v = 10^4$ ms^{-1} [204, 411], into Eq. (5.2), to get $\Delta T \simeq 45$ K [411]. This figure represents a very significant temperature increase of the grain surface, able to melt the polymer matrix ($T_m = 451$ K) in which it is embedded. For comparison, in identical experimental conditions, an aluminium grain would experience a temperature increase of only $\Delta T = 7.4$ K (**Exercise 5.2b**). In addition, it is worth noticing that the temperature increase attained in QC grains keeps large enough to properly melt the polymer matrix even in the presence of radiative losses (**Exercise 5.2c**).

5.4 GREEN ENERGY APPLICATIONS

5.4.1 Surface catalysis

The high brittleness of QCs turns to be an advantage as it allows one to prepare large amounts of ultra-fine powders through grinding, without losing chemical homogeneity, a feature of interest in the field of heterogeneous catalysis.[8] Thus, QCs may become superior to other metallic alloy phases as catalyst materials precisely due to their brittle nature [873]. The potential application of QCs for catalysis was first investigated by Nosaki and co-workers, who observed that the generation of hydrogen from the decomposition of methanol through the reaction $CH_3OH \rightarrow CO + 2H_2$, started at significantly lower temperatures on nano-sized i-AlPdTM QC surfaces than those reported when using conventional catalysts, such as Pd or Cu periodic crystals. The obtained results also showed that the highest quantity of hydrogen was formed on the catalysts under quasicrystalline form [656]. Later on, Tsai and collaborators found that QCs are promising precursors for the fabrication of high-activity catalysts when they are subjected to a dealloying treatment,[9] focusing on

---

[8]Heterogeneous catalysis is the science and technology aimed at modifying the course of chemical reactions by engineering the energy barriers between elementary steps of molecular transformations.

[9]Dealloying or selective leaching is a corrosion process of extracting one or more component elements preferentially from an alloy dissolving them in a liquid. Dealloyed materials generally take the form of

the catalytic properties of i-AlCuFe QCs with various compositions for the steam reforming reaction of methanol. This endothermic process takes place above 500 K between methanol and water steam, according to the reaction $CH_3OH + H_2O \rightarrow CO_2 + 3H_2$. This process is an effective way to produce hydrogen for fuel cells under mild reaction conditions. To this end, QCs are prepared as a powder, then crushed in order to promote catalytic activity by increasing the surface area, and finally leached in NaOH or Na_2CO_3 aqueous solution. It was observed, however, that for QCs powders leached with NaOH, a degradation of catalytic activity appears at high temperatures (above 613 K), due to a decrease in thermal stability. Quite on the contrary, i-AlCuFe QC particles leached in Na_2CO_3 exhibited an excellent catalytic activity [403, 981]. In addition to AlCuFe based QCs, another alloy systems, such as i-AlCuRu or AlNiCu, treated by the alkali leaching method, have been considered as suitable catalysts as well [334, 692]. More recently, the activity of an i-$Al_{63}Cu_{25}Fe_{12}$ QC leached using a 5 wt% Na_2CO_3 aqueous solution, was reported to be significantly improved by further calcination in air at 873 K, at variance with the behavior generally seen in typical catalysts, where calcination leads to sintering, which results in a serious loss of their catalytic activity [404].

Since the structural order of QCs is fundamentally different from that of both periodic and amorphous metallic alloys, it is tempting to think that a unique nanoporous pattern may result from the leaching process. In the case of Al-based QCs, this process selectively removes Al along with related Al oxides, leaving a network of nanoparticles consisting of concentric layers. At its center one has a nano-sized QC core, shrouded by a porous Al-OH shell that supports dispersed nanoparticles of Cu and Fe, so that the QC derived catalyst can be viewed as copper and copper oxide particles located on the QC surface. Therefore, from the viewpoint of heterogeneous catalysis this kind of QCs can be assimilated to the so-called Raney like catalysts, since a leaching treatment is required in order to create active metallic species on their surface. Similar results were reported for quaternary alloy phases in the systems i-AlCuFeB and d-AlCuFeCr. Whereas the d-AlCuFeCr QC derived alloys exhibited a high catalytic activity and high selectivity in H_2 and CO_2, the i-AlCuFeB QC derived alloy ones did not, albeit the presence of B atoms may conveniently inhibit the formation of coke (ultimately leading to the catalyst deactivation) during catalytic reforming [692].

In order to elucidate the origin of the high catalytic performance in the above mentioned quasicrystalline alloy systems, detailed microstructural analyses have been performed by several groups and the obtained results indicate that the cluster structure of QCs seems to play a crucial role. In fact, in the case of i-AlCuFe QCs the cluster structure is essentially preserved during leaching, which mainly results in a preferential dissolution of Al atoms in the outer cluster shells, leading to the emergence of a QP distribution of nanoparticles of Cu-Fe metals localized on the top surface atomic framework, thereby highlighting the importance of the original QP atomic scaffold in the resulting catalyst compound. Furthermore, since both Cu and Fe atoms keep their relative positions in the original cluster structure, they cannot agglomerate during a subsequent calcination process, avoiding inconvenient sintering processes, which typically occur in conventional periodic phases, ultimately degrading their catalytic function. Indeed, the calcination treatment produced a multilayered structure on the QC surface stemming from a self-assembly process. The first layer around the QC core is composed of dense Al_2O_3 with a thickness of about 50 nm, and a second layer of similar thickness grows around, mainly consisting of CuO and Fe nanoparticles with a small amount

---

millimeter-sized bodies with a microstructure comprised of nanoscale pores and metal ligaments, hence rendering structures exhibiting a large surface-to-volume ratio of interest in the areas of catalysis, sensing and nanomaterials.

of Al oxides with grain sizes less than 10 nm. The ability to form such a stratified structure seems to be closely related to the QP nature of the atomic layers sequence along the 5-fold axes in the i-AlCuFe, with atomic layers with more than about 80 at.% Al alternating with layers containing less than about 60 at.% Al, so that the dissolution rate is considerably slowed down in these layers, naturally leading to the observed multilayered structure. In fact, it is known that the surface atomic structure of Al-based i-QCs is bulk terminated (i.e., the surface layer planes are identical to the atomic planes in the bulk) and that the cluster shells mainly contain Al atoms (see Sec. 1.5.2). Thus, after a certain amount of these atoms have been removed from the cluster outer shells and dissolved into the electrolyte, the long-range connection of pseudo-Mackay icosahedra is broken in the original icosahedral structure, eventualy leading to the decomposition of the QC phase. The remaining atoms are then driven to agglomerate into a new atomic arrangement with enhanced catalytic function. This scenario has been recently confirmed from a detailed experimental study of the structure evolution of i-$Al_{70}Pd_{17}Fe_{13}$ grains to nanoporous Pd particles upon progressive dealloying in dilute NaCl aqueous solution [511].

The formation of Cu-enriched nano-regions on the leached surface of bulk AlCuFe iQCs has recently spurred the interest in the possible use of these materials as an antimicrobial agent. Indeed, the most commonly recognized mechanism proposed to explain the antimicrobial activity of Cu is that the membrane of bacteria ruptures when they make contact with Cu surface. The resulting surface leakage leads to the loss of cell content and eventually leads to cell death. On these basis, three types of QC alloy powders, with the nominal composition i-$Al_{62.5}Cu_{25}Fe_{12.5}$, i-$Al_{65}Cu_{23}Fe_{11}B$, and i-$Al_{65}Cu_{23}Fe_{11}Co$, were studied in detail [985]. The obtained results indicate that the diameter of inhibition zone in agar diffusion test of the leached quasicrystalline surfaces is comparable to that of copper and ampicillin[10], while it is less than that of kanamycin.[11] The antimicrobial ability of leached quasicrystalline surface may be attributed to the formation of uniformly dispersed nano-sized Cu/Cu_2O sites on the surface of Al–Cu–Fe–Co and Al–Cu–Fe–B QCs. Therefore, leached QCs can be promising materials to reduce the microbial burden from commonly touched surfaces such as bed rails, faucets and doorknobs in hospitals to reduce the number of hospital-acquired infections.

5.4.2 Hydrogen storage

Hydrogen atoms can be absorbed into either interstitial sites or on surfaces of materials. The ability to absorb hydrogen in metals or alloys is greatly dependent upon: (1) chemical affinity between hydrogen and host atoms, (2) type of interstitial sites, and (3) the total number and actual size of interstitial sites in the crystal. There are basically two types of interstitial sites in crystal lattices, namely, octahedral and tetrahedral interstices. The sites that the hydrogen atoms occupy first depend on the physical size of the interstices as well as the chemical affinities between hydrogen and the metal atoms surrounding the sites.

Following the discovery of the second largest group of thermodynamically stable QCs in the i-TiZrNi and i-TiZrHf systems, studies on the hydrogenation properties of such QCs started, searching for suitable samples for their possible application as hydrogen-storage materials [332, 409]. Diffraction studies of the i-TiZrNi phase indicated that it is based on Bergman atomic clusters as main building blocks (see Fig. 1.22). This cluster contains

---

[10]Ampicillin is an antibiotic used to prevent and treat a number of bacterial infections, such as respiratory tract infections, urinary tract infections, meningitis, salmonellosis, and endocarditis.

[11]Kanamycin, is an antibiotic used to treat severe bacterial infections and tuberculosis. It is not a first line treatment and it was removed from the World Health Organization's List of Essential Medicines in 2019.

a significantly large number of tetrahedral (and no octahedral) interstitials, clearly outnumbering the amount of interstices available in the case of common cubic lattices. Thus, one can reasonably expect that the more complex structure of quasicrystalline phases may provide suitable sites for hydrogen absorption by its own.

Abundant literature indicates that solid state hydrogen storage in the form of hydrides offers high storage capacity and practical operation pressures. Among the different considered hydrides, MgH_2 is regarded to be one of the more suitable candidates for hydrogen storage due to its abundance (Mg is the 8th most abundant element on the earth crust), low cost, non-toxicity, reversibility, and high gravimetric and volumetric hydrogen capacities (7.6 wt.% and 110 gl^{-1}, respectively) [230]. Notwithstanding these appealing features, the practical use of magnesium hydride as a hydrogen storage medium is hindered due to the high thermal stability of this molecule and the sluggish nature of the re-hydrogenation kinetics, which makes hydrogen release at moderate temperatures very difficult, thereby requiring high temperatures (above 400°C) for hydrogen de/re-hydrogenation processes to proceed. As a result, maintaining about 6 wt.% of storage capacity, coupled with reversibility for MgH_2 is still a challenging problem.

The hydrogen absorption properties of icosahedral phase powders has been compared to those of both amorphous and crystalline samples of interest in hydrogen energy industry (such as $LaNi_5$ or TiFe hydrides) in high pressure vessels experiments and by using electrochemical hydrogenation at room temperature as well [880]. The maximum hydrogen concentrations, obtained for i-$Ti_xZr_{83-x}Ni_{17}$ QCs (with $41 \leq x \leq 61$), ranged within 1.50−1.74 hydrogen-to-metal ratio. These figures are not as good as those reported for the most efficient materials discovered for hydrogen storage purposes (i.e., Mg and V, with an hydrogen-to-metal ratio of 2.0), but are certainly better than those obtained for the samples TiFe (0.98), $LaNi_5$ (1.00), and Mg_2Ni (1.33) of common use in H-batteries research. In addition, hydrogen in the icosahedral phase can be desorbed comparatively more easily than in the amorphous phase. Furthermore, after hydrogen desorption by heating the hydrogenated icosahedral phase powder to 800 K, the quasilattice constant of QCs completely returned to that before hydrogenation, showing good reversibility for the absorption and desorption of hydrogen. It was also observed that after a second gas-phase loading of hydrogen, QCs remained stable, though now coexisting with a $(Ti,Zr)H_2$ hydride phase.

Quite interestingly, i-$Al_{65}Cu_{20}Fe_{15}$ QC derived catalysts (see Sec. 5.4.1) have been recently reported to dramatically improve the de/re-hydrogenation behavior of MgH_2. It should be noted that i-$Al_{65}Cu_{20}Fe_{15}$ QC is harder (7.5 Mohs) than MgH_2 (4.0 Mohs), so that during the ball milling process followed to obtain these materials, the QCs powders pulverized the originally sized ∼300 nm hydride particles down to about 50 nm. Thus, the exposed area of MgH_2 for catalysis increases and diffusion pathways for hydrogen release get shortened. In this way, another characteristic property of QCs, namely, their hardness, is naturally exploited in a useful way. In addition, the catalytic activity related to the presence of QP Cu-Fe polyhedral frameworks located on top of the QC cores (see Sec. 5.4.1), results in an efficient weakening of the H-Mg-H bonds, thereby lowering the desorption temperature for hydrogen from MgH_2. The mechanism for absorption of hydrogen works in a similar but reverse way. In order to further investigate the details of catalytic activity, the change in the electronic state of Ca and Fe atoms was studied by X-ray photoemission spectroscopy (XPS) techniques and the obtained data clearly indicate a significant change in the valence value of both atomic species. In particular, it is important to note that since the electronegativity of Cu (1.90) and Fe (1.83) lies between that of Mg (1.31) and hydrogen (2.20), both copper and iron atoms can gain or lose electrons easily as compared to Mg and H atoms. Hence, because of the variable valence of Cu and Fe atoms they can efficiently act as intermediate charge carriers and enhance the electron transfer between Mg^{+2} and H^- ions during the de/re-hydrogenation reaction [669].

5.4.3   Rechargeable batteries

The working principle of a battery is relatively straightforward in its basic configuration, and it consists of a cell composed of two electrodes, each connected to an electric circuit, separated by an electrolyte that can accommodate charged ions. Usually, the electrodes are physically separated by a barrier material that prevents them from coming into physical contact to each other (short-circuit). In the discharge mode, when the battery serves to drive the electric current, an oxidation process takes place at the negative electrode (anode), resulting in electrons moving from the anode through the circuit to the positive electrode (cathode) where a complementary reduction process occurs. The cell voltage largely depends on the electrochemical potential difference of the electrodes, and the overall process is spontaneous. For rechargeable batteries the process can be reversed by using an external electricity source in order to produce complementary redox reactions at the electrodes.

Lead-acid and nickel-cadmium batteries have over 100 years successful history with low-cost and long-term operating stability.[12] Nevertheless, current heavy metals related environmental pollution concerns, are causing their market to be progressively shrinking. In view of this situation, the concept of nickel-metal hydride batteries was introduced due to its low-cost, good safety and relatively high energy to power density ratio. However, these batteries (first commercialized in 1989) are still open to improvements regarding their life-span and self-discharge effects at the anode. The concept of lithium-ion batteries was first introduced into the market by Sony in 1991. Most of the research performed on suitable cathode materials for rechargeable lithium-ion batteries have focused on the so-called intercalated compounds with layered, spinel or olivine type structures, where Li^+ ions occupy the octahedral sites of the inter-lamellar spaces. Advancement in the electrochemical energy storage technology has seen the development of next generation lithium-ion battery electrode materials, exemplified by the graphite anode and lithium iron phosphate $LiFePO_4$ cathode, which has recently attracted much attention as promising cathode material for lithium-ion batteries, due to its environmental benigness and potentially low cost, instead of the more expensive $LiCoO_2$ generally used compound. The overwhelming advantage of iron-based compounds is that, in addition to being inexpensive and naturally abundant, they are also less toxic than Co, Ni, and Mn.

The very possibility of using Ti-based iQCs in metal hydride rechargeable batteries was considered as a natural spin off from previous studies of hydrogen storage in Ti-based QCs (see Sec. 5.4.1), on the basis of: (1) their large number of tetrahedrally coordinated interstitial sites (where hydrogen atoms can accommodate) and (2) the large chemical affinity of Ti atoms with hydrogen [860]. In this way, research on electrochemical hydrogen storage performance was undertaken by considering i-TiZrNiCu and i-TiZrNiPd QCs [21]. Unfortunately, QCs belonging to this alloy system render relatively small maximum discharge capacities, within the range $80-120$ mAh/g after 25 discharge cycles (a figure well below that of commercial graphite anodes reported value of 370 mAh/g), thereby promoting the exploration of alternative Ti-based QCs to this end.[13] Thus, i-$(Ti_{1.6}NiV_{0.4})_{100-x}Sc_x$ QCs were synthesized and their electrochemical hydrogenation properties studied. The

---

[12] Lead-based batteries, still used as a starter battery in combustion engine cars, are based on two lead electrodes, at least one of them partially oxidized to PbO_2, separated by a sulfuric acid-containing electrolyte.

[13] Improved electrochemical storage properties of composites containing TiZrNi QCs with Pd and multiwalled carbon nanotubes were subsequently reported to exhibit a maximum discharge capacity of about

discharge capacity was observed to slightly decrease from 272 to 240 mAh/g with increasing the Sc content from $x = 0.5$ to 6, which however trade off for enhanced cycling stability and self-discharge property [348]. On the basis of the obtained results, the effect of Li atoms infiltration due to electroosmosis on the electrochemical performance of electrodes consisting of i-Ti$_{1.4}$NiV$_{0.6}$ QCs was subsequently investigated in a three-electrode set-up at room temperature. As a result of the infiltration of Li atoms in the QC structure the discharge capacity of the i-Ti$_{1.4}$NiV$_{0.6}$-Li material was improved as compared to that observed in i-Ti$_{1.4}$NiV$_{0.6}$. Thus, a maximum discharge capacity of 307 mAh/g was recorded for i-Ti$_{1.4}$NiV$_{0.6}$-Li at a current density of 30 mA/g. In addition, both high-rate dischargeability and cycling stability were improved as a result of the Li atoms infiltration [666].

Spurred by these promising results, i-Al$_{63}$Cu$_{25}$Fe$_{12}$ QCs have been studied as a possible anode material for lithium-ion batteries as well. It was reported that these compounds can store lithium atoms reversibly, although irreversible processes also take place during the first discharge process. As a result, although the first specific discharge capacity was of about 204 mAh/g, this capacity eventually drops to about 65 mAh/g after 50 charge-discharge cycles. X-ray diffraction analysis demonstrated that, during discharging process Li ions enter into the QC structure to form a solid solution, and a portion of these ions cannot fully leave the QC during the charging process, which induces an irreversible capacity [471].[14] On the other hand, a high electrochemical stability and reactivity of nanoporous Fe$_3$O$_4$/CuO/Cu composites synthesized by dealloying i-Al$_{65}$Cu$_{23}$Fe$_{12}$ QCs was reported [508], and charge/discharge tests of graphite/i-AlCuFe QC composites prepared by mechanical alloying show a high-stability capacity of 480 mAh/g after 20 cycles, which is larger than the sum of capacities of graphite and i-AlCuFe QCs (65 mAh/g) separately [944].

5.4.4 Thermoelectric devices

Thermoelectric devices are compact (a few mm thick by a few cm square) solid-state devices used in small scale power generation and refrigeration applications, where a thermal gradient generates an electrical current flow or, alternatively, a DC current is applied to remove heat from the cold side, respectively. The efficiency of a thermoelectric device is measured in terms of the so-called thermoelectric figure of merit, which is defined by the expression

$$ZT = \frac{\sigma S^2 T}{\kappa_e + \kappa_{ph}}, \qquad (5.3)$$

where T is the temperature, $\sigma(T)$ is the electrical conductivity, $S(T)$ is the Seebeck coefficient and $\kappa_e(T)$ and $\kappa_{ph}(T)$ are the thermal conductivities due to the electrons and lattice phonons, respectively [556]. The quest for good thermoelectric materials then entails the search for solids *simultaneously* exhibiting very low thermal conductivity values and both high electrical conductivity and Seebeck coefficient values. The low thermal conductivity requirement naturally led to consider complex enough lattice structures, generally including the presence of relatively heavy atoms within the unit cell. Attaining large values of the electrical conductivity and Seebeck coefficient, on the other hand, usually requires a precise doping control, as well as an accurate tailoring of the sample's electronic structure close to the Fermi level.

---

180 mAh/g [992]. This value has been recently improved up to 230 mAh/g by covering i-Ti$_{49}$Zr$_{26}$Ni$_{25}$ QCs fabricated via mechanical alloying followed by annealing with a porous polyaniline coating [507].

[14]At this point, since the theoretical specific capacity of AlLi alloy is 993 mAh/g (which is about three times that of graphite anodes), I wonder on the possible use of i-AlCuLi QCs as potential cathode materials in a full-QC-based batteries design, where both electrodes in the cell were based on QC compounds.

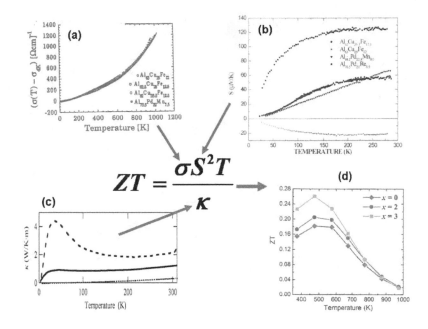

Figure 5.3 Collage picture illustrating the contribution of different transport coefficients (arrows) to the resulting ZT value: (a) temperature dependence of the electrical conductivity for i-AlCuFe and i-AlPdMn representatives up to 1000 K. (Reprinted figure with permission from Mayou D, Berger C, Cyrot-Lackmann F, Klein T, Lanco P, Phys. Rev. Lett. 1993, 70, 3915-3918. Copyright (1993) by the American Physical Society); (b) temperature dependence of the Seebeck coefficient for i-AlCuFe, i-AlPdMn, and i-AlPdRe samples (Courtesy of Roberto Escudero); (c) temperature dependence of the thermal conductivity for i-Al$_{74.6}$Re$_{17.4}$Si$_8$ (solid line) and its 1/1-cubic approximant (dashed line) (Courtesy of Tsunehiro Takeuchi); (d) ZT curves as a function of temperature for i-Al$_{71-x}$Ga$_x$Pd$_{20}$Mn$_9$ ($x = 0, 2, 3$) samples (Reprinted from Takagiwa Y, Kimura K, Sci. Technol. Adv. Mater. 2014, 15, 044802; doi:10.1088/1468-6996/15/4/044802, Creative Commons Atribution-NonCommercial-ShareAlike 3.0 License).

Quite interestingly, the temperature dependence of the transport coefficients appearing in Eq. (5.3) suggests that intermetallic QCs, exhibiting a more semiconductor-like than metallic character (see Chapter 3), may be considered for thermoelectric applications [553, 857]. In fact, we can list four basic features supporting iQCs (as well as their approximant phases) as potential thermoelectric materials:[562]

1. Their electrical conductivity steadily increases as the temperature increases up to the melting point (Fig. 5.3a).

2. Al-bearing iQCs show relatively large Seebeck coefficient values ($50-120\ \mu$VK^{-1}) as compared to those observed in usual metallic alloys ($1-10\ \mu$VK^{-1}) at room temperature, and the $S(T)$ curves usually deviate from the linear behavior characteristic of charge diffusion in ordinary metallic alloys in the temperature interval $100-300$ K (Fig. 5.3b).

TABLE 5.2 Room temperature values of the transport coefficients and figure of merit for iQCs belonging to different alloy systems. (a) After Ref. [200], (b) Estimated upper limit.

Sample	$\sigma(\Omega^{-1}\mathrm{cm}^{-1})$	$S\ (\mu\mathrm{V\ K}^{-1})$	$\kappa(\mathrm{Wm}^{-1}\mathrm{K}^{-1})$	ZT	Ref.
$\mathrm{Zn_{57}Mg_{34}Er_9}$	6170	+7	4.5	0.002	[285]
$\mathrm{Al_{65}Cu_{20}Ru_{15}}$	250	+27	1.8^b	0.003	[695]
$\mathrm{Cd_{84}Yb_{16}}$	5560	+14	9.4	0.004	[458]
$\mathrm{Ag_{42.5}In_{42.5}Yb_{15}}$	5140	+12	4.8	0.005	[459]
$\mathrm{Al_{62.5}Cu_{24.5}Fe_{13}}$	310	+44	1.8^a	0.010	[695]
$\mathrm{Al_{64}Cu_{20}Ru_{15}Si_1}$	390	+50	1.8^b	0.020	[695]
$\mathrm{Al_{71}Pd_{20}Re_9}$	450	+80	1.3	0.070	[855]
$\mathrm{Al_{71}Pd_{20}Mn_9}$	714	+90	1.5	0.120	[855]

3. In the case of i-AlCu(Fe,Ru,Os) and i-AlPd(Mn,Re) representatives, both $\sigma(T)$ and $S(T)$ increase as the temperature is increased over a broad temperature range ($T = 100-550$ K), which yields relatively high power factor (σS^2) values [706].

4. The thermal conductivity of most iQCs (within the range $\kappa = 1-5$ Wm^{-1}K^{-1} at room temperature) is about two orders of magnitude lower than that of ternary alloys over a wide temperature range (Fig. 5.3c). These unusually low thermal conductivity values are comparable to those observed for thermal insulators of extensive use in aeronautical industry [210].

The low thermal conductivity of QCs can be understood in terms of three main facts: (1) since the charge carrier concentration is severely reduced due to the presence of a pseudogap around the Fermi level (see Sec. 3.2.1), heat mainly propagates by means of phonons; (2) the high fragmentation of the frequency spectra in QCs leads to a small group velocity for most of these phonons (see Sec. 3.8); and (3) reciprocal space has a nearly fractal pattern in QCs, so that the transfer of momentum to the lattice is not bounded below, and the thermal current intensity is strongly reduced due to an enhancement of phonon–phonon scattering processes occurring at all scales in the reciprocal space. In a similar way, an increased number of umklapp electron—phonon processes in QCs may enhance the phonon drag contribution to the total Seebeck coefficient as well [147].

Inspired by all these features a series of experimental studies looking for promising iQCs as potential thermoelectric materials have been performed by several research groups [146, 458, 553, 705, 857]. To this end, the most promising QC alloy system was first determined by searching for those compounds exhibiting the lowest thermal conductivity value. Then, the high sensitivity of QCs transport coefficients to minor stoichiometric changes was exploited in order to properly enhance the power factor of the selected QCs, without seriously compromising their characteristic low thermal conductivities.

In Table 5.2, we list the transport coefficients for those iQC representatives yielding the best ZT values at room temperature. By inspecting this Table we see that isostructural i-Al$_{71}$Pd$_{20}$Re$_9$ and i-Al$_{71}$Pd$_{20}$Mn$_9$ samples exhibit promising ZT values, which are comparable to those reported for well-known thermoelectric materials, such as half-Heusler,[15]

---

[15]The term derives from the name of the German mining engineer and chemist Friedrich Heusler (1866–1947), who in 1903 studied such intermetallic compounds, which are ferromagnetic though their constituting elements are not.

skutterudites,[16] and clathrates[17] compounds at room temperature. Accordingly, the most promising QCs belong to the AlPd(Mn,Re) system. Subsequent search was then focused on refining their stoichiometries to further increase the corresponding power factor values. Indeed, ZT values differing by more than two orders of magnitude can be attained in a single QC system by slightly changing the sample's composition by a few atomic percent (hence preserving the QP lattice structure). We also note that both positive and negative values of the Seebeck coefficient can be obtained in this way, which allows for both the n- and p-type legs present in a typical thermoelectric cell to be fabricated from the same material. Furthermore, enhanced ZT values (0.2−0.25) are obtained at higher temperatures (450−500 K) for related quaternary iQCs, as it is shown in Fig. 5.3d for i-Al$_{71-x}$Ga$_x$Pd$_{20}$Mn$_9$ representatives[855]. The main reason for this ZT enhancement is the lowering of phonon thermal conductivity by the weakening of inter-cluster bonds, along with alloying effects [856].

At this point, however, we must highlight a main difference between QCs and other promising thermoelectric materials based on large unit cell inclusion compounds, such as Heusler phases, skutterudites, and clathrates, which exhibit a similar ZT value at room temperature [556]. This difference refers to the rapid increase of the thermal conductivity with the temperature in Al-based iQCs (and related approximants) starting at temperatures above $T \sim 300-400$ K [19, 863, 864], which currently hinders a competitive use of QCs in the thermoelectric industry. In fact, the $\kappa(T)$ increase leads to ZT curves which progressively decrease as the temperature is raised above the Debye temperature of the samples ($\Theta_D \approx$ 450−500 K, see Table 3.2), as can be seen in Fig. 5.3d. Conversely, the $ZT(T)$ curves of Heusler, skutterudites and clathrates inclusion compounds progressively *increase* within this temperature range, attaining peak ZT values within the interval 0.8−1.3 at temperatures around 800−900 K. In this regard, the recent obtention of dQCs with various compositions in the Ni-Mn-In alloy system, coexisting with an approximant phase and a Heusler compound, opens promising perspectives in the quest for QC based thermoelectric materials [490].

By all indications, the $\kappa(T)$ increase at temperatures above ~ 400 K is due to the charge carriers contribution $\kappa_e(T)$, while the phonon contribution $\kappa_{ph}(T)$ steadily decreases as T increases[18] [19, 66, 366, 549, 863, 864]. Within the framework of a two-band electronic structure model (see Sec. 3.6) the enhancement of the $\kappa_e(T)$ contribution in QCs may be accounted for in terms of a bipolar diffusion term [863], along with a possible dependence of the charge carrier density with the temperature at high enough temperatures [709]. Therefore, the high-temperature ZT value could be improved if the pseudogap width was widened

---

[16]The word "skutterudite" is derived from a town in Norway where minerals with this crystalline structure were first discovered. Binary skutterudites compounds crystallize in a body-centered-cubic (bcc) structure (space group $Im\bar{3}$) with $n_U = 32$ atoms in the unit cell, and obey the structural formula MX$_3$, where M is Fe or a Co group transition-metal and X is a pnicogen (group 15) atom. The skutterudite structure consists of six square planar rings formed by four pnicogen atoms each, with the rings oriented along the [100], [010], and [001] crystallographic directions, while the metal atoms complete a cubic lattice.

[17]In *ice clathrates* water molecules form a hydrogen-bonded framework where each water molecule is tetrahedrally bonded to four H$_2$O neighbors, such as in normal ice, but with a more open structure which forms different types of cavities that can enclose atoms or small molecules. Quite interestingly, the group 14 tetravalent atoms silicon, germanium, and tin also form similar clathrate structures, hosting alkaline atoms inside. In this case, the host lattice framework is based on strong covalent bonds with lengths comparable to those found in the diamond-like structures of these elements, and the resulting materials are referred to as *covalent clathrates*. Interestingly enough, a QP clathrate structural model for a 4 nm thick Sn layer grown atop the surface of an i-AlPdMn QC has been recently reported, indicating the existence of the first intrinsic − although apparently metastable − monoatomic realization of a QC [815].

[18]Quite remarkably, albeit the increase of the thermal conductivity of QCs with temperature reduces their potential as competitive thermoelectric materials in the high temperature regime, this unusual feature makes them attractive for the possible design of thermal rectifiers[864].

in order to reduce the unfavorable contribution related to bipolar transport effects. This inconvenient effect could also be alleviated by considering composite materials containing small sized QCs in order to reduce the thermal conductivity increase. In doing so, one will also benefit from a Seebeck coefficient value enhancement, arising from Al atoms loss during the calcination process applied during the composites preparation, thereby exploiting its extreme sensitivity to minor stoichiometric changes [350].

5.5 OPTICAL DEVICES

In this section, we will consider different ways of exploiting QPO in optical devices based on multilayered structures or aperiodic tilings of the plane. In fact, PQCs show up a number of characteristic properties that are not exhibited by their periodic counterparts (see Sec. 4.2). These properties can be broadly classified attending to (1) structural aspects, (2) the fractal nature of their frequency and Fourier spectra, and (3) the spatial distribution of their electromagnetic modes.

1. *Structural aspects*: A remarkable feature of multilayered PQCs is that they have less interfaces than periodic ones for a given system size, which is in general convenient to reduce losses and dispersion effects. This property is also useful to detect gases, liquids, or biological molecules by means of photonic multilayers based on porous silicon nanotechnology. In this case, the sensing mechanism relies on the refractive index change of porous silicon due to the partial substitution of the air in the pores on exposure to chemical substances. The resulting refraction index change is transduced, in turn, in a characteristic shift of the reflectivity spectrum. For the sake of comparison, the sensitivities of optical biochemical sensors, based on both periodic and Thue-Morse porous silicon multilayers, were measured. The obtained results clearly indicated that the aperiodic multilayer is more sensitive than the periodic one [624]. This improvement was traced back to both a higher filling capability (due to the presence of less interfaces in PQCs) and a greater number of narrower resonance transmittance peaks (due to the frequency spectrum high fragmentation in the aperiodic multilayers, see below), thereby increasing their spectral resolution. Another interesting structural aspect of PQCs is the presence of additional degrees of freedom that can be exploited in the design of photonic devices. These degrees of freedom are related to phasons, which can be readily visualized as distortions stemming from projection effects of periodic lattices described in a high-dimensional space (see Sec. 1.4.5).

2. *Fourier spectrum structure:* The presence of novel symmetries in PQCs allows for the existence of higher-order rotational axes, whereas the inflation symmetry gives rise to a denser Fourier spectrum pattern. The presence of higher rotational symmetries in PQCs leads to the formation of more isotropic and complete PBGs (i.e., for both TE and TM polarizations), by using lower refractive index materials, which is not possible with periodic photonic crystals. On the other hand, a denser Fourier spectrum results in a more plentiful spectrum structure than that of a periodic multilayer, thereby providing superior flexibility in achieving quasi-phase-matching conditions for frequency conversion processes in nonlinear optics, for instance (see Sec. 5.5.7) [802, 1001].

3. *Frequency spectrum structure:* The propagation of photons can be suppressed within a certain range of energies by the presence of PBGs in both photonic crystals and PQCs, but the number of tunable design parameters to this end is larger in QP structures. For example, the number of PBGs around the central frequency increases in

Thue-Morse multilayers according to the expression $N_n = 1 + (2^{n-1} \pm 1)/3$, where $n \geq 1$ is the generation order of the sequence, and the plus (minus) sign stands for even (odd) n values, respectively [714]. The availability of several frequencies for operation at the same time (multimode effect) becomes particularly useful for a number of applications such as optical spectroscopy and multimode biosensors [745], where the detection of multiple proteins is needed for unambiguous diagnosis, as well as in improved frequency selectivity in optical waveguides [516], which play a significant role in telecommunications and optoelectronics due to their ability to confine and guide light waves, keeping them spatially separated in coaxial QP designs [347, 391]. In fact, in contrast with conventional optical waveguides, based on the effect of the total internal reflection inside a high refraction index core, Bragg reflection based waveguides rely on their multilayered configuration, determined by a stack of alternating high- and low-index layers, leading to the formation of a series of PBGs [248]. On the other hand, the presence of more resonant frequencies due to multiple interference effects throughout the QP structure makes it possible, for instance, to fabricate multiwavelength narrow band optical filters of higher quality using PQCs, since their highly fragmented frequency spectra offer fuller transmission peaks than periodic ones in a given frequency range for a given system length. Vertical light extraction of a filtered color, working over all the visible spectral window, was demonstrated for both periodic and Thue-Morse based patterns of circular nanoparticles in 2D nanostructured devices, but in the case of Thue-Morse based devices, the extracted signals exhibited double energy and narrower full width at half maximum with respect to those of the periodic counterpart [742, 744]. On this basis, the enhanced omnidirectional light extraction of 2D PQC consisting of active nano-pillars arranged according to the golden angle Vogel spiral (see Sec. 1.3.1) has been proposed in the design of improved LEDs, whose main shortcoming is the low extraction efficiency associated to the trapping of radiation in the active device region [477]. In a similar vein, emission enhancement effects occurring at wavelengths corresponding to resonance states in light-emitting SiN/SiO_2 multilayered structures arranged according to the Thue-Morse sequence have been demonstrated as well [150].

4. *Nature of electromagnetic modes:* An important feature stemming from the presence of singular frequency spectra in PQCs is the existence of critical modes (see Sec. 2.5), which are spatially more complex than those related to both extended Bloch functions and exponentially localized defect modes in periodic and random photonic crystals, respectively. The presence of these critical states gives rise to the existence of either relatively localized or significantly extended electromagnetic intensity patterns depending on the considered frequency value. In this way, the existence of critical optical modes generally offers a higher design capability and tuning flexibility. Thus, on one hand, the occurrence of highly localized optical modes in the absence of disorder (i.e., in defect-free PQCs) is of interest for the fabrication of high-quality-factor resonators or coupled-resonator waveguides. On the other hand, compared with periodic arrays, where the resonance of a nanocavity can be achieved for a single frequency with a specific electric field pattern, in PQCs a *multifrequency* nanocavity array can exist, and for each frequency the electric field pattern has a different spatial behavior (see Fig. 4.8). Therefore, in contrast to periodic translation symmetry, spatial QPO naturally introduces a full set of non-equivalent sites in 2D and 3D PQCs. Consequently, a great diversity of electromagnetic spatial patterns, exhibiting both larger field enhancement and spatial density, can be obtained for different frequencies. This property paves the way for the design of useful platforms for multimode lasing [622, 787, 976], and

TABLE 5.3   Improved optical properties stemming from intrinsic QPO related properties, along with some resulting potential applications.

Intrinsic feature	Optical property	Application
Spectrum fragmentation	Multiple PBGs Very narrow peaks Size dependent transmission	Laser compression Multimode sensors Optical spectroscopy Multiband waveguides Polychromatic cavities Omnidirectional mirrors Photovoltaic solar cells Refractive index sensor Multiple narrow band filters
Critical localization	Faster emission rate Small group velocities Resonant standing waves Hierarchical EM patterns Light emission enhancement	Nonlinear optics Luminiscent devices Label-free biosensing Lasing pulse stretching Broadband plasmonics
Novel symmetries	More isotropic PBGs Denser reciprocal space Refractive index contrast Omnidirectional light extraction	LEDs technology Omnidirectional mirrors Omnidirectional mirrors Higher harmonic generation
Interfaces density Phason defects	Less losses and dispersion Selective defects	Optical sensing Guided resonances

multi-color optical sensing [485]. For example, the systematic use of critically localized modes has been proposed as a new approach for implementing label-free optical biosensing in aperiodic structures based on Rudin-Shapiro [86], and Thue-Morse multilayers [743], as well as dodecagonal 2D aperiodic tilings [745].

Other interesting design is provided by 2D aperiodic structures inspired by the Ammann-Beenker QC (see Sec. 4.3), since it lacks classically forbidden symmetries, and hence it belongs to a class of QCs which have spurred an increasing attention during the last decade [495, 229]. This PQC was generated by placing dielectric pillars at the maxima of a multi-beam interference pattern hologram and it demonstrated that light confinement can be induced into an air mirror-like cavity by the inherent symmetry of the spatial distribution of the dielectric scatterers forming the side walls of the open cavity, thereby introducing a new avenue for the design of photonic devices [1011].

These examples nicely illustrate the main point we will address through this section, namely, that carefully designed aperiodic structures can satisfactorily attain certain physical requirements necessary for the fabrication of improved devices of technological interest. For the sake of illustration, in Table 5.3, we summarize several advantages of QP designs over their periodic counterparts attending to the above described characteristic properties.

5.5.1 Photonic bandgap engineering

As we previously mentioned, the presence of adjustable multiple PBGs has been demonstrated in PQCs, both experimentally and by numerical calculations. The number of PBGs can be further increased by considering hybrid-order devices (see Sec. 4.2.1), which can be successfully used to construct broad omnidirectional reflectors by properly superposing PBGs extending over complementary spectral windows in the different units composing the hybrid-order PQC. Thus, it has been numerically shown that the omnidirectional reflection range of a FDM can be enlarged by combining the Fibonacci structure with a periodic one, so that the resulting modular structure exhibits a broader omnidirectional reflection region than that corresponding to each of the composing modules independently considered[457]. The main idea is that narrow gaps appearing in the highly fragmented QP units compensate for the relatively broad band regions appearing in the periodic ones. In this way, one can optimize engineered combinations of QP and periodic multilayered stacks in order to obtain not too thick photonic heterostructures (e.g., $[ABA]^4[BAABA]^2[BA]^7$), exhibiting large omnidirectional reflection [192].

Several methods to find optimal dielectric materials designs aimed at obtaining the widest PBG in 2D PQCs have been considered to date (see Sec. 4.3.1). For the case of TM polarization, nearly optimal decoration for a given QP pattern is obtained by placing identical optical elements centered at each QP lattice point and properly adjusting their size parameters in order to tune Mie and Bragg scattering effects, thereby leading to wide enough band gaps. On the other hand, to obtain optimal PBGs for TE polarization one must focus on tiles designs, which can be regarded as individual scattering objects supporting electromagnetic resonances, able to couple to each other when placed in the connected QP network [253, 733]. In this way, it has been reported that, at low dielectric contrasts, the higher symmetry quasicrystalline structures display greater band gaps than the crystalline ones, but, at higher contrasts, the hexagonal periodic structure yields the greater PBGs. Significantly, for all considered dielectric contrasts, optimized PQCs gaps are more isotropic than those of periodic structures due to the fact that their effective Brillouin zones are more circular than the Brillouin zones of periodic photonic crystals. This suggests to take a step further and investigate aperiodic structures exhibiting progressively higher-order rotational symmetries, going beyond 10- and 12-fold PQCs.

To this end, in the so-called inverse Fourier transform method one starts by choosing an n-sized polygon in the reciprocal space and subsequently one derives the optical elements positions required in real space. In this way, one can design PQCs with relatively wide PBGs possessing arbitrary rotational symmetries, such as those found in 18-, 40- and 120-fold QP lattices [484, 589]. Nonetheless, the best approximations to a "Bragg-ring" feature in reciprocal space has been obtained in spiral lattices based PQCs. By comparing the PBG corresponding to the spiral lattice shown in Fig. 1.16 with those corresponding to periodic hexagonal, on the one hand, and Stampfli dodecagonal PQCs (Fig. 4.7c), on the other hand, it was concluded that the spiral lattice design exhibits one of the widest complete TM gap ($\Delta\omega/\omega = 21\%$) for the relatively low contrast index value $\Delta n = 2$. In comparison the Stampfli lattice has $\Delta\omega/\omega = 14\%$ and the hexagonal lattice has $\Delta\omega/\omega = 5.6\%$ for the same Δn value. The sunflower's low-contrast wide PBG stems from long-range interactions allowing for the delocalized optical states to sampling the overall spiral structure, so that the resulting Bragg ring incorporates many reciprocal space contributions to further broaden the gap. In contrast, Bragg reflections from periodic crystals acts on a sparse pure-point Fourier space, thereby restricting the possible overlap of low-contrast spectral gaps [702].

Evolutionary optimization techniques, such as the particle-swarm optimization algorithm [410], have also been used in order to design plasmonic silver nanospheres arrays able to achieve broad-band field enhancement spanning the 400–900 nm spectral range [254]. The resulting optimized structure is aperiodic and exhibits a dense Fourier transform similar to those previously reported for deterministic aperiodic lattices with absolutely continuous Fourier spectra [87, 151].

5.5.2 Photovoltaic solar cells

A photovoltaic cell (PV) is a device that converts solar energy into electricity by means of the so-called photovoltaic effect. This effect consists in the generation of an electron–hole pair from the absorption of a photon with energy enough to promote an electron from the valence band to the conduction band in a semiconducting material. It is generally assumed that all incident photons with energy equal or greater than the band gap ε_g are absorbed, and they generate one electron–hole pair, whereas radiative recombination annihilates those pairs. The balance of electrons gives the current, which is delivered to an external load by the positive and negative contacts, yielding an electrical work qV. Non-radiative transitions (e.g., electron–phonon interactions) between any two bands are forbidden in ideal conditions.

The theoretical efficiency of ideal solar cells may be as high as 93%,[520] although in practice, it significantly depends on their electronic band structure, since the photon-induced electrical current originates from the difference between the rates of photon absorption and photon emission due to radiative recombination of the electron–hole pairs. Unfortunately, the band structure of typical single p–n junction solar cells is not well matched to the overall spectrum of solar radiation, and the ideal sunlight-to-electricity conversion efficiency of PV cells is consequently significantly reduced. Indeed, around 50% of the solar radiation provided to the Earth comes in form of infrared light, yet modern solar panels have been unable to efficiently convert this light into electric power. As a result, standard devices based on silicon p–n junctions yield a sunlight-to-electricity efficiency of just about 33%. This relatively low efficiency has spurred a flurry of research aimed at improving this figure, and numerous designs have been developed to enhance the performance of solar cells materials by properly *tailoring* their energy spectrum structure.

Earlier designs were based on periodic heterostructures, characterized by the existence of several minibands in the energy spectrum. In this way, the photovoltaic efficiency is enhanced due to the presence of a number of new possible excitation channels between the valence (conduction) bands and the intermediate bands, which allow for the absorption of additional solar energy coming from lower energy photons (Fig. 5.4a, where solid (dashed) arrows indicate absorption (recombination) processes respectively generating (annihilating) electron–hole pairs). As a consequence, an efficiency enhancement takes place due to additional photon-induced transitions between these intermediate levels. Thus, the photovoltaic efficiency limit progressively increases from 63% (for $\varepsilon_g \sim 2$ eV) for one intermediate band up to 76% (for $\varepsilon_g \sim 2.6$ eV) and 80% (for $\varepsilon_g \sim 2.3$ eV) for two and three intermediate bands, respectively (Fig. 5.4b). These results strongly suggest that the trench between the ideal thermodynamic limit and reported efficiencies in actual devices may gradually be filled by considering materials having highly fragmented energy spectra in photovoltaic cells. The quest for the optimum band structures to this end opens promising avenues in semiconductor based materials science engineering, which naturally leads one to the consideration of QP materials. In fact, one reasonably expects that devices based on aperiodic arrangements, including Fibonacci, Thue-Morse, or Cantor superlattices as representative examples, are a quite natural choice, for they properly combine the presence of highly fragmented miniband electronic structures (Fig. 5.4c), with a self-similar distribution of energy levels, hence further contributing to improve the photovoltaic efficiency by adding up the contributions

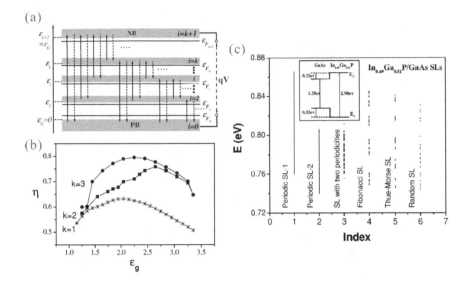

Figure 5.4 (a) Band diagram of solar cells based in a highly fragmented multiband spectrum structure. The model solar cell includes a negative-contact band (NB), a positive-contact band (PB), and a number of intermediate narrower bands. (b) Limiting efficiency for solar cells with one ($k = 1$), two ($k = 2$), and three ($k = 3$) intermediate bands as a function of the band gap ε_g. (c) Electronic miniband structures for several periodic and aperiodic In$_{0.49}$Ga$_{0.51}$P/GaAs superlattices whose band-edge diagram is shown in the inset. The structural parameters of the superlattices labeled by the index number read as follows: 1) $N = 21$, $a = b = 2.5$ nm, 2) $N = 21$, $a = b = 3.5$ nm, 3) is composed of two parts, one with $N = 10$, $a = b = 2.5$ nm, and the other with $N = 11$, $a = b = 3.5$ nm, 4) $N = 21$, $a = 2.5$ nm, $b = 3.5$ nm; 5) $N = 16$, $a = 2.5$ nm, $b = 3.5$ nm; 6) $N = 21$, $a = 2.5$ nm, $b = 3.5$ nm; where N is the number of layers, and a (b) are the thicknesses of blocks A (B). (Reprinted from Jing H, He J, Peng R W, Wang M, Aperiodic-order-induced multimode effects and their applications in optoelectronic devices, *Symmetry* **11** 1120 (12pp) (2019) doi:10.3390/sym11080988, Creative Commons Atribution-NonCommercial-ShareAlike 3.0 License).

coming from electronic transitions involving lower and lower energy scales. One reasonably expects this approach may produce high-performance PV devices based on electronic band gap engineering, and it could be used in the design of novel optoelectronic devices as well [346, 391, 680].

Thin film solar cells with thicknesses ranging from hundreds of nm to a few μm are attractive candidates for low-cost replacement of significantly thicker wafer-based devices, due to reduced costs of raw materials and processing. Light trapping photon management strategies rendering these cells optically thick enough, despite its limited physical thickness, is then crucial to attain high efficiency solar cells. To this end, the design of back-reflectors acting as diffraction gratings able to turn the incident light by an angle of 90° at specific frequencies, that can then be guided and thus absorbed along the plane of the cell, have proved very useful devices. In this regard, recent works have analyzed in detail the richness of QP geometries, demonstrating that a nano-grating grown according to the Fibonacci sequence used as back-reflector in a thin film solar cell guarantees higher absorption enhancement

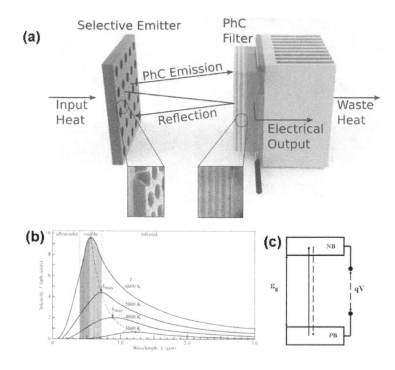

Figure 5.5 (a) Basic design of a second generation thermophotovoltaic cell for conversion of heat into electricity (Reprinted from Bermel P et al., Design and global optimization of high-efficiency thermophotovoltaic systems, Opt. Express 18 S3 A314–A334 (2010), Creative Commons Atribution-NonCommercial-Share Alike 3.0 License). (b) Spectrum and intensity of the thermal radiation emitted by a blackbody at thermal equilibrium (Planck's law). The peak intensity is given by the Wien's law. (c) Electronic structure of a simple PV cell characterized by a negative-contact band (NB) and a positive-contact band (PB) separated by the band gap ε_g.

with respect to the periodic counterpart, over the whole incidence angle range, providing a slightly better performance of about 5% for angles less than 20° and for angles greater than 45°[601]. This improvement stems from the denser Bragg spectrum of Fibonacci lattices, which enables a favorable phase match condition over more wavelengths, hence leading to a higher multiple resonance pattern for a given spectral range. This property also renders the QP back-reflector more robust with respect to the incoming light incidence angle variations, naturally arising from Sun's diurnal apparent motion.

5.5.3 Thermophotovoltaic devices

Thermophotovoltaic (TPV) devices transform heat from high-temperature sources, such as industrial furnaces used for oil refining or steal and glass making, as well as absorbers heated by concentrated sunlight (solar furnaces), into electricity via a two-step process, as is illustrated in Fig. 5.5a. Firstly, input heat generates thermal radiation at the surface of a selective emitter, generally a refractory material such as W, Cr, Ta, or Mo. Then, this thermal radiation is absorbed by a PV cell that converts the high enough energy photons into electron–hole pairs. These charge carriers are finally conducted to the metallic leads to produce an electrical current [42, 135]. Therefore, the bandgap width of the PV cell semiconducting material introduces a reference energy scale (usually within the 0.2−1.4 eV interval) in TPV devices design (Fig. 5.5c). This scale determines the energy range of suitable incoming thermal photons from the selective emitter. However, according to the Planck distribution law

$$\mathcal{P}(\lambda, T) = \frac{2hc^2}{\lambda^5} \frac{1}{\exp\left(\frac{hc}{\lambda k_B T}\right) - 1}, \tag{5.4}$$

which describes the spectrum and intensity of the thermal radiation (measured in Wm^{-3} units) emitted by an object at thermal equilibrium at temperature T (see Fig. 5.5b), where h is the Planck constant, c is the light velocity, and k_B is the Boltzmann constant, at working temperatures of interest (within the range 600−1600 K), most materials emit the vast majority of thermal photons with energies below the electronic bandgaps of typical PV cells (**Exercise 5.3**). These photons are thereby absorbed as waste heat, which substantially reduces the efficiency of TPV devices to figures close to just 1%.

Two main strategies have been addressed in order to circumvent this shortcoming in the design of second generation TPV devices. On the one hand, the emission spectra of terrestrial thermal sources can be tailored via optical engineering, so that the infrared photons can be returned to the emitter to be re-absorbed, keeping it at an elevated temperature, while the PV cell itself significantly reduces radiative overheating exposition. Thus, the recourse to photon recycling via reflection of low-energy photons with a thin multilayered mirror placed on top of the PV material (see encircled inset on the right in Fig. 5.5a) makes TPV devices an attractive platform for energy conversion that can potentially outperform solar PV cells (**Exercise 5.4**). In fact, theoretical TPV efficiencies are predicted to exceed 50% for an ideal p-n junction PV cell equipped with a perfect back-reflector when the blackbody emitter temperature reaches 1500 K [88]. In this context, it is worth noticing that the performance of periodic and Fibonacci metal nano-grating back-reflectors for solar light absorption in thin films (∼200 nm) amorphous silicon solar cell have been compared. The obtained results indicated that the Fibonacci grating exhibits a greater enhancement than the periodic one [601], hence opening a promising avenue in the design of improved TPV cells.

Another appealing approach, which also exploits the concept of spectral shaping, aims at: (1) directly suppress the emission of undesirable photons with energies below the PV cell bandgap coming from the selective emitter, and (2) to enhance the emission of useful photons with energies above the PV cell bandgap. A convenient way of getting this thermal emission control consists in coating the original emitter material with a properly designed multilayered thin film, in order to properly tailor the resulting thermal radiation spectrum of the composite structure (see Sec. 5.5.4) [141]. On the other hand, the photon density of states, and thus the thermal radiation spectrum, can be strongly modified by placing a 2D photonic crystal atop the original emitter in order to maximize the fraction of photons emitted within the useful energy range via PBG engineering [1000]. Alternatively, one may consider the very possibility of replacing the emitter in the basic TPV cell design by a 2D PQC with a tailored band gap structure (see Fig. 4.7) in order to improve the high energy photons emission. In addition, a properly designed 1D PQC coating may be included atop the PV cell to enhance the IR photons recycling (see Fig. 5.6). Finally, keeping in mind the results discussed in Sec. 5.5.2, one may further optimize the TPV output by considering a multiband PV material (see Fig. 5.4) instead of the simple PV band structure shown in Fig. 5.5c.

In fact, by properly combining both photon recycling and spectral shaping strategies, significantly higher efficiencies have been reported for micro-TPV reactors (26%) and solar thermal TPV devices (45%), respectively [52]. In doing so, one must simultaneously optimize the structural parameters of two separate optical structures, namely, the cylindrical hole depth, radius, and lattice parameter of the 2D photonic crystal atop the emitter, along with the refraction index values and widths of the layered infrared mirror on the PV cell. This task typically poses a so-called non-convex optimization problem, in which many local optima can exist, usually requiring carefully designed global optimization algorithms [52].

A simpler alternative approach is provided by a straightforward application of the so-called *harmony principle* in the design of optimized functional structures. This principle is inspired by the profusion of highly efficient designs found in nature under the driving force of genetic evolution, the overall arrangement of leaves and other botanical elements (phyllotaxis) in many plants being a very representative example [987]. When applied to design purposes harmony refers to combinations of parts to form an orderly whole. Accordingly, mathematical harmony describes the proportionality of the parts among themselves and with the whole [833]. In particular, the harmony principle can be readily applied in order to get natural generalizations of the golden ratio, as obtained from the division of a line at different points defining a series of segments, say x_j, such that all of them are related to each other according to a concatenated set of golden-like ratios. For instance, for a line which is divided into four of such segments of lengths $x_1 > x_2 > x_3 > x_4 \equiv 1$, we have

$$\frac{x_1 + x_2 + x_3 + 1}{x_1 + x_2 + x_3} = \frac{x_1 + x_2 + x_3}{x_1 + x_2}, \quad \frac{x_1 + x_2 + x_3}{x_1 + x_2} = \frac{x_1 + x_2}{x_1}, \quad \frac{x_1 + x_2}{x_1} = \frac{x_1}{1}. \quad (5.5)$$

The above set of nested ratios can be reduced to the algebraic equation $x^4 = x^3 + 1$, where we denote $x_1 \equiv x$. It can be iteratively shown that if the original line is divided into p segments following nested golden ratio-like relationships we obtain the general algebraic equation $x^p = x^{p-1} + 1$, which reduces to the well-known classical golden section for $p = 2$. For higher p values one gets the so-called generalized golden p-ratios, whose values are given by $\tau_{II} = 1.465\ldots$, $\tau_{III} = 1.380\ldots$, and $\tau_{IV} = 1.324\ldots$, for $p = 3$, 4, and 5, respectively [833]. As we see, one gets progressively smaller golden p-ratios as the p value is increased. It would be interesting to experimentally ascertain as to whether multilayered devices with a layer thickness distribution based on these generalized golden mean τ_p values may outperform (or not) the reflectance spectra reported in the literature for infrared mirrors of interest [557].

Such a QPO based designs can be readily extended to solar thermoelectric generators engineering as well. In this case, the key element is a thin film heat absorber which collects the solar radiation concentrated through a suitable lens. In order to optimize the device efficiency one requires a wavelength-selective surface absorptance for this film to precisely match the solar spectrum distribution in the near UV, visible and near IR domains. Whereas most experimental substrates considered up to now rely on specific materials, rather than engineered photonic crystals [169], one may expect significant improvement in the spectral response of absorbing films by using properly designed PQCs to this end.

5.5.4 Thermal emission control

During the last few years, the search for materials exhibiting quite specific thermal emission properties has been spurred by the necessity of attaining high-performance devices for light harvesting and thermal energy control. With this goal in mind, devices displaying either very low or very high emissivity values over relatively broad frequency windows are usually required. For instance, efficient daytime radiative coolers (see Sec. 5.5.5) must radiate heat to outer space through a transparency window in the Earth's atmosphere ranging between

8 and 13 μm, but they should also reflect most incoming solar radiation within the UV and visible wavelengths [723]. Now, even in the case of highly absorptive compounds, not all incident radiation is usually absorbed by a given material. This is because the material reflects some radiation at its surface. According to the Kirchoff's second law, the ratio of the thermal emittance to the absorptance is a constant, which equals unity only when the material is a perfect blackbody. Therefore, surfaces of real materials behave as the so-called gray-body when heated, exhibiting emittance values lower than unity, reduced by a factor $\epsilon(\lambda, T)$, known as spectral emissivity.

An ingenious way to circumvent this shortcoming, enhancing the thermal emissivity of a given material above that corresponding to its gray-body value *at certain* frequencies consists in coating a given bulk material with a properly designed multilayered thin film, in order to exploit the beneficial optical properties contributed by each component in a synergetic way. Within this approach the coat can act either as a passive filter (when the refractive indices of the layers material are all real valued) or as an active emitter/absorber element (when at least one of the layers material in the film has a complex valued refraction index). In both cases, the resulting thermal radiation spectrum of the composite structure can be substantially modified as compared to that corresponding to the original substrate alone, due to resonance effects within the multilayered coat [141]. These devices were early studied by considering a periodic sequence of layers in the thin film coat, and later on aperiodic designs were progressively considered [154, 152, 544, 548, 552, 554].

Within the framework of the transfer matrix technique the thermal optical power is given by $\mathcal{E}(\omega) = 1 - R(\omega) - T(\omega)$, where $R(\omega)$ and $T(\omega)$ are the reflection and transmission coefficients, respectively. This magnitude measures the ratio of the optical power emitted at frequency ω into a spherical angle element $d\Omega$ by a unit surface area of the thin film, to the power emitted by a blackbody with the same area at the same temperature. In this way, the power spectrum of a multilayer located in front of an emitting hot surface is given by the expression $\tilde{\mathcal{P}}(\omega, T) = \mathcal{E}(\omega)\mathcal{P}(\omega, T)$, where $\mathcal{P}(\omega, T)$ is the blackbody Planck's law [141]. Some illustrative examples are shown in Fig. 5.6. Let us first consider the case where a periodic film coating sits atop the heated substrate (Fig. 5.6Ia). As we see, the film significantly blocks heat radiation emitted by the substrate at the frequencies corresponding to the photonic crystal bandgap ($\omega/\omega_0 = 1$) as expected, but we also observe that the substrate's emission is enhanced from the gray-body level all the way up to the perfect blackbody rate at a number of frequencies corresponding to the pass-band transmission resonances of the multilayered film. This occurs because the thin film acts as an antireflective coating at these resonances. In this way, all the incident radiation tunnels through the multilayer structure into the substrate for these selected frequencies, so that the substrate effectively behaves as a perfectly absorbing blackbody in that case [170]. A similar enhancement of the substrate's thermal emittance at certain resonance frequencies, accompanied by the corresponding inhibition at the stop-bands, is observed in the aperiodically arranged thin film coatings as well. In the case of the Fibonacci coating (Fig. 5.6Ib) a characteristic trifurcation splitting can be clearly appreciated around a number of frequencies, and one finds a strong emittance within the spectral range corresponding to the midgap in the periodic case. An analogous pattern is observed in the thermal spectrum corresponding to the Thue-Morse thin film (Fig. 5.6Ic), whereas that corresponding to the period-doubling sequence (Fig. 5.6Id) is more similar to the periodic one around the $\omega/\omega_0 = 1$ spectral range. This feature illustrates the fact that the period-doubling sequence describes a limit-periodic crystal, whereas both Fibonacci and Thue-Morse multilayers can be regarded as QCs (see Sec. 2.2). In all the aperiodic arrangements one has a rich thermal emission spectrum, arising from the highly fragmented nature of their transmission profiles. Quite interestingly, these spiky thermal

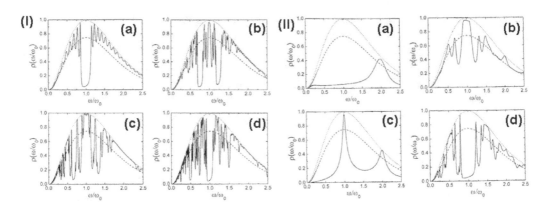

Figure 5.6 (I) Thermal radiation spectra for multilayered thin film coat as a function of the reduced frequency ($\lambda_0 = 700$ nm) under normal incidence conditions, with $n_A = 1.45$ (SiO$_2$) and $n_B = 1.0 + 0.01i$, where the layers are respectively arranged according to the following sequences (a) periodic, (b) Fibonacci ($N = 377$), (c) Thue-Morse ($N = 512$), and (d) period-doubling ($N = 512$). The perfect blackbody thermal spectrum is given by the dotted curve, whereas the dashed curve gives the thermal spectrum for the substrate, with refractive index $n_C = 3 + 0.03i$. The temperature is chosen so that the blackbody Wien's peak is aligned with the midgap frequency $\omega_0 = 2\pi c/\lambda_0$. All the curves are properly normalized by this peak power. (II) The same as (I) but considering that B layers are composed of a metamaterial with refraction index $n_B = -1.0 + 0.01i$. (Adapted from Ref. [170]. Courtesy of Eudenilson L. Albuquerque).

emission spectra can be substantially smoothed (hence obtaining broader spectral ranges with enhanced emittance) by using metamaterials in the composition of the aperiodic multilayer coats, as it is shown in Fig. 5.6II. On the other hand, it has been reported that the inclusion of graphene layers at the interfaces of successive dielectric layers, while preserving the general features of the thermal emittance spectra described above for the different kinds of aperiodic structures (i.e., Fibonacci, Thue-Morse, and period-doubling), also induces the presence of a characteristic photonic gap located at the $\omega/\omega_0 = 2$ relative frequency due to a specific resonance effect [143].

5.5.5 Daytime radiative coolers

A passive cooler is a device which decreases its temperature below that of the ambient air without any electricity (or any other form of energy) input. To achieve this goal the emitter's surface needs to be engineered to show high emissivity only in the atmospheric windows ranging between 8 and 13 μm to radiate heat to outer space, along with high reflectance outside these windows, to avoid absorption of both sunlight and thermal radiation from the atmosphere. In this way, the energy budget of the emitter becomes negative and its temperature cools down. Passive cooling was first demonstrated at night [739]. Daytime radiative cooling under direct sunlight was subsequently shown by properly extending the same fundamental technique using a photonic radiative cooler. To this end, a thin film multilayer composed of seven alternating layers of HfO$_2$ (high refraction index) and SiO$_2$ (low refraction index) of varying thicknesses were deposited on top of a 200 nm silver

TABLE 5.4   The second column lists the optimized thicknesses of the layers composing the solar reflector component in the radiative cooler studied in Ref. [558]. Note that their respective ratios (third column) can be expressed in terms of successive powers of the golden mean (fourth column).

n	d_n (nm)	$\frac{d_n}{d_1}$	τ^n
1	13	1.00	$\tau = 1.618\ldots$
2	34	2.615	$\tau^2 = 2.618\ldots$
3	54	4.154	$\tau^3 = 4.236\ldots$
4	73	5.615	$\tau^5/2 = 5.545\ldots$

substrate. The resulting coating was in turn fixed on top of a 750 μm thick, 200 mm diameter silicon wafer. The bottom four layers in the coating film have thicknesses within the range 10$-$70 nm, and assist in optimizing solar reflection. The top three layers are one order of magnitude thicker (within the range 200$-$700 nm) and are primarily responsible for thermal radiation from the cooler device. Thus, it is the presence of two different kinds of layer sizes which ultimately accomplishes the two functional tasks required for the coating, namely, high solar reflectance (due to the shorter layers stack) and high infrared emissivity within the 8$-$13 μm window (due to the longer layers stack). Both the total number and the proper thickness values of the alternating low/high refraction index layers was determined by using a numerical optimization method. Making use of this device a cooling of about 5°C below ambient air temperature, amounting to 40 Wm^{-2} cooling power, was reported under direct sunlight [723, 739].

These promising figures were obtained by properly combining suitable material properties and interference effects in an integrated structure that collectively achieves high solar reflectance (97% at normal incidence) and strong thermal emission in the selected infrared frequency window. Nevertheless, some room is still left for further improvement (say, full solar reflectance along with perfect transmission through the Earth's atmosphere infrared windows), and one may wonder if these goals may be achieved by considering alternative layer sequence orderings for the different layers. Indeed, although previously unnoticed, the fingerprints of the golden mean sequence can be found in the reported optimized structure for the daytime radiative passive cooler described above.

To see this, in Table 5.4 we explicitly list the optimized thicknesses of the four layers comprising the solar reflector component. Albeit these layers are deposited according to the periodic sequence $ABAB$ (where A stands for HfO$_2$ and B for SiO$_2$), we note that their thicknesses take on different values. Quite interestingly, we realize that three d_n values are Fibonacci numbers (allowing for the experimental ±1 nm tolerance), so that their respective ratios d_n/d_1 reasonably coincide with successive golden ratio powers, as it is prescribed by Eq. (1.3), and the 4th layer thickness is very close to the average value $(55+89)/2 = 144/2 = 72$, which involves two consecutive Fibonacci numbers. Thus, the thicknesses distribution obtained from the optimization algorithm approximately satisfies the harmony principle introduced in Sec. 5.5.3 in a natural way. By inspecting Table 5.4, it is appealing to consider the possible outcome of a multilayer where the thicker layer is replaced by one with $d_4 = 89$ nm, so as to fit the ratio $d_4/d_1 \simeq \tau^4$ [557].

The possibility of attaining a broadband reflectance, covering a nearly full-visible spectral window, has been recently demonstrated in a TiO$_2$ based photonic crystal whose lattice constant is gradually increased according to an arithmetic series of the form $d_\ell = d_0 + (\ell - 1)d$, $\ell = 1, \ldots N$, where d_0 is the thickness of the first layer and d is the common difference. In this way, a flattened reflectance peak reaching 100% intensity within the range $400 - 600$ nm was numerically obtained by properly assembling two different slabs containing $N = 8$ layers (with $d = 3.6$ nm, and $d_0 = 165.2$ nm or $d'_0 = 194$ nm, respectively) to form two $N = 16$ bilayers, which were in turn stacked to each other to get a $N = 32$ multilayer with a total length $L = 819.2$ nm [315]. Note that though the total length of this graded multilayer reflector amounts to about four times that corresponding to the solar reflector component used in the daytime cooler device discussed above (i.e., $L = 174$ nm) [739], its characteristic modular design can readily explain its ability to ultimately outperform the latter one. In fact, this property can be readily understood by regarding the overall structure as a modular stack of shorter photonic crystals with different lattice parameters, each one reflecting light with a peak located at its respective Bragg resonance condition. Thus, by properly selecting the location and width of each module main reflectance peaks, one may design a much broader, almost flat, reflectance spectrum profile, naturally stemming from superimposing each module's spectra.

Further elaborating on this physical picture it was numerically shown that an even wider full reflectance band (ranging from $380 - 780$ nm) could be obtained by considering hierarchical photonic crystals consisting in geometrically distributed layers, where the aperiodic order of the $n(z)$ profile is given in terms of a geometric series rather than an arithmetic one [7]. The influence of the angle of incidence and light polarization was carefully analyzed, and the refraction index dispersion effects in the layers was explicitly taken into account, hence providing a quite realistic description. Geometric series naturally introduce a scale factor in the resulting refraction index profile, a feature they share with fractal structures characterized by self-similar symmetry. Accordingly, it is reasonable to expect that additional improvement in both solar reflectance and infrared thermal emission properties could be attained in radiative cooling devices able to exploit the additional design degree of freedom provided by the scalability property characteristic of fractal and QPO based multilayer geometries [274, 1003].

One may also consider QP structures combining self-similar features typical of PQCs with graded refraction index profiles. A suitable example includes two kinds of layers: one has a constant refraction index (n_A) and the other has a linearly varying refraction index given by the expression, $n_B = n_A + (n_B - n_A)z/d_B$, where d_B is the B layer thickness [812]. Of course, instead of a simple linear relation one may consider an exponential one, or any other refraction index gradation laws at will [813]. In turn, layers A and B could be arranged according to a sequence prescribed by a given substitution rule (i.e., Fibonacci, Thue-Morse, period-doubling) in order to fully exploit QPO related freedom degrees in these modular designs.

5.5.6 Fibonacci lenses

In photonics technology, the so-called diffractive optical elements have found a large number of new applications in many different areas, covering the whole electromagnetic spectrum from X-ray microscopy to THz imaging. In particular, diffractive lenses, such as conventional Fresnel zone plates, are essential in many focusing and image forming systems, albeit they have some inherent limitations. Fresnel zone plates comprise a series of concentric circular pupils, alternating transparent (say A) and opaque (say B) zones, which can be arranged according to either periodic or aperiodic (Cantor, Fibonacci, Thue-Morse) sequences (Fig. 5.7a). In the later case, one obtains more than one main focal point. For

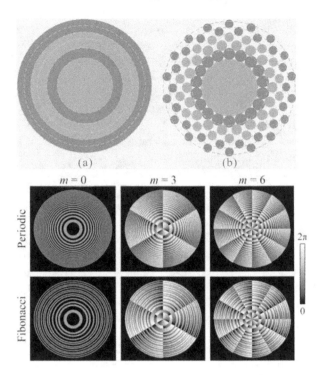

Figure 5.7 (a) A QP zone plate, where the zones are arranged according to the generalized Fibonacci sequence with three transparent zones (dark gray), and three opaque zones (light gray). (b) A phase-only generalized Fibonacci photon sieve. (Reprinted from Optics Communications 368, Ke J, Zhang J, Focusing properties of phase-only generalized Fibonacci photon sieves, 34–38, Copyright (2016), with permission from Elsevier). Phase distribution of vortex zonal plates based on a 8th generation Fibonacci sequence (bottom panels) and that of equivalent periodic lenses with the same number of zones (middle panels). (Reprinted from Calatayud A, Ferrando V, Remón L, Furlan W D, Monsoriu J A, Twin axial vortices generated by Fibonacci lenses, Optics Express 21, 10234-10239 (2013) Creative Commons Atribution-NonCommercial-ShareAlike 3.0).

instance, in Fibonacci lenses one has two twin (i.e., with the same intensity) foci along the optical axis: one it is located in front, the other one behind, the focus of an equivalent periodic Fresnel zone plate with the same number of circular zones. In addition, the axial positions of these foci are given by the Fibonacci numbers and the ratio of the two focal distances equals the golden ratio [247, 615]. In a similar way, Thue-Morse zone plates produce a pair of self-similar and equal-intensity foci, which have low chromatic aberration. Extensions of this basic design, encompassing a periodic concatenation of short order aperiodic sequences give rise to similar results, exhibiting two tighter self-similar and much higher intensity foci, and providing more design parameters for focusing optimization purposes. [966]

In order to overcome the disadvantages of traditional zone plates, the very concept of photon sieve, which is a Fresnel zone plate with transparent zones replaced by a great number of completely separated pinholes (Fig. 5.7b), was introduced and applied to the study of both periodic and aperiodic designs [408].

Diffractive optical elements have also been designed to fabricate the so-called optical vortices, of high value as optical traps because, in addition to trap microparticles, they are able to set these particles into rotation due to their inherent orbital angular momentum. In this regard, vortex lenses able to simultaneously generate two optical vortices along the axial coordinate whose diameters are related by the golden mean have been demonstrated. This sort of Fibonacci vortex lattice is defined by a zone phase distribution given by $\Phi(r, \theta) = \mathrm{mod}_{2\pi}(m\theta + \Phi_j(r))$, where $m \in \mathbb{N}$, θ is the azimuthal angle about the optical axis at the pupil plane, and the radial phase distribution $\Phi_j(r)$ is defined in the domain $[0, 1]$, and this interval is partitioned in F_{j+1} intervals of length $d_j = 1/F_{j+1}$. Then, $\Phi_j(r)$ can take on two different values (0 or π) in the considered interval depending on the zone being labeled as A or B in the corresponding Fibonacci sequence. Typical examples of Fibonacci vortex lenses for different values of the parameter m are illustrated in Fig. 5.7, where they are also compared with periodic counterparts [106]. A modified precious mean based zone plate has been recently proposed to generate a twin equal-intensity foci which have the same resolution [967]. On the other hand, the possibility of designing flat lenses by using QP arrays of steel cylinders embedded in an air background has been theoretically discussed. In this case, the motivation for using the high-symmetry PQC is to maintain an efficient interference of waves (long-range order), while reducing the orientational order of the system (crystallographical restriction theorem is relaxed) to get a more isotropic propagation [997].

5.5.7 High-harmonics generation

Clear experimental results indicating certain advantages of QP systems over periodic ones were early reported in the nonlinear optics field, where it was shown that second harmonic generation processes were more efficient in QP multilayers, due to their richer Fourier spectrum structure. In fact, due to their higher space group symmetry, QP multilayers can provide more reciprocal vectors to the so-called quasi-phase-matching optical process, and this ultimately results in a more plentiful spectrum structure than that of a periodic multilayer [802, 1001]. The importance of the role played by the quasiperiodicity of the substrate is further highlighted when considering third harmonic generation processes, where it has been shown that the conversion efficiency in a QP multilayer is increased by a factor of eight in comparison with the two-step process required for a usual third-harmonic generator based on periodic systems [128, 668, 1002]. In a similar way, emission enhancement effects occurring at wavelengths corresponding to resonance states in light-emitting SiN/SiO_2 multilayered structures arranged according to the Thue-Morse sequence have been demonstrated as well [150]. Recently, the second-harmonic enhancement efficiency in FDM containing ZnS, SiO_2, and MoS_2 layers has been theoretically studied, confirming that the efficiency of FDMs is increased with respect to periodic multilayers with the same number of MoS_2 layers [24].

The nonlinear properties of optical heterostructures can also be used to fabricate a compact-sized compressor for laser pulse. This compression is physically determined by the group velocity dispersion in the material, so one can expect that by adding more layers to a periodic multilayer one should obtain narrower optical bands and the compression effect will be increased. However, this is inevitably accompanied by an increase of the total thickness of the structure, which is undesirable. In this context, the recourse to QP structures, exhibiting a significantly larger fragmentation of their optical spectrum for similar system sizes, appears as a natural choice. Inspired by this principle, the laser compression performance of both periodic and FDMs made of high-index ZnS and low-index Na_3AlF_6 layers was experimentally compared. As expected, the Fibonacci based structure exhibits a compression enhancement due to its larger group velocity dispersion [576]. Recently, the strong nonlinear response of graphene layers inserted in a FDM has been used to analyze the bistability[19] and multistability properties in the THz frequency range by exploiting the rich variety of resonances provided by the underlying QPO [635, 926, 998].

5.6   EXERCISES WITH SOLUTIONS

5.1 *Making use of the contact angle values given in Fig. 5.2d and Fig. 5.2g, determine the adhesion energy of the studied i-AlCuFe based coating (a) after annealing; (b) after polishing the annealed coating.*

As a first approximation the adhesion energy of a liquid on a solid surface is given by the relationship [46],

$$W_L = (1 + \cos\theta)\gamma_L, \tag{5.6}$$

where θ is the contact angle and γ_L is the surface tension of the liquid. Plugging the contact angle values given in Fig. 5.2d and Fig. 5.2g, respectively, and adopting the value $\gamma_L = 75$ mJm^{-2} for water, we get:

(a) $W_L = (1 + \cos\frac{3\pi}{4}) \times 75 = 21.97$ mJm^{-2},

for the i-AlCuFe based coating after annealing, and

(b) $W_L = (1 + \cos\frac{23\pi}{36}) \times 75 = 43.30$ mJm^{-2},

for the annealed coating after polishing the surface. Thus, the obtained adhesion energy with water is always below that of Teflon (44 mJm^{-2}) and just one-third of that of window glass under realistic working conditions.

5.2 *(a) Obtain Eq. (5.2) making use of Eq. (5.1). (b) Determine the temperature increase of an aluminium grain with $\alpha = 0.14$, $\rho_m = 2690$ kgm^{-3}, $C_p = 895$ JK^{-1}kg^{-1}. (c) Determine the temperature increase of a quasicrystalline grain in the powder-polymer composites described in the text by taking into account radiative losses.*

(a) Neglecting both convective and radiative losses terms in Eq. (5.1) one gets the first order differential equation

$$\frac{dT}{dt} = \frac{6\alpha P_L}{\pi \rho_m D D_L^2 C_P} \equiv A, \tag{5.7}$$

which can be readily integrated to obtain

$$T - T_b \equiv \Delta T_0 = A\Delta t = \frac{6\alpha P_L}{\pi \rho_m D D_L C_P v}. \tag{5.8}$$

(b) Plugging into Eq. (5.8) the physical parameters for aluminum grains given above, along with the laser experimental data given in the main text, we have

$$\Delta T = \frac{6 \times 0.14 \times 5 \times 10^4 \text{ J s}^{-1}}{2690 \text{ kg m}^{-3} \times 75 \times 10^{-6} \text{ m} \times 10^{-3} \text{ m} \times 895 \text{ JK}^{-1}\text{kg}^{-1} \times 10^4 \text{ ms}^{-1}\pi} = 7.4 \text{ K},$$

which is significantly lower than the value obtained when using i-AlCuFe QC grains, namely

$$\Delta T = \frac{6 \times 0.9 \times 5 \times 10^4 \text{ J s}^{-1}}{4370 \text{ kg m}^{-3} \times 75 \times 10^{-6} \text{ m} \times 10^{-3} \text{ m} \times 581 \text{ JK}^{-1}\text{kg}^{-1} \times 10^4 \text{ ms}^{-1}\pi} = 45.1 \text{ K}.$$

---

[19]Optical bistability is a nonlinear phenomenon in which the transmission characteristics of a system depend on the input light intensity, so that an input intensity corresponds to two stable resonant outputs.

(c) If we neglect the convective term, Eq. (5.1) can be written in the form

$$\frac{dT}{dt} = A - \frac{6\sigma\epsilon}{\rho_m DC_P}T^4 \equiv A - BT^4, \qquad (5.9)$$

where we have used Eq. (5.7) and introduced the constant parameter B. This first order differential equation can be solved separating variables to get

$$\Delta t = \int_{T_b}^{T} \frac{dT'}{A - BT'^4} = B^{-1}\int_{T_b}^{T} \frac{dT'}{C^4 - T'^4}, \qquad (5.10)$$

where

$$C^4 \equiv \frac{A}{B} = \frac{\alpha P_L}{\epsilon\sigma\pi D_L^2}. \qquad (5.11)$$

Making use of the integral

$$\int \frac{dx}{a^4 - x^4} = \frac{1}{2a^3}\tan^{-1}\frac{x}{a} + \frac{1}{4a^3}\ln\left(\frac{x+a}{x-a}\right),$$

Eq. (5.10) can be expressed as

$$2C^3 B\Delta t = \tan^{-1}\left(\frac{C\Delta T}{TT_b + C^2}\right) + \ln\sqrt{\frac{TT_b - C^2 - C\Delta T}{TT_b - C^2 + C\Delta T}}, \qquad (5.12)$$

where we have used the relationship $\tan^{-1} A - \tan^{-1} B = \tan^{-1}\frac{A-B}{AB+1}$. Plugging the physical parameters given in the text, along with $\epsilon = 0.05$ (see Sec. 5.2.3), and $\sigma = 5.67 \times 10^{-8}\,\mathrm{W\,m^{-2}\,K^{-4}}$, into Eq. (5.11) we obtain

$$C^2 = \sqrt{\frac{\alpha P_L}{\sigma\epsilon\pi D_L^2}} = \sqrt{\frac{0.9 \times 5 \times 10^4\,\mathrm{W}}{0.05\pi \times 5.67 \times 10^{-8}\,\mathrm{W\,m^{-2}\,K^{-4}} \times 10^{-6}\,\mathrm{m^2}}} = 2.25 \times 10^9\,\mathrm{K^2},$$

which is significantly larger than $TT_b = 500 \times 440 = 2.2 \times 10^5\,\mathrm{K^2}$, in the experimental range of temperatures considered [411]. Therefore, $C^2 \gg TT_b$, and Eq. (5.12) can be approximated as[20]

$$2C^3 B\Delta t \simeq \tan^{-1}\left(\frac{\Delta T}{C}\right) + \frac{1}{2}\ln\left(\frac{1 + \frac{\Delta T}{C}}{1 - \frac{\Delta T}{C}}\right) \simeq 2\frac{\Delta T}{C} + \frac{2}{5}\left(\frac{\Delta T}{C}\right)^5, \qquad (5.13)$$

which, making use of Eqs. (5.8) and (5.11), can be rearranged in the form

$$\Delta T^5 + 5C^4\Delta T - 5C^4\Delta T_0 = 0. \qquad (5.14)$$

Finally, solving numerically the quintic Eq. (5.8) we obtain $\Delta T = 44.30$ K, indicating that the radiative losses effects are below 2%.

5.3 (a) *Determine the peak emission wavelength, λ_m, of a blackbody at equilibrium temperature T, and calculate $\lambda_m(600\,\mathrm{K})$ and $\lambda_m(1600\,\mathrm{K})$. (b) The melting temperatures of tantalum hafnium carbide (Ta_4HfC_5) and tungsten refractory compounds are $T_m(\mathrm{THC}) =*

---

[20]We use the Taylor series $\arctan x = x - \frac{1}{3}x^3 + \frac{1}{5}x^5 + O\left(x^7\right)$ and $\ln\left(\frac{1+x}{1-x}\right) = 2x + \frac{2}{3}x^3 + \frac{2}{5}x^5 + O\left(x^6\right)$.

4488 K and $T_m(\text{W}) = 3695$ K, *respectively. Obtain their corresponding $\lambda_m(T_m)$ values and determine if photons with these wavelength values have enough energy to activate a PV cell with a band gap width of $\varepsilon_g = 1.5$ eV.*

(a) Introducing the reduced energy $x \equiv \frac{hc}{\lambda k_B T}$, Eq. (5.4) can be expressed in the form $P(x) = Ax^5(e^x - 1)^{-1}$, where $A \equiv \frac{2(k_B T)^5}{h^4 c^3}$. Imposing the extreme condition $\frac{dP}{dx} = 0$, and taking into account that $x \neq 0$, we get the equation

$$\frac{xe^x}{e^x - 1} = 5,$$

which can be numerically solved to obtain $x_m \simeq 4.9651 \equiv \frac{hc}{k_B T \lambda_m}$. Therefore,

$$\lambda_m T \simeq 4.9651^{-1}\frac{hc}{k_B} = \frac{6.626 \times 10^{-34}\,\text{J s} \times 2.998 \times 10^8\,\text{m s}^{-1}}{4.9651 \times 1.381 \times 10^{-23}\,\text{J K}^{-1}} = 2.897 \times 10^{-3}\,\text{K m} \quad (5.15)$$

This expression describes the so-called Wien's law. Making use of Eq. (5.15), we get

$$\lambda_m(600\ \text{K}) = \frac{2.897 \times 10^{-3}\,\text{K m}}{600\ \text{K}} = 4.83 \times 10^{-6}\,\text{m}$$

$$\lambda_m(1600\ \text{K}) = \frac{2.897 \times 10^{-3}\,\text{K m}}{1600\ \text{K}} = 1.81 \times 10^{-6}\,\text{m}$$

so that the maximum emission takes place in the infrared window of the electromagnetic spectrum.

(b) Plugging the indicated melting temperatures into Eq. (5.15), we obtain

$$\lambda_m(\text{THC}) = \frac{2.897 \times 10^{-3}\,\text{K m}}{4488\ \text{K}} = 6.455 \times 10^{-7}\,\text{m}$$

$$\lambda_m(\text{W}) = \frac{2.897 \times 10^{-3}\,\text{K m}}{3695\ \text{K}} = 7.840 \times 10^{-7}\,\text{m}$$

and the related photons energies are respectively given by

$$E_{\text{THC}} = \frac{hc}{\lambda_m(\text{THC})} = \frac{6.626 \times 10^{-34}\,\text{J s} \times 2.998 \times 10^8\,\text{m s}^{-1}}{6.455 \times 10^{-7}\,\text{m}}\frac{10^{19}\ \text{eV}}{1.602\ \text{J}} = 1.92\ \text{eV}$$

$$E_{\text{W}} = \frac{hc}{\lambda_m(\text{W})} = \frac{6.626 \times 10^{-34}\,\text{J s} \times 2.998 \times 10^8\,\text{m s}^{-1}}{7.84 \times 10^{-7}\,\text{m}}\frac{10^{19}\ \text{eV}}{1.602\ \text{J}} = 1.58\ \text{eV}$$

Accordingly, the photons corresponding to the emissivity peak have enough energy to activate the TPV cell in both cases.

5.4 *An electron is trapped in a quantum dot modeled in terms of a one-dimensional square well of width $a = 1.5$ nm and infinite walls. (a) Determine the wavelength of the photon required to excite the electron from the fundamental level up to the fourth energy level. (b) Determine the temperature of a blackbody whose Wien peak equals the wavelength of the photon obtained in (a).*

(a) The energy spectrum for a 1D quantum well with infinite walls is given by:

$$E_n = \frac{h^2 n^2}{8 m_e a^2}, \quad n = 1, 2, \ldots$$

where h is the Planck constant, m_e is the electron mass, and a is the well width. The wavelength of the photon required to excite the electron from the fundamental level $(n = 1)$ up to the fourth energy level $(n = 4)$ is:

$$\lambda = \frac{hc}{E_4 - E_1} = \frac{8m_e c a^2}{15h}$$

$$= \frac{8 \times 9.109 \times 10^{-31}\,\text{kg} \times 2.998 \times 10^8\,\text{m s}^{-1} \times \frac{9}{4} \times 10^{-18}\,\text{m}^2}{15 \times 6.626 \times 10^{-34}\,\text{J s}} = 494.6\,\text{nm}$$

As we see, the photon belongs to the visible spectral range.

(b) By plugging the wavelength of the photon obtained in (a) into Eq. (5.15), we get

$$T = \frac{2.897 \times 10^{-3}\,\text{m K}}{4.946 \times 10^{-7}\,\text{m}} = 5857.6\,\text{K}.$$

This temperature value is comparable to that of Sun's photosphere, significantly higher than the melting temperatures of the benchmark refractory compounds considered in **Exercise 5.3b**.

CHAPTER 6

New Frontiers in Quasicrystals Science

6.1   EXTENDING THE QUASICRYSTAL REALM

> *"The world of quasicrystals opened by Daniel Shechtman extends its reach into the twenty-first century chemistry".* (Tomonari Dotera, 2011) [195]

As we learnt in Secs.1.1.1–1.1.3, first generation QCs were discovered serendipitously. In this regard, it is instructive to compare the list of the first QCs reported in the literature with the list of main alloys used in industry at that time. In doing so, we realize that the main non-ferrous alloys are based on aluminum, magnesium, copper, nickel, cobalt, and titanium [29], which are also the main elements composing the most studied QCs belonging to the systems i-AlCuLi, i-GaMgZn, i-ZnMg(RE), and i-TiZrNi. Subsequently, the search for QC forming alloy systems was guided by the electron per atom criterion, based on the Hume-Rothery alloy concept, leading to the discovery of the second generation high-quality QCs i-AlCu(Fe,Ru,Os), i-AlPd(Mn,Re), d-AlCu(Co,Ni), and i-Cd(Ca,Yb), along with their ternary relatives i-(Ag,Au)In(Ca,Yb) or i-CdMg(RE), and more recently to the synthesis of the third generation i-CuAlSc or i-GaNaAu QCs via detailed electronic band engineering.

Although there is no fundamental reason preventing the spontaneous emergence of long-range QPO in non-metal based systems, it took two decades after Shechtman's announcement of the first metallic QC in an AlMn alloy, until the first representative of the so-called *soft* QCs[1] was found in nature in a solution of wedge-shaped macromolecules that form micelles with radii of about 10 nm [988]. A micelle can be roughly envisaged as a stiff core surrounded by a brush-like corona of flexible tethers.[2] Micelles form readily in solution of molecules having hydrophilic and hydrophobic components (the so-called amphiphiles) and they are easily deformable under thermal motion. Accordingly, their thermodynamic equilibrium phases are characterized by a competition between elastic forces favoring spherical shapes and interactions that tend to minimize the contact area among neighboring micelles. Thus, crystallization as an ordering principle is not restricted to atoms alone [291].

Since 2004, a number of different materials have successively joined the family of soft QCs exhibiting 10-, 12-, or 18-fold diffraction symmetries, including organic dendrimer liquid crystals,[3] star terpolymers, block co-polymer micelles, DNA oligomers, mesoporous

---

[1]The term "soft QC" was coined by the polymer chemist Virgil Percec in 2005 [195].

[2]The term micelle was coined in nineteenth century scientific literature as the diminutive of the Latin word mica (particle), conveying a new word for "tiny particle".

[3]Liquid crystals are a state of matter, which has properties between those of conventional liquids and

silica, colloidal inorganic metallic nanoparticles,[4] or densely packed Cd(Se,S) nanometer-sized tetrahedral-shaped particles, thereby demonstrating that quasicrystallinity can be considered a universal form of ordering. Thus, soft matter QC science emerges as a cross discipline subject connecting physics and chemistry, where plenty of room is left for the search of new kinds of QP arrangements in condensed matter.

6.1.1 Oxide quasicrystals

"We rise the question of what the structure of a silicate quasicrystal may be like. The question is not totally meaningless, since the plenty of quasicrystalline samples obtained in various laboratories of the world from various alloys can give the impression that quasiperiodicity is reserved by Nature for alloys only". (Gábor Gévay, 1990) [280]

To the best of my knowledge the possible existence of oxide based QCs was earlier proposed by Gávor Gévay by considering the silica tetrahedron $(SiO_4)^{-4}$, which is common to all silicates, as the basic building block of a hierarchical structure including five levels: (1) five tetrahedra are joined by the vertices forming a regular pentagonal loop (displaying D_{5h} symmetry), (2) these pentagonal loops are considered as faces of a regular pentagonal dodecahedron, in such a way that the Si atoms are located at the vertices and three of the O atoms around each Si atom are located at the mid-points of edges, whereas the fourth O atom is directed outwards, resulting in a cage with formula $(Si_{20}O_{50})^{-20}$, (3) these dodecahedra are properly linked by face sharing to obtain a golden rhombus[5] containing 16 silico-dodecahedrane cages each, (4) these rhombuses are regarded as the faces of two different rhombohedra, with acute (obtuse) angle meeting at the poles, and (5) these rhombohedra play the role of the basic building blocks of a 3D Penrose tile, which constitutes the silicate QC itself [280]. By the light of current knowledge on the cluster structure in metallic QCs, one may alternatively consider to make use of 10 prolate and 10 oblate rhombohedra in order to build up a rhombic triacontahedron instead [513], since this polyhedron is a well-known basic atomic cluster in the architecture of Tsai-type QCs (see Sec. 1.5.2).

Gévay's proposal introduced three inspiring ideas in non-metallic based QC design, namely, (1) the possible role of Si atoms as QC forming elements, (2) the use of tetrahedral clusters as possible building blocks in the route towards icosahedral symmetry, and (3) the consideration of oxide compounds. In this regard it is worth to note that:

1. The occurrence of a quasicrystalline decagonal phase of silicon, stabilized through annealing (300°C to 400°C) of silicon films, probably composed of icosahedron-like Si_{28} clusters prepared by evaporation of Si under helium ambient ($\sim$200 torr), was first reported in 2000 [402]. Subsequently, the possible existence of a single component Si

---

those of solid crystals. For instance, a liquid crystal may flow like a liquid, but its molecules may be oriented in a crystal-like way. There are many different types of liquid-crystal phases, which can be distinguished by their different optical properties. Examples of liquid crystals can be found both in the natural world and in technological applications. For example, many proteins and cell membranes are liquid crystals. Other well-known examples of liquid crystals are solutions of soap and various related detergents, as well as the tobacco mosaic virus, and some clays.

[4]Colloid is a short synonym for colloidal system where molecules or polymolecular particles dispersed in a medium have at least one dimension between approximately 1 nm and 1 μm.

[5]In geometry, a golden rhombus is a rhombus whose diagonals are in the golden ratio. The internal supplementary angles of the golden rhombus are: acute angle $\alpha = 2\tan^{-1}(\tau^{-1}) \simeq 63.434°$, obtuse angle $\beta = 2\tan^{-1}(\tau) = \pi - \tan^{-1}(2) \simeq 116.565°$. The golden rhombus should be distinguished from the two rhombi of the Penrose tiling, which are both related in other ways to the golden ratio but have shapes different to the golden rhombus. Several notable polyhedra have golden rhombi as their faces, including the acute and obtuse golden rhombohedra (with six faces each), the rhombic icosahedron (with 20 faces), and the rhombic triacontahedron (with 30 faces).

ddQC was predicted on the basis of numerical simulations [394], and the existence of ddQC silicene monolayers with 3- and 4-fold coordination has been recently predicted via molecular dynamics simulations [713].

2. The potential use of tetrahedral building blocks in QC structural design was demonstrated by the discovery of 10-fold arrangements resulting from the self-assembly of nanometer-sized truncated tetrahedrons of CdSe and CdS particles at a liquid-air interface [633, 963]. To this end, two different types of surfaces were covered with oleic acid and octadecylphosphonic acid, respectively, promoting selective facet-to-facet attachment of neighboring tetrahedrons. The preferred alignment of tetrahedral nanocrystals led to the formation of decagon units.

3. The first observation of an oxide based QC was reported for a barium titanate ($BaTiO_3$) structure,[6] which formed a high-temperature interface-driven 12-fold symmetric array on the 3-fold symmetric (111) surface of a Pt single crystal substrate [255]. The resulting QC structure was grown by annealing a several monolayer (14–30 Å) thick $BaTiO_3$ (111) layer on Pt(111). Upon annealing to 1250 K in ultra-high vacuum, the film restructures into a QC thin film of 2 to 4 Å thickness in between a few thicker $BaTiO_3$ (111) islands. The building blocks of this dodecagonal structure assemble in the theoretically predicted Stampfli tiling (see Fig. 4.7c), exhibiting arrangements of triangles, squares and 30° rhombs, with a fundamental length-scale of 0.69 nm. Later on, an approximant structure was found with an atomic pattern described in terms of a $3^2.4.3.4$ Archimedean tiling (see Fig. 4.6), where Ti atoms reside at the corners of each tiling element along with 3-fold coordinated oxygen atoms, and Ba atoms intercalate between TiO_3 tetrahedra, leading to a fundamental edge length of the tiling $a = 0.67$ nm [256]. The fundamental mechanisms involved in the growth process were subsequently disclosed using *in situ* low energy electron microscopy studies indicating that the oxide QC formation proceeds in two steps via an amorphous 2D precursor wetting layer, which eventually develops long-range QPO above 1170 K. Furthermore, the reversible isothermal conversion of the oxide QC into $BaTiO_3$ islands and bare Pt(111) can be induced by annealing in a controlled partial pressure O_2 atmosphere [257]. The observation of a similar oxide ddQC and its related approximant in the $SrTiO_3$ system was reported shortly after these findings [785]. The valence band structure of dd-$BaTiO_3$ was recently studied by photoemission using momentum spectroscopy. In this way, a dispersive band was identified and assigned to a combination of in-plane O_{2p} orbitals. In addition, the signature of Ti_{3d} states near the Fermi level was observed as well, which results in a metallic character of this oxide ddQC [133].

Mesoporous silica are one kind of fascinating colloids, which allow the formation of dense and close-packed assembly architectures by using self-assembly of surfactant micelles, thereby leading to a solid replica of a micellar aggregate. Since their discovery, these special silica particles have been of great interest as building blocks for diverse application fields. In this regard, it is remarkable the growth of mesoporous silicas which exhibit 12-fold symmetry in both electron diffraction and morphology (Fig. 6.1), hence providing dodecagonal analogs of the axial dQCs shown in Fig. 1.8a. By all indications, these structures form through non-equilibrium growth processes wherein the competition between different micellar configurations has a central role in tuning the structure [968].

---

[6]$BaTiO_3$ is one of the best investigated perovskite oxide systems which is also widely used in thin film applications and oxide heterostructures.

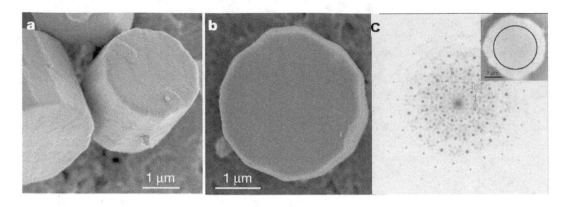

Figure 6.1 (a,b) Scanning electron images taken from 1 μm sized mesoporous particles exhibiting a dodecagonal prism growth habit, (c) Electron diffraction pattern using the aperture size indicated in the circle in the inset picture. (Reprinted from Xiao C, Fujita N, Miyasaka K, Sakamoto Y, Terasaki O, Dodecagonal tiling in mesoporous silica. *Nature* **487**, 349–353 (2012). https://doi-org.bucm.idm.oclc.org/10.1038/nature11230. With permission from Macmillan Publishers Limited. Copyright 2012).

6.1.2 Organic quasicrystals

> *"We feel that the quasicrystalline order is not a privilege of metallic phases. It is expected that, perhaps one of these days, non-metallic, even organic quasicrystals may be discovered, thus widening further the scope of crystallography and related areas of science".* (Gábor Gévay, 1993) [281]

Icosahedral morphologies are often observed in diamond crystals grown by chemical vapor deposition methods. Among different considered growth scenarios, a molecular mechanism wherein dodecahedrane ($C_{20}H_{20}$) molecules[7] might act as nuclei for the growth of icosahedral diamond crystals was put forward in 1983 [590]. The feasibility of this proposal was further explored in 2015 by considering a growth process involving the breaking of C=H bonds and the formation of new C=C bonds, typically by addition of CH_3 radicals to the resulting surface reactive sites. As soon as methyl groups replace all the original H atoms in dodecahedrane, an embryonic (111) plane will develop orthogonal to each of the original C=H bonds. Continued substitution of H atoms by CH_3 groups would eventually lead to the emergence of an icosahedral crystal bounded by 20 (111) planes. The number of atoms included in the different shells of a defect free icosahedral diamond grain is $n = 20 \sum (1 + k)^2$, where $k = 0, 1, \ldots$ is the shell number [956]. However, the resulting structures cannot properly be regarded as QCs, but nano-sized crystals. In order to obtain full-fledged organic iQCs based on dodecahedrane molecules building blocks one may follow a hierarchical assembly process completely analogous to that described in Sec. 6.1.1 to construct a Si-based oxide iQC, by simply replacing the $(SiO_4)^{-4}$ molecules by $C_{20}H_{20}$ ones [281].

---

[7]The hydrocarbon $C_{20}H_{20}$ molecule consists of 20 carbon atoms sitting at the vertices of a regular pentagonal dodecahedron, such that the edges correspond to C=C sigma bonds. The fourth valence of each C atom (sp^3 hybridization geometry) is directed outwards linking H atoms. This molecule was first synthesized in 1981 [671].

The possible existence of iQCs based on a QP stacking of C_{60} or C_{70} fullerenes as possible molecular building blocks has also been explored, with no success so far [251, 602]. Recently, the formation of a ddQC arrangement of self-assembled fullerenes adsorbed on a $Pt_3Ti(111)$ surface at 320 K has been reported [677]. The resulting 2D monolayer can be described in terms of a square-triangle tiling with vertex configurations corresponding to $3^2.4.3.4$, $3^3.4^2$ and 3^6 Schläfli symbols (see Fig. 4.6), along with interpenetrating and fused dodecagons. Since no higher order tiling patterns, showing the characteristic inflation factor $2 + \sqrt{3}$ were observed, it was concluded that this ddQC should be regarded as a random tiling array. The tiling also contains a small number of polygonal defects, indicating that the QC is not in its ground state. Indeed, the origin of this structure can be assigned to interfacial interactions between the self-assembled monolayer of C_{60} molecules and the platinum alloy surface, both of which otherwise form periodic structures. Furthermore, the ddQC can be decomposed into different local areas corresponding to several types of approximant crystals, the most frequent being the 8/3 one. Careful inspection of diffraction data reveal that the reflections are not located in the positions that correspond to an ideal ddQC, but exhibit slight spot shifts signaling the presence of considerable phason strains. Furthermore, it is deduced that optimization of the intermolecular interactions between neighboring fullerenes hinders the realization of translational periodicity in the fullerene monolayer on Pt-terminated $Pt_3Ti(111)$ surface.

Other possible organic molecules have been considered in order to obtain quasicrystalline phases as well. According to numerical calculation one potential candidate may be the (still hypothetical compound) coronene $C_{30}H_{10}$, which exhibits a D_{5h} symmetry at the molecular electronic level [999]. On the other hand, ferrocenecarboxylic acid (FcCOOH)[8] self-assembled monolayers were studied using scanning tunnelling microscopy, and it was found that, rather than producing dimeric or linear structures typical of carboxylic acids, FcCOOH forms highly unusual cyclic hydrogen-bonded pentamers, which combine with simultaneously formed FcCOOH dimers to form 2D quasicrystallites that exhibit local 5-fold symmetry and maintain translational and rotational order (without periodicity) for distances of more than 40 nm [955].

6.1.3 Soft quasicrystals

"I want to emphasize that the origin of quasiperiodicity is already much better understood for mesoscopic quasicrystals than for intermetallic quasicrystals". (Walter Steurer, 2012) [841]

6.1.3.1 Supramolecular dendrimers

Dendrimers are highly branched macromolecules with nanometer-scale dimensions made of shorter dendron (tree in Greek) molecules. Dendrimers are defined by three components: a hard central core, an interior dendritic structure (the branches), and an exterior soft corona composed of flexible functional groups. In the case that dendrimers adopt a conical shape, they can assemble into spherical nm-sized objects called spherical supramolecular dendrimers. These assemblies further self-organize into 3D arrays with long-range order, which may be either periodic or aperiodic in nature. The first soft QC was discovered by investigating a library of supramolecular dendrimers in the late 1990s and early 2000s, and was published in 2004 [988]. Subsequently, numerous new libraries of supramolecular

---

[8]Ferrocene (Fc) is an organometallic compound with the formula $Fe(C_5H_5)_2$. The molecule consists of two cyclopentadienyl rings bound on opposite sides of a central iron atom with their planes parallel and mutually orthogonal to the joining line through the Fe atom.

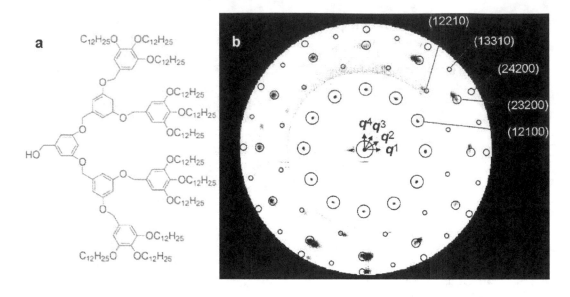

Figure 6.2 (a) The dendron $((3,4,5)\text{-}(3,5)_2)12G3\text{-}CH_2OH$, the first to be confirmed as forming a soft ddQC (b) Experimental and simulated (circles) X-ray diffraction patterns along the 12-fold axis (at room temperature) of a soft ddQC based on the spherical dendrimers obtained from the dendron shown in a). The intensities of outer reflections are scaled up by 100. (Reprinted from Ungar G, Percec V, Zeng X, Leowanawat P, Liquid quasicrystals, Isr. J. Chem. **51** 1206-1215 (2011). With permission from Wiley-VCH Verlag GmbH & Co. KGaA Weinheim. Copyright 2011).

dendrimers have been investigated, and in all of them soft QCs have been discovered as being self-organized from spherical supramolecular dendrimers [920]. The first soft QC was based on the organic molecule dendron shown in Fig. 6.2a, that was amenable to growth of large monodomains exhibiting 12-fold rotational symmetry. The obtained X-ray diffraction patterns were indexed using five Miller indices, four of them required in order to describe the QPO in 2D planes, the remaining one describing the periodic stacking of QP planes along the orthogonal direction (Fig. 6.2b). Accordingly, the first soft QC should be classified as an axial ddQC in 3D, analogous to the dQCs previously reported in metallic alloys.

The standard cut and project method from a 4D space generates tilings with 12-fold symmetry in 2D, formed by squares, triangles and $\pi/6$ rhombuses. However, rhomboidal units are seldom found in high-resolution transmission electron micrographs of ddQCs. In order to generate a dodecagonal tiling consisting of only squares and equilateral triangles, the acceptance domain in the perpendicular space must be allowed to be fractally shaped, which makes the structural analysis difficult. An alternative way of generating QP tilings relies on their self-similar property. A parent patch is first inflated by a given inflation factor, then the inflated tiles are decomposed into tiles of original size and a new pattern is thus produced. The first such inflation rule to generate square-triangle tilings with 12-fold symmetry was found by Stampfli (see Fig. 4.7c) [834], other was discovered by Schlottmann [333], both having inflation factor $\lambda = 2 + \sqrt{3}$. Consequently, after one inflation–decomposition step, the unit cell parameter of the tiling increases by a factor of ~ 4, and the area of the unit cell by a factor of ~ 14 [989].

The calculated diffraction pattern based on the structural model shown in Fig. 6.3 renders a good match with the experimental one shown in Fig. 6.2b. The distribution of

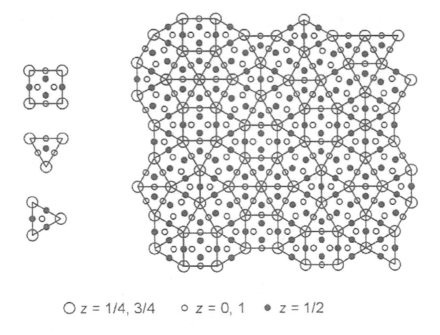

$\bigcirc$ z = 1/4, 3/4 $\circ$ z = 0, 1 $\bullet$ z = 1/2

Figure 6.3 Packing of supramolecular spheres in a dodecagonal arrangement based on equilateral triangles and squares. The symbols indicate the relative elevations of the spheres forming the denser nets. The lattice constant is 8.14 nm. (Reprinted from Ungar G, Percec V, Zeng X, Leowanawat P, Liquid quasicrystals, Isr. J. Chem. **51** 1206-1215 (2011). With permission from Wiley-VCH Verlag GmbH & Co. KGaA Weinheim. Copyright 2011).

equilateral triangles and squares is locally similar to that corresponding to the dodecagonal Stampfli tiling and their relative frequency agrees well with the expected value $4\sqrt{3}/3$ (see Sec. 4.3). However, the inflation scale factor $\lambda = 2 + \sqrt{3}$, characteristic of metallic ddQCs (**Exercise 1.3**), can not be readily derived from the obtained diffraction patterns, but one obtains the value $\lambda = 1 + \sqrt{3}$ instead (**Exercise 6.1**).

6.1.3.2 Block copolymers

Another class of soft QCs was found as an spin-off from studies of Archimedean tiles (see Sec. 4.3) formed from block copolymers assemblies. Block copolymers are composed of chemically distinct polymers linked together and joining at a common vertex. Using these copolymers as building blocks, both simulations and earlier experiments demonstrated that the so-called ABC star-shaped terpolymers melts may be organized according to the Archimedean tiles 6^3, 4.8^2, and $4.6.12$ (see Fig. 4.6). In 2005, an experiment with an ABC star-shaped terpolymer revealed a more complex phase corresponding to the Archimedean tile $3^2.4.3.4$ (the same arrangement observed in 2D oxide crystals, see Sec. 6.1.1), composed of equilateral triangles and squares, whose edge length is about 80 nm. By finely tuning the ratio of polymer components in the block copolymers a tiling pattern possessing some typical features of a dodecagonal QP structure was obtained in 2007 [327]. Thus, a mesoscopic tiling pattern with 12-fold symmetry was observed in a three-component polymer system composed of polyisoprene, polystyrene, and poly(2-vinylpyridine) star-shaped terpolymer, and a polystyrene homopolymer blend. Transmission electron microscopy images revealed a nonperiodic tiling pattern covered with equilateral triangles and squares, with a triangle/square number ratio of 2.3, which agrees well with the ideal ratio $4\sqrt{3}/3 = 2.3094\ldots$ (see Sec. 4.3).

In 2011, phases with both 12- and 18-fold diffraction symmetry in small-angle synchrotron X-ray diffraction pattern were reported for a colloidal system of self-assembled spherical micelles in aqueous solution, composed of block copolymers consisting of hydrophobic polyisoprene (PI) at the core, surrounded by a relatively large shell of hydrophilic polyethylene oxide (PEO) in the outer shell [250]. The 18-fold diffraction patters were observed for micelles with composition $PI_{30}\text{-}PEO_{120}$, $PI_{32}\text{-}PEO_{120}$, and $PI_{30}\text{-}PEO_{124}$. In order to rule out possible twinning effects the experimental diffraction patterns were compared with simulated patterns for both a fcc twin (consisting of non-rotated, 20°-rotated, and 40°-rotated (111) layers along the [111] stacking direction) and a suitable QP model based on rhombohedric tiles with an scaling factor $\lambda \simeq 0.536$. In so doing, one sees that QC model exhibits reflections that are not present in the diffraction pattern of the fcc twinned structure. Therefore, the appearance of these characteristic reflections in the experimental diffraction patterns provides strong indication for the true formation of a novel QC phase.[9] Such a possibility is further supported by the observation of phase transitions between the fcc phase and the 12- and 18-fold phases, which can be easily followed in situ by time-resolved diffraction experiments [250]. The origin of 18-fold symmetry has not been fully clarified so far, and it is an open question whether the related QC has enneagonal (9-fold) or 18-fold rotational symmetry in real space (see Sec. 6.3.3) [291]. If so, this will be the first report of a QC with a novel symmetry in about 30 years. In addition, the discovery of this 18-fold QC constitutes the first account of a novel quasicrystalline structure on the nanoscale with no equivalent known for metallic QCs, thereby inspiring the search for atomic based analogues of this soft matter representative [291].

6.1.3.3 Nanoparticles

Nanoparticles of different metals, semiconductors and magnetic materials can self-assemble from colloidal solutions into superlattices. In 2009, the formation of ddQCs was reported in different binary nanoparticle systems, namely: (1) 13.4 nm Fe_2O_3 and 5 nm Au; (2) 12.6 nm Fe_3O_4 and 4.7 nm Au; and (3) 9 nm PbS and 3 nm Pd nanocrystals capped with oleic and dodecanethiol surfactant molecules [854]. These molecules introduced short-range steric repulsion counterbalancing the attractive van der Waals force and preventing uncontrollable aggregation of nanocrystals in the colloidal solution. The observed arrangement of nanocrystals fits a dodecagonal equilateral triangle-square tiling, with an average ratio 2.3 ± 0.2, which is close to the ideal value $4/\sqrt{3} = 2.3094\ldots$. Because of material flexibility the formation of this assembly does not require a unique combination of interparticle interactions, but can be regarded as a general sphere-packing phenomenon, generally governed by entropy and simple interparticle potentials, hence suggesting that other nanocrystal based QCs may be possible via fine-tuning of size-ratio and composition [195].

In summary, contrary to the structure of intermetallic QCs, soft QCs basic construction building blocks are supramolecular aggregates, micelles or other colloidal particles instead of individual atoms or atomic clusters about 1 nm in size. Soft QCs follow other assembly mechanisms as well, demonstrating that quasicrystalline order is universal over different

---

[9]At the time being, we cannot completely rule out that a yet unknown structure with a very large unit cell, or the coexistence of a yet unknown complex periodic phase, could similarly explain the observed diffraction patterns [250].

length scales in condensed matter. Thus, whereas chemical bonding is the key factor in intermetallics, specific pair potentials, three-body interactions, and the shape entropy play the decisive roles in the case of soft-matter based QCs, where 12-fold symmetry is the prevailing one.

In fact, in order to go beyond dense sphere packing arrays, soft QCs building blocks must exhibit complex enough interaction potentials providing two length scales at least [724, 841]. These scale lengths, say $\ell_1 < \ell_2$, are related to the minima of suitable pair interaction potentials, so that ℓ_2 sets advantageous positions for non-crystallographic configurations leading to QC formation. To this end, isotropic Lennard-Jones-Gauss pair potentials of the form

$$U_{\mathrm{LJG}}(r) = \epsilon_0 \left[\left(\frac{r_0}{r} \right)^{12} - 2 \left(\frac{r_0}{r} \right)^6 - \epsilon \exp \left(-\frac{(r - r_G)^2}{2 r_0^2 \sigma^2} \right) \right], \tag{6.1}$$

where ϵ_0 and r_0 set the units of energy and distance, respectively, and r_G, ϵ, and σ determine the position, depth and width of the minimum at ℓ_2, have been extensively studied in the recent literature [277, 278, 587]. Another possible approach is to consider patchy particles with attractive regions at the surface. In this case, the interaction between two particles at a distance r is composed of an isotropic Lennard-Jones-like contribution, along with an *anisotropic* angular $V(\theta)$ term, in the form

$$U(r, \theta) = \epsilon_0 \left[\left(\frac{r_0}{r} \right)^{2n} - 2 \left(\frac{r_0}{r} \right)^n \right] V(\theta), \tag{6.2}$$

for $r > r_0$, whereas the interaction potential reduces to the Lennard-Jones form for $r < r_0$. Accordingly, in these systems the ordering preferred by the isotropic parts of the potential (which possess a local minimum at $r = r_0$) competes with structures that possess the binding angles determined by the attractive patches, described by the angular term

$$V(\theta_k, \theta_l) = e^{-\frac{\theta_k^2 + \theta_l^2}{2\sigma_1^2}} + e^{-\frac{\theta_k^2 + \theta_l^2}{2\sigma_2^2}}, \tag{6.3}$$

where θ_k and θ_l respectively denote the angles between the nearest patch of particle i (respectively j) to the bond r_{ij} between particles i and j, and $\sigma_{1,2}$ indicate the patch widths [278]. A system of particles with an angular dependence such that each particle has five attractive patches was studied by considering the role of the patches width $\sigma_{1,2}$ in the resulting structures. For particles with fairly narrow patches the system forms a periodic crystal, where each particle has a coordination number of 5, albeit in an arrangement that deviates from perfect 5-fold symmetry. For sufficiently wide patches, a competition is set up between such a non-uniform pentavalent coordination and the hexagonal coordination characteristic of 2D crystals of isotropic pair potentials. These hexa- and penta-valent environments form triangle-square tilings, which are well known precursors of dodecagonal QCs. Therefore, by properly tuning the σ_1 and σ_2 values one can eventually obtain 12-fold QC phases [734].

6.1.4 Biological quasicrystals

"Biology is filled with symmetry, where regularity of form leads to functional features. This takes the study of biomolecular symmetry both challenging and appealing, since there always seem to be interesting exceptions in the gray area between perfect symmetry and complete asymmetry". (David S. Goodsell, 2019) [303]

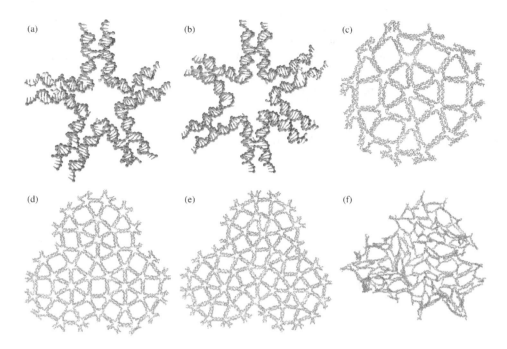

Figure 6.4 DNA representation of (a) a five-arm tile, (b) a six-arm tile, (c)–(e) QC-like motifs of increasing size absorbed onto a surface and (f) the largest motive in solution. The simulations of these structures were performed at 22°C and a salt concentration of 0.5 mol dm^{-3}. (Reprinted from Reinhardt A, Schreck J S, Romano F, Doye J P K, Self-assembly of two-dimensional binary quasicrystals: a possible route to a DNA quasicrystal, J. Phys.: Condens. Matter **29** 014006 (12pp) 2017; doi10.1088/0953-8984/29/1/014006. Creative Commons Atribution License CC BY 3.0).

Many of the biomolecules are symmetrical complexes of identical subunits, including point group symmetries in the enzymes or helical symmetry in several structural proteins (e.g., actin and collagen fibers) and DNA macromolecules. Collagen fibers provide an illustrative example of aperiodic arrangement of amino acid building blocks, adopting the configuration of the so-called Boerdijk-Coxeter helix containing $1 + \sqrt{3}$ edges per turn [763]. Both types of symmetries can be observed in the virus capsids. In particular, icosahedral symmetry is found in many biological nano-containers. In addition, because of the enantiomorphic nature of most biomolecules, point groups are restricted to groups of rotational symmetries, with all of the axes passing through a single point. These include cyclic symmetries with a single rotation axis, dihedral groups with a cyclic axis and perpendicular 2-fold axes, along with octahedral and icosahedral groups [303].

6.1.4.1   DNA nanomotifs

DNA provides an excellent model system for studying rational QC design. Indeed, during the last 35 years, DNA has been exploited to construct a wide range of nanostructures because of its programming capability. Thus, a suitable candidate for the ideal patchy particles considered in the previous section is provided by DNA multi-arm motifs which can be regarded as star-shape tiles (Fig. 6.4), able to self-assemble in a range of 2D regular

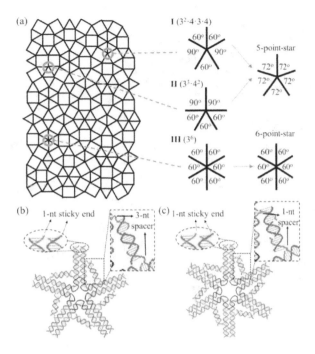

Figure 6.5 (a) The three different vertices in a Stampfli dodecagonal tiling are highlighted and the basic geometry of the DNA motifs which occupy them are indicated. The corresponding detailed molecular designs are shown in (b) and (c). (Reprinted with permission from Liu L, Li Z, Li Y, Mao C, Rational design and self-assembly of two-dimensional dodecagonal DNA quasicrystals, J. Am. Chem. Soc. **141** 4248-4251 (2019). Creative Commons Atribution License CC BY 3.0).

structures. These DNA motifs are complementary, so that the bulk of the star tile is fully bonded, but the end of each of the arms includes unpaired strands which can bond with other star tiles. In this way, the bonding can be chosen to be as generic or as specific as we wish by simply selecting appropriate DNA end sequences, and the resulting motifs have a well defined valence, determined by the number of arms. Accordingly, the DNA star tiles can be properly modeled in terms of the pair interaction potentials given by Eqs. (6.2)–(6.3) with a narrow patch width. In so doing, Monte Carlo simulations and free-energy techniques showed that binary solutions of penta- and hexa-valent patchy particles could form thermodynamically stable ddQCs, provided that their patch interactions are chosen in an appropriate way. Such patchy particles may be thought of star-shaped DNA multi-arm motifs, chosen to bond with one another very specifically by tuning the DNA sequences of the protruding arms [734].

Accordingly, the possibility of engineering ddQCs based on DNA motifs was explored by considering three basic blocks, namely two 5-branch and one 6-branch star-shaped tiles, respectively occupying the three different vertices one has in a dodecagonal Stampfli tiling (Fig. 6.5a). All branches in a motif are identical to each other and each one is a 4-turn long DNA double strand. Between any two adjacent branches, a bridge is introduced to control the interbranch angle and the overall rigidity of the motif (Fig. 6.5b,c). The bridge lengths range from 7.7 to 10.2 nm. Without the bridges the motifs are too flexible and could not maintain their designed geometry during the assembly process, which was conducted in two

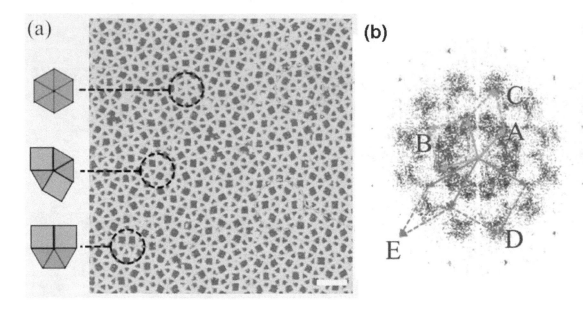

Figure 6.6 (a) Dodecagonal quasicrystalline networks assembled from b5PS and b6PS motifs at a ratio of 80:20 in bulk solution. (b) Fast Fourier transform pattern of the arrangement shown in (a). (Adapted with permission from Liu L, Li Z, Li Y, Mao C, Rational design and self-assembly of two-dimensional dodecagonal DNA quasicrystals, J. Am. Chem. Soc. **141** 4248-4251 (2019). Creative Commons Atribution License CC BY 3.0).

steps: (1) assembly of the individual DNA motifs separately in solution; (2) surface-assisted self-assembly of DNA networks absorbed onto mica surfaces. Depending on the molar ratio of the penta- and hexa-valent DNA motifs (referred to as b5PS and b6PS, respectively), one can obtain periodic or QP arrangements. The ddQC phase was obtained for molar concentrations (b5PS:b6PS) ranging from 90:10 to 80:20 [509].

Fig. 6.6a illustrates an atomic force microscopy image (covering a 4 μm × 4 μm surface) of a DNA-based ddQC where b5PS motifs adopt either configuration I ($3^2.4.3.4$) or II ($3^3.4^2$), and b6PS motifs adopt configuration III (3^6). In the AFM image there are 695 triangles and 301 squares, yielding a triangle to squares ratio 2.309, pretty close to the ideal value $4/\sqrt{3}$ for a dodecagonal Stampfli tiling [509]. In the fast Fourier pattern shown in Fig. 6.6b diffraction reflections can be seen on five concentric rings labelled A, B, C, D, and E from the center outwards. On each ring the diffraction spots exhibit 12-fold rotational symmetry and spots on the different circles are mutually correlated as indicated by the arrow vectors. For example, diffraction peaks in circle E can be derived from vector addition of two adjacent reflections in circle B.

6.1.4.2 Icosahedral viruses geometrical design

"The probability of formation of a highly complex structure from its basic constituents is increased if that structure can be broken down into a finite series of successively inclusive sub-structures". (John D Bernal, 1960) [53]

a   *b*

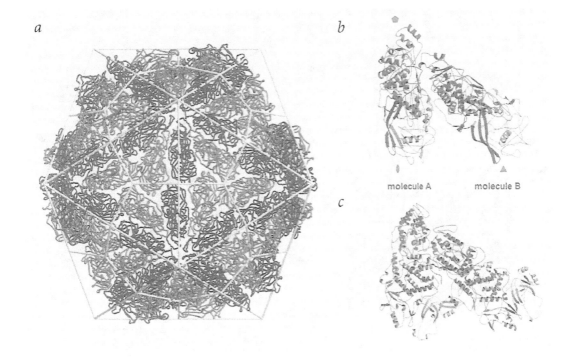

c

molecule A   molecule B

Figure 6.7   a) Icosahedral structure of L-A virus capsid ($T = 1$)   inner core   viewed down to a 2-fold axis. The capsid (diameter 40 nm) is made of 60 protein subunit dimers (shown in b) grouped in capsomers containing five dimers each, which are located at the 5-fold vertices. Icosahedral symmetry elements are indicated by symbols located at specific protein regions. c) Protein subunit dimer of the closely structurally related blue tongue virus. (Reprinted by permission from Springer Nature, Nature Structural Biology, L-A virus at 3.4 Å resolution reveals particle architecture and mRNA decapping mechanism, Naitow H, Tang J, Canady M, Wickner R B, Johnson J E, 9 725–728, 2002. Copyright 2002).

As we mentioned in Sec. 1.1.2 the first evidence of icosahedral symmetry in condensed matter was obtained from X-ray diffraction measurements of certain virus crystals. From those studies it was concluded that icosahedral virus capsids are structured according to a hierarchical *modular design* architecture.[10] The first two levels are constituted by amino acids linked in proteins (typically containing 100–200 amino acids), which play the role of basic building blocks. Proteins can appear either single or, more generally, grouped in sets including two, three, four, or even more proteins each, to give rise to the so-called protein subunits (Fig. 6.7b–c). The proteins belonging to these subunits can be either identical (i.e., they have the same primary amino acid sequence) or different among them. In the former case, proteins usually differ in their spatial conformation (Fig. 6.7b). The protein subunits then associate among them to form the third level in the structural hierarchy, namely, the capsomers (Fig. 6.7a and Fig. 6.8b).[11] Within the capsomers, protein subunits

---

[10]Strong evidence for a modular structure was provided by chemically induced disintegration of virus particles into triangular and pentagonal fragments exhibiting well resolved protein substructures [961].

[11]André Lwoff (1902–1994), T F Anderson, and François Jacob (1920–2013) proposed in 1959 the terms

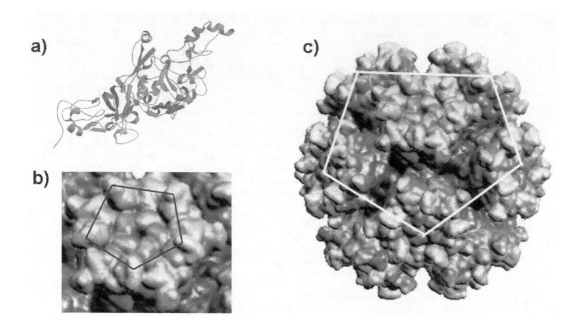

Figure 6.8 Adenovirus A (dsDNA, average diameter 27.6 nm, $T = 1$) structure at 0.33 nm resolution. a) X-ray crystal structure of the protein subunit. b) A close up view of one capsomers which is composed of five protein subunits arranged themselves into a regular pentagon geometry. c) Virus capsid showing the arrangement of the capsomers into a global pentagonal dodecahedron geometry. (Adapted from VIPER EMdb virus capsid repository [937]).

can associate in groups of three (trimers), five (pentamers), or six (hexamers) subunits each. Capsomers generally adopt polygonal shapes arranging themselves according to pentagonal and hexagonal local patterns (Fig. 6.8b). Finally, in the fourth hierarchical level, the capsomers organize themselves throughout the space, endowing the virus capsid with a global polyhedral symmetry as a whole (Fig. 6.8c). Thus, while protein subunits mainly exhibit triangular or trapezoidal geometries, the resulting capsomers can display either pentagonal or hexagonal *local* arrangements of protein subunits, which in turn organize themselves so that the whole array of capsids exhibits a *global* icosahedral symmetry (Figs. 6.7a and 6.8c).

There exist a number of biologically relevant aspects favoring the preference for polyhedra exhibiting icosahedral symmetry, namely:

- They provide the maximum number of possible symmetry elements available for distributing protein subunits through the capsid.

- They have the largest volume-to-surface ratio (Table 6.1), thereby optimizing the encapsulation of nucleic acid genome within the capsid.

- These polyhedra also exhibit large perimeter-to-surface ratios (Table 6.1), hence

---

"capsid" and "capsomers" to represent, respectively, the protein shell and the *units* comprising it, and the term "virion" to denote the complete potentially infective virus particle (i.e., a capsid enclosing the nucleic acid) [522].

allowing for the required mechanical stability of the capsid via chemical bonding among capsomers.

Table 6.1 Volume (V), area (S), effective perimeter $(P = Ea)$, where E is the number of edges, and the related surface-to-volume S^3/V^2 and perimeter-to-surface P^2/S dimensionless ratios of the five Platonic solids and some Archimedean solids with edge a. $\tau = (1 + \sqrt{5})/2$ is the golden ratio and $\psi \equiv \sqrt{2 + \tau}$. The data for the sphere (radius r) are given for the sake of comparison.

Solid	$V(a^3)$	$S(a^2)$	$P(a)$	S^3/V^2	P^2/S
Tetrahedron	$\frac{\sqrt{2}}{12}$	$\sqrt{3}$	6	374.12	20.79
Cube	1	6	12	216.00	24.00
Octahedron	$\frac{\sqrt{2}}{3}$	$2\sqrt{3}$	12	187.06	41.57
Dodecahedron	$\frac{4+7\tau}{2}$	$3\sqrt{5}\tau\psi$	30	149.86	43.59
Icosahedron	$\frac{5}{6}\tau^2$	$5\sqrt{3}$	30	136.46	103.92
Icosidodecahedron	$\frac{14+17\tau}{3}$	$5\sqrt{3}+3\sqrt{5}\tau\psi$	60	131.49	122.85
Triacontahedron	$4\tau\psi$	$12\sqrt{5}$	60	127.48	134.16
Truncated icosahedron	$\frac{41+43\tau}{2}$	$30\sqrt{3}+3\sqrt{5}\tau\psi$	90	125.22	111.56
Sphere	$\frac{4}{3}\pi r^3$	$4\pi r^2$	–	113.10	–

In this way, one gets a highly stable closed protein shell with an optimal volume to surface ratio composed of capsomers containing a minimal number of identical protein subunits. Indeed, there exists a sound biological argument for construction of small viruses out of identical protein subunits packed together in a regular manner to provide a protective shell for the nucleic acid, namely, that coding of these relatively small identical coat proteins is an efficient use of the limited information contained in the relatively short virus nucleic acid [112]. Accordingly, in the case of a viral capsid the regular polyhedra requirement expresses a principle of *genetic economy* allowing for regular and repetitive *chemical interactions* among capsid proteins, which are arranged according to spatial patterns providing maximal contact, via polygonal shapes exhibiting the largest perimeter to area ratio (Table 6.1).

In Fig. 6.9, we plot the volume-to-surface ratio versus the perimeter-to-surface ratio for the polyhedra listed in Table 6.1. By inspecting this figure several conclusions can be drawn. First, Platonic solids lacking icosahedral symmetry are not well suited to optimize virus designs. Second, dodecahedral capsid shapes reasonably optimize the volume-to-surface ratio, but their perimeter-to-surface ratio falls too short. Accordingly, one should expect that dodecahedral virus shapes would be observed for relatively small capsids only. Third, icosahedral symmetry polyhedra other than the pentagonal dodecahedron substantially improve the perimeter-to-surface ratio (dashed line jump in Fig. 6.9) and further decrease the surface-to-volume ratio. Among the considered polyhedra the optimal ratios are those corresponding to the truncated icosahedron (best S^3/V^2 ratio) and the rhombic triaconta-hedron (best P^2/S ratio), respectively.

6.1.4.3 Chemical interactions and virus symmetries

The main structural problem in the field of physical virology can be stated as follows: how to construct a regular protein array with icosahedral symmetry formed by multiple copies of (nearly) identical asymmetric (i.e., chiral) proteins interacting among them in (nearly) identical environments?

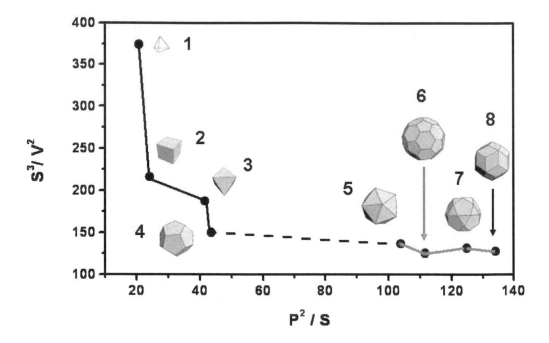

Figure 6.9 Surface-to-volume (S^3/V^2) versus perimeter-to-surface (P^2/S) plot for different polyhedra: 1, tetrahedron; 2, cube; 3, octahedron; 4, dodecahedron; 5, icosahedron; 6, truncated icosahedron; 7, icosidodecahedron; and 8, rhombic triacontahedron.

If just one kind of chemical interaction would occur between each pair of interacting protein subunits, then arranging identical units in identical environments would necessarily produce a symmetrical structure, and there would be only a geometrically limited number of kinds of symmetry available. In particular, icosahedral symmetry demands that the maximum number of asymmetric protein subunits that can be arranged on a surface with spherical topology, such that each has exactly the same environment, be 60.[12] Nevertheless, construction of icosahedral capsids containing just 60 protein subunits severely restricts the size of the genome that can be packaged inside, and it soon became evident that most viruses have considerably more than 60 protein subunits in their capsids. This fact provides a nice illustration of the balance between the amount of required biological information (i.e., biocomplexity) and geometrical size in biological systems.[13] Therefore, we must face the problem of how to form a closed shell exhibiting icosahedral symmetry using more than 60 protein subunits.

In this regard, the presence of different chemical *interaction centers* in the complex structure of a folded protein provides the basis for a straightforward matematization of such a protein in terms of suitable geometrical entities, where these *chemical centers* play the role of geometrical *points*. In this way, proteins exhibiting two different interaction centers may be assimilated to segments, those having three interaction centers to triangles

---

[12]The chiral nature of proteins and their related capsomers imply the absence of both inversion and mirror symmetry elements in the resulting icosahedral virus capsid, hence halving the number of the icosahedral group symmetry elements from 120 (I_h group) to 60 (I group).

[13]Note that this geometrical constraint does not operate in the case of helical viruses, which are open along the helical axis and can accommodate a nucleic acid of any length inside their capsids.

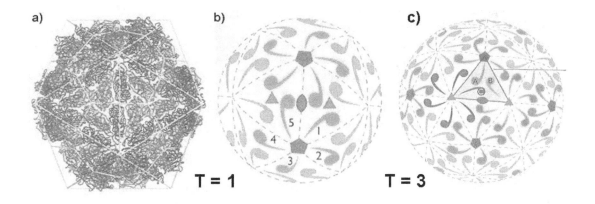

Figure 6.10 a) The capsid of the L-A virus particle (viewed down to an icosahedral 2-fold axis) is made of 60 protein subunit dimers (see Fig. 6.7) grouped in capsomers containing five dimers each. (Adapted by permission from Springer Nature, Nature Structural Biology, L-A virus at 3.4 Å resolution reveals particle architecture and mRNA decapping mechanism, Naitow H, Tang J, Canady M, Wickner R B, Johnson J E, 9 725-728. Copyright 2002). b) Sketch showing an ideal icosahedral packing in simple icosahedral virus structures with triangulation number $T = 1$. A comma represents a single protein chiral molecule which forms the basic structural unit by itself, and every molecule is related to its neighbors through head-to-head, back-to-back and tail-to-tail interactions, respectively corresponding to the 2-, 3-, and 5-fold axes characteristic of icosahedral symmetry. c) In the $T = 3$ structure, able to accommodate 180 proteins, there are three modes of packing protein subunits to obtain the basic structural unit. The A units are present in pentamers, formed by tail-to-tail interactions. The B and C subunits are arranged in rings of six molecules via tail-to-tail interactions as well. In addition the A units interact in rings of three with B and C subunits (via back-to-back interactions), and in pairs (via head-to-head interactions). Despite their packing differences, the bonding interactions among A, B, and C subunits are considered to be similar. (Adapted from [252]).

(not necessarily equilateral), and so on. In fact, in order to account for most morphological units reported from microscopic imaging of virus particles one must assume the existence of at least *three* kinds of chemical interactions among neighboring proteins subunits involving three different interactions centers, which are generically referred to as *tail* (t), *head* (h), and *back* (b) ones (Fig. 6.10b). These interactions are related to a specific symmetry element of the icosahedral group, namely:

- $h-h$ interactions, involving two proteins in front each other (related to 2-fold axis),

- $b-b-b$ interactions, involving three proteins in a triangular configuration (related to 3-fold axis), and

- $t-t-t-t-t$ interactions, involving five proteins in a pentagonal arrangement (related to 5-fold axis).

Furthermore, it is assumed that, although tail, back and head interactions involve different protein locations, all of them exhibit the *same* intensity and directional properties (i.e., they are effectively *isotropic*), so that one can reasonably assume that $t - t \simeq h - h \simeq b - b$ to all effects.[14] This is the main tenet of the so-called *quasi-equivalence principle* (QEP), originally introduced in 1962 by Donald L. D. Caspar and Aaron Klug (1926–2018) on the basis that the lowest energy capsid structure is expected to have the maximum number of most stable bonds formed [113]. According to the QEP, protein subunits can be positioned on a *triangulated* icosahedral surface lattice such that the environments of the individual proteins could be similar although not identical.[15] The size of the capsid shell can then be properly increased, still preserving the basic icosahedral design shown in Fig. 6.10b, by partitioning the surface of this icosahedron into a larger number of triangles, which are subsequently properly scaled up (Fig. 6.10c). But, now the facet triangles will no longer all have the protein subunits located in equivalent positions, so that the three protein subunits originally assigned to each triangular face will not generally occupy equivalent environments. In particular, the presence of hexagonal rings containing six protein subunits, which apparently are not compatible with icosahedral symmetry, naturally occur. In this way, the resulting polyhedra will be able to accommodate $12 \times 5 + 10(T - 1) \times 6 = 60T$ protein subunits.

Accordingly, the assembly of viral capsids exhibiting six 5-fold, ten 3-fold, and fifteen 2-fold icosahedral symmetry axis can be physically realized, albeit containing *local* hexagonal patterns as well. In order to predict the locations and relative orientations of the different capsomers Caspar and Klug considered embeddings of the surface of an icosahedron into an hexagonal periodic planar lattice, such that the vertices of the equilateral triangular faces met centers of the hexagons in the underlaying lattice (Fig. 6.11). In this way, by replacing each involved hexagon by six equilateral triangles, triangulations of the plane are obtained that are compatible with the overall icosahedral symmetry. By decorating each triangular facet in such a triangulation net with a protein subunit in each corner one obtains clusters of five proteins (pentamers) at the 5-fold axis of icosahedral symmetry, as well as clusters of six protein subunits (hexamers) otherwise, in a natural way.

It resulted that the problem of arranging units in a hexagonal close packing on an icosahedral surface, with local pentagonal arrangements at certain vertices, had earlier been solved by Michael Goldberg (1902–1990) in 1937, who introduced a type of polyhedra particularly well suited to this end [296]. Indeed, the so-called Goldberg polyhedra are convex polyhedra defined by the following properties: (1) each face is either a hexagon or pentagon, (2) exactly three faces meet at each vertex, and (3) they have full *rotational* icosahedral symmetry (I group). Icosahedral symmetry ensures that the pentagons must be regular ones, although many of the hexagons may not be. Goldberg polyhedra can be derived from a hexagonal plane periodic lattice by converting some of the hexagons to pentagons. Once an origin is defined, every hexagon in the lattice can be uniquely identified by the number of steps along the **h** and **k** directions in the hexagonal net (Fig. 6.11). To construct a given Goldberg polyhedron, one face of the desired scaled up polyhedron is drawn in the hexagonal net. The origin is replaced with a pentagon, as well as the hexagon located at the (h, k) position. The third replaced hexagon is then obtained by completing an equilateral triangle. These substitutions have the effect of folding the plane into a closed surface. To this end, since an icosahedron has 6 axes with 5-fold symmetry, twelve pentagons must be introduced in the lattice in substitution to hexagons to form a closed structure with icosahedral symmetry [392]. In the simplest case, all the hexagons around a given origin are converted to

---

[14]Note that no $t-b$, $h-t$, or $h-b$ interactions are allowed to occur.

[15]Triangulation refers to the description of triangular faces of a larger icosahedral structure in terms of its subdivision into smaller triangles termed facets.

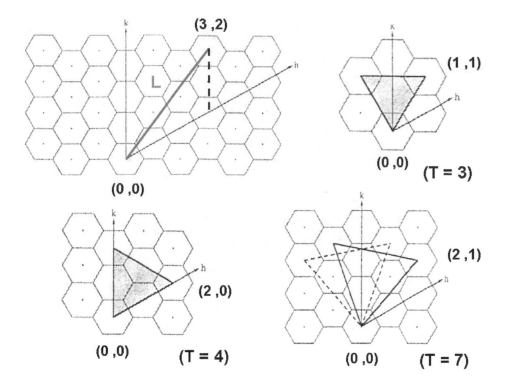

Figure 6.11 Geometric procedure for generating Goldberg polyhedra from a hexagonal planar lattice. The vertices of regular triangular faces of the Goldberg polyhedron hexagons coincide with the 6-fold axes of the plane hexagonal lattice. The coordinates of the hexagons centers are referred to the axes **h** and **k**.

pentagons and a pentagonal dodecahedron in obtained. Therefore this Platonic solid is but a particular case of the more general family formed by Goldberg polyhedra.

In the more general cases the locations of the substituted pentagons are given by the so-called Goldberg indices (h, k), where $h, k \in \mathbb{N}$, are the indices of the hexagonal translation mapped onto the edge. In this way, the so-called *triangulation number* naturally appears in the process of obtaining a Goldberg polyhedron from a hexagonal net, and it is given by the ratio between the area of the polyhedron equilateral triangle face of size L on the hexagonal lattice and the area of a reference equilateral triangle of size $L_0 = 1$, that is (see Fig. 6.11)

$$T = \frac{\frac{\sqrt{3}}{4}L^2}{\frac{\sqrt{3}}{4}} = L^2 = h^2 + k^2 - 2hk \cos \frac{2\pi}{3} = h^2 + k^2 + hk. \qquad (6.4)$$

The triangulation number T then gives the number of facets per face.[16] The significance of this number is that it introduces a selection rule, so that only certain numbers of protein subunits can be expected to form closed polyhedral shells. In fact, since an icosahedron has 20 faces, and each facet is assumed to have three protein subunits, one gets $60T$ protein subunits in total in a virus capsid characterized by a given T number. We note that, in terms

---

[16]We note that T is invariant under the exchange $h \leftrightarrow k$.

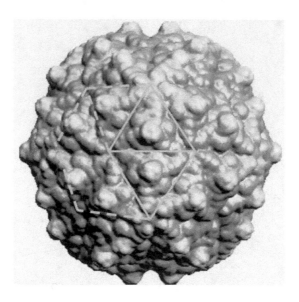

Figure 6.12 Capsid protein structure of flock house virus nodavirus ($T = 3$), which exhibits the shape of a rhombic triacontahedron. (Adapted from VIPER EMdb virus capsid repository [937]).

of the triangulation number one can express the perimeter, surface and volume of a regular polyhedron of edge a in the form $P = ET^{1/2}$, $S = S_0 T$, and $V = V_0 T^{3/2}$, respectively, where S_0 and V_0 values depend on the considered polyhedron. As a consequence, one can readily check that both the surface-to-volume ratio and the perimeter-to-surface ratios are independent of the adopted T value. Attending to their related T number, Goldberg polyhedra can be classified into three families, namely: (1) the $G(n, 0)$ family for which $T = n^2$, (2) the $G(n, n)$ family for which $T = 3n^2$, and (3) the general $G(n, m)$ $n \neq 0$, $n \neq m$ family. These polyhedron families can be related to three virus capsid classes according to a one-to-one correspondence [580]. Therefore, the T number allows for a systematic classification of the virus capsids arranged according to the (h, k) indices (**Exercise 6.2**).

Accordingly, the T number can be properly regarded as a parameter giving a measure of the structural *complexity* of a virus capsid. In the simplest $T = 1$ case ($h = 0, k = 1$) the interactions of all protein subunits with their neighbors are identical, leading to a pentagonal dodecahedron (i.e., G(1,0)) overall arrangement of the capsid shell. For the next structure in complexity degree ($T = 3$) two main capsid geometries have been reported. The most common capsid geometry exhibits the basic arrangement of 12 vertices with 5-fold symmetry clusters along with 20 hexagonal facets interposed between them present in a G(1,1) polyhedron, which coincides with a truncated icosahedron, an Arquimedean solid exhibiting the best S^3/V^2 ratio in Table 6.1. On the other hand, there also exist $T = 3$ class virus capsids (e.g., nonadavirus, tomato bushy stunt virus, bean mosaic virus) which *cannot* be described in terms of Goldberg polyhedra.

Indeed, in this case the capsid shape is described by a rhombic triacontahedron, exhibiting the best P^2/S ratio in Table 6.1, which is a Catalan solid but not a Goldberg polyhedron, since it lacks hexagonal facets (Fig. 6.12). In fact, it is comprised of 30 golden rhombs forming a dihedral angle of $2\pi/5$, which are joined at 60 edges and 32 vertices, of which twelve are 5-fold and twenty are 3-fold ones.[17] Thus, rhombic triacontahedron capsids do not form true hexamers but, rather, a series of trimers of dimers about each of the icosahedral 3-fold axes [392]. One of such dimers is shown in Fig. 6.12 (solid line) as a part of the pseudo-hexagon. Accordingly, the $T = 3$ capsids exhibiting a rhombic triacontahedron shape are structurally simpler than $G(1,1)$ ones, since they lack hexagonal arrangements but low symmetry rhombic ones, and we only require one type of geometrical motive (golden rhombs) instead of the pentagons and hexagons present in truncated icosahedra.

From a chemical point of view, however, they must be considered more complex since triacontahedral capsids belonging to the $T = 3$ class viruses result from two distinctly *different interactions* between identical subunits at 5- and 2-fold axis. The variations in protein conformation required to this end, involve an alteration between order and disorder of the flexible regions located near the N- and C-termini, which are commonly referred to as molecular switches. When in an ordered state, these protein arms interdigitate with their neighbors serving as wedges between proteins that are arranged in a hexagonal pattern, so that the angle between the wedged proteins along the 6-fold symmetry axis is relatively flat, whereas the angle between non-wedged proteins along the 5-fold symmetry axis is relatively sharp [392].

6.1.4.4 Quasi-equivalence principle limitations

The great success of the Caspar and Klug model was its ability to explain the vastly different capsid structures of many viruses in terms of an underlying arrangement of the protein subunits. To this end, the QEP requires a defined valence about each capsomer, that is, pentamers should have five neighboring proteins and hexamers should have six. Nonetheless, as one considers large T values it is reasonable to expect a significant level of non-equivalence in capsid assembly. Indeed, although in theory quasi-equivalent shells of any size can be formed, in practice the largest capsids formed of a single subunit type are $T = 7$. By relaxing the single subunit type condition, that is, allowing the capsid to contain more than one subunit protein type one can get $T > 7$ capsids. Accordingly, capsids with large T numbers appear to require a relatively large number of different protein subunits. For instance, adenoviruses capsid is composed of a variety of proteins with 12 pentamers on the vertices of an icosahedron and 240 pseudo hexamers covering the faces. Hence, adenovirus achieves a pseudo $T = 25$ shell by making a variety of approximations to the original concepts proposed, yet the result is a lattice that places pentamers and pseudo hexamers on the rotation axis predicted by the Caspar and Klug model. This example illustrates that the surface lattice notion can be utilized even when the original hypothesis for its existence, namely the presence of a single type of subunit in the capsid that is capable of forming both hexamers and pentamers, does not hold [393].

Indeed, during the 1980s, several viruses with relatively large T values were observed to show some features not compatible with the standard Caspar and Klug model and there are now a growing number of experimentally resolved structures which do not satisfy this model predictions about local protein arrangements. For the sake of illustration let us consider the case of papovaviruses, which are oncogenic viruses that include the polyoma and papilloma virus subfamilies. The papovaviruses have a common capsid structure composed of 72 capsomers, and they were originally classified as belonging to the $T = 7d$ class[18] on

---

[17]The short diagonals of the rhombi form the edges of a pentagonal dodecahedron, and the long diagonals that of an icosahedron. Accordingly, the rhombic triacontahedron is the dual of the icosidodecahedron. The rhombic triacontahedron forms the hull of the projection of a 6D hypercube to 3D (see [840]).

[18]For $T \geq 7$ the capsid can appear in two different enantiomorphic forms, obtained by permuting the Goldberg indices $h \leftrightarrow k$, respectively. An example is shown in Fig. 6.11 for the case $h = 2$, $k = 1$ (solid

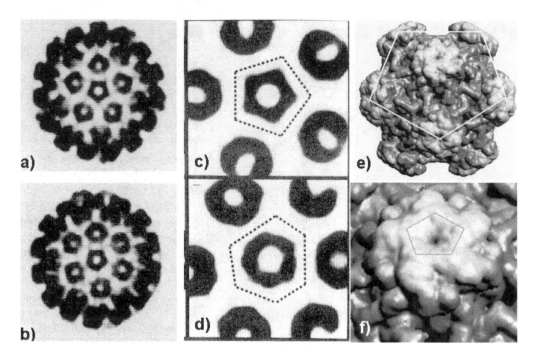

Figure 6.13 Electron microscopic views of half the capsid of bovine papilloma virus ($T = 7$, about 60 nm in diameter) down the 5-fold a) and 6-fold b) axes. Electron density sections showing the substructure of the pentavalent c) and hexavalent d) morphological units indicate they are pentamers formed by five interacting proteins. (Reprinted by permission from Springer Nature, Nature (London) 295 110–115, Polyoma virus capsid structure at 22.5 Å resolution, Rayment I, Baker T S, Caspar D L D, and Murakami W T, Copyright 1982). e) Recombinant human papilloma virus HPV16 (from VIPER EMdb virus capsid repository [937]) exhibiting a series of protrusions (capsomers) arranged into an overall pentagonal dodecahedral pattern ($T = 1$). f) A close up view of one capsomer, which is composed of five L1 protein subunits arranged themselves into a regular pentagon geometry.

the basis of low-resolution micrographs, thereby assuming their capsids were made of 420 protein subunits. Later on, the structure of polyoma virus was determined at a higher resolution by single crystal X-ray diffraction in 1982, and the 5-fold symmetry present in the diffraction data was interpreted as indicating that *all* 72 capsomers present on the virus surface were built from five identical or very similar protein *pentamers* coordinated to their neighbors in *both* hexavalent and pentavalent configurations (Fig. 6.13) [731]. Thus, in these virus capsids, there exist 12 *pentavalent* pentamers making edge-to-edge contacts with their neighbors and 60 *hexavalent* pentamers, instead of 12 pentamers and $10(T-1)$ hexamers as expected. Then, the capsids of papovaviruses contain $12 \times 5 + 60 \times 5 = 360$ protein subunits, which would correspond to a hypothetical $T = 6$ value that is *not allowed* by the triangulation number selection rule (**Exercise 6.2**). Therefore, the notion of hexamer and pentamer capsomers must be replaced with that of hexavalent (pentavalent) capsomers, respectively.

---

triangle) and $h = 1$, $k = 2$ (dashed triangle). The corresponding polyhedra are mirror images to each other, and are respectively labeled as $T = 7d$ (dextro form) and $T = 7l$ (levo form).

In fact, one can reasonably expect that the coordination value of a given capsomer in a typical modularly designed biological structure is not merely an intrinsic property of such a capsomer, but also depends on external conditions, and may be very sensitive to them when the difference in free energy of possible associations of different kinds is small [54].

Thus, while the Caspar and Klug model predicts the correct location of capsomers by triangulating the capsid surfaces according to a $T = 7$ pattern,[19] it erroneously assumes 420 protein subunits rather than the 360 subunits organized in pentamers that are actually observed. Many more examples of anomalous virus capsids have been reported. For instance, the capsid of L–A virus is classified as $T = 2$, which is impossible at integers h and k after Eq. (6.4). On the other hand, the capsid of dengue virus and the capsids of other flaviviridae (Zika virus, yellow fever virus, and Western Nile virus) do not contain hexamers at all. The capsids of nucleocytoplasmic large viruses of asfarviridae, ascoviridae, iridoviridae, poxviridae, mimiviridae, and phycodnaviridae viral families do not contain any hexamers either, but instead have trimers and pentamers as their structural units. The violation of the Caspar and Klug model points to the fact that the assumption of QEP is too restrictive for many viruses.

This drawback has spurred alternative approaches, which consider the entire virus capsid as an extended chemical network involving local protein interaction centers located at specific amino acid residues as playing the role of effective *chemical lattice* points. On the basis of these chemical points (vertices) one may introduce segments (edges), planes (polygonal tiles) and polyhedra in order to describe the modular design characteristic of virus capsids. In this way, the hexagonal lattice picture in terms of triangulations is no longer necessary, permitting tesselations in terms of other shapes, such as rhombs, trapezoids, or pentagons. Due to the requirement of 5-fold symmetry axis, some of these tiles need to have some corner with an angle of $2\pi/5$. These criteria then represent a generalization of the QEP in the context of a tiling approach [915]. In accordance with this generalized QEP, protein subunits (represented by dots) are located precisely at those corners of the tiles which subtend the same angle. This assumption ensures that identical types of protein subunits can only occupy structurally equivalent sites on the tiles. From a mathematical point of view, this generalization manifests itself as a requirement on the *decoration* of the tiles. As we see, the proposed tiles have more vertices where the different protein subunits may be allocated than in the triangular tiles case, so that certain criteria must be adopted in order to properly assign which vertices are related to a protein subunit and which ones are not.

To this end, the knowledge gained from aperiodic tilings can be fruitfully used [915]. Since pentagons can not tessellate the plane without gap nor overlaps, a straightforward extension of the Goldberg construction is not possible. The use of aperiodic tiling notions, originally introduced in the study of quasicrystalline alloys, has been successfully borrowed to properly classify the protein structure of certain virus families during the last decade. Thus, the possible design of suitable generalized lattices via projection from higher-dimensional lattices with icosahedral symmetry was studied in detail [916]. On the other hand, it was demonstrated that the location of different protein subunits on the virus capsid can be precisely determined by properly blending rotational and invariance scale symmetries in a 6D crystallographical space, thereby introducing an elegant unified approach [374].

Alternatively, one can nicely fit the bovine papilloma virus electron density map shown in Fig. 6.14a by dividing a sphere into 12 spherical pentagons instead (Fig. 6.14b). The

---

[19]Accordingly, the nomenclature based on the triangulation number can still be used for these viruses, for the sake of convenience, with the proviso that the $60T$ rule is no longer valid.

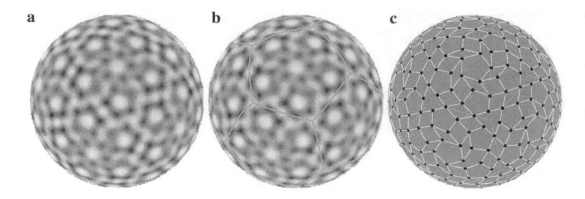

Figure 6.14 a) Structure of the bovine papilloma virus determined by image reconstruction of cryoelectron micrographs at a radius of 23 nm. The intensity maxima (corresponding to the positions of proteins centers) are located at the centers of clear circles. b) Superimposition of the protein density map with a dodecahedral tessellation of the sphere. c) The idealized quasilattice of protein density maxima. (Reprinted figure with permission from Konetsova O V, Rochal S B, and Lorman V L, Chiral quasicrystalline order and dodecahedral geometry in exceptional families of viruses, Phys. Rev. Lett. 108 038102 2012 Copyright (2012) by the American Physical Society).

vertices of these pentagons correspond to the vertices of a dodecahedron inscribed in the sphere. Accordingly, the capsid can be *pentagulated* instead of triangulated [114]. Inside each of the resulting spherical pentagons, the protein center positions can be connected by lines of approximately equal length which define the nodes of an ordered tiling characterized by a local pentagonal symmetry. The tiling is composed of three types of tiles, namely, regular pentagons, and thin and thick rhombuses (Fig. 6.14c). In addition, there also exist some defects in the form of equilateral triangles with their centers located at the dodecahedron vertices [442]. This approach has been successfully applied to other kinds of virus capsids as well [443, 699].

6.1.4.5   Virus quasicrystals?

In the previous sections, we have focused on the study of viruses as macromolecular protein aggregates. In this section, we will zoom out our perspective by considering virus particles themselves as the building blocks of giant unit cell supramolecular virus crystals, as it is illustrated in Fig. 6.15. The first virus to be crystallized was the tobacco mosaic virus (TMV) by Wendell M Stanley (1904–1971) in 1935, followed by tomato bushy stunt virus (TBSV), crystallized by Norman W Pirie (1907–1997) and Frederick Bawden in 1936, and the turnip yellow mosaic virus (TYMV), which crystallizes from salt solutions as isotropic octahedra. Poliomyelitis virus was the first animal crystal virus, obtained by F L Schaffer and C E Schwerdt in 1955. In Table 6.2, we list the crystallographical data of different crystal virus representatives, including all possible classical crystallographical systems. It is interesting to note that, depending upon the adopted growth conditions a given virus can crystallize in different systems. For instance, GFLV virus crystals corresponding to spatial groups P1 (1, triclinic), P2 (3, monoclinic), and $P2_13$ (198, cubic) have been reported [784].

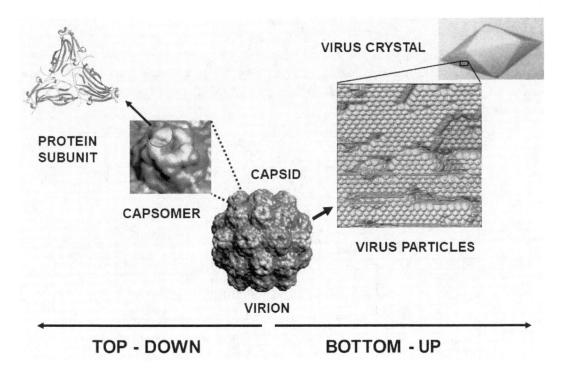

VIRUS CRYSTAL

PROTEIN SUBUNIT

CAPSID

CAPSOMER

VIRUS PARTICLES

VIRION

TOP - DOWN

BOTTOM - UP

Figure 6.15   (Right top) At a macroscopic scale the turnip yellow mosaic virus crystal shows a hexagonal bipyramid habit (McPherson A, DeLucas L, Microgravity protein crystallization, Microgravity 1 15010 (2015) doi: 10.1038/npjmgrav.2015.10 Creative Commons Attirbution 4.0 International License). This organic supramolecular crystal is composed of polyhedral virus particles (adapted from [293]), which are in turn made of protein subunits grouped in ordered capsomers [937].

Virus crystallization is ruled by a combination of chemical and geometrical factors, where the size and relative orientation of the capsids along the resulting crystal symmetry axes play a fundamental role. In this regard, it was a fortunate fact that all TBSV virus particles will adopt the same spatial configuration in the final crystalline array, hence providing X-ray diffraction patterns amenable to be easily interpreted. This virus crystallizes into a fcc lattice (spatial group I23), so that two symmetry axes of the virus capsid are oriented along crystal directions, namely, a 2-fold axis along [100] and a 3-fold axis along the [111] directions, respectively. The remaining capsid symmetry axes are therefore effaced from the resulting virus crystal. On the contrary, in TYMV virus crystals the capsids lie with their centres at the lattice points of a bcc unit cell of side 35 nm, but are not all of them in the same orientation, so that the true crystallographic unit cell has a side of 70 nm (8 times larger than that of TBSV) and contains 16 virus particles [262].

Chemical aspects play a key role when different viruses crystallize into similar growth habits. This is the case, for instance, of poliovirus and rhinovirus (both members of the picornavirus family) isomorphic crystals (see Table 6.2), stemming from a fine correlation among the chemical interaction centers of all the involved capsids [234]. Detailed studies aimed at evaluating the balance between chemical and purely geometrical factors in the resulting growth habit of virus crystals, have disclosed that the relative size of capsids along their different symmetry axes plays a significant role in the emergence of local ordering

Table 6.2 Virus crystals crystallographic data including the triangulation number (T), the capsid diameter, the spatial group (SG), and the lattice parameters. Key: Black beetle virus (BBV), tomato bushy stunt virus (TBSV), cowpea mosaic virus (CPMV), flock house virus (FHV), southern bean mosaic virus (SBMV), HK97 bacteriophage, Nodamura virus (NOV), Coxsackie virus (CVB3), grapevine fanleaf virus (GFLV).

VIRUS	T	$\varnothing$ (nm)	SG	a, b, c (nm)	α, β, γ (°)	REF
BBV	3	33.2	$P4_232$	36.2	90.0	[636]
TBSV	3	34.6	$I23$	38.6	90.0	[323]
Polioma	$7d$	49.5	$I23$	57.2	90.0	[234][731]
CPMV	1	30.6	$P6_122$	45.1 45.1 103.8	90.0 90.0 120.0	[503]
FHV	3	33.4	$R3$	32.36	67.7	[636]
SBMV	3	30.2	$R32$	31.79	63.435	[636]
HK97	$7l$	60.0	$P4_32_12$	101.0 101.0 73.0	90.0 90.0 90.0	[271]
Poliovirus	1	32.0	$P2_12_12$	32.4 35.9 38.1	90.0 90.0 90.0	[732]
Rhinovirus	1	31.4	$P2_12_12$	32.3 35.8 38.0	90.0 90.0 90.0	[234]
NOV	3	34.8	$P2_1$	56.21 35.41 61.28	90.0 110.9 90.0	[636]
CVB3	1	34.6	$P2_1$	57.46 30.21 52.16	90.0 107.7 90.0	[636]
GFLV	1	30.6	$P1$	27.84 29.83 29.16	102.0 116.4 108.2	[784]

effects in global low-symmetry crystals, such as those belonging to monoclinic and triclinic systems [636].

Is it possible that an ensemble of icosahedral virus particles may eventually pile up to give rise to a virus iQC? To the best of my knowledge the existence of such sort of soft matter biological QC has not yet been reported, and its search would be a very interesting goal for the time to come.

6.2 THE QUEST FOR PERFECT QUASICRYSTALS

> "In all these cases, the limiting shape or number was simpler and more symmetrical than its finite approximators. This is the allure of infinity. Everything becomes better there". (Steven Strogatz, 2019) [845]

In previous chapters, we have learnt that self-similar features in the diffraction patterns of both metallic and soft QCs, as well as PQCs, stem from the presence of scale invariant (inflation/deflation) symmetry, along with the local isomorphism property of their underlying building blocks in physical space. In turn, these symmetries determine the presence of typical fractal-like features in the resulting energy and frequency spectra, ultimately determining their physical properties. Thus, an appealing motivation for the experimental study of QP solids was the theoretical prediction that they should exhibit peculiar quantum states, described in terms of power law decaying QP wavefunctions, belonging to highly fragmented energy or frequency spectra displaying fractal self-similar patterns in the ideal QP limit $N \to \infty$ (see Secs.2.6.2 and 2.7) [8, 9, 532, 544].

In fact, only in the thermodynamic limit it can be properly said that the highly fragmented spectrum of a QP lattice consists of a self-similar Cantor set with zero Lebesgue measure. Nevertheless, although mathematically derived results deal with ideal, arbitrarily large systems, actual systems are always finite in length. Fortunately, the degree of fragmentation of energy and frequency spectra progressively increases as the size of the considered system is increased. This implies that, strictly speaking, the so-called approximant phases should exhibit the characteristic physical properties of QCs in a lesser degree, as it is experimentally observed (see Sec. 3.3.2). Therefore, in order to experimentally check most theoretical results one must follow a systematic study, looking for the possible presence (or not) of certain trends in the physical properties as the size of the considered system is progressively increased. In this way, one may expect that intrinsic properties of QCs should progressively grab the limelight as the inflation symmetry of the underlying structure properly extends through all scale lengths, ranging from the atomic to the macroscopic domains in a perfect QC. Therefore, the quest for perfect QCs is motivated by the pursuit to obtain experimental evidences of long-range QPO effects in the physical properties of QCs.

Accordingly, large size samples of very high structural quality are needed to this end. However, perfect structural order in both QP and periodic crystals is an abstract concept that is rarely (if ever) realized in real samples, which unavoidably exhibit an equilibrium quantity of thermal vacancies and some dislocations [372]. Furthermore, inflation symmetry underlying QCs geometry imposes very restrictive constraints on the possible presence of any sort of defects in the spatial structure of the considered solid. On the other hand, most information about long-range QPO is in the very weak reflections with large perpendicular-space components of the diffraction vectors. This would correspond to rational approximants with lattice parameters of the order of micrometers and billions of Bragg reflections (**Exercise 6.3**) [838]. Thus, one may certainly expect that a direct observation of fully intrinsic properties of QP arrangements of matter would be a very difficult experimental task.

Consequently, one may reasonably ask as to whether the quality of the quasicrystalline samples studied to date is high enough to properly display some of the expected intrinsic properties of perfect QP orderings of matter, even in those samples whose structural quality is comparable to that achieved for the best periodic crystals. Certainly, structural order can be improved by thermal annealing at a temperature high enough that atomic diffusion can relax the structure. In fact, an insufficient annealing time may result in poor reproducibility of the measurements, since most physical properties are quite sensitive to the metallurgical state of the sample. Thus, special care has to be taken by annealing samples for various lengths of time, and subsequently measuring their properties until the results converge and are reproducible. Only proceeding in such a careful and systematic way it has been possible to clearly confirm that the counter intuitive increase of resistivity upon annealing is directly related to an improvement of the long-range QPO in metallic QCs, for instance [107]. In a similar vein, a detailed comparison of the optical reflectivity properties of mineral QCs found in certain meteorites and laboratory synthesized metallic QC analogs, has disclosed significant differences between their reflectance spectra (see Sec. 3.4.3). Since mineral samples exhibit a better structural quality than the synthetic ones (**Exercise 1.5**), this result properly illustrates the importance of studying very high quality QCs in order to be able to disclose intrinsic properties stemming from the QPO itself, and naturally spurs the interest in further comparative analysis of other transport properties of both sorts of compounds. In this regard, the study of physical properties of QCs grown under micro-gravity conditions will be of great interest as well.

In the case of high quality QCs belonging to the AlPdRe system, the possible existence of a metal-insulator transition at very low temperatures was considered, though earlier works assumed this behavior as arising from quantum interference effects, rather than signaling the emergence of a specific property of long-range QPO in these materials. In that case, contrarily to the usual situation in amorphous metallic systems the presence of quantum interference effects would indicate an insulating-metallic transition rather than a metallic-insulating one. This scenario can be pictured in terms of interference effects among delocalized electronic wavefunctions leading to the formation of resonant standing waves over all the spatial scales involved in the QP structure, albeit the experimental confirmation of this feature still remains an open question in the field [371, 756]. It is important to highlight that this localization phenomenon in QCs can be properly regarded as the *opposite* effect to the well-known Anderson localization induced by defects in periodic ones. Indeed, ideally one may think of a metallic QC as a hierarchical superstructure based on a self-similar network of nested atomic clusters. The chemical bonding in such a superstructure can then be envisioned as resulting from a superposition of resonant states related to each nested cluster unit. At $T = 0$, the resulting network of self-similar standing waves can be regarded as a new kind of chemical bonding mechanism whereby originally metallic-like (i.e., extended) sp states eventually collapse into a covalent-like structure due to strong sp-d hybridization effects with the QP network of transition metal atoms located inside different cluster shells. In that case, ideal QCs at zero temperature will be perfect insulators. As the temperature is progressively increased electron–phonon interactions will destroy the electronic standing waves, hence increasing electronic conductivity, as experimentally observed (see Sec. 3.4.1). Further increasing the temperature may lead to some sort of dynamical localization of the eigenstates. In this scenario, an activated hopping mechanism among localized states takes over and the self-similarity determined by the spatial cluster distribution naturally leads to a variable range hopping process. Eventually, as the temperature is further increased the activation of high energy vibration normal modes of atomic clusters could give rise to a significant release of charge carriers from atomic clusters, thereby enhancing both electrical and thermal conductivities.

Translation invariance provides a useful tool to determine the stoichiometry of a perfect periodic crystal from the knowledge of the number and chemical nature of the atoms contained in its unit cell. Nevertheless, since QCs lack a well-defined unit cell (alternatively, one may consider that their unit cell has an effectively infinite size) such a procedure can not be applied, and the question about the proper way to determine the stoichiometric proportion of a perfect self-similar structure naturally arises. In fact, scale invariance symmetry now replaces the periodic boundary conditions, and the criteria used to count the atoms participation ratio at the unit cell borders in periodic crystals are no longer applicable in their standard form. Taking into account the dominant role played by the golden mean in the basic geometry of both dQCs and iQCs one may reasonably expect this ratio to explicitly appear in the resulting theoretical stoichiometries of these alloys. Thus, by assuming that the stoichiometric ratios in ternary iQCs of the form $A_x B_y C_z$, with $x > y > z$, obey the nested relationships $y = \tau^n z$, $x = \tau^n(y + z)$, along with $x + y + z = 1$, one obtains compositions which are relatively close to the experimentally reported ones for i-AlCuFe ($n = 1$) and i-AlPdMn ($n = 2$) representatives, respectively (**Exercise 6.4a**) [376]. As compared with the ideal obtained compositions the AlPdMn real QCs have slightly less aluminum and slightly more palladium, while the AlFeCu ones contains 2% less iron and 2% more copper. Quite interestingly, these experimental compositions are very close to that corresponding to the 5/3 and 3/2 periodic approximants, respectively, when calculated making use of Eqs. (6.11)–(6.12), but with 5/3 or 3/2 instead of τ.

Ideal stoichiometries stemming from a long-range extended chemical network of alternating weak and strong covalent bonds was discussed for i-AlPdRe QC on the basis of detailed *ab initio* band structure calculations of 1/1 and 2/1 crystal approximants [449]. The exact obtained expression reads $Al_{37+2}Pd_{\frac{2\tau+1}{2}}Re_{\frac{\tau+1}{3}}$, which can be approximated as $Al_{69.62}Pd_{21.51}Re_{8.86}$, leading to the ideal $e/a = 2.4261\ldots$value for the electron per atom ratio in this alloy system (**Exercise 6.4b**). On the other hand, in the case of ddQCs the resulting stoichiometric ratios are related to the irrational number $\sqrt{2 + \sqrt{3}}$, directly related to the geometric structure of the underlying quasi-lattice, as expected [454]. Similar results have been obtained in the case of approximant crystal structures (which exhibit giant unit cells), illustrating the connection between crystal structure in nature and the mathematical notion of rational approximants of an irrational number. These results naturally rise the question as to whether the QCs synthesized in laboratories so far are really close to the ideal stoichiometry stemming from inflation symmetry in irrational fractal solids, or they should be rather considered as just good approximants from a stoichiometric viewpoint. If so, the quest for better *irrational stoichiometry* QCs in the phase diagrams of suitable alloy systems would be a quite appealing task. In this regard, it is interesting to consider that most metallic QCs were originally found when studying alloy systems which were mainly explored for industrial applications.[20] This suggests that some room is left for the search of possible QCs based on more suitable atomic compositions and elemental atoms choices, aimed at attaining a much better long-range QP structure to enhance the emergence of intrinsic QPO related physical properties.

6.3 FROM QUASICRYSTALS TO HYPERCRYSTALS

> *"The influence of aperiodic order on physical and chemical properties is still under debate, but aperiodic crystals cannot be ignored any longer, nor can they be considered as 'bad crystals'!".* (Marc de Boissieu, 2019) [166]

6.3.1 On the quasicrystal notion and definition

> *"I was a member of the commission when this definition was hammered out, and I argued strongly in favor of it. It wasn't a cop-out; it was designed to stimulate research. (...) That the commission still retains this definition today suggests the difficulty of the question we deliberately but implicitly posed".* (Marjorie Senechal, 2006) [794]

Certainly, solid state physics community was not prepared for the advent of QCs. The development of theoretical crystallography naturally led to a magnificent

---

[20]For instance, Al-Mn (3xxx) and Al-Cu-Li (2xxx) alloys were studied in order to improve mechanical properties of materials used in aeronautical industry. Al-Mg-Zn (7xxx, 7xx) alloys are also of interest due to their mechanical properties.

classification scheme introducing the 32 crystallographic point groups and the 230 space groups that exhausted (as it was thought from 1890 on) the pure crystalline possibilities in 3D.[21] The resulting intellectual confidence prevented us from seeking other theoretical possibilities of long-range orderings of matter, non-periodic, albeit regular arrangements of atoms able to bond and grow and eventually crystallize. This was a very unfortunate outcome, for the knowledge of such a possibility might have prepared us to expect the very existence of additional crystals forms, and not to overlook them when their fingerprints were serendipitously found in empirical observations [288]. For that knowledge may have probably avoided that our superb theory was ultimately reduced to a humble, provisional definition of what a crystal is...

This book opens with a quotation introducing the very QC *notion*, and this section begins with a quotation commenting on the difficult task aimed at introducing a proper *definition* of what a QC is. This fundamental issue is still fully open currently. For instance, one of the members of the first Commission on Aperiodic Crystals (established in 1991) stated in 2006 *"a QC is a crystal with forbidden symmetry"*, [794] albeit the existence of forbidden symmetries is not actually required nowadays, according to the *Online Dictionary of Crystallography* of the International Union of Crystallography (see Sec. 1.1.5). Indeed, it has been argued that the prevailing definition of QCs, requiring them to contain an axis of symmetry that is forbidden in periodic crystals, is inadequate, since this definition is too restrictive in that it excludes an important and interesting collection of structures that exhibit all the well-known properties of QCs without possessing any forbidden symmetry [495]. In this vein, we can read *"QCs are long-range ordered solids in which the diffraction pattern exhibits symmetries that are incompatible with translational symmetry"*. This is illustrated by the 'cubic' quasicrystalline phase of $Al_{69}Pd_{22}Mn_9$, that is, a non periodic structure of cubic symmetry [163], or by non-periodic square-triangle tilings of the plane with 6-fold symmetry [672]. This debate has naturally led to a somewhat solomonic attitude expressed in the terms *"A quasicrystal is thus considered to be characterized by the absence of a Bravais lattice and/or the observation of 'forbidden' crystallographic symmetry* [888].

Other QC definition proposals focus on their spectral properties instead. For instance, Freeman J. Dyson (1923-2020) stated that

> *"A quasi-crystal is a distribution of discrete point masses whose Fourier transform is a distribution of discrete point frequencies. Or to say it more briefly, a quasi-crystal is a pure point distribution that has a pure point spectrum. This definition includes as a special case the ordinary crystals, which are periodic distributions with periodic spectra."* [218]

This definition nicely fits with the spectral classification chart illustrated in Fig. 2.2, and it also highlights the conceptual primacy of QPO over that of periodic order, which is just a particular case within the even larger set of almost periodic functions (see Fig. 1.17).

In the earlier times of the field, QCs were mainly denoted in terms of negative attributes rather than positive ones. Thus, it was routinely said that QCs exhibit both *forbidden* symmetries (in reference to those symmetries allowed for periodic crystals) and *anomalous* transport properties (in comparison to those observed in both periodic and amorphous metallic alloys). As a consequence, one was naturally tempted to consider QCs as some sort of defective, flawed materials. An alternative, more appealing approach should focus on the characteristic features of QCs by their own, for these novel features endow quasicrystalline materials with new, previously unsuspected properties. For example, QCs display

---

[21]The interested reader is referred to the link https://www.iucr.org/__data/assets/pdf_file/0020/749/ fedorov.pdf

long-range orderings of atoms through the space which can be properly described in terms of a systematic application of *inflation* symmetry operations. This symmetry had not been previously considered in classical crystallography, and it is directly related to the emergence of self-similar, hierarchical patterns embodying atomic cluster aggregates. Therefore, from a structural viewpoint QCs can be regarded as self-similar arrays of atoms, where the translation symmetry, characteristic of periodic crystals, is *replaced* by a scale invariance one. Accordingly, QCs are representatives of *fractal solids*, with scale factors given by *irrational numbers*.

In an analogous way, rather than trying to fit the electronic structure of QCs within the conceptual schemes introduced in order to describe classical periodic solids (namely, metals, semimetals, semiconductors or insulators) we should expand these categories to properly deal with their highly-fragmented energy and frequency spectra. Indeed, depending on the precise location of the Fermi energy, the very notion of metal, semiconductor of even insulator may be applied to a fractal solid characterized by a fractal-like electronic energy spectrum, as it can be easily grasped by inspecting Fig. 2.4a, for instance.

What is the influence of the long-range QPO on physical properties of QCs? Are there new and characteristic signatures of this long-range QPO on physical properties? As we saw in Sec. 1.5.1, nD dimensional crystallography allows one to describe QCs in an Euclidean space of dimension $n > 3$, large enough for recovering periodicity in terms of the usual Euclidean vector space representation [306]. Accordingly, the long-range QPO present in QCs can be envisioned as a *distortion* effect stemming from the projection of a periodic lattice in more than three dimensions onto physical space. Thus, a more appropriate term for fractal solids characterized by the presence of QP distributions of atoms throughout the space, would probably be *hypercrystals* [544, 548]. Notwithstanding the generalized use of projections from higher dimensional spaces to describe QP structures, it has not been possible to derive their physical properties in a similar way. For instance, since Bloch's theorem can be applied in the higher dimensional space, in principle one may think of obtaining the electronic structure of a given QC in nD, and then project it to the physical $3D$, although to the best of author's knowledge this has not yet been reported.[22]

About a year before Daniel Shechtman's discovery of an actual QC, Alan L. Mackay laid down the expansion of crystallography, that he dubbed *generalized crystallography*, specially emphasizing the importance of self-similarity, hierarchy, and information, as well as the convenience of higher dimensional and non-Euclidean approaches to develop a unified way of dealing with a broad collection of spatial structures and their changes, growth, evolution, and transformations. Accordingly, generalized crystallography was born as a project aimed to unify all the disciplines in which crystals play a role, not only in mineralogy, solid state physics, structural chemistry and material sciences, but also in biochemistry, molecular biology, genetics, virology, medicine, information theory, or mathematics. Within this context, the notion of hypercrystal mentioned above may be regarded as a step forward towards a broader crystal notion, encompassing almost periodic functions as well. In the route towards more general hierarchical designs one may also think of multilayered structures where the slabs of a QP superlattice are themselves made of quasicrystalline materials. In this case, we have two different spatial scales where QPO is present in the same piece of matter. By all indications, the possible fabrication of this sort of two-level QPO structures depends on the very possibility of finding suitable epitaxial growth techniques, able to preserve the QC long-range order during the superlattice synthesis process [689, 712].

---

[22]By exploiting the analogy between the Schrödinger and Helmholtz equations, it was shown that the reflectivity distribution of an incommensurate 1D optical cavity is related to the DOS of a tight-binding Hamiltonian in a 2D triangular lattice [642]. In the same vein, critical phenomena leading to the 4D to 5D phase transition in the aperiodic composite n-nonadecane/urea was interpreted in terms of a low frequency excitation along the aperiodic direction that condenses at the transition temperature [583, 584].

6.3.2 High n-fold quasicrystals?

The most commonly occurring rotational symmetries in QCs are 5-, 8-, 10-, and 12-fold. In contrast, rotational symmetries with orders 7-, 9-, 11-, 13-fold, or higher have rarely been observed. The minimal embedding dimension of an n-fold symmetric structure is given by the so-called Euler totient function $\Phi(n)$,[23] and read $\Phi = 4$ for $n = 5, 8, 10$, and 12, and $\Phi = 6$ for $n = 7, 9, 14$, and 18. Accordingly, all known QCs have a minimal embedding dimension of four, which may be indicating that they are easier to form than other possible QCs with larger D values [291]. For the case of 2D structures, it has been shown that only hypothetical QCs based on quadratic irrational numbers of the form, $a + b\sqrt{c}$ ($a, b, c \in \mathbb{Q}$), should be energetically stable [487], hence justifying why only QCs with 5-, 8-, 10-, and 12-fold symmetries would be allowed. In the case of 3D QCs, however, symmetries based on cubic irrationalities, such as 7- and 9-fold would also be possible according to these thermodynamics-based considerations [824]. The possible symmetry of 3D heptagonal structures, with diffraction symmetry $14/mmm$, can be described in analogy to the decagonal case [738]. Whereas no regular or semi-regular polyhedra (Platonic or Archimedean solids) exist with rotational symmetry larger than 5-fold, polyhedra with only axial n-fold symmetry (e.g., pyramids) are possible for arbitrary n. The 2D QP part of a QC with axial n-fold symmetry can be described as cut of a 4D hypercrystal in the case of quadratic irrationalities and of a 6D hypercrystal in the case of cubic irrationalities. Consequently, in the case of 7-fold symmetry we need a 6D embedding space. Then, the dimension of the perpendicular space (4D) is higher than that of the physical space (2D). This has the consequence that there are two types of 2D phason fields instead of just one for the known QCs [344].

To date, many different variations of self-similar tilings with 7-fold symmetry in 2D have been proposed, most of which are derived from two main approaches, namely, projection from higher dimensions and inflation/deflation techniques. Heptagonal rhomb tilings consist of three different rhombic prototiles with equal edge lengths and acute angles $\pi/7$, $2\pi/7$, and $3\pi/7$. The heptagonal rhomb tiling is invariant under scaling by a factor λ^2, with $\lambda = 1 + 2\cos(2\pi/7) \simeq 2.247$, and λ results as one of the solutions of the cubic equation $x^3 - 2x^2 - x + 1 = 0$ [649]. Alternative procedures to construct a rigorous self-similar 7-fold rhombic tiling containing repeatedly appearing patches with 7-fold rotational symmetry have been devised recently [574], exploiting a global hierarchical framework to orchestrate the local and global 7-fold symmetry [6].

Contrary to pentagonal symmetry, the heptagonal one is rarely observed in nature; examples are some representatives of sea weeds and of ascidians [260]. More frequently observed are 7-membered rings in organic heterocyclic molecules and in inorganic borides and borocarbides, such as $YCrB_4$, ThB_4, $ThMoB_4$, and Y_2ReB_6. There are ternary borides that can be regarded as approximants of heptagonal QCs [839, 663], however, neither stable nor metastable intermetallic QCs with this symmetry have been reported to date [663, 841]. 7-fold QC structures were assembled using holographic optical trapping technique in which computer-generated holograms are projected through a high-numerical-aperture microscope objective lens to create large three-dimensional arrays of optical traps [750]. Colloidal particles in laser fields were investigated in order to ascertain the geometric constrains that affect the formation of 7- and 9-fold QP symmetries, and it was concluded that colloidal particles on quasicrystalline laser fields favor symmetries with rank $\Phi = 4$ compared to $\Phi = 6$, in agreement with atomic systems. This behavior is attributed to large differences in the

---

[23] In number theory, Euler's totient function $\Phi(n)$ counts the positive integers up to a given integer n that are relatively prime to (i.e., do not contain any factor with) n.

number density of highly symmetric sites where quasicrystalline order first originates, so that the laser intensities needed to induce high rotational symmetry are much higher than their lower counterparts [603].

6.3.3 Quasicrystals and number theory

As soon as one realizes that the regular structure of a perfect lattice is suitable for comparison with regularities among natural numbers the way is paved for a fruitful relationship between mathematics, crystallography, and physics [790]. Thus, a promising field where the science of QCs may prove useful is number theory, which is primarily concerned with the properties of integer numbers. The algorithmic generation of aperiodic distributions is central to number theory, and it has motivated many engineering applications in cryptography and coding theory, related to the difficulty of certain arithmetic problems, such as the factoring of large numbers, the invertibility of certain functions, or the distribution of prime numbers [154]. Aperiodicity in number theory can be seen in the erratic behavior of simple arithmetic functions, such as the divisor function, which provides the number $d(n)$ of distinct divisors of n, the Euler's totient function (see Sec. 6.3.2), or the prime-counting function $\pi(x)$, which gives the number of prime numbers smaller than or equal to x. Closely related to the study of this function one encounters the so-called Riemann hypothesis (RH) [561].

The RH is a long standing mathematical problem, included in the celebrated Hilbert's problems listed in 1900,[24] and it states that the nontrivial zeros[25] of the Riemann zeta function (RZF) $\zeta(s)$, have the form $s_n = 1/2 + i\gamma_n$, where the γ_n are all real, so that all these zeros arrange along the so-called *critical line* in the complex plane, which is the axis of symmetry of $\zeta(s)$. It is expected that the exact knowledge of the zeta function nontrivial zeros will unlock the (aperiodic) mathematical pattern hidden within the apparently random distribution of prime numbers, via the Euler-Riemann series

$$\zeta(s) \equiv \sum_{n=1}^{\infty} \frac{1}{n^s} = \prod_p \frac{1}{1-p^{-s}},$$

where the product runs through the set of all prime numbers p.[26]

It has been conjectured that the distribution of the RZF zeros along the critical line can be used to define an aperiodic 1D distribution of points, hence providing a novel example of 1D QC structure [154, 218]. To this end, the density distribution function

$$\rho_{\mathrm{R}} = \sum_{\gamma_n \in X} \delta(\gamma - \gamma_n), \tag{6.5}$$

where γ_n is the imaginary part of the nth nontrivial RZF zero, should be able to diffract [154]. According to detailed numerical calculations [659], the Fourier transform of this array of points is given by a pure point spectrum characterized by a distribution of Dirac

---

[24] The RH is one of the seven "Millennium Problems" stated by the Clay Institute of Mathematics in 2000, and rewarded with a US$ 1,000,000 prize each (http://www.claymath.org/millennium-problems/).

[25] The so-called *trivial* zeros are located at each even negative integer, as it can be deduced from the Riemann's functional equation $\zeta(s) = 2(2\pi)^{s-1}\sin(\pi s/2)\Gamma(1-s)\zeta(1-s)$, for $\mathrm{Re}\,s < 1$, where $\Gamma(s) = \int_0^\infty x^{s-1}e^{-x}dx$ ($\mathrm{Re}\,s > 0$).

[26] The expression n^s for $s \in \mathbb{C}$ is defined as $n^s = e^{s\ln n}$, and $e^z = \sum_{k=0}^{\infty} \frac{z^k}{k!}$ for $z \in \mathbb{C}$. Consequently, $n^s = \sum_{k=0}^{\infty} \frac{s^k(\ln n)^k}{k!}$ (a similar expression stands for p^s).

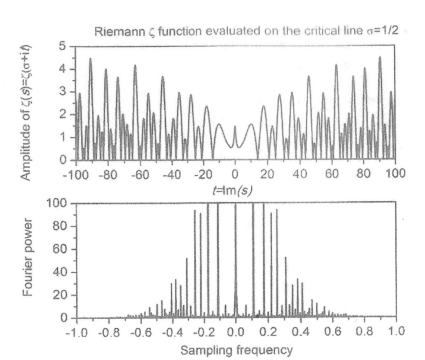

Figure 6.16 (Top) Amplitude of the Riemann zeta function calculated on the critical line for the absolute values of the imaginary part of s less than 100. (bottom) calculated Fourier power spectrum of the Riemann zeta function on the critical line highlighting the self-similar background peaks with small amplitudes. To avoid aliasing, the signal has been sampled at a high sampling frequency (15 Hz) for a total of 7500 sample points. The data shown is limited to a 2 Hz range of sampling frequency (Courtesy of Luca Dal Negro) [154].

δ-functions, one at each of the logarithms of ordinary prime numbers and prime-power numbers, according to the expression

$$\mathcal{F}\left\{\sum_{\gamma_n \in X} \delta(\gamma - \gamma_n)\right\} = \sum_{k_m \in X^*} F_m \delta(k - k_m), \qquad (6.6)$$

where k is a reciprocal space vector, $k_m = \log p^m$, with p prime and $m \in \mathbb{N}$, and F_m are real coefficients. The top panel of Fig. 6.16 shows the amplitude of the RZF calculated along the critical line for $|\gamma_n| \leq 100$, illustrating its symmetry about the real axis and displaying the distribution of the first 29 nontrivial zeros. We note that the zeros become denser as one traverses higher up the critical line.[27] In the bottom panel we show the Fourier power spectrum of the RZT plotted in the top panel. The spectrum exhibits a complex hierarchy of aperiodically distributed discrete peaks, showing a characteristic self-similar pattern, which bears remarkable similarities with the distinctive behavior of QP functions [154]. These results point toward a possible connection between the theory of 1D QCs and the Riemann

---

[27]The average spacing between consecutive zeros at a given height z on the critical line is given by $\Delta(z) = 2\pi / \log(z/2\pi)$ for large enough z [659].

hypothesis. Actually, Eq. (6.6) only guarantees that we are dealing with an aperiodic crystal, since they are characterized by having an essentially discrete diffraction spectrum [357]. To properly classify the density distribution given by Eq. (6.5) as a 1D QC we require a mathematical proof of RZF quasiperiodicity on the critical line.[28] The electronic transport properties of a 1D lattice where atoms are arranged according to the distribution function given by Eq. (6.5) were recently addressed by comparing them with those corresponding to both Fibonacci and Anderson lattice models [922]. To this end, a tight-binding Hamiltonian of the form given by Eq. (2.7) was considered, assuming a fixed lattice spacing, $V_n = 0$ $\forall n \in \mathbb{N}$ (transfer model), and taking transfer integral values determined by average spacing between consecutive RFZ zeros in the form $t_{n,n+1} = \Delta(\gamma_{n+1})/(\gamma_n - \gamma_{n+1})$. In doing so, it was concluded that, in contrast to the FQC, the "Riemann" lattice behaves generically as an insulator due to the localization trend of the electronic eigenstates.

A different approach to study the RH is to consider the $\{\gamma_n\}$ series as describing the eigenvalues of a suitable self-adjoint operator, since this condition guarantees all γ_n values would necessarily be real. This idea, known as the Pólya-Hilbert conjecture, is supported by much evidence, notably that the nontrivial zeros appear to closely follow the Gaussian unitary ensemble statistics of random matrix theory and quantum chaos. Indeed, most studies have considered the required self-adjoint operator to be related to a Hamiltonian describing the dynamics of a suitable quantum system. Notwithstanding this, the possible connection between classical systems and the RH has been explored as well, by considering the Fourier transform of opto-mechanical waves propagating in a suitable substrate or far-field diffraction radiation profiles [55].

The distribution of prime numbers in 1D and 2D arrays has also been considered in several studies. Regarding the distribution of prime numbers themselves recent results indicate they exhibit a rich correlated structure, and that their structure factor

$$S(k) = \left| \sum_p e^{-ikp} \right|^2 ,$$

exhibits well-defined Bragg-like peaks along with a small 'diffuse' contribution. This indicates that primes are appreciably more correlated than anyone had previously conceived. Numerical results definitively suggest an explicit formula for the locations and heights of the peaks. This formula predicts infinitely many peaks in any non-zero interval. Although such a behavior is similar to that of QCs, primes differ from the latter in that the ratio between the locations of any two predicted peaks is rational [994]. It was also shown numerically that the diffuse part decays slowly as the sample size increases. This suggests that the diffuse part vanishes in an appropriate infinite-system-size limit, thereby the Fourier spectrum of prime numbers may be classified as pure point. Furthermore, their limiting structure factor is that of a union of an infinite number of periodic systems and is characterized by a dense set of Dirac delta functions (Bragg peaks), similar to but different from the dense Bragg peaks that arise in QCs and standard limit-periodic systems (see Fig. 2.3) [890, 891]. Recently, a novel class of deterministic aperiodic structure based on elliptic curves was proposed and the light scattering properties of particle arrays designed according to them were investigated. These aperiodic structures are characterized by diffuse diffraction spectra, similar to those observed in homogeneous and isotropically disordered systems [157]. In this way, representatives of all possible pure Lebesgue spectra have been found in the study of lattice points based on number theory based distributions.[29]

---

[28]To the best of my knowldge such a proof has not been reported to date (July 2020).

[29]It has been shown that the tight-binding of the Fibonacci chain and the Aubry-André or Harper model

6.4 EXERCISES WITH SOLUTIONS

6.1 *Determine the characteristic inflation scale factor of a soft ddQC making use of the Bragg reflections arranged along the 15° radial lines in the diffraction pattern shown in Fig. 6.2b.*

The main peaks along the 15° radial lines have Miller indices (12210) and (13310) outwards from the center. The average distances of the model peaks (circles), measured along each of the twelve radial lines, read $d_{12210} = 22.7 \pm 0.1$ mm, and $d_{13310} = 31.0 \pm 0.1$ mm, respectively. Thus, we get the ratio

$$\lambda_{\text{exp}} = \frac{d_{13310}}{d_{13310} - d_{12210}} = \frac{31.0}{31.0 - 22.7} = \frac{310}{83},$$

that is, $\lambda \simeq 3.73 \pm 0.06$, which is pretty close to the exact value $\lambda_{\text{dd}} = 2 + \sqrt{3} = 3.732\ldots$, within the experimental accuracy. Now, by closely inspecting the experimental X-ray pattern shown in Fig. 6.2b, we realize that most reflections are concentrated in the inner dodecagonal ring, as well as along the 30° radial lines, whereas most small circles along 15° radial lines are lacking experimental reflections inside, thereby indicating that the long-range inflation symmetry is not so well established in soft QCs as it is in metallic ddQCs (see **Exercise 1.3**). Accordingly, we will take into account the main peaks along the 30° radial lines, having Miller indices (12100), (23200) and (24200) outwards from the center, whose average distances read $d_{12100} = 16.0 \pm 0.1$ mm, $d_{23200} = 27.7 \pm 0.1$ mm, and $d_{24200} = 32.0 \pm 0.1$ mm. In this way, we obtain two relevant ratios:

$$\lambda_\alpha = \frac{d_{24200}}{d_{23200} - d_{12100}} = \frac{32.0}{27.7 - 16} = \frac{320}{117} \simeq 2.735,$$

$$\lambda_\beta = \frac{d_{24200}}{d_{24200} - d_{23200}} = \frac{32.0}{32.0 - 27.7} = \frac{320}{43} \simeq 7.442,$$

which are relatively close to the values $\lambda_\alpha = 1 + \sqrt{3} = 2.732\ldots$, and $\lambda_\beta = \lambda_\alpha^2 = 2(2 + \sqrt{3}) = 2\lambda_{\text{dd}} = 7.464$, hence providing evidence on the presence of a relevant inflation factor, λ_α, in the structure [840].

6.2 *Some of the largest virus capsids reported in the VIPERdb repository belong to the faustovirus ($T = 277$, average diameter 260 nm) and phycodnavirus ($T = 169d$, average diameter 179 nm) families [427, 242]. Determine their Goldberg indices and the expected number of protein subunits in their capsids.*

Since $169 = 13^2$ one may consider that the phycodnavirus belongs to the $G(h, 0)$ class, with $h = 13$. However, its capsid exhibits a dextro chiral signature, so that it must be a member of the general Goldberg polyhedra class $G(h, k)$ with $h \neq k \neq 0$. The same can be said of faustovirus capsid, since 277 is neither a square number nor a multiple of 3.

Accordingly, we must cope with the cumbersome task of calculating the triangular number series by systematically increasing h and k integer numbers from $h = 1, k = 0 \rightarrow T = 1$ up to $h = 12, k = 7 \rightarrow T = 277$, making explicit use of Eq. (6.4). In so doing, we obtain

$$\begin{aligned}
T = \{ & 1, 3, 4, 7, 9, 12, 13, 16, 19, 21, 25, 27, 28, 31, 36, 37, 39, 43, 48, 49, 52, 57, 61, 63, 64, 67, \\
& 73, 75, 76, 79, 81, 84, 91, 93, 97, 100, 103, 108, 109, 111, 112, 117, 121, 124, 127, 129, \\
& 133, 139, 144, 147, 148, 151, 156, 157, 163, \mathbf{169}, 171, 172, 175, 181, 189, 192, 193, 196, \\
& 201, 208, 217, 219, 223, 229, 243, 247, 244, 252, 271, 273, 277 \}
\end{aligned}$$

---

are in fact topologically equivalent, since both systems are characterized by the same Chern numbers. This property seems to be closely related with the gap labelling theorem [450].

In the list above there are five triangular numbers which can be obtained from two different combinations of Goldberg indices, namely $T = 49 \rightarrow (5,3)$ or $(7,0)$, $T = 91 \rightarrow (6,5)$ or $(9,1)$, $T = 133 \rightarrow (9,4)$ or $(11,1)$, $T = 147 \rightarrow (7,7)$ or $(11,2)$, and $T = 169 \rightarrow (8,7)$ or $(13,0)$. We also note that $h = 12, k = 7$ are triangular numbers themselves.

Finally, making use of the $60T$ rule the expected number of subunit proteins in faustovirus (phycodnavirus) will be $60 \times 277 = 16620$ and $60 \times 169 = 10140$, respectively.

6.3 *The smallest observed distance between Bragg reflections of d-AlCuNi QC is $\ell_{\mathrm{m}} \simeq 0.008$ Å^{-1}, which in the case of a cubic approximant corresponds to a lattice parameter of $a = 12.5$ nm [838]. Determine the full number of Bragg reflections data set for standard resolution* (MoKα, $\theta_{\max} = 30°$).

Plugging the diffraction data into Bragg formula

$$d_{\mathrm{m}} = \frac{\lambda_{\mathrm{MoK}\alpha}}{2 \sin \theta_{\max}} = \frac{\lambda_{\mathrm{MoK}\alpha}}{2 \sin 30°} = \lambda_{\mathrm{MoK}\alpha} = 0.70926 \text{ Å}.$$

Keeping in mind that the total number of possible Bragg reflections is given by

$$n_{\mathrm{T}} = \frac{2\pi V}{3\, d_{\mathrm{m}}^3} = \frac{2\pi}{3} \left(\frac{a}{d_{\mathrm{m}}} \right)^3 = \frac{2\pi}{3} \left(\frac{125}{0.70926} \right)^3 \simeq 11464952.$$

The number of unique reflections of a cubic primitive approximant would amount to $n_{\mathrm{T}}/48 = 11464952/48 \simeq 238853$, and for a related iQC with primitive 6D hypercubic lattice to $n_{\mathrm{T}}/120 = 11464952/120 \simeq 95541$. This figure exceeds by more than two orders of magnitude the values generally employed in QC structure analysis [838].

6.4 *(a) Determine the ideal stoichiometry of i-AlPdMn and i-AlCuFe QCs making use of the nested relationships given in the text. (b) Obtain the approximate stoichiometry of the ideal $\mathrm{Al}_{3\tau+2}\mathrm{Pd}_{\frac{2\tau+1}{2}}\mathrm{Re}_{\frac{\tau+1}{3}}$ QC, and its related electron per atom ratio.*

(a) By properly combining the relationships

$$y = \tau^n z, \quad x = \tau^n(y + z), \quad x + y + z = 1, \tag{6.7}$$

we have

$$y + z = 1 - x \Rightarrow x = \tau^n(1 - x) \Rightarrow x = \tau^n(1 + \tau^n)^{-1}. \tag{6.8}$$

Analogously, after Eqs. (6.7) and (6.8) we get

$$z = 1 - x - y = 1 - \tau^n(1 + \tau^n)^{-1} - y, \tag{6.9}$$

and plugging Eq. (6.9) into $y = \tau^n z$ in Eq. (6.7) we finally obtain

$$y = \tau^n \frac{1 - \tau^n(1 + \tau^n)^{-1}}{1 + \tau^n} = \tau^n(1 + \tau^n)^{-2}, \quad z = 1 - x - y = (1 + \tau^n)^{-2}. \tag{6.10}$$

Therefore,

$$n = 1 \rightarrow x = \frac{\tau}{1 + \tau} = \tau^{-1}, y = \frac{\tau}{(1 + \tau)^2} = \tau^{-3}, z = \frac{1}{(1 + \tau)^2} = \tau^{-4}, \tag{6.11}$$

$$n = 2 \rightarrow x = \frac{\tau^2}{1 + \tau^2}, \ y = \frac{\tau^2}{(1 + \tau^2)^2}, \ z = \frac{1}{(1 + \tau^2)^2}, \tag{6.12}$$

leading to the ideal compositions i-Al$_{\tau-1}$Cu$_{\tau-3}$Fe$_{\tau-4}$ and i-Al$_{\frac{\tau^2}{1+\tau^2}}Pd_{\frac{\tau^2}{(1+\tau^2)^2}}Mn_{\frac{1}{(1+\tau^2)^2}}$, which yield the approximate stoichiometries i-Al$_{61.80}$Cu$_{23.61}$Fe$_{14.59}$ and i-Al$_{72.36}$Pd$_{20.00}$Mn$_{7.64}$, respectively [376]. Making use of Eqs. (6.8) and (6.10), one can deduce the approximate compositions of the hypothetical ideal QCs corresponding to the cases $n > 2$, which are given by A$_{80.90}$B$_{15.45}$C$_{3.65}$ for $n = 3$ and A$_{87.27}$B$_{11.11}$C$_{1.62}$ for $n = 4$.

(b) Making use of the golden mean powers relationships given by Eq. (1.3), we have $\tau + 1 = \tau^2$, $2\tau + 1 = \tau^3$, and $3\tau + 2 = \tau^4$, so that the ideal i-AlPdRe QC composition can be alternatively written Al$_{\tau^4}$Pd$_{\frac{\tau^3}{2}}$Re$_{\frac{\tau^2}{3}}$. On the other hand, we get

$$S = \tau^4 + \frac{\tau^3}{2} + \frac{\tau^2}{3} = \frac{\tau^2}{6}\left(6\tau^2 + 3\tau + 2\right) = \frac{\tau^4}{6}\left(6 + \tau^2\right) = \frac{\tau^4}{6}\left(7 + \tau\right) = \frac{\tau^4}{12}(15 + \sqrt{5}),$$

where we used $\tau = (1 + \sqrt{5})/2$. Therefore, we finally obtain the atomic percentages

$$\frac{\tau^4}{S} = \frac{12}{15 + \sqrt{5}} = 3\frac{15 - \sqrt{5}}{55} \simeq 0.6962$$

$$\frac{\tau^3}{2S} = \frac{3}{5 + 4\sqrt{5}} = \frac{12\sqrt{5} - 15}{55} \simeq 0.2151$$

$$\frac{\tau^2}{3S} = \frac{4}{25 + 9\sqrt{5}} = \frac{25 - 9\sqrt{5}}{55} \simeq 0.0886$$

leading to the stoichiometric formula Al$_{69.62}$Pd$_{21.51}$Re$_{8.86}$. Making use of the valence values listed in Fig. 3.1 we obtain

$$e/a = 3\frac{15 - \sqrt{5}}{55} \times 3.01 + \frac{12\sqrt{5} - 15}{55} \times 0.96 + \frac{25 - 9\sqrt{5}}{55} \times 1.4 = \frac{3121}{1100} - \frac{1011}{5500}\sqrt{5} \simeq 2.4261.$$

For the sake of comparison, data reported in the literature range from $e/a = 2.4280$ (Al$_{70}$Pd$_{22.5}$Re$_{7.5}$) to $e/a = 2.4426$ (Al$_{70.5}$Pd$_{21}$Re$_{8.5}$).

Bibliography

[1] Abe E, Yan Y, Pennycook S J, Quasicrystals as cluster aggregates, *Nature Materials* **3** 759 (2004).

[2] Achim C V, Schmiedeberg M, Löwen H, Growth modes of quasicrystals, *Phys. Rev. Lett.* **112** 255501 (5pp) 2014.

[3] Adam J, Rich J B, The crystal structure of WAl_{12}, $MoAl_{12}$ and $(Mn,Cr)Al_{12}$, *Acta Cryst.* **7** 813 (1954).

[4] Agosta D S, Leisure R G, Adams J J, Shen Y T, Kelton K F, Elastic moduli of a Ti-Zr-Ni i-phase quasicrystal as a function of temperature, *Phil. Mag.* **87** 1–10 (2007).

[5] Ahn S J, Moon P, Kim T H, Kim H W, Shin H C, Kim E H, Cha H W, Kagng S J, Kim P, Koshino M, Son Y W, Yang C W, Ahn J R, Dirac electrons in a dodecagonal graphene quasicrystal, *Science* **361** 782 (2018).

[6] Ajlouni R, A seed-based structural model for constructing rhombic quasilattice with 7-fold symmetry, *Struct. Chem.* **29** 1875–1883 (2018).

[7] Alagappan G, Wu P, Geometrically distributed one-dimensional photonic crystals for light-reflection in all angles, *Opt. Express* **17** 11550—11557 (2009).

[8] Albuquerque E L, Cottam M G, Theory of elementary excitations in quasiperiodic structures, *Phys. Rep.* **376** 225–337 (2003).

[9] Albuquerque E L, Cottam M G, *Polaritons in Periodic and Quasiperiodic Structures* (Elsevier, Amsterdam, 2004).

[10] Albuquerque E L, Sesion Jr. P D, Band gaps of acoustic waves propagating in a solid/liquid phononic Fibonacci structure, *Physica B* **405** 3704–3708 (2010).

[11] Aliev G N, Goller B, Quasi-periodic Fibonacci and periodic one-dimensional hypersonic phononic crystals of porous silicon: Experiment and simulation, *J. Appl. Phys.* **116** 094903 (10pp) 2014.

[12] Amazit Y, Fisher M, Perrin B, Zarembowitch A, de Boissieu M, Pressure and temperature dependence of the elastic properties in AlMnPd quasicrystal, *Europhys. Lett.* **25** 441 (1994).

[13] Anand K, Fournée V, Prévot G, Ledieu J, Gaudry É, Nonwetting behavior of Al-Co quasicrystalline approximants, *ACS Appl. Mater. Interfaces* **12** 15793–15801 (2020).

[14] Anderson B C, Bloom P D, Baikerikar K G, Sheares V V, Mallapragada S K, Al-Cu-Fe quasicrystal/ultra-high molecular weight polyethylene composites as biomaterials for acetabular cup prosthetics, *Biomaterials* **23** 1761–1768 (2002).

[15] Anselmo D H A L, Dantas A L, Albuquerque E L, Localization and fractal spectra of optical phonon modes in quasiperiodic structures, *Physica A* **349** 259 (2005).

[16] Aragón J L, Dávila F, Gómez A, Prediction of the external shape of ideal icosahedral quasicrystals, *Phys. Rev. B* **51** 857–863 (1995).

[17] Aragón J L, Naumis G G, Gómez-Rodríguez A, Twisted graphene and quasicrystals: A cut and projection approach, *Crystals* **9** 519 (2019). doi: 10.3390/cryst9100519.

[18] Aragón J L, Torres M, Gil D, Barrio R A, Maini P K, Turing patterns with pentagonal symmetry, *Phys. Rev. E* **65** 051913 (2002).

[19] Archambault P, Janot C, Thermal conductivity of quasicrystals and associated processes, *MRS Bulletin* **22** 48–53 November 1997.

[20] Archambault P, Plaindoux Ph, Belin-Ferré E, Dubois J M, Pseudo-gap in quasicrystals: A key to understand their stability and properties, in *Quasicrystals: Preparation, Properties and Applications,* J M Dubois, P A Thiel, A P Tsai et al. eds., (Materials Research Society, Warrendale, PA, 1999).

[21] Arige Y, Takasaki A, Kimijima T, Swierczek K, Electrochemical properties of $Ti_{49}Zr_{26}Ni_{25-x}Pd_x$ ($x = 0-6$) quasicrystal electrodes produced by mechanical alloying, *J Alloy Comp.* **645** S152–S154 (2015).

[22] Armstrong N M R, Mortimer K D, Kong T, Bud'ko S L, Canfield P C, Basov D N, Timusk T, Quantum diffusion of electrons in quasiperiodic and periodic approximant lattices in the rare earth-cadmium system, *Phil. Mag.* **96** 1122–1130 (2016).

[23] Artacho E, Machado M, Sánchez-Portal D, Ordejón P, Soler J M, Electrons in dry DNA from density functional calculations, *Molecular Phys.* **101** 1587 (2003).

[24] Asadollah Zarif F, Khazaei Nezhad M, Rastegrar Moghaddam Rezaieun H, Enhancement of efficiency of second-harmonic generation from MoS monolayers in 1D Fibonacci photonic crystals, *Photon. Nanostruct. Fundam. Appl.* **36** 100726 (8pp) 2019.

[25] Asao T, Endo J, Tamura R, Takeuchi S, Shibuya T, Electrical resistivities of Al-Pd-Os and Al-Pd-Os-Re icosahedral quasicrystals, *Mat. Sci. Engineering* **294–296** 604–606 (2000).

[26] Ashraff J A, Stinchcombe R B, Exact decimation approach to the Green's functions of the Fibonacci-chain quasicrystal, *Phys. Rev. B* **37** 5723 (1988).

[27] Ashcroft N W, Mermin N D, *Solid State Physics* (Saunders College Publishing, Orlando, FL 1976).

[28] Asimow P D, Lin C, Bindi L, Ma C, Tschauner O, Hollister L S, Steinhard P J, Shock synthesis of quasicrystals with implications for their origin in asteroid collisions, *Proc. Natl. Acad. Sci. USA* **113** 7077—7081 (2016).

[29] Askeland D R, Wright W J, *The Science and Engineering of Materials* (7th Cengage Learning, 2016).

[30] Aynaou H, Velasco V R, Nougaoui A, El Boudouti E H, Bria D, Properties of elastic waves in quasiregular structures with planar defects, *Superlatt. Microstruct.* **32** 35 (2002); Aynaou H, Velasco V R, Nougaoui A, El Boudouti E H, Djafari-Rouhani B, Bria D, Application of the phase time and transmission coefficients to the study of transverse elastic waves in quasiperiodic systems with planar defects, *Surf. Sci.* **538** 101 (2003).

[31] Azhazha V, Khadzhay G, Malikhin S, Merisov B, Pugachov A, The electrical resistivity of Ti-Zr-Ni quasicrystals in the interval 1.3–300 K, *Phys. Lett. A* **319** 5–10 (2003).

[32] Baake M, Grimm U, *Aperiodic Order I: A Mathematical Invitation*, Encyclopedia of Mathematics and Its Applications 149 (Cambridge University Press, Cambridge, 2013).

[33] Baake M, Damanik D, Grimm U, What is ... aperiodic order?, *Notices AMS* **63** (6) 647–650 June/July 2016.

[34] Baake M, Grimm U, Diffraction of a binary non-Pisot inflation tiling, *J. Phys. Conf. Series* **809** 012026 (2017).

[35] Baake M, Frank N P, Grimm U, Three variations on a theme by Fibonacci, ArXiv: 1910.00988v1 [math.DS] (2019).

[36] Bailey A C, Yates B, Anisotropic thermal expansion of pyrolytic graphite at low temperatures, *J. Appl. Phys.* **41** 5088 (1970).

[37] Bandres M A, Rechtsman M C, Segev M, Topological photonic quasicrystals: Fractal topological spectrum and protected transport, *Phys. Rev. X* **6** 011016 (12pp) 2016.

[38] Barrio R A, Aragón J L, Vera C, Torres M, Jiménez I, Montero de Espinosa F, Robust symmetric pattern in the Faraday experiment, *Phys. Rev E* **56** 4222—4230 (1997).

[39] Barthes-Labrousse M G, Dubois J M, Quasicrystals and complex metallic alloys: trends for potential applications, *Phil. Mag.* **88** 2217–2225 (2008).

[40] Basov D N, Timusk T, Barakat F, Greedan J, Grushko B, Anisotropic optical conductivity of decagonal quasicrsytals, *Phys. Rev. Lett.* **72** 1937–1940 (1994).

[41] Bassiri S, Papas C H, Engheta N, Electromagnetic wave propagation through a dielectric-chiral interface and through a chiral slab, *J. Opt. Soc. Am. A* **15** 1450–1459 (1998).

[42] Bauer T, *Thermophotovoltaics: Basic Principles and Critical Aspects of System Design.* Springer Science & Business Media (2011).

[43] Bayindir M, Cubukcu E, Bulu I, Ozbay E, Photonic band-gap effect, localization, and waveguiding in the two-dimensional Penrose lattice, *Phys. Rev. B* **63** 161104(R) (2001).

[44] Belin E, Dankhazi Z, Sadoc A, Calvayrac A, Klein T, Dubois J M, Electronic distributions of states in crystalline ans quasicrystalline Al-Cu-Fe and Al-Cu-Fe-Cr alloys, *J. Phys. Condens. Matter* **4** 4459 (1992).

[45] Belin-Ferré E, Electronic structure of quasicrystalline compounds, *J. Non-Cryst. Solids* **334&335** 323 (2004).

[46] Belin-Ferré E, Dubois J M, Wetting of aluminium-based complex metallic alloys, *Int. J. Mat. Res.* **97** 1–11 (2006).

[47] Bellissard J, Iochum B, Scoppola E, Testard D, Spectral properties of one dimensional quasi-crystals, *Commun. Math. Phys.* **125** 527 (1989).

[48] Bellissard J, Bovier A, Ghez J M, Spectral properties of a tight binding Hamiltonian with period doubling potential, *Commun. Math. Phys.* **135** 379 (1991).

[49] Bellissard J, Bovier A, Ghez J M, Gap labelling theorems for one dimensional discrete Schrödinger operators, *Rev. Math. Phys.* **4** 1–37 (1992).

[50] Bendersky L, Quasicrystal with one-dimensional translation symmetry and a tenfold rotation axis, *Phys. Rev. Lett.* **55** 1461 (1985).

[51] Bergman G, Waugh J L T, Pauling L, Crystal structure of the intermetallic compound $Mg_{32}(Al,Zn)_{49}$ and related phases, *Nature (London)* **169** 1057 (1952); The crystal structure of the metallic phase $Mg_{32}(Al, Zn)_{49}$, *Acta Cryst.* **10** 254 (1957).

[52] Bermel P. et al., Design and global optimization of high-efficiency thermophotovoltaic systems, *Opt. Express* **18** A314–A334 (2010).

[53] Bernal J D, The scale of structural units in Biopoesis in *Aspects on the Origin of Life*, Ed. Florkin M (Pergamon Press, Oxford, 1960).

[54] Bernal J D, The range of generalised crystallography, *Kristallografiya* **13** 927–951 (1968). [Transl. in *Sov. Phys. Cryst.* **13** 811–831, 1969].

[55] Berry M V, Riemann zeros in radiation patterns: II. Fourier transforms of zeta, *J. Phys. A* **48** 385203 (8pp) 2015.

[56] Bert F, Bellessa G, Tunneling states in Al-Li-Cu quasicrystals, *Phys. Rev. B* **65** 014202 (2001).

[57] Besicovitch A S, *Almost Periodic Functions* (Cambridge University Press, Dover Publications Inc., Cambridge,1954).

[58] Bianchi A D, Bommeli F, Chernikov M A, Gubler U, Degiorgi L, Ott H R, Electrical, magneto- and optical-conductivity of quasicrystals in the Al-Re-Pd system, *Phys. Rev. B* **55** 5730 (1997).

[59] Bienenstock A, Ewald P P, Symmetry of Fourier space, *Acta Cryst.* **15** 1253–1261 (1962).

[60] Biggs B D, Poon S J, Munirathnam N R, Stable Al-Cu-Ru icosahedral quasicrystals: A new class of electronic alloys, *Phys. Rev. Lett.* **65** 2700–2703 (1990).

[61] Biggs B D, Li Y,Poon S J, Electronic properties of icosahedral, approximant, and amorphous phases of an Al-Cu-Fe alloy, *Phys. Rev. B* **43** 8774–8751 (1991).

[62] Bihar Z, Bilušić A, Lukatela J, Smontara A, Leglić P, McGuiness P J, Dolinšek J, Jagličić Z, Jamovec J, Demange V, Dubois J M, Magnetic, electrical and thermal transport properties of Al-Cr-Fe approximant phases, *J. Alloys Compd.* **407** 65–73 (2006).

[63] Bilušic A, Pavuna D, Smontara A, Figue of merit of quasicrystals: the case of Al-Cu-Fe, *Vacuum* **61** 345 (2001); Bilušic A, Smontara A, Lasjaunias J C, Ivkov J, Calvayrac Y, Thermal and thermoelectric properties of icosahedral $Al_{62}Cu_{25.5}Fe_{12.5}$ quasicrystal, *Mater. Sci. Eng. A* **294–296** 711–714 (2000).

[64] Bilušić A, Budrović Ž, Smontara A, Dolinšek J, Canfield P C, Fisher I R, Transport properties of icosahedral quasicrystal $Al_{72}Pd_{19.5}Mn_{8.5}$, *J. Alloys Compd.* **342** 413–415 (2002).

[65] Bilušić A, Smontara A, Dolinšek J, McGuiness P J, Ott H R, Phonon scattering in quasicrystalline i-AlPdMn: A study of the low-temperature thermal conductivity, *J. Alloys Compd.* **432** 1–6 (2007).

[66] Bilušić A, Smiljanić I, Bihar Ž, Stanić D, Smontara A, Heat conduction in complex metallic alloys, *Croat. Chem. Acta* **83** 21–25 (2010).

[67] Bindi L, Steinhardt P J, Yao N, Lu P J, Natural quasicrystals, *Science* **324** 1306—1309 (2009).

[68] Bindi L, Steinhardt P J, Yao N, Lu P J, Icosahedrite, $Al_{63}Cu_{24}Fe_{13}$, the first natural quasicrystal, *Am. Mineral.* **96** 928—931 (2011).

[69] Bindi L, Eiler J M, Guan Y, Hollister L S, MacPherson G, Steinhardt P J, Yao N, Evidence for the extraterrestrial origin of a natural quasicrystal, *Proc. Natl. Acad. Sci. USA* **109** 1396–1401 (2012).

[70] Bindi L, Yao N, Lin C, Hollister L S, Andronicos C L, Distler V V, Eddy M P, Kostin A, Kryachko V, MacPherson G J, et al., Natural quasicrystal with decagonal symmetry, *Sci. Rep.* **5** 9111 (2015) doi:10.1038/srep09111.

[71] Bindi L, Yao N, Lin C, Hollister L S, Andronicos C L, Distler V V, Eddy M P, Kostin A, Kryachko V, MacPherson G J, Decagonite, $Al_{71}Ni_{24}Fe_5$, a quasicrystal with decagonal symmetry from the Khatyrka CV3 carbonaceous chondrite, *Am. Mineral.***100** 2340—2343 (2015).

[72] Bindi L, Lin C, Ma C, Steinhardt P J, Collisions in outer space produced an icosahedral phase in the Khatyrka meteorite never observed previously in the laboratory, *Sci. Rep.* **6** 38117 (2016) doi:10.1038/srep39117.

[73] Bindi L, Chapuis G, *Aperiodic mineral structures*, EMU Notes in Mineralogy, Vol. 19 in *Mineralogical Crystallography* (eds. J. Plasil, J. Majzian, S. Kriovichev) Chapter 5, 213–254 (2017).

[74] Bindi L, Pham J, Steinhardt P J, Previously unknown quasicrystal periodic approximant found in space, *Sci. Rep.* **8** 16271 (2018) doi:10.1038/s41598–018–34375–x.

[75] Bindi L, Steinhardt P J, How impossible crystals came to Earth. A short history, *Rocks & Minerals* **93** 50–57 (2018).

[76] Bindi L, Stanley C J, Natural versus synthetic quasicrystals: analogies and differences in the optical behavior of icosahedral and decagonal quasicrystals, *Rindiconti Lincei. Scienze Fisiche e Naturali* 2019. doi:10.1007/s12210–019–00859–9.

[77] Bloch F, Reminiscences of Heisenberg and the early days of quantum mechanics, *Physics Today* **29** 23 (1976).

[78] Blumberg J, Shoufie Ukhtary M, Saito R, Enhancement of the electric field and diminishment of the group velocity of light in dielectric multilayer systems: A general description, *Phys. Rev. Appl.* **10** 064015 (2018).

[79] Bobnar M, Vrtnik S, Jagličić Z, Wencka M, Cui C, Tsai A P, Dolinšek J, Electrical, magnetic, and thermal properties of the single-grain $Ag_{42}In_{42}Yb_{16}$ icosahedral quasicrystal: Experiment and modeling, *Phys. Rev. B* **84** 1342015 (2011).

[80] Boguslawski M, Lučić N M, Diebel F, Timotijević D V, Denz C, Jović Savić D M, Light localization in optically induced deterministic aperiodic Fibonacci lattices, *Optica* **3** 711–717 (2016).

[81] Bohr H, Zur theorie der fast periodischen funktionen, *Acta Math.* **45** 29 (1924); **46** 101 (1925); **47** 237 (1926).

[82] Bohr H, *Collected Mathematical Works. II Almost periodic functions* (Copenhagen: Dansk Matematisk Forening, 1952).

[83] Boisen M B Jr., Gibbs G V, Mathematical Crystallography, in *Reviews in Mineralogy*, Ribbe P H, Ed.; Mineralogical Society of America: Chantilly, VA, USA, 1985; Volume 15, pp. 191—194.

[84] Bombieri E, Taylor J E, Quasicrystals, tilings, and algebraic number theory: some preliminary connections, *Contemp. Math.* **64** 241 (1987).

[85] Borgstahl G, Goldman A I, Thiel P A, Aperiodic order coming to age: from inorganic materials to dynamic protein structures, *Acta Cryst. A* **75** 212–213 (2019).

[86] Boriskina S V, Dal Negro L, Sensitive label-free biosensing using critical modes in aperiodic photonic structures, *Opt. Express* **16** 12511–12522 (2008).

[87] Boriskina S V, Gopinath A, Dal Negro L, Optical gap formation and localization properties of optical modes in deterministic aperiodic photonic structures, *Opt. Express* **16** 18813–18826 (2008).

[88] Boriskina S V, Heat is the new light, *Optics & Photonics News* **28–33** November 2017.

[89] Boudard M, Klein H, de Boissieu M, Audier M, Vincent H, Structure of quasicrystalline approximant phase in the Al Pd Mn system, *Philos. Mag. A* **74** 939 (1996).

[90] Boudard M, de Boissieu M, in *Physical Properties of Quasicrystals*, pp. 91—126, Ed. Z. M. Stadnik, Springer series in Solid-State Sciences **126** (Springer, Berlin 1999).

[91] Bovier A, Ghez J M, Remarks on the spectral properties of tight-binding and Kronig-Penney models with substitution sequences, *J. Phys. A* **28** 2313–2324 (1994).

[92] Brand R A, Dianoux A J, Calvayrac Y, Vibrational density of states in the archetypical quasicrystal i-$Al_{62}Cu_{25.5}Fe_{12.5}$: Neutron time-of-flight results, *Phys. Rev. B* **62** 8849–8861 (2000).

[93] Brandao E R, Costa C H, Vasconcelos M S, Anselmo D H A L, Mello V D, Octonacci photonic quasicrystals, *Opt. Mats.* **46** 378–383 (2015).

[94] Braun O M, Kivshar Y S, *The Frenkel-Kontorova Model: Concepts. Methods and Applications*, (Springer-Verlag, Berlin 2004); *Phys. Rep.* **306** 1 (1998).

[95] Braun O M, Vanossi A, Tosatti E, Incommesurability of a confined system under Shear, *Phys. Rev. Lett.* **95** 026102 (2005).

[96] Bruin H, Clark A, Fokkink R, The Pisot substitution conjecture, *Topol. Appl.* **205** 1–3 (2016).

[97] Burkov S E, Timusk T, Ashcroft N W, Optical conductivity of icosahedral quasi-crystals, *J. Phys.: Condens. Matt.* **4** 9447 (1992).

[98] Burlak G, Koshevaya S, Sánchez-Mondragón J, Grimalsky V, Electromagnetic oscillations in a multilayer spherical stack, *Opt. Commun.* **180** 49 (2000).

[99] Burlak G, Díaz-de-Anda A, Optical fields in a multilayered microsphere with a quasiperiodic spherical stack, *Opt. Commun.* **281** 181 (2008).

[100] Burlak G, Díaz-de-Anda A, Santaolaya Salgado R, Pérez Ortega J, Narrow transmittance peaks in a multilayered microsphere with a quasiperiodic left-handed stack, *Opt. Commun.* **283** 3569–3577 (2010).

[101] Bursill L A, Lin P J, Xudong F, Close-packing of growing discs, *Mod. Phys. Lett. B* **1** 195 (1987).

[102] Butler W H, Williams R K, Electron-phonon interaction and thermal conductivity, *Phys. Rev. B* **18** 6483 (1978).

[103] Cahn J W, Quasicrystals, *J. Res. Natl. Inst. Stand. Technol.* **106** 975–982 (2001).

[104] Cahn J W, A celebration of the pioneering work on quasicrystals in France and the expansion of crystallography, *C. R. Physique* **15** e1–e5 (2014).

[105] Cai T C, Ledieu J, MacGrath R, Fournée V, Lograsso T A, Ross A R, Thiel P A, Pseudomorphic starfish: nucleation of extrinsic metqal atoms on a quasicrystalline substrate, *Surf. Sci.* **526** 115 (2003).

[106] Calatayud A, Ferrando V, Remón L, Furlan W D, Monsoriu J A, Twin axial vortices generated by Fibonacci lenses, *Opt. Express* **21** 10234–10239 (2013).

[107] Calvayrac Y, Quivy A, Bessière M, Lefebvre S, Cornier-Quinquandon M, Gratias D, Icosahedral AlCuFe alloys: towards ideal quasicrystals, *J. de Physique* **51** 417–431 (1990).

[108] Canfield P C, Caudle M L, Ho C-S, Kreyssig A, Nandi S, Kim M G, Lin X, Kracher A, Dennis K W, McCallum R W, Goldman A I, Solution growth of a binary icosahedral quasicrystal of $Sc_{12}Zn_{88}$, *Phys. Rev. B* **81** 020201 (4pp) 2010.

[109] Cao H, Dal Negro L, Noh H, J. Trevino, Lasing in deterministic aperiodic nanostructures, in L. Dal Negro (Ed.), *Optics of Aperiodic Structures: Fundamentals and Device Applications* (Pan Stanford, USA, 2014) pp. 160–168.

[110] Cao L S, Peng R W, Li D, Wu X, Qi D X, Gao F, Wang M, Tunable phonon resonances and thermal conductance in weakly nonlinear disordered systems with short-range correlations, *Europhys. Lett.* **83** 66001 (6pp) 2008.

[111] Cartwright J H E, Mackay A L, Beyond crystals: the dialectic of materials and information, *Phil. Trans. R. Soc. A* **370** 2807–2822 (2012).

[112] Caspar D L D, Structure of tomato bushy stunt virus, *Nature (London)* **177** 475–7 (1956).

[113] Caspar D L D, Klug A, Physical principles in the construction of regular viruses, *Cold Spring Harbor Symp. Quant. Biol.* **27** 1–24 (1962).

[114] Caspar D L D, Fontano E, Five-fold symmetry in crystalline quasicrystal lattices, *Proc. Natl. Acad. Sci. USA* **93** 14271–14278 (1996).

[115] Cataldo H M, Tejero C F, Stability of the hard-sphere icosahedral quasilattice, *Phys. Rev. B* **52** 13629–13273 (1995).

[116] Cerny R, Francois M, Yvon K, Jaccard D, Walker E, Petricek V, Cisarova I, Nissen H-U, Wessiken R, A single-crystal x-ray and HRTEM study of the heavy-fermion compound $YbCu_{4.5}$, *J. Phys.: Condens. Matter* **8** 4485 (1996).

[117] Chakrabarti A, Karmakar S N, Renormalization-group method for exact Green's functions of self-similar lattices: Application to generalized Fibonacci chains, *Phys. Rev. B* **44** 896 (R) (1991).

[118] Chakrabarti A, Karmakar S N, Moitra R K, On the nature of eigenstates of quasiperiodic lattices in one dimension, *Phys. Lett. A* **168** 301 (1992).

[119] Chakrabarti A, Karmakar S N, Moitra R K, Renormalization-group analysis of extended electronic states in one-dimensional quasiperiodic lattices, *Phys. Rev. B* **50** 13276 (1994).

[120] Chakrabarti A, Karmakar S N, Moitra R K, Role of a new type of correlated disorder in extended electronic states in the Thue-Morse lattice, *Phys. Rev. Lett.* **74** 1403 (1995).

[121] Chakrabarti A, Electronic states and charge transport in a class of low dimensional structured systems, *Physica E* **114** 113616 (18pp) 2019.

[122] Chan L, Grimm U, Substitution-based sequences with absolutely continuous diffraction, *J. Phys.; Conf. Series* **809** 012027 (2017).

[123] Charrier B, Ouladdiaf B, Schmitt D, Observation of quasimagnetic structures in rare-earth-based icosahedral quasicrystals, *Phys. Rev. Lett.* **78** 4637 (1997).

[124] Chatterjee R, Kanjilal A, Swift-heavy-ion irradiation on $Al_{62}Cu_{25.5}Fe_{12.5}$ quasicrystals, *J. Non-Cryst. Solids* **334&335** 431–435 (2004).

[125] S. Chattopadhyay S, Chakrabarti A, Hidden dimers and the matrix maps: Fibonacci chains revisited, *Phys. Rev. B* **65** 184204 (2002).

[126] Chen K H, Su T I, Fang H C, Lee S C, Lin S T, Anomalous Hall effect in the hopping regime of the insulating $Al_{70}Pd_{22.5}Re_{7.5}$ quasicrystal, *J. Alloys Compd.* **342** 352–354 (2002).

[127] Chen L, Chen X, A study of a stable Al-Cu-Fe quasicrystal in solid and liquid state, *Phys. stat. sol. (b)* **169** 15–21 (1992).

[128] Chen Y B, Zhang C, Zhu Y Y, Zhu S N, Wang H T, Ming N B, Optical harmonic generation in a quasi-phase-matched three-component Fibonacci superlattice $LiTaO_3$, *Appl. Phys. Lett.* **78** 577 (2001).

[129] Chernikov M A, Bernasconi A, Beeli C, Schilling A, Ott H R, Low-temperature magnetism in icosahedral $Al_{70}Pd_{21}Mn_9$, *Phys. Rev. B* **48** 3058–3066 (1993).

[130] Chernikov M A, Bianchi A, Ott H R, Low-temperature thermal conductivity of icosahedral $Al_{70}Pd_{21}Re_9$, *Phys. Rev. B* **51** 153–157 (1995).

[131] Chernikov M A, Paschen S, Felder E, Vorburger P, Ruzicka B, Degiorgi D, Ott H R, Fisher I R, Canfield P C, Low-temperature transport, thermal, and optical properties of single-grain quasicrystals of icosahedral phases in the Y-Mg-Zn and Tb-Mg-Zn alloy systems, *Phys. Rev. B* **62** 262–273 (2000).

[132] Chew W, *Waves and Fields in Inhomogeneous Media* (IEEE Press, New York 1996).

[133] Chiang C T, Ellguth M, Schumann F O, Tusche C, Kraska R, Förster S, Widdra W, Electronic band structure of a two-dimensional oxide quasicrystal, *Phys. Rev. B* **100** 125149 (2019).

[134] Christofi A, Pinheiro F A, Dal Negro L, Probing scattering resonances of Vogel's spirals with the Green's matrix spectral method, *Optics Lett.* **41** 1933–1936 (2016).

[135] Chubb D, *Fundamentals of Thermophotovoltaic Energy Conversion* (Elsevier 2007).

[136] Coates S, Smerdon J A, McGrath R, Sharma H R, A molecular overlayer with the Fibonacci square grid structure, *Nature Commun.* **9** 3435 (2018), doi:10.1038/s41467-018-05950-7.

[137] Coelho I P, Vasconcelos M S, Bezerra C G, Effects of mirror symmetry on the transmission fingerprints of quasiperiodic photonic multilayers, *Phys. Lett. A* **374** 1574–1578 (2010).

[138] Collins L C, Witte T G, Silverman R, Green D B, Gomes K K, Imaging quasiperiodic electronic states in a synthetic Penrose tiling, *Nature Commun.* **8** 15961 (2017).

[139] Conrad M, Krumeich F, Harbrecht B, A dodecagonal quasicrystalline chalcogenide, *Angew, Chem. Int. Ed.* **37** 1383–1386 (1998).

[140] Coppolaro M, Castaldi G, Galdi V, Aperiodic order induced enhancement of weak nonlocality in multilayered dielectric metamaterials, *Phys. Rev B* **98** 195128 (12pp) 2018.

[141] Cornelius C M, Dowling J P, Modification of Planck blackbody radiation by photonic band-gap structures, *Phys. Rev. A* **59** 4736—4746 (1999).

[142] Costa C H, Pereira L F C, Bezerra C G, Light propagation in quasiperiodic dielectric multilayers separated by graphene, *Phys. Rev. B* **96** 125412 (9pp) 2017.

[143] Costa C H, Vasconcelos M S, Fulco U L, Albuquerque E L, Thermal radiation in one-dimensional photonic quasicrystals with graphene, *Opt. Mat.* **72** 756–764 (2017).

[144] Crick F H C, Watson J D, The structure of small viruses, *Nature (London)* **177** 473–5 (1956).

[145] Cui C, Shimoda M, Tsai A P, Studies on icosahedral Ag-In-Yb: a prototype for Tsai-type quasicrystals, *RSC Adv.* **4** 46907–46921 (2014). doi;10.1039/c4ra07980a

[146] Cyrot-Lackmann F, Grenet T, Berger C, Fourcaudot G, Gignoux C, Couxhes minces d'alliages quasi-cristallins, leur preparation et leurs utilisations, French patent 9503938 (1995).

[147] Cyrot-Lackmann, F. Quasicrystals as potential candidates for thermoelectric materials, *Mater Sci. Eng. A* **294–296** 611–612 (2000).

[148] Dahr S K, Palenzola A, Manfrinetti P, Pattalwar S M, Physical properties of the quasicrystal $YbCd_{5.7}$ and its approximant $YbCd_6$, *J. Phys.: Condens. Matter* **14** 517–522 (2002).

[149] Dal Negro L, Oton C J, Gaburro Z, Pavesi L, Johnson P, Lagendijk A, Righini R, Colocci M, Wiersma D S, Light transport through the band-edge states of Fibonacci quasicrystals, *Phys. Rev. Lett.* **90** 055501 (2003).

[150] Dal Negro L, Yi J H, Nguyen V, Yi Y, Michel J, Kimerling L C, Spectrally enhanced light emission from aperiodic photonic structures, *Appl. Phys. Lett.* **86** 261905 (2005).

[151] Dal Negro L, Feng N N, Gopinath A, Electromagnetic coupling and plasmon localization in deterministic aperiodic arrays, *J. Opt. A: Pure Appl. Opt.* **10** 064013 (2008).

[152] Dal Negro L, Boriskina S V, Deterministic aperiodic nanostructures for photonics and plasmonics applications, *Laser Photon. Rev.* **6** 178–218 (2012).

[153] Dal Negro L, Lawrence J, Trevino J, Analytical light scattering and orbital angular momentum spectra of arbitrary Vogel spirals, *Opt. Express* **20** 18209 (2012).

[154] Dal Negro L, Lawrence J, Trevino J, Walsh G, Aperiodic order for nanophotonics, in L. Dal Negro (Ed.) *Optics of Aperiodic Structures: Fundamentals and Device Applications* (Pan Stanford, USA, 2014) pp. 1–55.

[155] Dal Negro L, J. Lawrence, J. Trevino, Engineering aperiodic spiral order in nanophotonics: Fundamentals and device applications, in P. Bettoti (Ed.), *Nanodevices for Photonics and Electronics. Advances and Applications* (Pan Stanford, USA, 2016) pp. 57–125.

[156] Dal Negro L, Wang R, Pinheiro A, Structural and spectral properties of deterministic aperiodic optical structures, *Crystals* **6** 161 (35 pp) 2016; doi:10.3390/cryst6120161.

[157] Dal Negro L, Chen Y, Sgrignuoli F, Aperiodic photonics of elliptic curves, *Crystals* **9** 482 (28pp) 2019.

[158] Dallapiccola R, Gopinath A, Stellacci F, Dal Negro L, Quasi-periodic distribution of plasmon modes in two-dimensional Fibonacci arrays of metal nanoparticles, *Opt. Express* **16** 5544 (2008).

[159] Damanik D, Embree M, Gorodetski A, Spectral properties of Schrödinger operators arising in the study of quasicrystals, *Prog. Math. Phys.* **309** 307–370, in *Mathematics of Aperiodic Order*, J. Kellendonk, D. Lenz, J. Savinien (eds.) (Springer, Berlin 2015).

[160] Dana I, Topologically universal spectral hierarchies of quasiperiodic systems, *Phys. Rev. B* **89** 205111 (2014).

[161] Dareau A, Levy E, Bosch Aguilera M, Bouganna R, Akkermans E, Gerbier F, Beugnon J, Revealing the topology of quasicrystals with a diffraction experiment, *Phys. Rev. Lett.* **119** 215304 (2017).

[162] Davydov D N, Mayou D, Berger C, Gignoux C, Neumann A, Jansen A G M, Wyder P, Density of states in quasicrystals and approximants: tunneling experiment on bare and oxidized surfaces, *Phys. Rev. Lett.* **77** 3173–3176 (1996).

[163] de Boissieu M, Boudard M, Ishimasa T, Elkaim E, Laurait J P, Letoublon A, Audier M, Duneau M, Reversible transformation between an icosahedral Al-Pd-Mn phase and a modulated structure of cubic symmetry, *Philos. Mag.* **A78** 305—326 (1998).

[164] de Boissieu M et al., Lattice dynamics of the Zn-Mg-Sc icosahedral quasicrystal and its Zn-Sc periodic 1/1 approximant, *Nature Mater.* **6** 077–094 (2007).

[165] de Boissieu M, Phonons, phasons and atomic dynamics in quasicrystals, *Chem. Soc. Rev.* **41** 6778–6786 (2012).

[166] de Boissieu M, Ted Janssen and aperiodic crystals, *Acta Cryst.* **A75** 273–280 (2019).

[167] de Bruijn N, Algebraic theory of Penrose´s non-periodic tilings of the plane. I, *Ned. Akad. Wet. Proc. Ser. A* **84** 39 (1981)

[168] de Lange C, Janssen T, Electrons in incommensurate crystals: Spectrum and localization, *Phys. Rev. B* **28** 195–209 (1983).

[169] De Leob M T, Chong H, Kraft M, Solar thermoelectric generators fabricated on a silicon-on-insulator substrate, *J. Micromech. Microeng.* **24** 085011 (2014).

[170] De Medeiros F F, Albuquerque E L, Vasconcelos M S, Mauriz P W, Thermal radiation in quasiperiodic photonic crystals with negative refractive index, *J. Phys. Condens. Matter* **19** 496212 (2007).

[171] del Rio R, Jitomirskaya S, Last Y, Simon B, What is localization?, *Phys. Rev. Lett.* **75** 117 (1995).

[172] de Wolff P M, The pseudo-symmetry of modulated crystal structures, *Acta Crystallogr. A* **30** 777 (1974); Symmetry operation for displacively modulated structures, *Acta Crystallogr. A* **33** 493 (1977).

[173] Degiorgi L, Chernikov M A, Beeli C, Ott H R, The electrodynamic response of the icosahedral quasicrystal $Al_{70}Pd_{21}Mn_9$, *Solid State Commun.* **87** 721–726 (1993).

[174] Delahaye J, Berger C, Scaling of the conductivity in icosahedral Al-Pd-Re metallic samples, *Phys. Rev. B* **64** 094203 (2001).

[175] Delahaye J, Berger C, Fourcaudot G, Thouless and critical regimes in insulating icosahedral AlPdRe ribbons, *J. Phys. Condens. Matter* **15** 8753 (2003).

[176] Delahaye J, Berger C, The question of intrinsic origin of the metal-insulator transition in i-AlPdRe quasicrystal, *Eur. Phys. J. B* **88** 102 (17 pp.) 2015. doi: 10.1140/epjb/e2015-50720-7.

[177] Della Villa A, Enoch S, Tayeb G, Pierro V, Galdi V, Capolino F, Band gap formation and multiple scattering in photonic quasicrystals with a Penrose-type lattice, *Phys. Rev. Lett.* **94** 183903 (2005).

[178] Demange V, Milandri A, de Weerd M C, Machizaud F, Jeandel G, Dubois J M, Optical conductivity of Al-Cr-Fe approximant compounds, *Phys. Rev. B* **65** 144205 (11pp) 2002.

[179] Deng X H, Liu J T, Huang J H, Zou L, Liu N H, Omnidirectional bandgaps in Fibonacci quasicrystals containing single-negative materials, *J. Phys. Condens. Matter* **22** 055403 (5pp) 2010.

[180] Depine A, Martínez-Ricci M L, Monsoriu J A, Silvestre E, Andrés P, Zero permeability and zero permittivity band gaps in 1D metamaterial photonic crystals, *Phys. Lett. A* **364** 352 (2007).

[181] Desideri J P, Macon L, Sornette D, Observation of critical modes in quasiperiodic systems, *Phys. Rev. Lett.* **63** 390–393 (1989).

[182] Dian-lin Z, Shao-chun C, Yun-ping W, Li L, Xue-mei W, Ma X L, Kuo K H, Anisotropic thermal conductivity of the 2D single quasicrystals: $Al_{65}Ni_{20}Co_{15}$ and $Al_{62}Si_3Cu_{20}Co_{15}$, *Phys. Rev. Lett.* **66** 2778 (2006).

[183] Dieter M P, Mattheus H B, Jeffcoat R A, Museman R F, Comparison of lead bioavailability in F344 rats fed lead acetate, lead oxide, lead sulfide, or lead ore concentrate from Skagway, Alaska, *J. Toxicol. Environ. Health* **39** 79 (1993).

[184] Díez E, Domínguez-Adame F, Maciá E, Sánchez A, Dynamical phenomena in Fibonacci semiconductor superlattices, *Phys. Rev. B* **54** 16792–16798 (1996).

[185] Djemia P, Ganot F, Dugautier C, Quilichini M, Brillouin scattering from the icosahedral quasicrystal $Al_{70.4}Mn_{8.4}Pd_{21.2}$, *Solid State Comm.* **106** 459 (1998).

[186] Dolinšek J, Klanjšek M, Apih T, Smontara A, Lasjaunias J C, Dubois J M, Poon S J, Searching for sharp features in the pseudogap of icosahedral quasicrystals by NMR, *Phys. Rev. B* **62** 8862 (2000).

[187] Dolinšek J, Jeglič P, McGuiness P J, Jagličić Z, Bilušić A, Bihar Ž, Smontara A, Landauro C V, Feuerbacher M, Grushko B, Urban K, Magnetic, electrical, thermal transport, and thermoelectric properties of the ξ' and ψ complex metallic alloy phases in the Al-Pd-Mn system, *Phys. Rev. B* **72** 064208 (2005).

[188] Dolinšek J, Jagličić Z, Smontara A, Physical properties of the complex metallic alloy phases in the Al-Pd-Mn system, *Phil. Mag.* **86** 671–678 (2006).

[189] Dolinšek J, Vrtnik S, Klanjšek M, Jagličić Z, Smontara A, Smiljanić I, Bilušić A, Yokoyama Y, Inoue A, Landauro C V, Intrinsic electrical, magnetic, and thermal properties of single-crystalline $Al_{64}Cu_{23}Fe_{13}$ icosahedral quasicrystal: Experiment and modeling, *Phys. Rev. B* **76** 054201 (9pp) 2007.

[190] Dolinšek J, Smontara A, Anisotropic physical properties of complex metallic alloys, in *Complex Metallic Alloys Fundamentals and Applications* (Wiley-VCH Verlag, Weinheim 2011).

[191] Dolinšek J, Electrical and thermal transport properties of icosahedral and decagonal quasicrystals, *Chem. Soc. Rev.* **41** 6730–6744 (2012).

[192] Dong J W, Han P, Wang H Z, Broad omnidirectional reflection band forming using the combination of Fibonacci quasi-periodic and periodic one-dimensional photonic crystals, *Chin. Phys. Lett.* **20** 1963 (2003).

[193] Dong J W, Chang M L, Huang X Q, Hang Z H, Zhong Z C, Chen W J, Huang Z Y, Chan C T, Conical dispersion and effective zero refractive index in photonic quasicrystals, *Phys. Rev. Lett.* **114** 163901 (5pp) 2015.

[194] Donnadieu P, Wang K, Degand C, Garoche P, Improving the Al_6Li_3Cu quasicrystal by annealing treatments, *J. Non-Cryst. Solids* **183** 100—108 (1995).

[195] Dotera T, Quasicrystals in soft matter, *Isr. J. Chem.* **51** 1197–1205 (2011).

[196] Dotera T, Bekku S, Ziherl P, Bronze-mean hexagonal quasicrystal, *Nature Mats.* **16** 987–992 (2017).

[197] Du G, Burns A, Prigodin V N, Wang C S, Joo J, Epstein A J, Anomalous Anderson transition in carbonized ion-implanted polymer p-phenylenebenzobisoxazole, *Phys. Rev. B* **61** 10142–10148 (2000).

[198] Duan F, Guojun J, *Introduction to Condensed Matter Physics* Vol. 1 (World Scientific, Singapore 2005) pp. 493–494.

[199] Dubois J M, Weinland P, Matériaux de revêtement pour alliages métalliques et métaux, Organisation Mondiale de la Propriété Intellectuale brevet no. WP 90/01567, European patent registry number EP0356297 (1990); original French patent application number 8810559 (1988).

[200] Dubois J M, Kang S S, Archambault P, Colleret B, Thermal diffusivity of quasicrystalline and related crystalline alloys, *J. Mat. Res.* **8** 38 (1993).

[201] Dubois J M, The applied physics of quasicrystals, *Physica Scripta* **T49** 17–23 (1993).

[202] Dubois J M, Kang S S, Perrot A, Towards applications of quasicrystals, *Mat. Sci. Eng A* **170–180** 122–126 (1994).

[203] Dubois J M, *New Horizons in Quasicrystals: Research and Applications*, Eds. Goldman A I, Sordelet D J, Thiel P A, and Dubois J M (World Scientific, Singapore 1997) p 208.

[204] Dubois J M, Archambault P, Colleret B, Quasicrystalline aluminium heat protection element and thermal spray method to form elements, US Patent 5888661 (1999).

[205] Dubois J M, Brunet P, Belin-Ferré E, *Quasicrystals: Current Topics* edited by Belin-Ferré E, Berger C, Quiquandon M and Sadoc A (World Scientific, Singapore 2000) p 498.

[206] Dubois J M, Quasicrystals, *J. Phys. Condens. Matter* **13** 7753 (2001).

[207] Dubois J M, Machizaud F, Devices for absorbing infrared radiation comprising a quasicrystalline element, US Patent 6,589,370 B1 (2003).

[208] Dubois J M, Brunet P, Costin W, Merstallinger A, Friction and fretting on quasicrystals under vacuum, *J. Non-Cryst. Solids* **334 & 335** 475–480 (2004).

[209] Dubois J M, Fournée V, Belin-Ferré E, Wetting and friction on quasicrystals and related componds, *Mater. Res. Soc. Symp. Proc.* **805** LL8.6.1 (2004).

[210] Dubois J M, *Useful Quasicrystals* (World Scientific, Singapore 2005)

[211] Dubois J M, So useful, those quasicrystals, *Isr. J. Cham.* **51** 1169–1175 (2011).

[212] Dubois J M and Belin-Ferré E (Eds.) *Complex Metallic Alloys Fundamentals and Applications* (Wiley-VCH Verlag, Weinheim 2011).

[213] Dubost B, Lang J M, Tanaka M, Sainfort P, Audier M, Large AlCuLi single quasicrystals with triacontahedral solidification morphology, *Nature (London)* **324** 48–50 (1986).

[214] Dulea M, Severin M, Riklund R, Transmission of light through deterministic aperiodic non-Fibonaccian multilayers, *Phys. Rev. B* **42** 3680–36897 (1990).

[215] Dunlap D H, Wu H L, Phillips P W, Absence of localization in a random-dimer model, *Phys. Rev. Lett.* **65** 88 (1990).

[216] Dutra R F, Messias D, Mendes C V C, Ranciara Neto A, Sales M O, de Moura A B F, Electronic dynamics in 2D aperiodic systems under effect of electric field, *Phys. StatusSolidi B* 1900782 (6pp) 2020.

[217] Dyachenko P N, Miklyaev Yu. V, Dmitrienko V E, Three-dimensional photonic quasicrystal with a complete band gap, *JETP Letters* **86** 240 (2007).

[218] Dyson F, Birds and frogs, *Notices of AMS* **56** (2) 212–223 (2009).

[219] Ebert Ph, Feuerbacher M, Tamura N, Wollgarten M, Urban K, Evidence for a cluster-based structure of AlPdMn single quasicrystals, *Phys. Rev. Lett.* **77** 3827 (1996).

[220] Ebert Ph, Importance of bulk properties in the structure and evolution of cleavage surfaces of quasicrystals, *Prog. Surf. Sci.* **75** Quasicrystals 109 (2004).

[221] Edagawa K, Chernikov M A, Bianchi A D, Felder E, Gubler U, Ott H R, Low-temperature thermodynamic and thermal-transport properties of decagonal $Al_{65}Cu_{20}Co_{15}$, *Phys. Rev. Lett.* **77** 1071 (1996).

[222] Edagawa K, Kajiyama K, Tamura R, Takeuchi S, High-temperature specific heat of quasicrystals and a crystal approximant, *Mater. Sci. Eng. A* **312** 293–298 (2001).

[223] Edagawa K, Kajiyama K, High temperature specific heat of Al-Pd-Mn and Al-Cu-Co quasicrystals, *Mater. Sci. Eng. A* **294–296** 646–649 (2000).

[224] Eisenhammer T, Lazarov M, Radiation converter containing quasi-crystalline material, German Patent No. 4425140 (1994).

[225] Eisenhammer T, Quasicrystal films: numerical optimization as a solar selective absorber, *Thin Solid Films* **270** 1–5 (1995).

[226] Eisenhammer T, Mahr A, Haugeneder A, Assmann W, Selective absorbers based on AlCuFe thin films, *Solar Energy Mats. Solar Cells* **46** 53–65 (1997).

[227] Eisenhammer T, Haugeneder A, Mahr A, High-temperature optical properties and stability of selective absorbers based on quasicrystalline AlCuFe *Solar Energy Mats. Solar Cells* **54** 379–386 (1998).

[228] Eksioglu Y, Vignolo P, Tosi M P, Matter-wave interferometry in periodic and quasi-periodic arrays, *Opt. Commun.* **243** 175 (2004).

[229] Elcoro L, Pérez-Mato J M, Cubic superspace symmetry and inflation rules in metastable MgAl alloy, *Eur. Phys. J. B* **7** 85–89 (1999).

[230] El-Eskandarany M S, Shaban E, Ali N, Aldakheel F, Alkandary A, In-situ catalyza-tion approach for enhancing the hydrogenation/dehydrogenation kinetics of MgH_2 powders with Ni particles, *Sci. Rep.* **6** 1–13 (2016).

[231] Elser V, Indexing problems in quasicrystal diffraction, *Phys. Rev. B* **32** 4892 (1985).

[232] Elser V, The diffraction pattern of projected structures, *Acta Cryst. A* **42** 36–43 (1986).

[233] Endou R, Niizeki K, Fujita N, Universalities in one-electron properties of limit quasiperiodic lattices, *J. Phys. A: Math. Gen.* **37** L151 (2004).

[234] Erickson J W, Frankenberger E A, Rossmann M G, Shay Fout G S, Medappa K C, Rueckert R R, Crystallization of a common cold virus, human rhinovirus14: "isomorphism" with poliovirus crystals, *Proc. Natl. Acad. Sci USA* **80** 931 (1983).

[235] Esaki L, Tsu R, Superlattice and negative differential conductivity in semiconduc-tors, *IBM J. Res. Develop.* **14** 61 (1970).

[236] Escorcia-García J, Mora-Ramos M E, Study of optical propagation in hybrid periodic/quasiregular structures based on porous silicon, *Progress in Electromag-netics Research Symposium (PIERS)* online **5(2)** 167–170 (2009).

[237] Escorcia-García J, Mora-Ramos M E, Propagation and confinement of electric field waves along one-dimensional porous silicon hybrid periodic/quasiperiodic structure, *Opt. Photon. J* **3** 1–12 (2013).

[238] Escudero R, Lasjaunias J C, Calvayrac Y, Boudard M, Tunnelling and point contact spectroscopy of the density of states in quasicrystalline alloys, *J. Phys. Condens. Matter* **11** 383–404 (1999).

[239] Escudero R, Morales F, Point contact tunneling spectroscopy of the density of states in Tb-Mg-Zn quasicrystals, *J. Non-Cryst. Solids* **439** 46–50 (2016).

[240] Estevez J O, Arriage J, Méndez-Blas A, Robles-Cháirez M G, Contreras-Solorio D A, Experimental realization of the porous silicon optical multilayers based on the 1–s sequence, *J. Appl. Phys.* **111** 013103 (6pp) 2012.

[241] Evangelou S N, Multi-fractal spectra and wavefunctions of one-dimensional quasi-crystals, *J. Phys. C: Solid State Phys.* **20** L295 (1987).

[242] Fang Q, Zhu D, Agarkova I, Adhikari J, Klose T, Liu Y, Chen Z, Sun Y, Gross M L, Van Etten J L, Zhang X, Rossmann M G, Near-atomic structure of a giant virus, *Nat Commun.* 10(1) 388 (2019). doi: 10.1038/s41467–019–08319–6.

[243] Feeman T G, *The Mathematics of Medical Imaging. A Beginner's Guide* (Springer, Berlin 2010).

[244] Fehrenbacher L, Sputtering technique forms versatile quasicrystalline coatings, *MRS Bulletin* **36** 581 (2011).

[245] Fernández-Álvarez L, Velasco V R, Sagittal elastic waves in Fibonacci superlattices, *Phys. Rev. B* **57** 14141 (1998).

[246] Ferralis N, Szmodis A W, Diehl R D, Diffraction from one- and two-dimensional quasicrystalline gratings, *Am. J. Phys.* **72** 1241 (2004).

[247] Ferrando V, Calatayud A, Andrés P, Torroba R, Furlan W D, Monsoriu J A, Imaging properties of kinoform Fibonacci lenses, *IEEE Photon. J.* **6** 6500106 (2014).

[248] Fesenko V I, Tuz V R, Dispersion blue-shift in an aperiodic Bragg reflection waveguide, *Opt. Commun.* **365** 225–230 (2016).

[249] Fisher I R, Islam Z, Panchula A F, Cheon K O, Kramer M J, Canfield P C, Goldman A I, Growth of large-grain R-Mg-Zn quasicrystals from ternary melt (R = Y, Er, Ho, Dy, and Tb), *Phil. Mag. B* **77** 1601–1615 (1998).

[250] Fischer S, Exner A, Zielske K, Perlich J, Deloudi S, Steurer W, Lindner P, Förster S, Colloidal quasicrystals with 12-fold and 18-fold diffraction symmetry, *Proc. Natl. Acad. Sci. USA* **108** 1810–1014 (2011).

[251] Fleming R M, Kortan A R, Siegrist T, Thiel F A, Marsh P, Haddon R C, Tycko R, Dabbagh G, Kaplan M L, Mujsce A M, Pseudotenfold symmetry in pentane-solvated C_{60} and C_{70}, *Phys. Rev. B* **44** 888(R) (1991).

[252] Flint S J, Enquist L W, Racaniello V R, Skalka A M, *Principles of virology, Molecular biology, Pathogenesis, and Control of Animal Viruses,* 2nd Ed. (ASM Press, Washington 2004).

[253] Florescu M, Torquato S, Steinhardt P J, Complete band gaps in two-dimensional photonic quasicrystals, *Phys. Rev. B* **80** 155112 (2009).

[254] Forestiere C, Donelli M, Walsh G F, Zeni E, Miano G, Dal Negro L, Particle-swarm optimization of broadband nanoplasmonic arrays, *Opt. Lett.* **35** 133 (2010).

[255] Förster S, Meinel K, Hammer R, Trautmann M, Widdra W, Quasicrystalline structure formation in a classical crystalline thin-film system, *Nature* **502** 215–218 (2013).

[256] Förster S, Trautmann M, Roy S, Adeagbo W A, Zollner E M, Hammer R, Schumann F O, Meinel K, Nayak S K, Mohseni K, Hergert W, Meyerheim H L, Widdra W, Observation and structure determination of an oxide quasicrystal approximant, *Phys. Rev. Lett.* **117** (5pp) 2016.

[257] Förster S, Flege J I, Zollner E M, Schumann F O, Hammer R, Bayat A, Schindler K M, Falta J, Widdra W, Growth and decay of a two-dimensional oxide quasicrystal: High-temperature in situ microscopy, *Ann. Phys. (Berlin)* **529** 1600250 (2017).

[258] Foster K, Leisure R G, Shaklee J B, Kim J Y, Kelton K F, Elastic moduli of a Ti-Zr-Ni icosahedral quasicrystal and the 1/1 bcc crystal approximant, *Phys. Rev. B* **59** 11132–11135 (1999).

[259] Fournee V, Belin-Ferre E, Pecheur P, Tobola J, Dankhazi Z, Sadoc A, Müller H, Electronic structure of Al-Pd-Mn crystalline and quasicrystalline alloys, *J. Phys. Condens. Matter* **14** 87 (2002).

[260] Franco B J O, 3rd-order Fibonacci sequence associated to a heptagonal quasiperiodic tiling of the plane, *Phys. Lett. A* **178** 119–122 (1993).

[261] Francoual S, Livet F, de Boissieu M, Yakhou F, Bley F, Létoublon A, Caudron R, Gastaldi J, Dynamics of phason fluctuations in the i-AlPdMn quasicrystal, *Phys. Rev. Lett.* **91** 225501 (2003).

[262] Franklin R E, Caspar D L D, Klug A, The structure of viruses as determined by x-ray diffraction, in *Plant Pathology: Problems and Progresses, 1908– 1958*, Holton C S ed., (University of Wisconsin Press, Madison, Wisconsin 1959) pp. 447–461.

[263] Friedenberg J, The perceived beauty of regular polygonal tessellations, *Symmetry* **11** 984 (17pp) 2019.

[264] Fu X, Liu Y, Zhou P, Sritrakool W, Perfect self-similarity of energy spectra and gap-labeling properties in one-dimensional Fibonacci-class quasilattices, *Phys. Rev. B* **55** 2882–2889 (1997).

[265] Fujimori M, Kimura K, The bonding nature of icosahedral clusters of the group III elements. *J. Solid State Chem.* **133** 310–313 (1997).

[266] Fujita N, Quasiperiodic canonical-cell tiling with pseudo icosahedral symmetry, *Ann. Phys.* 385 225–286 (2017).

[267] Fujiwara T, Yokokawa T, Universal pseudogap at the Fermi energy in quasicrystals, *Phys. Rev. Lett.* **66** 333–336 (1991).

[268] Fujiwara T, Yamamoto S, Trambly de Laissardiére G, Band structure effects of transport properties in icosahedral quasicrystals, *Phys. Rev. Lett.* **71** 4166 (1993); Trambly de Laissardiére G, Fujiwara T, Electronic structure and conductivity in a model approximant of the icosahedral quasicrystal AlCuFe, *Phys. Rev. B* **50** 5999 (1994); Electronic structure and transport in a model approximant of the decagonal quasicrystal AlCuCo, *ibid.* **50** 9843 (1994).

[269] Fung K K, Yang C Y, Zhou Y Q, Zhao J G, Zhan W S, Shen B G, Icosahedrally related decagonal quasicrystal in rapidly cooled Al-14–at.%-Fe alloy, *Phys. Rev. Lett.* **56** 2060 (1986).

[270] Gähler F, Klitzing R, in *The Mathematics of Long-Range Aperiodic Order*, ed. R. V. Moody (Kluwer, Dordrecht 1995) pp. 141–174. (Nato Science Series C: Mathematical and Physical Sciences, vol. 489).

[271] Gan L, Johnson J E, An optimal exposure strategy for cryoprotected virus crystals with lattice constants greater than 1000 Å, *J. Synchrotron Rad.* **15** 223–226 (2008).

[272] Gantmacher F R, *The Theory of Matrices* **2** (Chelsea, New York 1974).

[273] Gantmakher V F, Chemical localization, *Physics-Uspekhi* **45** 1165–1174 (2002).

[274] Gaponenko S V, *Introduction to Nanophotonics* (Cambridge University Press, Cambridge 2010).

[275] Gardner M, Extraordinary nonperiodic tiling that enriches the theory of tiles, *Sci. Am.* **236** 110–117 (1977).

[276] Gellermann W, Kohmoto M, Sutherland B, Taylor P C, Localization of light waves in Fibonacci dielectric multilayers, *Phys. Rev. Lett.* **72** 633 (1994).

[277] Gemeinhardt A, Martinsons M, Schmiedeberg M, Growth of two-dimensional dodecagonal colloidal quasicrystals: Particles with isotropic pair interactions with two length scales vs, patchy colloids with preferred binding angles, *Eur. Phys. J. E* **41** 126 (2018).

[278] Gemeinhardt A, Martinsons M, Schmiedeberg M, Stabilizing quasicrystals composed of patchy colloids by narrowing the patch width, *EPL* **126** 38001 (2019).

[279] Gévay G, Szederkény T, Quasicrystals and their spontaneous formation possibilities in the nature. *Acta Miner. Petrogr.* **29** 5—12 (1988).

[280] Gévay G, An icosahedral silicate quasicrystal model, *Acta Miner. Petrogr.* **31** 5–11 (1990).

[281] Gévay G, Non-metallic quasicrystals: Hypothesis or reality? *Phase Trans.* **44** 47–50 (1993).

[282] Ghosh A, Karmakar S N, Trace map of a general aperiodic Thue-Morse chain: Electronic properties, *Phys. Rev. B* **58** 2586 (1998).

[283] Ghosh A, Dynamical properties of three component Fibonacci quasicrystals, *Eur. Phys. J. B* **21** 45–51 (2001).

[284] Ghulinyan M, Oton C J, Dal Negro L, Pavesi L, Sapienza R, Colocci M, Wiersma D S, Light-pulse propagation in Fibonacci quasicrystals, *Phys. Rev. B* **71** 094204 (2005).

[285] Giannò K, Sologubenko A V, Chernikov M A, Ott H R, Fisher I R, Canfield P C, Electrical resistivity, thermopower, and thermal conductivity of single grained (Y, Tb, Ho, Er)-Mg-Zn icosahedral quasicrystals, *Mater. Sci. Eng. A* **294–296** 715 (2000).

[286] Giannò K, Sologubenko A V, Chernikov M A, Ott H R, Fisher I R, Canfield P C, Low-temperature thermal conductivity of a single-grain Y-Mg-Zn icosahedral quasicrystal, *Phys. Rev. B* **62** 292–300 (2000).

[287] Gignoux C, Berger C, Fourcaudot G, Grieco J C, Rakoto H, Indications for a metal-insulator transition in quasicrystalline i-AlPdRe, *Europhys. Lett.* **39** 171 (1997).

[288] Gilead A, Further light on the philosophical significance of Mackay's theoretical discovery of crystalline pure possibilities, *Found. Chem.* **21** 285–296 (2019).

[289] Giroud F, Grenet T, Berger C, Lindqvist P, Gignoux C, Fourcaudot G, Resistivity, Hall effect and thermopower in AlPdMn and AlCuFe quasicrystals, *Czech. J. Phys.* **46** Suppl. S5, 2709–2710 (1996).

[290] Giulietti D, Lucchesi M, Emissivity and absorptivity measurements on some high-purity metals at low temperature, *J. Phys. D: Appl. Phys.* **14** 877–881 (1981).

[291] Glotzer S C, Engel M, Complex order in soft matter, *Nature* **471** 309–310 (2011).

[292] Glotzer S C, Assembly engineering: Materials design for the 21st century (2013 P V Danckwerts lecture), *Chem. Eng. Sci.* **121** 3–9 (2015).

[293] Goddard T http://www.rbvi.ucsf.edu/Research/afm/stmv/afmaverage.html

[294] Godrèche C, Luck J M, Vallet F, Quasiperiodicity and types of order: a study in one dimension, *J. Phys. A* **20** 4483 (1987).

[295] Gögebakan M, Avar B, Uzun O, Quasicrystalline phase formation in the conventionally solidified Al-Cu-Fe system, *Mat Sci. Poland* **27** 919 (2009).

[296] Goldberg M, A class of multi.symmetric polyhedra, *Tohoku Math. J.* **43** 104–108 (1937).

[297] Goldman A I, Kong T, Kreyssig A, Jesche A, Ramazanoglu M, Dennis K W, Budko S L, Canfield P C, A family of binary magnetic icosahedral quasicrystals based on rare earths and cadmium, *Nature Mats.* **12** 714–718 (2013).

[298] Goldman A I, Magnetism in icosahedral quasicrystals: current status and open questions, *Sci. Techol. Adv. Mater.* **15** 044801 (15pp) 2014.

[299] Goldstein H, Poole C, Safko J, *Classical Mechanics* (3rd Ed.) (Addison Wesley, New York 2002).

[300] Gollup J P, Langer J S, Pattern formation in nonequilibrium physics, *Rev. Mod. Phys.* **71** S396–S403 (1999).

[301] Gómez-Urrea H A, Escorcia-García J, Duque C A, Mora-Ramos M E, Analysis of light propagation in quasiregular and hybrid Rudin-Shapiro one-dimensional photonic crystals with superconducting layers, *Photon. Nanostruct. Fund. Appl.* **27** 1–10 (2017).

[302] González J E, Cruz-Irisson M, Sánchez V, Wang C, Thermoelectric transport in poly(G)-poly(C) double chains, *J. Phys. Chem. Solids* **136** 109136 (7 pp) 2020.

[303] Goodsell D S, Symmetry at the cellular mesoscale, *Crystals* **11** 1170 (11pp) 2019.

[304] Graebner J E, Chen H S, Specific heat of an icosahedral superconductor, $Mg_3Zn_3Al_2$, *Phys. Rev. B* **58** 1945–1948 (1987).

[305] Gratias D, Calvayrac Y, Devaud-Rzepski J, Faudot F, Harmelin H, Quivy A, Bancel P A, The phase diagram and structures of the ternary AlCuFe system in the vicinity of the icosahedral region, *J. Non-Cryst. Sol.* **153–154** 482 (1993).

[306] Gratias D, Quiquandon M, Structures of quasicrystals: where are the atoms?, *Phil. Mag.* **88** 1887 (2008).

[307] Gratias D, Quiquandon M, Discovery of quasicrystals: The early days, *C. R. Physique* **20** 803–816 (2019).

[308] Grenet T, in *Quasicrystals Current Topics*, Eds. Belin-Ferré E, Berger C, Quiquandon M, and Sadoc A (World Scientific, Singapore 2000).

[309] Grimm U, Schreiber M, *Quasicrystals: An introduction to Structure, Physical Properties, and Applications*, Eds. Suck J B, Schreiber M, and Häussler P (Springer, Berlin 2002) p. 49.

[310] Grimm U, Electrons in quasicrystals, *Isr. J. Chem.* **51** 12571262 (2011).

[311] Grimm U, Aperiodic crystals and beyond, *Acta Cryst.* **B71** 258–274 (2015).

[312] Guedes de Lima B A S, Medeiros Gomes R, Guedes de Lima S J, Dragoe D, Barhtes-Labrousse M G, Kouitat-Njiwa R, Dubois J M, Self-lubricating, low-friction, wear-resistant Al-based quasicrystalline coatings, *Sci. Tech. Adv. Mats.* **17** 71–79 (2016).

[313] Guidoni L, Triché C, Verkek P, Grynberg G, Quasiperiodic optical lattices, *Phys. Rev. Lett.* **79** 3363 (1997)

[314] Guo Q, Poon S J, Metal-insulator and localization in quasicrystalline $Al_{70.5}Pd_{21}Re_{6.5}Mn_2$ alloys, *Phys. Rev. B* **54** 12793–12797 (1996).

[315] Guo M, Xie K, Wang Y, Zhou L, Huang H, Aperiodic TiO_2 nanotube photonic crystal: Full-visible-spectrum solar light harvesting on photonic devices. *Sci. Rep.* **4** 6442 (2014).

[316] Guyot P, News on five-fold symmetry, *Nature (London)* **326** 640 (1987).

[317] Haberken R, Fritsch G, Härting M, Semi-metallic behaviour of i-AlCuFe quasicrystals, *Appl. Phys. A* **57** 431–435 (1993).

[318] Haberken R, Khedhri K, Madel C, Häussler P, Electronic transport properties of quasicrystalline thin films, *Mater. Sci. Eng.* **294–296** 475–480 (2000).

[319] Han I, Xiao X, Sun H, Shahani A J, A side-by-side comparison of the solidification dynamics of quasicrystalline and approximant phases in the Al-Co-Ni system, *Acta Cryst. A* **75** 281–296 (2019).

[320] Hardy H K, Silcock J M, The phase sections at 500 C and 350 C of aluminium-rich aluminium-copper-lithium alloys, *J. Inst. Metals* **84** 423 (1956).

[321] Hargittai I, *The Chemical Intelligencer* October, pp. 25–48 (1997).

[322] Hargittai I, Structures beyond crystals, *J. Molec. Struct.* **976** 81–86 (2010).

[323] Harrison S C, Structure of tomato bushy stunt virus: Three-dimensional X-ray diffraction analysis at 30 Å resolution, *Cold Spring Harbor Symp. Quant. Biol.* **36** 495–501 (1971).

[324] Hasegawa J, Tamura R, Takeuchi S, Tokiwa K, Watanabe T, Electronic transport of the icosahedral Zn-Mg-Sc quasicrystal and its 2/1 and 1/1 cubic approximants, *J. Non Cryst. Solids* **334–335** 368–375 (2004).

[325] Hattori Y, Fukamichi K, Suzuki K, Niikura A, Tsai A P, Inoue A, Masumoto T, Electronic specific heat coefficient and magnetic entropy of icosahedral Mg-RE-Zn (RE = Gd, Tb and Y) quasicrystals, *J. Phys.: Condens. Matter* **7** 4183–4191 (1995).

[326] Hattori Y, Fukamichi K, Suzuki K, Aruga-Katori H, Goto T, Spin-glass-like behaviour and low-temperature specific heat of amorphous Er_xNi_{100--x} random magnetic anisotropy system, *J. Phys. Condens. Matter* **7** 4193 (1995).

[327] Hayashida K, Dotera T, Takano T, Matsushita Y, Polymeric quasicrystal: mesoscopic quasicrystalline tiling in ABC star polymers, *Phys. Rev. Lett.* **98** 195502 (2007).

[328] He G, Muser M H, Robbins M O, Adsorbed layers and the origin of static friction, *Science* **284** 1650 (1999).

[329] He X L, Li X Z, Zhang Z, Kuo K H, One-dimensional quasicrystal in rapidly solidified alloys, *Phys. Rev. Lett.* **61** 1116 (1988).

[330] He S, Maynard J D, Eigenvalue spectrum, density of states, and eigenfunctions in a two-dimensional quasicrystal, *Phys. Rev. Lett.* **62** 1888 (1989).

[331] Hellner E, Koch E, Cluster or framework considerations for the structure of Tl_7Sb_2, α-Mn, Cu_5Zn_8, and their variants $Li_{22}Si_{51}$, $Cu_{41}Sn_{11}$, $Sm_{11}Cd_{45}$, Mg_6Pd and Na_6Tl with octuple unit cells, *Acta Cryst. A* **37** 1–6 (1981).

[332] Hennig R G, Majzoub E H, Kelton K F, Location and energy of interstitial hydrogen in 1/1 approximant W-TiZrNi of the icosahedral TiZrNi quasicrystal: Rietveld refinement of x-ray and neutron diffraction data and density-functional calculations, *Phys. Rev. B* **73** 184205 (2006).

[333] Hermisson J, Richard C, Baake M, A guide to the symmetry structure of quasiperiodic tiling classes, *J. Phys. I France* **7** 1003 (1997).

[334] Highfield J, Liu T, Koo Y S, Grushko B, Borgna A, Skeletal Ru/Cu catalysts prepared from crystalline and quasicrystalline ternary precursors: characterization by X-ray absorption spectroscopy and CO oxidation, *Phys. Chem. Chem. Phys.* **11** 1196–1208 (2009).

[335] Hill C A, Chang T C, Wu Y, Poon S J, Pierce F S, Stadnik Z M, Temperature-dependent NMR features of the $Al_{65}Cu_{20}Ru_{15}$ icosahedral alloy, *Phys. Rev. B* **49** 8615 (1994).

[336] Hippert F, Simonet V, Trambly de Laissardière G, Magnetism of quasicrystals, in *Quasicrystals: Current Topics*, Eds. Belin-Ferré E, Berger C, Quiquandon M, and Sadoc A (World Scientific, Singapore 2000) p. 475.

[337] Hippert F, Audier M, Préjean J J, Sulpice A, Lhotel E, Simonet V, Calvayrac Y, Magnetic properties of icosahedral Al-Pd-Mn quasicrystals, *Phys. Rev. B* **68** 134402 (2003).

[338] Hofstadter D R, Energy levels and wave functions of Bloch electrons in rational and irrational magnetic fields, *Phys. Rev. B* **14** 2239 (1976).

[339] Holzer M, Three classes of one-dimensional, two-tile Penrose tilings and the Fibonacci Kronig-Penney model as a generic case, *Phys. Rev. B* **38** 1709 (1988).

[340] Homes C C, Timusk T, Wu X, Altounian Z, Sahnoune A, Ström-Olsen J O, Optical conductivity of the stable icosahedral quasicrystal $Al_{63.5}Cu_{24.5}Fe_{12}$, *Phys. Rev. Lett.* **67** 2694–2696 (1991).

[341] Honda Y, Edagawa K, Takeuchi S, Tsai A P, Inoue A, Electrical transport properties of Al-Cu-Os icosahedral quasicrystal, *Jpn. J. Appl. Phys.* **34** 2415–2417 (1995).

[342] Hou J, Hu H, Sun K, Zhang C, Superfluid-quasicrystal in a Bose-Einstein condensate, *Phys. Rev. Lett.* **120** 060407 (6pp) 2018.

[343] Hsueh W J, Chen C T, Chen C H, Omnidirectional band gaps in Fibonacci photonic crystals with metamaterials using a band-edge formalism, *Phys. Rev. A* **78** 013836 (6pp) 2008.

[344] Hu C Z, Ding D H, Yang W G, Wang R H, Possible two-dimensional quasicrystal structures with a six-dimensional embedding space, *Phys. Rev. B* **49** 9423 (1994).

[345] Hu J, Asimow P D, Ma C, Bindi L, First synthesis of a unique icosahedral phase from the Khatyrka meteorite by shock-recovery experiment, *IUCrJ* **7** (11pp) 2020, doi:10.1107/S2052252520002729.

[346] Hu Q, Zhao J Z, Peng R W, Gao F, Zhang R L, Wang M, "Rainbow" trapped in a self-similar coaxial optical waveguide, *Appl. Phys. Lett.* **96** 161101 (2010).

[347] Hu Q, Xu D H, Peng R W, Zhou Y, Yang Q L, Wang M, Tune the "rainbow" trapped in a multilayered waveguide, *Europhys. Lett.* **99** 57007 (4pp) 2012.

[348] Hu W, Yi J, Zheng B, Wang L, Icosahedral quasicrystalline $(Ti_{1.6}NiV_{0.4})_{100--x}Sc_x$ alloys: Synthesis, structure and their application in Ni-MH batteries, *J. Solid State Chem.* **202** 1–5 (2013).

[349] Huang C, Ye F, Chen X, Kartashov Y V, Konotop V V, Torner L, Localization-delocalization wavepacket transition in Pythagorean aperiodic potentials, *Sci. Rep.* **6** 32546 (8pp) 2016.

[350] Huang J R, Takagiwa Y, Tsai A P, Preparations and properties of nano-structured Al_xO/i-AlCuFe as quasicrystalline thermoelectric composites, *J. Phys. Conference Ser.* **1458** 012011 (4pp) 2020.

[351] Huang X, Gong C, Property of Fibonacci numbers and the periodiclike perfectly transparent electronic states in Fibonacci chains, *Phys. Rev. B* **58** 739 (1998).

[352] Huang X Q, Jiang S S, Peng R W, Hu A, Perfect transmission and self-similar optical transmission spectra in symmetric Fibonacci-class multilayers, *Phys. Rev. B* **63** 245104 (2001).

[353] Hubert H, Devouard B, Garvie L A J, O'Keeffe M, Buseck P R, Petuskey W T, McMillan P F, Icosahedral packing of B_{12} icosahedra in boron suboxide (B_6O), *Nature* **391** 376–378 (1998).

[354] Hui X, Yu Ch, Photonic bandgap structure and long-range periodicity of a cumulative Fibonacci lattice, *Photn Res.* **5** 11 (2017) doi: 2327–9125/17/010011–04.

[355] Hung J, Kok M H, Tam W Y, Complete photonic bandgaps in the visible range from spherical layer structures in dichromate gelatin emulsions, *Appl. Phys. Lett.* **94** 014102 (2009).

[356] Hurd C M, *Electrons in Metals* (Robert E. Krieger Publishing Company, FL 1981).

[357] ICrU Report of the Executive Committee for 1991 *Acta Cryst. A* **48** 922 (1992). See also http://www.iucr.ac.uk/iucr-top/comm/capd/terms.html.

[358] Iguchi K, A new class of invariant surfaces under the trace maps for nary Fibonacci lattices, *J. Math. Phys.* **35** 1008 (1994).

[359] Inaba A, Tsai A P, Shibata K, Vibrational properties of quasicrystals of AlCuRu, AlPdRe and AlPdMn deduced from heat capacities, *Proceedings of the 6th International Conference on Aperiodic Crystals*, Eds. Takeuchi S and Fujiwara T (World Scientific Singapore, 1998).

[360] Inaba A, Takakura H, Tsai A P, Gischer I R, Canfield P C, Heat capacities of icosahedral and hexagonal phases of Zn-Mg-Y system, *Mats. Sci. Eng.* **294–296** 723–726 (2000).

[361] Inaba A, Lortz R, Meingast C, Guo J Q, Tsai A P, Heat capacity and thermal expansion of a decagonal Al-Co-Ni quasicrystal, *J. Alloys Compd.* **342** 302–305 (2002).

[362] Inagaki K, Suzuki S, Ishikawa A, Tsugawa T, Aya F, Yamada T, Tokiwa K, Takeuchi T, Tamura R, Ferromagnetic 2/1 quasicrystal approximants, *Phys. Rev. B* **101** 180405 (5pp)-R 2020.doi:10.1103/PhysRevB.101.180405.

[363] Iochum B, Testard D, Power law growth for the resistance in the Fibonacci model, *J. Stat. Phys.* **65** 715 (1991).

[364] Ishii Y, Fujiwara T, Hybridization mechanism for cohesion of Cd-based quasicrystals, *Phys. Rev. Lett.* **87** 206408 (2001).

[365] Ishikawa R et al., Thermophysical properties of the melts of AlPdMn icosahedral quasicrystal *Phil. Mag.* **87** 2965 (2007).

[366] Ishikawa A, Takagiwa Y, Kimura K, Tamura R, Probing of the pseudogap via thermoelectric properties in the Au-Al-Gd quasicrystal approximant. *Phys. Rev. B* **95** 104201 (5pp) 2017.

[367] Ishimasa T, Nissen H U, Fukano Y, New ordered state between crystalline and amorphous in Ni-Cr particles, *Phys. Rev. Lett.* **55** 511–513 (1985).

[368] Ishimasa T, Kasano Y, Tachibana A, Kashimoto S, Osaka K, Low-tempersture phase of the Zn-Sc approximant, *Philos. Mag.* **87** 2887 (2007).

[369] Ishimasa T, Icosahedral quasicrystal, 1/1 and 2/1 approximants in Zn-based ternary alloys containing Au and Yb/Tb, *Phil. Mag. Lett.* **99** 351 (2019) doi:10.1080/09500839.2019.1695068.

[370] Islam Z, Fisher I R, Zarestky J, Canfield P C, Stassis C, Goldman A I, Reinvestigation of long-range magnetic ordering in icosahedral Tb-Mg-Zn, *Phys. Rev. B* **57** R11047 (1998).

[371] Jaiswal A, Rawat R, Lalla N P, Insulator-like electrical transport in structurally ordered $Al_{65}Cu_{20+x}Ru_{15-x}$ ($x = 1.5, 1.0, 0.5, 0.0$ and -0.5) icosahedral quasicrystals, *J. Non Cryst. Solids* **351** 239–250 (2005).

[372] Jagličić Z, Jagodič M, Grushko B, Zijlstra E S, Weber Th, Steurer W, Dolinšek J, The effect of thermal treatment on the magnetic state and cluster-related disorder of icosahedral Al-Pd-Mn quasicrystals, *Intermetallics* **18** 623–632 (2010).

[373] Janner A, Which symmetry will an ideal quasicrystal admit? *Acta Cryst. A* **47** 577–590 (1991).

[374] Janner A, Towards a classification of icosahedral viruses in terms of indexed polyhedra, *Acta Cryst. A* **62** 319–330 (2006).

[375] Janot C, de Boissieu M, Quasicrystals as a hierarchy of clusters, *Phys. Rev. Lett.* **72** 1674–1677 (1994).

[376] Janot C, Conductivity in quasicrystals via hierarchically variable-range hopping, *Phys. Rev. B* **53** 181–191 (1996).

[377] Janot C, Atomic clusters, local isomorphism, and recurrently localized states in quasicrystals, *J. Phys. Condens. Matter* **9** 1493–1508 (1997).

[378] Janot C, Loreto L, Farinato R, Bloch oscillations in quasicrystals? *Phys. Lett. A* **276** 291–295 (2000).

[379] Janot C, Dubois J M, in *Quasicrystals: An Introduction to Structure, Physical Properties and Applications*, Eds. Suck J B, Shreiber M, Haüssler P, Springer Series in Materials Science **55** (Springer, Berlin 2002) p. 183

[380] Janssen T, Radukescu O, Rubtsov A N, Phasons, sliding modes and friction, *Eur. Phys. J. B* **29** 85 (2002).

[381] Janssen T, Chapuis G, de Boissieu M, *Aperiodic Crystals. From Modulated phases to Quasicrystals* (Oxford University Press, Oxford 2007).

[382] Janssen T, Fifty years of aperiodic crystals. *Acta Cryst. A* **68** 667–674 (2012).

[383] Jazbec S, Vrtnik S, Jagličić Z, Kashimoto S, Ivkov J, Popcevic P, Smontara A, Kim H J, Kim J G, Dolinšek J, Electronic density of states and metastability of icosahedral Au-Al-Yb quasicrystal, *J. Alloys Compd.* **586** 343–348 (2014).

[384] Jazbec S, Kashimoto S, Koželj P, Vrtnik S, Jagodič M, Jagličić Z, Dolinšek J, Schottky effect in the i-Zn-Ag-Sc-Tm icosahedral quasicrystal and its 1/1 Zn-Sc-Tm approximant, *Phys. Rev. B* **93** 054208 (1–14) 2016.

[385] Jenks C J, Thiel P A, Quasicrystals: a short review from a surface science perspective, *Langmuir* **14** 1392 (1998).

[386] Jenks C J, Thiel P A, Comments on quasicrystals and their potential use as catalysts, *J Mol Catal A Chem* **131** 301 (1998); Jenks C J, Lograsso T A, Thiel P A, Surface reactivity of a sputter-annealed Al-Pd-Mn quasicrystal, *J Am Chem Soc* **120** 12668 (1998).

[387] Jeon S Y, Kwon H, Hur K, Intrinsic photonic wave localization in a three-dimensional icosahedral quasicrystal, *Nature Phys.* **13** 363–369 (2017).

[388] Jiang H, Chen H, Li H, Zhang Y, Omnidirectional gap and defect mode of one.dimensional photonic crystals containing negative-index materials, *Appl. Phys. Lett.* **83** 5386–5388 (2003).

[389] Jiang J Z, Jensen C H, Rasmussen A R, Gerward L, Evidence of a stable binary CdCa quasicrystalline phase, *Appl. Phys. Lett.* **78** 1856–1857 (2001).

[390] Jin G J, Wang Z D, Are self-similar states in Fibonacci systems transparent?, *Phys. Rev. Lett.* **79** 5298 (1997).

[391] Jing H, He J, Peng R W, Wang M, Aperiodic-order-induced multimode effects and their applications in optoelectronic devices, *Symmetry* **11** 1120 (13pp) 2019. doi:10.3390/sym11091120.

[392] Johnson J E, Functional implications of protein-protein interactions in icosahedral viruses, *Proc. Nat. Acad. Sci. USA* **93** 27–33 (1996).

[393] Johnson J E, Speir J A, Quasi-equivalent viruses: A paradigm for protein assemblies, *J. Mol. Biol.* **269** 665–675 (1997).

[394] Johnston J C, Phippen S, Molinero V, A Single-Component Silicon Quasicrystal, *J. Phys. Chem. Lett.* **2, 5** 384–388 (2011).

[395] Kalman D, Mena R, The Fibonacci numbers - Exposed, *Mathematics Magazine* **76** 167 (2003).

[396] Kalozoumis P A, Morfonios C, Diakonos F K, Schmelcher P, Invariants of broken discrete symmetries, *Phys. Rev. Lett.* **113** 050403 (5 pp) 2014.

[397] Kalozoumis P A, Richoux O, Diakonos F K, Theocharis G, Schmelcher P, Invariant currents in lossy acoustic waveguides with complete local symmetry, *Phys. Rev. B* **92** 014303 (10 pp) 2015.

[398] Kalozoumis P A, Morfonios C, Diakonos F K, Schmelcher P, Local symmetries in one-dimensional quantum scattering, *Phys. Rev. A* **87** 032113 (10pp) 2013.

[399] Kalozoumis P A, Morfonios C, Palaiodimopoulos N, Diakonos F K, Schmelcher P, Local symmetries and perfect transmission in aperiodic photonic multilayers, *Phys. Rev. A* **88** 033857 (9pp) 2013.

[400] Kalugin P A, Chernikov M A, Bianchi A, Ott H R, Structural scattering of phonons in quasicrystals, *Phys. Rev. B* **53** 14145–14151 (1996).

[401] Kalugin P, Katz A, Electrons in deterministic quasicrystalline potentials and hidden conserved quantities, *J. Phys. A: Math. Theor.* **47** 315206 (27pp) 2014.

[402] Kamalakaran R, Singh A K, Srivastava O N, Quasicrystalline decagonal phase of Si clusters evaporated in helium and annealed, *Phys. Rev. B* **61** 12686 (2000).

[403] Kameoka S, Tanabe T, Tsai A P, Al-Cu-Fe quasicrystals for steam reforming of methanol: a new form of copper catalysts, *Catalysis Today* **93–95** 23 (2004).

[404] Kameoka S, Tanabe T, Satoh F, Terauchi M, Tsai A P, Activation of Al-Cu-Fe quasicrystalline surface: fabrication of a fine nanocomposite layer with high catalytic performance, *Sci. Techol. Adv. Mater.* **15** 014801 (7pp) 2014.

[405] Kamiya K, Takeuchi T, Kabeya N, Wada N, Ishimasa T, Ochiai A, Deguchi K, Imura K, Sato N K, Discovery of superconductivity in quasicrystal, *Nat. Commun.* **9** 154 (2018); doi: 10.1038/s41467–017–02667–x.

[406] Karpus V, Tuménas S, Suchodolskis A, Arwin H, Assmus W, Optical spectroscopy and electronic structure of the face-centered icosahedral quasicrystals Zn-Mg-RE (R = Y, Ho, Er), *Phys. Rev. B* **88** 094201 (12pp) 2013.

[407] Kashimoto S, Motomura S, Francoual S, Matsuo S, Ishimasa T, *Phil. Mag.* **86** 725 (2006).

[408] Ke J, Zhang J, Focusing properties of phase-only generalized Fibonacci photon sieves, *Opt. Commun.* **368** 34–38 (2016).

[409] Kelton K F, Kim Y J, Stroud R M, A stable Ti-based quasicrystal, *Appl. Phys. Lett.* **70** 3230 (1997).

[410] Kennedy J, Eberhart R C, Shi Y, *Swarm Intelligence* (Morgan Kaufmann 2001).

[411] Kenzari S, Bonina D, Degiovanni A, Dubois J M, Fournée V, Quasicrystal-polymer composites for additive manufacturing technology, *Acta Phys. Polonica A* **126** 449–452 (2014).

[412] Kenzari S, Bonina D, Dubois J M, Fournée V, Quasicrystal-polymer composites for selective laser sintering technology, *Mats. Design* **35** 691–695 (2012).

[413] Keys A S, Glotzer S C, How do quasicrsytals grow?, *Phys. Rev. Lett.* **99** 235503 (2007).

[414] Khamzin A A, Nigmatullin R R, Groshev D E, Analytical investigation of the specific heat for the Cantor energy spectrum, *Phys. Kett. A* **379** 928–932 (2015).

[415] Kim S K, Lee J H, Kim S H, Hwang I K, Lee Y H, Photonic quasicrystal single-cell cavity mode, *Appl. Phys. Lett.* **86** 031101 (2005).

[416] Kim T H, Lee G W, Gangopadhyay A K, Whyers R, Rogers J R, Goldman A I, Kelton K F, Structural studies of a Ti–Zr–Ni quasicrystal-forming liquid, *J. Phys.: Condens. Matter* **19** 455212 (2007).

[417] Kimura K, Iwahashi H, Hashimoto T, Takeychi S, Mizutani U, Ohashi S, Itoh G, Electronic properties of the single-grained icosahedral phase of Al-Li-Cu, *J. Phys. Soc. Jpn.* **58** 2472–2481 (1989).

[418] Kiorpelidis I, Diakonos F K, Theocharis G, Pagneux V, Richoux O, Schmelcher P, Kalozoumis P A, Duality of bound and scattering wave systems with local symmetries, *Phys. Rev. A* **99** 012117 (8pp) 2019.

[419] Kirihara K, Kimura K, Covalency, semiconductor-like and thermoelectric properties of Al-based quasicrystals: icosahedral cluster solids, *Science and Technology Ad. Materials* **1** 227–236 (2000).

[420] Kirihara K, Nakata T, Takata M, Kubota Y, Nishibori E, Kimura K, Sakata M, Covalent bonds in AlMnSi icosahedral quasicrystalline approximant. *Phys. Rev. Lett.* **85** 3469–3471 (2000).

[421] Kirihara K, Kimura K, Sign of covalency in AlPdRe icosahedral quasicrystals obtained from atomic density and quasilattice constant. *Phys. Rev. B* **64** 212201 (4pp) 2001.

[422] Kirihara K, Nagata T, Kimura K, Kato K, Takata M, Kubota Y, Nishibori E, Sakata M, Covalent bonds and their crucial effects on pseudogap formation in α-Al(Mn,Re)Si icosahedral quasicrystalline approximant. *Phys. Rev. B* **68** 014205 (12pp) 2003.

[423] Klein T, Gozlan A, Berger C, Cyrot-Lackmann F, Calvayrac Y, QuivyA, Anomalous transport properties in pure AlCuFe icosahedral phases of high structural quality, *Europhys. Lett.* **13** 129 (1990).

[424] Klein T, Berger C, Mayou D, Cyrot-Lackmann F, Proximity of a metal-insulator transition in icoshaedral phases of high structural quality, *Phys. Rev. Lett.* **66** 2907–2010 (1991).

[425] Klein T, Symko O G, Does the quasicrystal AlCuFe follow Ohm's law? *Phys. Rev. Lett.* **73** 2248–2251 (1994).

[426] Klein T, Symko O G, Davydov D N, Jansen A G M, Observation of a narrow pseudogap near the Fermi level of AlCuFe quasicrystalline thin films, *Phys. Rev. Lett.* **74** 3656–3659 (1995).

[427] Klose T, Reteno D G, Benamar S, Hollerbach A, Colson P, La Scola B, Rossmann M G, Structure of faustovirus, a large dsDNA virus, *Proc. Natl. Acad. Sci. USA.* **113(22)**.6206–11 (2016).

[428] Klug A, Finch J T, Franklin R E, The structure of turnip yellow mosaic virus: X-ray diffraction studies, *Biochim. et Biophys. Acta* **25** 242 (1957).

[429] Klug A, From virus structure to chromatin: X-ray diffraction to three dimensional electron microscopy, *Annu. Rev. Biochem.* **79** 1–35 (2010).

[430] Klyueva M, Shulyatev D, Adreev N, Tabachkova N, Sviridova T, Suslov A, New stable quasicrystal in the system Al-Cu-Co-Fe, *J. Alloys Compd.* **801** 478–482 (2019).

[431] Kohmoto M, Kadanoff L P, Tang C, Localization problem in one-dimension: mapping and escape, *Phys. Rev. Lett.* **50** 1870 (1983).

[432] Kohmoto M, Metal-insulator transition and scaling for incommensurate systems, *Phys. Rev. Lett.* **51** 1198 (1983).

[433] Kohmoto M, Banavar J R, Quasiperiodic lattice: Electronic properties, phonon properties, and diffusion, *Phys. Rev. B* **34** 563 (1986).

[434] Kohmoto M, Localization problem and mapping of one-dimensional wave equations in random and quasiperiodic media, *Phys. Rev. B* **34** 5043 (1986).

[435] Kohmoto M, Sutherland B, Iguchi K, Localization of optics: Quasiperiodic media, *Phys. Rev. Lett.* **58** 2436 (1987).

[436] Kohmoto M, Sutherland B, Tang C, Critical wave functions and a Cantor-set spectrum of a one-dimensional quasicrystal model, *Phys. Rev. B* **35** 1020 (1987).

[437] Kohmoto M, Entropy function for multifractals, *Phys. Rev. A* **37** 1345 (1988).

[438] Kok M H, Lu W, Tam W Y, Wong G K L, Lasing from dye-doped icosahedral quasicrystals in dichromate gelatin emulsions, *Opt. Express* **17** 7275 (2009).

[439] Kolář M, New class of one-dimensional quasicrystals, *Phys. Rev. B* **47** 5489(R) (1993).

[440] Kolář M, Iochum B, Raymond L, Structure factor of 1D systems (superlattices) based on two-letter substitution rules. I. delta (Bragg) peaks, *J. Phys. A: Math. Gen.* **26** 7343 (1993).

[441] Kondo R, Hashimoto T, Eadagawa K, Takeuchi S, Takeuchi T, Mizutani U, Electrical properties of Zn-Mg-RE (RE = Y, Fd) icosahedral quasicrystals, *J. Phys. Soc. Jpn.* **66** 1097–1102 (1997).

[442] Konetsova O V, Rochal S B, Lorman V L, Chiral quasicrystalline order and do-decahedral geometry in exceptional families of viruses, *Phys. Rev. Lett.* **108** 038102 (2012).

[443] Konevtsova O V, Pimonov V V, Lorman V L, Rochal S B, Quasicrystalline and crystalline types of local protein order in capsids of small viruses, *J. Phys.: Condens. Matter* **29** 284002 (8pp) 2017.

[444] Kong T, Bud'ko S L, Jesche A, McArthur J, Kreyssig A, Goldman A I, Canfield P C, Magnetic and transport properties of i-R-Cd icosahedral quasicrystals (R=Y, Gd-Tm), *Phys. Rev. B* **90** 014424 (13pp) 2014.

[445] Korotaev P Yu, Vekilov Yu Kh, Kaputkina N E, Electronic properties of aperiodic quantum dot chains, *Physica E* **44** 1580–1584 (2012).

[446] Korotaev P Yu, Vekilov Yu Kh, Kaputkina N E, Electronic spectrum and localization of electronic states in aperiodic quantum dot chains, *J. Exp. Theor. Phys.* **118**, 304–310 (2014).

[447] Krajčí M and Hafner J, Fermi surfaces and electronic transport properties of qua-sicrystalline approximants, *J. Phys. Condens. Matter* **13** 3817–3830 (2001).

[448] Krajčí M, Hafner J, Covalent bonding and bandgap formation in intermetallic com-pounds: a case study for Al_3V, *J. Phys. Condens. Matter* **14** 1865 (2002).

[449] Krajčí M, Hafner J, Pseudomorphic quasiperiodic alkali metal monolayers on an i-Al-Pd-Mn surface, *Phys. Rev. B* **75** 224205 (2007).

[450] Kraus Y E, Zilbergerg O, Quasiperiodicity and topology trascend dimensions, *Nature Phys.* **12** 624–626 (2016).

[451] Kreiner G, Franzen H F, The crystal structure of λ-Al_4Mn, *J. Alloys Compd.* **261** 83 (1997).

[452] Kroon L, Lennholm E, Riklund R, Localization-delocalization in aperiodic systems, *Phys. Rev. B* **66** 094204 (2002).

[453] Kroon L, Riklund R, Renormalization of aperiodic model lattices: spectral properties, *J. Phys. A: Math. Gen.* **36** 4519 (2003).

[454] Krumeich F, Müller E, Wepf R A, Conrad M, Reich C, Harbrecht B, Nesper R, The structure of dodecagonal (Ta,V)Te imaged by phase-contrast scanning transmission electron microscopy, *J. Solid State Chem.* **194** 106–112 (2012).

[455] Kumar V, Ananthakrishna G, Electronic structure of a quasiperiodic superlattice, *Phys. Rev. Lett.* **59** 1476 (1987).

[456] Kumar V, Extended electronic states in a Fibonacci chain, *J. Phys. Condens. Matt.* **2** 1349 (1990).

[457] Kumar Singh S, Thapa K B, Ojha S P, Large omnidirectional reflection range using combination of periodic and Fibonacci quasiperiodic structures, *Optoelectron. Adv. Mater.-Rapid Comm.* **1** 49 (2007).

[458] Kuo Y K, Lai H H, Huang C H, Ku W C, Lue C S, Lin S T, Thermoelectric properties of binary Cd-Yb quasicrystal and Cd_6Yb, *J. Appl. Phys.* **95** 1900–1905 (2004).

[459] Kuo Y K, Sivakumar K M, Lai H H, Ku C N, Lin S T, Kaiser A B, Thermal and electrical transport properties of Ag-In-Yb quasicrystals: an experimental study, *Phys. Rev. B* **72** 054202 (1–6) 2005.

[460] Kuo Y K, Lai J R, Huang C H, Lue C S, Lin S T, Transport properties of icosahedral $Al_{70}Pd_{22.5}Re_{7.5}$ quasicrystals, *J. Phys. Condens. Matter* **15** 7555–7562 (2003).

[461] Kuo Y. K., Kaurav N, Syu W K, Sivakumar K M, Shan U T, Lin S T, Wang Q, Dong C, Transport properties of Ti-Zr-Ni quasicrystalline and glassy alloys, *J. Appl. Phys.* **104** 063705 (7pp) 2008.

[462] Labib F, Fujita N, Ohhashi S, Tsai A P, Icosahedral quasicrystals and their cubic approximants in the Cd-Mg-Re (RE = Y, Sm, Gd, Tb, Dy, Ho, Er, Tm) systems, *J. Alloys Compd.* **822** 153541 (12pp) 2020.

[463] Labib F, Okuyama D, Fujita N, Yamada T, Ohhashi S, Sato T J, Tsai A P, Magnetic properties of icosahedral quasicrystals and their cubic approximants in the Cd-Mg-RE (RE = Gd, Tb, Dy, Ho, Er, and Tm) systems, *J. Alloys Compd.* **822** 153541 (2020).

[464] Laborde O, Frigerio J M, Rivory J, Pérez A, Plenet J C, Anomalous Hall effect in Al-Mn quasicrystals, *Solid State Commun.* **71** 711 (1989).

[465] Lai Y, Zhang Z Q, Chan C H, Tsang L, Anomalous properties of the band-edege states in large two-dimensional photonic quasicrystals, *Phys. Rev. B* **76** 165132 (2007).

[466] Lalla N P, Tiwari R S, Srivastava O N, Investigation on the synthesis, characterization and electronic behaviour of $Al_{65}Cu_{20+x}Ru_{15-x}$ (x = 2,1,0 and .1) quasicrystalline alloys, *J. Phys. Condens. Matter* **7** 2409 (1995).

[467] Lamb J S W, Wijnands F, From multi-site to on-site transfer matrix models for self-similar chains, *J. Stat. Phys.* **90** 261 (1998).

[468] Lambropoulos K, Simserides C, Electronic structure and charge transport properties of atomic carbon wires, *Phys. Chem. Chem. Phys.* **19** 26890–26897 (2017).

[469] Lambropoulos K, Simserides C, Spectral and transmission properties of periodic 1D tight-binding lattices with a generic unit cell: an analysis within the transfer matrix approach, *J. Phys. Commun.* **2** 03013 (2018).

[470] Lambropoulos K, Simserides C, Periodic, quasiperiodic, fractal, Kolakoski, and random binary polymers: Energy structure and carrier transport, *Phys. Rev. E* **99** 032415 (17pp) 2019.

[471] Lan X, Wang H, Sun Z, Jiang X, Al-Cu-Fe quasicrystals as the anode for lithium ion batteries, *J. Alloys Compd.* **805** 942–946 (2019).

[472] Lanco P, Klein T, Berger C, Cyrot-Lackmann F, Fourcaudot G, Sulpice A, High resistivity and diamagnetism in AlPdMn icosahedral phase, *Europhys. Lett.* **18** 227 (1992).

[473] Landauro C V, Solbrig H, Temperature dependence of the electronic transport in Al-Cu-Fe phases, *Mater. Sci. Eng. A* **294–296** 600 (2000).

[474] Landauro C V, Solbrig H, Modeling the electronic transport properties of Al-Cu-Fe phases, *Physica B* **301** 267 (2001).

[475] Landauro C V, Maciá E, Solbrig H, Analytical expressions for the transport coefficients of icosahedral quasicrystals, *Phys. Rev. B* **67** 184206 (12pp) 2003.

[476] Langsdorf A, Assmus W, Growth of large single grains of the icosahedral quasicrystals ZnMgY, *J. Cryst. Growth* **192** 152–156 (1998).

[477] Lawrence N, Trevino J, Dal Negro L, Aperiodic arrays of active nanopillars for radiation engineering, *J. Appl. Phys.* **111** 113101 (9pp) 2012.

[478] Lay Y Y, Jan J C, Chiou J W, Tsai H M, Pong W F, Tsai M H, Pi T W, Lee J F, Ma C I, Tseng K L, Wang C R, Lin S T, Observation of metal-insulator transition in Al-Pd-Re quasicrystal by x-ray absorption and photoemission spectroscopy, *Appl. Phys. Lett.* **82** 2035–2039 (2003).

[479] Lazo E, Humire F R, Saavedra E, Generation of extended states in diluted transmission lines with distribution of inductances according to Galois sequences: Hamiltonian map approach, *Physica B* **452** 74–81 (2014).

[480] Lazo E, Localization properties of non-periodic electrical transmission lines, *Symmetry* **11** 1257 (2019). doi:10.3390/sym11101257.

[481] Ledermann A, Cademartiri L, Hermatschweiler M, Toninello C, Czin G A, Wiersma D S, Wegener M, von Freymann G, Three-dimensional silicon inverse photonic quasicrystals for infrared wavelengths, *Nature Mater.* **5** 942 (2006).

[482] Ledieu J, Hoeft J T, Reid D E, Smerdon J A, Diehl R D, Lograsso T A, Ross A R, McGrath R, Pseudomorphic growth of a single element quasiperiodic ultrathin film on a quasicrystal surface, *Phys. Rev. Lett.* **92** 135507 (2004).

[483] Ledieu J, Fournée V, Surfaces of quasicrystals, *C. R. Physique* **15** 48–57 (2014).

[484] Lee T D M, Parker G J, Zoorob M E, Cox S J, Charlton M D B, Design and simulation of highly symmetric photonic quasi-crystals, *Nanotechnology* **16** 2703 (2005).

[485] Lee S Y, Amsden J J, Boriskina S V, Gopinath A, Mitropolous A, Kaplan D L, Omenetto F G, Dal Negro L, Spatial and spectral detection of protein monolayers with deterministic aperiodic arrays of metal nanoparticles, *Proc. Natl. Acad. Sci. U.S.A.* **107** 12086–12090 (2010).

[486] Levine D, Steinhardt P J, Quasicrystals: A new class of ordered structures, *Phys. Rev. Lett.* **53** 2477–2480 (1984).

[487] Levitov L S, Why only quadratic irrationalities are observed in quasi-crystals, *Europhys. Lett.* **6** 517–522 (1988).

[488] Levy E, Barak A, Fisher A, Akkermans E, Topological properties of Fibonacci quasicrystals: A scattering analysis of Chern numbers, arXiv;1509.04028 (2015).

[489] Li J, Zhou L, Chan C T, Sheng P, Photonic band gap from a stack of positive and negative index materials, *Phys. Rev. Lett.* **90** 083901 (2003).

[490] Li X Z, Zhang W Y, Sellmyer D J, Quasicrystalline phase and crystalline approximant in Ni-Mn-In Heusler alloy system, *Intermetallics* **119** 106703 (5pp) 2020.

[491] Liang X, Hamid I, Duan H, Dynamic stabilities of icosahedral-like clusters and their ability to form quasicrystals, *AIP Advances* **6** 065017 (9pp) 2016.

[492] Lifshitz R, Theory of color symmetry for periodic and quasiperiodic crystals, *Rev. Mod. Phys.* **69** 1181 (1997).

[493] Lifshitz R, Petrich D M, Theoretical model for Faraday waves with multiple-frequency forcing, *Phys. Rev. Lett.* **79** 1261 (1997).

[494] Lifshitz R, The square Fibonacci lattice, *J. Alloys Compd.* **342** 186–190 (2002).

[495] Lifshitz R, Quasicrystals: A matter of definition, *Foundations of Physics* **33** 1703–17011 (2003).

[496] Lin C, Steinhardt P J, Torquato S, Light localization in local-isomorphism classes of quasicrystals, *Phys. Rev. Lett.* **120** 247401 (5pp) 2018.

[497] Lin C R, Chou S L, Lin S T, The metal-insulator transition in quasicrystals, *J. Phys. Condens. Matter* **8** L725–L730 (1996).

[498] Lin C R, Lin S T, Wang C R, Chou S L, Horng H E, Cheng J M, Yao Y D, Lai S C, J. Electron transport and magnetic properties of the icosahedral Al-Pd-Re quasicrystals, *J. Phys. Condens. Matter* **9** 1509–1519 (1997).

[499] Lin Q, Corbett J D, New stable icosahedral quasicrystalline phase in the Sc-Cu-Zn system, *Phil. Mag. Lett.* **83** 755–762 (2003).

[500] Lin Q, Corbett J D, Electronic tuning of MgCuGa. A route to crystalline approximant and quasicrystalline phases, *J. Am. Chem. Soc.* **127** 12786–12787 (2005).

[501] Lin Q, Corbett J D, Development of the Ca-Au-In icosahedral quasicrystal and two crystalline approximants: practice via pseudogap electronic tuning, *J. Am. Chem. Soc.* **129** 6789–6797 (2007).

[502] Lin Q, Corbett J D, A chemical approach to the discovery of quasicrystals and their approximant crystals, *Struct. Bonding (Berlin)* **133** 1–40 (2009).

[503] Lin T, Chen Z, Usha R, Stauffacher C V, Dai J B, Schmidt T, Johnson J E, The refined crystal structure of Cowpea Mosaic Virus at 2.8 Å resolution, *Virology* **265** 20–34 (1999).

[504] Lindqvist P, Berger C, Klein T, Lanco P, Cyrot-Lackmann F, Calvayrac Y, Role of Fe and sign reversal of the Hall coefficient in quasicrystalline Al-Cu-Fe, *Phys. Rev. B* **48** 630–633 (1993).

[505] Lindqvist P, Lanco P, Berger C, Jansen A G M, Cyrot-Lackmann F, Magnetoconductance of quasicrystals and their approximants: a study of quantum interference effects, *Phys. Rev. B* **51** 4796 (1995).

[506] Lisiecki F, Rychly J, Kuswik P, Glowinski H, Klos J, Groß F, Träger N, Bykova I, Weigand M, Zelent M, Goering E J, Schütz G, Krawczyk M, Stobiecki F, Dubowik J, Gräfe J, Magnons in a quasicrystal: Propagation, extintion and localization of spin waves in Fibonacci structures, *Phys. Rev. Appl.* **11** 054061 (10pp) 2019.

[507] Liu H, Liu W, Sun Y, Chen P, Zhao J, Guo X, Su Z, Preparation and electrochemical hydrogen storage properties of $Ti_{49}Zr_{26}Ni_{25}$ alloy covered with porous polyaniline, *Int. J. Hydrogen energy* **45** 11675–11685 (2020).

[508] Liu H, Wang X, Wang J, Xu H, Yu W, Dong X, Zhang H, Wang L, High electrochemical performance of nanoporous Fe_3O_4/CuO/Cu composites synthesized by dealloying Al-Cu-Fe quasicrystal, *J. Alloys Compd.* **729** 360–369 (2017).

[509] Liu L, Li Z, Li Y, Mao C, Rational design and self-assembly of two-dimensional dodecagonal DNA quasicrystals, *J. Am. Chem. Soc.* **141** 4248–4251 (2019).

[510] Liu N H, Propagation of light waves in Thue-Morse dielectric multilayers, *Phys. Rev. B* **55** 3543–3547 (1997).

[511] Liu X, Wang Y, Qiang J, Wan B, Ma D, Zhang W, Dong Ch., Preparation and electro-catalytic activity of nanoporous palladium by dealloying rapidly-quenched $Al_{70}Pd_{17}Fe_{13}$ quasicrystalline alloy, *Trans. Nonferrous Met. Soc. China* **29** 785–790 (2019).

[512] Liu Y, Riklund R, Electronic properties of perfect and nonperfect one-dimensional quasicrystals, *Phys. Rev. B* **35** 6034 (1987).

[513] Lord E A, Mackay A L, Raganathan S, *New Geometries for New Materials* (Cambridge University Press, Cambridge 2006).

[514] Lu J P, Birman J L, Electronic structure of a quasiperiodic system, *Phys. Rev. B* **36** 4471–4474(R) 1987.

[515] Lu J P, Birman J L, Acoustic-wave propagation in quasiperiodic, incommensurate and random systems, *Phys. Rev. B* **38** 8067 (1988).

[516] Lučić N M, Savić D M J, Piper A, Grujić D Ž, Vasiljevi J M, Pantelić D V, Jelenković B M, Timotijević D V, Light propagation in qausi-periodic Fibonacci waveguide arrays, *J. Opt. Soc. Am. B* **32** 1510–1513 (2015).

[517] Luck J M, Cantor spectra and scaling of gap widths in deterministic aperiodic systems, *Phys. Rev. B* **39** 5834 (1989).

[518] Luck J M, Godrèche C, Janner A, Janssen T, The nature of the atomic surfaces of quasiperiodic self-similar structures, *J. Phys. A: Math. Gen.* **26** 1951 (1993).

[519] Lukas K C, Liu W S, Joshi G, Zebarjadi M, Dresselhaus M S, Ren Z F, Chen G, Opeil C P, Experimental determination of the Lorenz number in $Cu_{0.01}BiTe_{2.7}Se_{0.3}$ and $Bi_{0.88}Sb_{0.12}$, *Phys. Rev. B* **85**, 205410 (2012).

[520] Luque A, Martí A, Increasing the efficiency of ideal solar cells by photon induced transitions at intermediate levels, *Phys. Rev. Lett.* **78** 5014 (1997).

[521] Lüscher R, Flückiger T, Erbudak M, Kortan A R, Debye temperature of the pentagonal surface of the quasicrystal Al-Pd-Mn, *Surf. Sci.* **532–535** 8–12 (2003).

[522] Lwoff A, Anderson T F, Jacob F, Remarques sur les characteristiques de la particule virale infectieuse, *Ann. Inst. Pasteur* **97** 281–289 (1959).

[523] Macé N, Jagannathan A, Kalugin P, Mosseri R, Piéchon F, Critical eigenstates and their properties in one and two dimensional quasicrystals, *Phys. Rev. B* **96** 045138 (17pp) 2017.

[524] Maciá E, Domínguez-Adame F, Sánchez A, Effects of the electronic structure on the dc conductance of Fibonacci superlattices, *Phys. Rev. B* **49** 9503 (1994).

[525] Maciá E, Domínguez-Adame F, Sánchez A, Energy spectra of quasiperiodic systems via information entropy, *Phys. Rev. E* **50** 679 (1994).

[526] Maciá E, Domínguez-Adame F, Three-dimensional effects on the electronic structure of quasiperiodic systems, *Physica B* **216** 53 (1995).

[527] Maciá E, Domínguez-Adame F, Physical nature of critical wave functions in Fibonacci systems, *Phys. Rev. Lett.* **76** 2957 (1996).

[528] Maciá E, Electronic transport in the Koch fractal lattice, *Phys. Rev. B* **57** 7661 (1998).

[529] Maciá E, Physical nature of critical modes in Fibonacci quasicrystals, *Phys. Rev. B* **60** 10032 (1999).

[530] Maciá E, Thermal conductivity of one-dimensional Fibonacci quasicrystals, *Phys. Rev. B* **61** 6645 (2000).

[531] Maciá E, Modeling the electrical conductivity of icosahedral quasicrystals, *Phys. Rev. B* **61** 8771–8777 (2000).

[532] Maciá E, Domínguez-Adame F, *Electrons, Phonons, and Excitons in Low Dimensional Aperiodic Systems* (Editorial Complutense, Madrid 2000).

[533] Maciá E, May quasicrystals be good thermoelectric materials?, *Appl. Phys. Lett.* **77** 3045–3047 (2000).

[534] Maciá E, Theoretical prospective of quasicrystals as thermoelectric materials, *Phys. Rev. B* **64** 094206 (2001).

[535] Maciá E, Exploiting quasiperiodic order in the design of optical devices, *Phys. Rev B* **63** 205421 (2001).

[536] Maciá E, Dubois J M, Thiel P A, *Quasicrystals* entry in *Ullmann's Encyclopedia of Industrial Chemistry*, Sixth Edition, 2002 January Release on CD-ROM; Wiley-VCH, Winheim

[537] Maciá E, Universal features in the electrical conductivity of icosahedral Al-transition-metal quasicrystals, *Phys. Rev. B* **66** 174203 (12pp) 2002.

[538] Maciá E, Do quasicrystals follow Wiedemann-Franz's law?, *Appl. Phys. Lett.* **81** 88–90 (2002).

[539] Maciá E, Thermoelectric properties of icosahedral quasicrystals: A phenomenological approach, *J. Appl. Phys.* **93** 1014–1022 (2003).

[540] Maciá E, Modeling the thermopower of icosahedral AlCuFe quasicrystal: Spectral fine structure, *Phys. Rev. B* **69** 132201 (4pp) 2004.

[541] Maciá E, Takeuchi T, Otagiri T, Modeling the spectral conductivity of Al-Mn-Si quasicrystalline approximants: A phenomenological approach, *Phys. Rev. B* **72** 174208 (2005).

[542] Maciá E, Hierarchical description of phonon dynamics on finite Fibonacci superlattices, *Phys. Rev. B* **73** 184303 (8pp) 2006.

[543] Maciá E, Rodríguez-Oliveros R, Renormalization transformation of periodic and aperiodic lattices, *Phys. Rev. B* **74** 144202 (2006).

[544] Maciá E 2006, The role of aperiodic order in science and technology, *Rep. Prog. Phys.* **69** 397–441

[545] Maciá E, Rodríguez-Oliveros R, Theoretical assessment on the validity of the Wiedemann-Franz law for icosahedral quasicrystals, *Phys. Rev. B* **75** 104210 (2007).

[546] Maciá E, Dolinšek J, Anomalous electronic transport in ξ'-Al-Pd-Mn complex metallic alloy studied by spectral conductivity analysis, *J. Phys.: Condens. Matter* **19** 176212 (2007).

[547] Maciá E, Charge transfer in DNA: effective Hamiltonian approaches, *Z. Kristallograph.* **224** 91–95 (2009).

[548] Maciá E, *Aperiodic Structures in Condensed Matter: Fundamentals and Applications* (CRC Press, Boca Raton, FL 2009).

[549] Maciá E, Thermal conductivity in complex metallic alloys: Beyond Wiedemann-Franz law, *Phys. Rev. B* **79** 245112 (10pp) 2009.

[550] Maciá E, Theoretical aspects of thermal transport in complex metallic alloys: A generalization of the Wiedemann-Franz law, *Croat. Chem. Acta* **83** 65–68 (2010).

[551] Maciá E, de Boissieu M, Properties of CMAs: theory and experiments, in *Complex Metallic Alloys Fundamentals and Applications* (Wiley-VCH Verlag, Weinheim 2011).

[552] Maciá E, Exploiting aperiodic designs in nanophotonic devices, *Rep. Prog. Phys.* **75** 036502 (2012).

[553] Maciá E, Quasicrystals and the quest for next generation thermoelectric materials, *Rev. Sol. State Mats. Sci.* **37** 215–242 (2012).

[554] Maciá E, The importance of being aperiodic: Optical devices, in *Optics of Aperiodic Structures: Fundamentals and Device Applications*; Dal Negro, L. Ed. (Pan Stanford, USA, 2014) pp. 57–90.

[555] Maciá E, On the nature of electronic wave functions in one-dimensional self-similar and quasiperiodic systems, *ISRN Condensed Matter Phys.* **2014** 165943 (2014). doi: 10.1155/2014/165943.

[556] Maciá-Barber E, *Thermoelectric Materials: Advances and Applications* (Pan Stanford Publishing Pte. Ltd.; Singapore 2015).

[557] Maciá E, Thermal emission control via bandgap engineering in aperiodically designed nanophotonic devices, *Nanomaterials* **5** 814–825 (2015). doi:10.3390/nano5020814.

[558] Maciá E, Improving the efficiency of thermophotovoltaic devices: golden ratio based designs, in *Current Trends in Energy and Sustainability 2015 Edition* (eds. R Gómez-Calvet and J M Martínez-Duart) (Real Sociedad Espanola de Física, Madrid 2015) pp .83–94.

[559] Maciá E, Spectral classification of one-dimensional binary aperiodic crystals: An algebraic approach, *Ann. Phys.* **529** 1700079 (17pp) 2017.

[560] Maciá E, Clustering resonance effects in the electronic energy spectrum of tridiagonal Fibonacci quasicrystals, *Phys. Status Solidi B* **254** 1700078 (12pp) 2017.

[561] Maciá Barber E, Quo vadis quasicrystals? *Crystals* **7** 64 (2017). doi:10.3390/cryst7030064.

[562] Maciá Barber E, Chemical bonding and physical properties in quasicrystals and their related approximants: Known facts and current perspectives, *Appl. Sci.* **9** 2132 (2019).

[563] Mackay A L, A dense non-crystallographic packing of equal spheres, *Acta Crystallogr.* **15** 916 (1962).

[564] Mackay A L, Generalized structural geometry, *Acta Cryst. A* **30** 440–447 (1974).

[565] Mackay A L, De nive quinquangula: On the pentagonal snowflake, *Sov. Phys. Crystallogr.* **26** 517 (1981).

[566] Mackay A L, Crystallography and the Penrose pattern *Physica A* **114** 609–613 (1982).

[567] Mackay A L, Generalised crystallography, *Comps. Math. Appls.* **12B** 21–37 (1986).

[568] Mackay A L, What has the Penrose tiling to do with the icosahedral phases? Geometrical aspects of the icosahedral quasicrystal problem, *J. Micros.* **146** 233 (1987).

[569] Mackay A L, Generalised crystallography, *J. Mol. Struct. (Theochem)* **336** 293–303 (1995).

[570] Mackay A L, Some are less equal than others, *Nature* **391** 334–335 (1998).

[571] Mackay A L, Generalized crystallography, *Struct. Chem.* **13** 215–220 (2002).

[572] Macon L, Desideri J P, Sornette D, Surface acoustic waves in a simple quasiperiodic system, *Phys. Rev. B* **40** 3605–3615 (1989).

[573] Mäder R, Widmer R, Gröning P, Steurer W, Gröning O, Correlating scanning tunneling spectroscopy with the electrical resistivity of Al-based quasicrystals and approximants, *Phys. Rev. B* **87** 075425 (9pp) 2013.

[574] Madison A E, Tiling approach for the description of the sevenfold symmetry in quasicrystals, *Struct. Chem.* **28** 57–62 (2017).

[575] Mahan G D, *Many Particle Physics*, (Plenum, New York 1990).

[576] Makarava L N, Nazarov M M, Ozheredov I A, Shkurinov A P, Smirnov A G, Zhukovsky S V, Fibonacci-like photonic structure for femtosecond pulse compression, *Phys. Rev. E* **75** 036609 (2007).

[577] Man W, Megens M, Steinhardt P J, Chaikin P M, Experimental measurement of the photonic properties of icosahedral quasicrystals, *Nature (London)* **436** 993–996 (2005).

[578] Mancinelli C, Jenks C J, Thiel P A, Gellman A J, Tribological properties of a B2–type Al-Pd-Mn quasicrystal approximant, *J. Mater. Res.* **18** 1447 (2003).

[579] Mandel E D, Lifshitz R, Electronic energy spectra of square and cubic Fibonacci quasicrystals, *Philos. Mag.* **88** 2261–2273 (2008).

[580] Mannige R V and Brooks C L III 2010, Periodic table of virus capsids: Implications for natural selection andd design, *PLoS ONE* **5** (3) e9423.

[581] Mantela M, Lambropoulos K, Theodorakou M, Simserides C, Quasi-periodic and fractal polymers: Energy structure and carrier transfer, *Materials* **12** 2177 (30pp) 2019.

[582] Marcoux C, Socolar J E S, Sparse phonon modes of a limit-periodic structure, *Phys. Rev. B* **93** 174102 (2016).

[583] Mariette C, Guérin L, Rabiller P, Ecolivet C, García-Orduna P, Bourges P, Bosak A, de Sanctis D, Hollingsworth M D, Janssen T, Toudic B, Critical phenomena in higher dimensional spaces: The hexagonal-to-orthorhombic phase transition in aperiodic n-nonadecane/urea, *Phys. Rev. B* **87** 104101 (5pp) 2013.

[584] Mariette C, Guérin L, Rabiller P, Odin C, Verezhak M, Bosak A, Bourges P, Ecolivet C, Toudic B, High spatial resolution studies of phase transitions within organic aperiodic crystals, *Phys. Rev B* **101** 184107 (7pp) 2020.

[585] Martin S, Hebard A F, Kortan A R, Thiel F A, Transport properties of $Al_{65}Cu_{20}Co_{15}$ and $Al_{70}Ni_{15}Co_{15}$ decagonal quasicrystals, *Phys. Rev. Lett.* **67** 719 (1991).

[586] Martínez-Gutiérrez D, Velasco V. R, Transverse acoustic waves in piezoelectric ZnO/MgO and GaN/AlN Fibonacci-periodic superlattices, *Surf. Sci.* **624** 58 (2014).

[587] Martinsons M, Schmiedeberg M, Growth of two-dimensional decagonal colloidal quasicrystals, *J. Phys.: Condens. Matter* **30** 255403 (13pp) 2018.

[588] Matarazzo V, De Nicola S, Zito G, Mormile P, Rippa M, Abbate G, Zhou J, Petti L, Spectral characterization of two-dimensional Thue-Morse quasicrystals realized with high resolution lithography, *J. Opt.* **13** 015602 (2011).

[589] Matsui T, Agrawal A, Nahata A, Vardeny Z V, Transmission resonances through aperiodic arrays of subwavelength apertures, *Nature (London)* **446** 517 (2007).

[590] Matsumoto S, Matsui T, Electron microscopic observation of diamond particles grown from the vapour phase, *J. Mat. Sci.* **18** 1785–1793 (1983).

[591] Matsuo S, Nakano H, Ishimasa T, Fukano Y, Magnetic properties and the electronic structure of a stable Al-Cu-Fe icosahedral phase, *J. Phys.: Condens. Matter* **1** 6893 (1989).

[592] Maynard J D, Colloquium: Acoustical analogs of condensed matter problems, *Rev. Mod. Phys.* **73** 401–417 (2001).

[593] Mayou D, Berger C, Cyrot-Lackmann F, Klein T, Lanco P, Evidence for unconventional electronic transport in quasicrystals, *Phys. Rev. Lett.* **70** 3915 (1993).

[594] Mayou D Generalized Drude formula for the optical conductivity of quasicrystals 2000 *Phys. Rev. Lett.* **85** 1290

[595] Mayou D, in *Quasicrystals Current Topics*, Eds. Belin-Ferré E, Berger C, Quiquandon M, and Sadoc A (World Scientific, London 2000) p. 445.

[596] Mei M, Yessen W, Tridiagonal substitution Hamiltonians, *Math. Model Nat. Phenom.* **9** 204–238 (2014).

[597] Meier M M M, Bindi L, Heck P R, Neander A I, Spring N H, Riebe M 1 E, Maden C, Baur H, Steinhardt P J, Wieler R, Busemann H, Cosmic history and a candidate parent asteroid for the quasicrystal-bearing meteorite Khatyrka, *Earth & Planet. Sci. Lett.* **490** 122–131 (2018).

[598] Merlin R, Bajema K, Clarke R, Juang F Y, Bhattacharya P K, Quasiperiodic GaAs-AlAs heterostructures, *Phys. Rev. Lett.* **55** 1768 (1985).

[599] Merlin R, Clarke R, Quasi-periodic layered structures, U.S. Patent No. 4.955.692 (1990).

[600] Mermin N D, The space groups of icosahedral quasicrystals and cubic, orthorhombic, monoclinic and triclinic crystals, *Rev. Mod. Phys.* **64** 3–49 (1992).

[601] Micco A, Ricciardi A, Pisco M, La Ferrara V, Mercaldo L V, Delli Veneri P, Cutolo A, Cusano A, Light trapping efficiency of periodic and quasiperiodic back-reflectors for thin film solar cells: A comparative study. *J. Appl. Phys.* **114** 063103 (2013).

[602] Michaud F, Barrio M, Toscani S, López D O, Tamarit J Ll, Agafonov V, Szwarc H, Céolin R, Solid-state studies on single and decagonal crystals of C_{60} grown from 1,2–dichloroethane, *Phys. Rev. B* **57** 10351 (1998).

[603] Mikhael J, Schmiedeberg M, Rausch S, Roth J, Stark H, Bechinger C, Proliferation of anomalous symmetries in colloidal monolayers subjected to quasiperiodic light fields, *Proc. Natl. Acad. Sci.* **107** 7214–7218 (2010).

[604] Mizutani U, Sakabe Y, Matsuda T, Electronic properties of icosahedral quasicrystals in Mg-Al-Ag, Mg-Al-Cu and Mg-Zn-Ga alloys systems, *J. Phys.: Condens. Matter* **2** 6153–6167 (1990).

[605] Mizutani U, Sakabe Y, Shibuya T, Kishi K, Kimura K, Takeuchi S, Electron transport properties of thermodynamically stable Al-Cu-Ru icosahedral quasicrystals, *J. Phys.: Condens. Matter* **2** 6169–6178 (1990).

[606] Mizutani U, Role of the pseudogap in the electron transport of quasicrystals and their approximants *J. Phys. Condens. Matter* **10** 4609–4623 (1998).

[607] Takeuchi T, Onogi T, Banno E, Mizutani U, Direct evidence of the Hume-Rothery stabilization mechanism in Al-Mn-Fe-Si Mackay-type 1/1 cubic approximants, *Mater. Trans.* **42** 933 (2001); Mizutani U, Takeuchi T, Banno E, Fourneé V, Takata M, Sato H, Determination of spatially hybridized charge distribution and its effect on electron transport in the Al-Cu-Ru-Si 1/1 approximant-Theoretical basis for the Hume-Rothery rule, *Mat. Res. Soc. Symp. Proc.* **643** K13.1.1 (2001).

[608] Mizutani U, *Introduction to the Electron Theory of Metals*, Cambridge University Press, Cambridge 2001.

[609] Mizutani U, Takeuchi T, Sato H, Interpretation of the Hume-Rothery rule in quasicrystals and their approximants, *J. Non-Cryst. Solids* **334&335** 331 (2004).

[610] Mizutani U, *Hume-Rothery Rules for Structurally Complex Alloy Phases* (CRC Press, Taylor & Francis, Boca Raton 2011).

[611] Mizutani U, Sato H, The physics of the Hume-Rothery electron concentration rule, *Crystals* **7** 9 (2017).

[612] Mo Y, Huang W Q, Huang G F, Hu W, Wang L L, Pan A, Ballistic phonon transport through a Fibonacci array of acoustic nanocavities in a narrow constriction, *Phys. Lett. A* **375** 2000–2006 (2011).

[613] Monsoriu J A, Depine R A, Martínez-Ricci M L, Silvestre E, Interaction between non-Bragg band gaps in 1D metamaterial photonic crystals, *Opt. Express* **14** 12958 (2006).

[614] Monsoriu J A, Depine R A, Silvestre E, Non-Bragg band gaps in 1D metamaterial aperiodic multilayers, *J. Eur. Opt. Soc.* **2** 07002 (2007).

[615] Monsoriu J A, Calatayud A, Remón L, Furlan W D, Saavedra G, Andrés P, Bifocal Fibonacci diffractive lenses, *IEEE Photon J.* **5** 3400106 (2013).

[616] Montalbán A, Velasco V R, Tutor J, Fernández-Velicia F J, Phonon confinement in one-dimensional hybrid periodic/quasiregular structures, *Phys. Rev B* **70** 132301 (4pp) 2004.

[617] Montalbán A, Velasco V R, Tutor J, Fernández-Velicia F J, Selective spatial localization of the atom displacements in one-dimensional hybrid quasi-regular (Thue-Morse/Rudin-Shapiro)/periodic structures, *Surf. Sci.* **601** 2538 (2007).

[618] Montalbán A, Velasco V R, Tutor J, Fernández-Velicia F J, Selective confinement of vibrations in composite systems with alternate quasi-regular sequences, *Physica B* **387** 36 (2007).

[619] Montalbán A, Velasco V R, Tutor J, Fernández-Velicia F J, Phonons in hybrid Fibonacci/periodic multilayers, *Surf. Sci.* **603** 938 (2009).

[620] Monzón J J, Felipe A, Sánchez-Soto L L, Lempel-Ziv complexity of photonic quasicrystals, *Crystals* **7** 183 (12pp) 2017.

[621] Mooij J H, Electrical conduction in concentrated disordered transition metal alloys, *Phys. Status Solidi a* **17** 521 (1973).

[622] Moon S K, Jeong K Y, Noh H, Yang J K, Lasing in an optimized deterministic aperiodic nanobeam cavity, *Appl .Phys. Lett.* **109** 241106 (2016).

[623] Moretti L, Mocella V, Two-dimensional photonic aperiodic crystals based on Thue-Morse sequence, *Opt. Express* **15** 15314–15323 (2007).

[624] Moretti L, Rea I, De Stefano L, Rendina I, Periodic versus aperiodic: Enhancing the sensitivity of porous silicon based optical sensors, *Appl. Phys. Lett.* **90** 191112 (2007).

[625] Moretti L, Mocella V, The square Thue-Morse tiling for photonic application, *Phil. Mag.* **88** 2275 (2008).

[626] Morfonios C, Schmelcher P, Kalozoumis P A, Diakonos F K, Local symmetry dynamics in one-dimensional aperiodic lattices: a numerical study, *Nonlinear. Dyn.* **78** 71 (2014).

[627] Morfonios C, Kalozoumis P A, Diakonos F K, Schmelcher P, Nonlocal discrete continuity and invariant currents in locally symmetric effective Schrödinger arrays, *Ann. Phys.* **385** 623–649 (2017).

[628] Mott N F, Twose W D, The theory of impurity conduction, *Adv. Phys.* **10** 107 (1961).

[629] Mott N F, Davis F A, *Electronic processes in non-crystalline materials*, (Clarendon Press, Oxford 1979).

[630] Mugabassi S, Vourdas A, Localized wavefunctions in quantum systems with multiwell potentials, *J. Phys. A: Math. Theor.* **43** 325304 (2010).

[631] Muro Y, Sasakawa T, Suemitsu T, Takabatake T, Tamura R, Takeuchi S, Thermoelectric properties of binary Cd-Yb quasicrystal and its approximant, *Jpn. J. Appl. Phys.* **41** 3787–3790 (2002).

[632] Nagao K, Inuzuka T, Nishimoto K, Edagawa K, Experimental observation of quasicrystal growth, *Phys. Rev. Lett.* **115** 075501 (2015).

[633] Nagaoka Y, Zhu H, Eggert D, Chen O, Single-component quasicrystalline nanocrystal superlattices through flexible polygon tiling rule, *Science* **362** 1396–1400 (2018).

[634] Nakakura J, Ziherl P, Matsuzawa J, Dotera T, Metallic-mean quasicrystals as aperiodic approximants of periodic crystals, *Nature Commun.* **10** (2019).

[635] Nasari H, Sadegh Abrishamiam M, Terahertz bistability and multistability in graphene/dielectric Fibonacci multilayer, *Appl. Opt.* **19** 5313–5322 (2017).

[636] Natarajan P, Johnson J E, Molecular packing in virus crystals: geometry, chemistry, and biology, *J. Struct. Biol.* **121** 295–305 (1998).

[637] Naumis G G, Use of the trace map for evaluating localization properties, *Phys. Rev. B* **59** 11315 (1999).

[638] Naumis G G, Wang C, Thorpe M F, Barrio R A, Coherency of phason dynamics in Fibonacci chains, *Phys. Rev. B* **59** 14302 (1999).

[639] Naumis G G, Aragón J L, Torres M, Average lattice and the long-wave length behavior of quasicrystals, *J. Alloys Compd.* **342** 210 (2002).

[640] Naumis G G, Salazar F, Wang C, Phonon diffusion in harmonic and anharmonic one-dimensional quasiperiodic lattices, *Phil. Mag.* **86** 1043 (2006).

[641] Naumis G G, Phason hierarchy and electronic stability of quasicrystals, *Phys. Rev. B* **71** 144204 (2005).

[642] Naumis G G, Bazán A, Torres M, Aragón J L, Quintero-Torres R, Reflectance distribution in optimal transmittance cavities: The remains of a higher dimensional space, *Physica B* **203** 3179–3184 (2008).

[643] Naumis G G, López-Rodríguez F J, The electronic spectrum of a quasiperiodic potential: From the Hofstadter butterfly to the Fibonacci chain, *Physica B* **403** 1755–1762 (2008).

[644] Naumis G G, Topological map of the Hofstadter butterfly: Fine structure of Chern numbers and Van Hove singularities, *Phys. Lett. A* **380** 1772–1780 (2016).

[645] Naumis G G, Higher-dimensional quasicrystalline approach to the Hofstadter butterfly topological-phase band conductances: Symbolic sequences and self-similar rules at all magnetic fluxes, *Phys. Rev. B* **100** 165101 (2019).

[646] Naumovic D et al., in *New Horizons in Quasicrystals*, Eds. Goldman A I, Sordelet D J, Thiel P A and Dubois J M (World Scientific, Singapore 1997) p. 86.

[647] Nava R, Tagüena-Martínez J, del Río J A, Naumis G G, Perfect light transmission in Fibonacci arrays of dielectric multilayers, *J. Phys.: Condens. Matter* **21** 155901 (2009).

[648] Nayak J, Maniraj M, Rai A, Singh S, Rajput P, Gloskovskii A, Zegenhagen J, Schlagel D L, Lograsso T A, Horn K, Barman S R, Bulk electronic structure of quasicrystals, *Phys. Rev. Lett.* **109** 216403 (4pp) 2012.

[649] Niizeki K, Self-similarity of quasilattices in two-dimensions. I. The n-gonal quasilattice, *J. Phys. A* **22** 193 (1989).

[650] Niizeki K, Akamatzu T, Special points in the reciprocal space of an icosahedral quasi-lattice and the quasi-dispersion relation of electrons, *J. Phys.: Condens. Matt.* **2** 2759 (1990).

[651] Niizeki K, Fujita N, Superquasicrystals: self-similar-ordered structures with non-crystallographic point symmetries, *J. Phys. A: Math. Gen.* **38** L199 (2005).

[652] Nishino, Y, Electronic structure and transport properties of pseudogap system Fe_2VAl, *Materials Transactions* **42** 902–910 (2001).

[653] Niu Q, Nori F, Renormalization-group study of one-dimensional quasiperiodic systems, *Phys. Rev. Lett.* **57** 2057 (1986).

[654] Niu Q, Nori F, Spectral splitting and wave-function scaling in quasicrystalline and hierarchical structures, *Phys. Rev. B* **42** 10329 (1990).

[655] Noguchi K, Kawanishi S, Nishimoto K, Tamura R, Synthesis of single-grained $Zn_{88}Sc_{12}$ quasicrystal and its electrical resistivity, *Acta Cryst. A* **67** C625 (2011).

[656] Nosaki K, Masumoto T, Inoue A, Yamaguchi T, Ultrafine particle of quasi-crystalline aluminium and process for producing aggregates thereof , US patent No. 5,800,638 (1998).

[657] Notomi M, Suzuki H, Tamamura T, Edagawa K, Lasing action due to the two-dimensional quasiperiodicity of photonic quasicrystals with a Penrose lattice, *Phys. Rev. Lett.* **92** 123906 (2004).

[658] Notomi M, Manipulating light with strongly modulated photonic crystals, *Rep. Prog. Phys.* **73** 096501 (2010).

[659] Odlyzko A M, On the distribution of spacings between zeros of the zeta function, *Math. Comput.* **48** 273–308 (1987).

[660] Ohashi W, Spaepen F, Stable GaMgZn quasi-periodic crystals with pentagonal dodecahedral solidification morphology, *Nature (London)* **330** 555–556 (1987).

[661] Oppenheim J, Ma C, Hu J, Bindi L, Steinhardt P J, Asimow P D, Shock synthesis of decagonal quasicrystals, *Sci. Rep.* **7** 15628 (2017).

[662] Oppenheim J, Ma C, Hu J, Bindi L, Steinhardt P J, Asimow P D, Shock synthesis of five-component icosahedral quasicrystals, *Sci. Rep.* **7** 15629 (2017).

[663] Orsini-Rosenberg H, Steurer W, Ab initio investigations on the stability of seven-fold approximants, *Philos. Mag.* **91** 2567–2578 (2011).

[664] Ostlund S, Pandit R, Rand D, Schellnhuber H J, Siggia E D, One-dimensional Schrödinger equation with an almost periodic potential, *Phys. Rev. Lett.* **50** 1873 (1983).

[665] Ostlund S, Pandit R, Renormalization-group analysis of the discrete quasiperiodic Schrödinger equation, *Phys. Rev. B* **29** 1394 (1984).

[666] Ouyang Z, Liu K, Dhang J, Xing C, Liu W, Wang L, Effect of Li on electrochemical properties of $Ti_{1.4}V_{0.6}Ni$ quasicrystal alloy produced by rapid quenching, *Intermetallics* **62** 50–55 (2015).

[667] Oviedo-Roa R, Pérez L A, Wang C, AC conductivity of the transparent states in Fibonacci chains, *Phys. Rev. B* **62** 13805–13808 (2000).

[668] Pan S D, Yuan Y, Zhao L N, Liu Y H, Zhu S N, Tuning visible light generation based on multi-channel quasiperiodically poled $LiTaO_3$, *Opt. Commun.*, **284** 429–431 (2011).

[669] Pandey S K, Bhatnagar A, Mishra S S, Yadav T P, Shaz M A, Srivastava O N, Curious catalytic characteristics of Al-Cu-Fe quasicrystal for de/rehydrogenation of MgH_2, *J. Phys. Chem. C* **121** 24936–24944 (2017).

[670] Panova G Kh, Zemlyanov M G, Parshin P P, Shikov A A, Brand R A, Low-energy lattice excitations in the decagonal Al-Ni-Fe and icosahedral Al-Cu-Fe quasicrystals and the (Al,Si)-Cu-Fe cubic phase, *Phys. Solid State* **52** 771–775 (2010).

[671] Paquette L A, Balogh D W, Usha R, Kountz D, Christoph G G, Crystal and molecular structure of a pentagonal dodecahedrane, *Science* **211** 575–576 (1981).

[672] Paredes R, Aragón J L, Barrio R A, Nonperiodic hexagonal square-triangle tilings, *Phys. Rev. B* **58** 11990–11995 (1998).

[673] Park M J, Kim H S, Lee S B, Emergent localization in dodecagonal bilayer quasicrystals, *Phys. Rev. B* **99** 245401 (5pp) 2019.

[674] Park J Y Ogletree D F, Salmeron M, Jenks C J, Thiel P A., Friction and adhesion properties of clean and oxidized Al-Ni-Co decagonal quasicrystals: a UHV atomic force microscopy/scanning tunneling microscopy study, *Tribology Lett.* **17** 629 (2004).

[675] Park J Y, Ogletree D F, Salmeron M, Ribeiro R A, Canfield P C, Jenks C J, Thiel P A, High frictional anisotropy of periodic and aperiodic directions on a quasicrystal surface, *Science* **309** 1354 (2005); Atomic scale coexistence of periodic and quasiperiodic order in a 2-fold Al-Ni-Co decagonal quasicrystal surface, *Phys. Rev. B* **72** 220201(R) 2005.

[676] Parsamehr H, Chen T S, Wang D S, Leu M S, Han I, Xi Z, Tsai A P, Shahani A J, Lai C H, Thermal spray coating of Al-Cu-Fe quasicrystals: Dynamic observations and surface properties, *Materalia* **8** 100432 (2019).

[677] Paβens M, Caciuc V, Atodiresei N, Feuerbacher M, Moors M, Dunin-Borkowski R E, Blügel S, Waser R, Karthäuser S, Interface-driven formation of a two-dimensional dodecagonal fullerene quasicrystal, *Nature Commun.* **8** 15367 (7pp) 2017.

[678] Paβens M, Karthäuser S, Rotational switches in the two-dimensional fullerene quasicrystal, *Acta Cryst.* **A75** 41–49 (2019).

[679] Pauling L, So-called icosahedral and decagonal quasicrystals are twins of an 820–atom cubic crystal, *Phys. Rev. Lett.* **58** 365 (1987).

[680] Peng R W, Hu A, Jiang S S, Zhang C S, Feng D, Structural characterization of three-component Fibonacci Ta/Al multilayer films, *Phys. Rev. B* **46** 7816 (1992); Pan F M, Jin G J, Wu X S, Wu X L, Hu A, Jian S S, Raman scattering by longitudinal acoustic phonons in a three-component TaAl Fibonacci superlattice, *Phys. Lett. A* **228** 301 (1997).

[681] Peng R W, Huang X Q, Qiu F, Wang M, Hu A, Jiang S S, Mazzer M, Symmetry-induced perfect transmission of light waves in quasiperiodic dielectric multilayers, *Appl. Phys. Lett.* **80** 3063–3065 (2002).

[682] Peng R W, Liu Y M, Huang X Q, Qiu F, Wang M, Hu A, Jiang S S, Feng D, Ouyang L Z, Zou J, Dimerlike positional correlation and resonant transmission of electromagnetic waves in aperiodic dielectric multilayers, *Phys. Rev B* **69** 165109 (2004).

[683] Penrose R, The role of aesthetics in pure and applied mathematical research, *Bull. Inst. Math. Appl.* **10** 266 (1974).

[684] Penrose R, in *Tilings and Quasicrystals: a non-local growth problem?* ed. M. V. Jaric (Academic Press Inc, London 1989) vol. 3, Ch 2, pp. 53–79.

[685] Pérez K S, Estevez J O, Méndez-Bals A, Arriaga J, Palestino G, Mora-Ramos M E, Tunable resonance transmission modes in hybrid heterostructures based on porous silicon, *Nanoscale Res. Lett.* **7** 302 (8pp) 2012.

[686] Pérez-Álvarez R, García-Moliner F, Trallero-Giner C, Velasco V R, Polar optical modes in Fibonacci heterostructures, *J. Raman Spectrosc.* **31** 421 (2000).

[687] Perkins C L, Trenary M, Tanaka K, Direct observation of $(B_{12})(B_{12})_{12}$ supericosahedra as the basic structural element in YB_{66}, *Phys. Rev. Lett.* **77** 4772 (1996).

[688] Perrot A, Dubois J M, Heat diffusivity in quasicrystals and related alloys, *Ann. Chim. Fr.* **18** 501–511 (1993).

[689] Petucci J, Karimi M, Huang Y T, Curtarolo S, Diehl R D, Ordering and growth of rare gas films (Xe, Kr, Ar, and Ne) on the pseudo-ten-fold quasicrystalline approximant $Al_{13}Co_4$ (100) surface, *J. Phys.: Condens. Matter* **26** 095003 (10pp) 2014.

[690] Pham J, Meng F, Lynn M J, Ma T, Kreyssig A, Kramer M J, Goldman A I, Miller G J, From quasicrystals to crystals with interpenetrating icosahedra in Ca-Au-Al: In situ variable-temperature transformation, *J. Am. Chem. Soc.* **140** 1337–1347 (2018).

[691] Phillips P W, Wu H L, Localization and its absence: a new metallic state for conducting polymers, *Science* **252** 1805 (1991).

[692] Phung Ngoc B, Geantet C, Aouine M, Bergeret G, Raffy S, Marlin S, Quasicrystal derived catalyst for steam reforming of methanol, *Int. J. Hydrogen Energy* **33** 1000–1007 (2008).

[693] Phung Ngoc B, Geantet C, Dalmon J A, Aouine M, Bergeret G, Delichere P, Raffy S, Marlin S, Quasicrystalline structures as catalyst precursors for hydrogenation reactions, *Catal. Lett.* **131** 59–69 (2009).

[694] Phung Ngoc B, Geantet C, Aouine M, Bergeret G, Raffy S, Marlin S, Quasicrystal derived catalyst for steam reforming of methanol, *Int J Hydrogen Energ* **33** 1000 (2008).

[695] Pierce F S, Bancel P A, Biggs B D, Guo Q, Poon S J, Composition dependence of the electronic properties of Al-Cu-Fe and Al-Cu-Ru-Si semimetallic quasicrystals, *Phys. Rev. B* **47** 5670–5676 (1993).

[696] Pierce F S, Poon S J, Biggs B D, Band-structure gap and electron transport in metallic quasicrystals and crystals, *Phys. Rev. Lett.* **70** 3919–3922 (1993).

[697] Pierce F S, Poon S J, Guo Q, Electron localization in metallic quasicrystals, *Science* **261** 737–739 (1993).

[698] Pierce F S, Guo Q, Poon S J, Enhanced insulatorlike electron transport behavior of thermally tuned quasicrystalline states of Al-Pd-Re alloys, *Phys. Rev. Lett.* **73** 2220–2223 (1994).

[699] Pimonov V V, Konevtsova O V, Rochal S B, Anomalous small viral shells and simplest polyhedra with icosahedral symmetry: the rhombic triacontahedron case, *Acta Cryst.* **A75** 135—14 (2019).

[700] Pina C M, López-Acevedo V, Quasicrystals and other aperiodic structures in mineralogy, *Crystals* **6** 137 (2016). doi:10.3390/cryst6110137.

[701] Poddar A, Das S, Schnelle W, Gmelin E, Plachke D, Carstanken H D, Electrical transport, magnetic and thermal properties of icosahedral Al-Pd-Mn quasicrystals, *J. Mag. Mag. Mat.* **300** 263–272 (2006).

[702] Pollard M E, Parker G J, Low-contrast bandgaps of a planar parabolic spiral lattice, *Opt. Lett.* **34** 2805–2807 (2009).

[703] Poon S J, Electronic properties of quasicrystals: an experimental review, *Adv. Phys.* **41** 303–363 (1992).

[704] Pope A L, Tritt T M, Chernikov M A, Feuerbacher M, Legault S, Gagnon R, Strom-Olsen J, Observation of the interplay of microstructure and thermopower in the $Al_{71}Pd_{21}Mn_{8--x}Re_x$ quasicrystalline system, *Mat. Res. Soc. Symp. Proc.* **545** 413 (1999).

[705] Pope A L. Tritt T M, Gagnon R, Strom-Olsen J, Electronic transport in Cd-Yb and Y-Mg-Zn quasicrystals, *Appl. Phys. Lett.* **79** 2345–2347 (2001).

[706] Pope A L, Tritt T M, Chernikov M A, Feuerbacher M, Thermal and electrical transport properties of the single-phase quasicrystalline material $Al_{70.8}Pd_{20.9}Mn_{8.3}$, *Appl. Phys. Lett.* **75** 1854–1856 (1999).

[707] Préjean J J, Lasjaunias J C, Berger C, Sulpice A, Resistive and calorimetric investigations of an insulating quasicrystal i-AlPdRe, *Phys. Rev. B* **61** 9356–9364 (2000).

[708] Prekul A F, Kuz'min N Yu, Shchegolikhina N J, Electronic structure of icosahedral quasicrystals: role of deffects, *J. Alloys Compd.* **342** 405–409 (2002).

[709] Prekul A F, Shchegolikhina N, Correlations between electric, magnetic, and galvano-magnetic quantities in stable icosahedral phases based on aluminium, *Crystallogr. Rep.* **52** 996–1005 (2007).

[710] Prekul A F, Shalaeva E V, Shchegolikhina N, Calorimetric investigation of electronic and lattice excitations of the icosahedral quasicrystals in the range of moderate temperatures, *Phys. Solid State* **52** 1797–1802 (2010).

[711] Prekul A F, Shchegolikhina N, Two-level electron excitations and distinctive physical properties of Al-Cu-Fe quasicrystals, *Crystals* **6** 119 (2016). doi:10.3390/cryst6090119.

[712] Pussi K, Gierer M, Diehl R D, The uniaxially aperiodic structure of a thin Cu film on fivefold i-Al-Pd-Mn, *J. Phys.: Condens. Matter* **21** 474213 (7pp) 2009.

[713] Qian C, Wang J, Dodecagonal quasicrystal silicene: preparation, mechanical property, and friction behaviour, *Phys. Chem. Chem. Phys.* **22** 74–81 (2020).

[714] Qiu F, Peng R W, Huang X Q, Liu Y M, Wang M, Hu A, Jiang S S, Resonant transmission and frequency trifurcation of light waves in Thue-Morse dielectric multilayers, *Europhys. Lett.* **63** 853–859 (2003).

[715] Quiquandon M, Quivy A, Devaud L, Faudot F, Lefebvre S, Bessière M, Calvayrac Y, Quasicrystal and approximant structures in the Al-Cu-Fe system, *J. Phys.: Condens. Matter* **8** 2487 (1996).

[716] Quivy A, Quiquandon M, Calvayrac Y, Faudot F, Gratias D, Berger C, Brand R A, Simonet V, Hippert F, A cubic approximant of the icosahedral phase in the (Al,Si)-Cu-Fe system *J. Phys. Condens. Matter* **8** 4223–4235 (1996).

[717] Quotane I, El Boudouti E H, Djafari-Rouhani B, El Hassouani Y, Velasco V R, Bulk and surface acoustic waves in solid-fluid Fibonacci layered materials, *Ultrasonics* **61** 40–51 (2015).

[718] Radin C, The pinwheel tilings of the plane, *Ann. Math.* **139** 661–702 (1994).

[719] Rahimi H, Near infrared ultra compact phase retarders based on photonic superlattices containing fullerene layers, *Opt. Quant. Electron.* **47** 3627–3635 (2015).

[720] Rahimi H, Investigation of mode localization in hybrid periodic/quasiperiodic heterostructures containing nano-scale fullerene layers, *Optik* **126** 4676–4678 (2015).

[721] Rahimi H, TM wave resonant peaks in defect-free multilayer structures containing nano-scale Y123 superconductors, *K. Supercond. Nov. Magn.* **29** 1767–1772 (2016).

[722] Rajasekaran E, Zeng X C, Diestler D J, *Micro/Nanotribology and Its Applications*, Ed. Bhushan B (Kluwer Academic Publishers, Dordrecht 1997, The Netherlands).

[723] Raman A P, Anoma M A, Zhu L, Rephaeli E, Fan S, Passive radiative cooling below ambient air temperature under direct sunlight, *Nature* **415** 540—545 (2014).

[724] Ratliff D J, Archer A J, Subramanian P, Rucklidge A M, Which wave numbers determine the thermodynamic stability of soft matter quasicrystals? *Phys. Rev. Lett.* **123** 148004 (6pp) 2019.

[725] Rapp Ö, in *Physical Properties of Quasicrystals*, Ed. Stadnik Z M, Springer Series in Solid-State Physics **126** (Springer-Verlag, Berlin 1999) p 127.

[726] Rapp Ö, Electronic transport properties of quasicrystals: the unique case of magnetoresistance, *Mats. Sci. Engineer.* **204–206** 458–463 (2000).

[727] Rapp Ö, Karkin A A, Goshchitskii B N, Voronin V I, Srinivas V, Poon S J, Electronic and atomic disorder in icosahedral AlPdRe, *J. Phys. Condens. Matt.* **20** 114120 (6pp) 2008.

[728] Rapp Ö, Generalization of the Mooij correlation to quasicrystals, *J. Phys. Condens. Matt.* **25** 065701 (7pp) 2013.

[729] Rapp Ö, Reinterpretation of the zero-temperature conductivity in icosahedral AlPdRe, *Phys. Rev. B* **94** 024202 (4 pp) 2016. doi: 10.1103/PhysRevB.94.024202.

[730] Rattier M, Benisty H, Schwoob E, Weisbuch C, Krauss T F, Smith C J M, Houdré R, Oesterle U, Omnidirectional and compact guided light extraction from Archimedean photonic lattices, *Appl. Phys. Lett.* **83** 1283–1285 (2003).

[731] Rayment I, Baker T S, Caspar D L D, Murakami W T, Polyoma virus capsid structure at 22.5 Å resolution, *Nature (London)* **295** 110–115 (1982).

[732] Rayment I, Animal virus structure, in *Biological Macromolecules and Asemblies*, vol. **1** ed. McPherson (John Wiley & Sons 1984).

[733] Rechtsman M C, Jeong H C, Chaikin P M, Torquato S, Steinhardt P J, *Phys. Rev. Lett.* **101** 073902 (2008).

[734] Reinhardt A, Schreck J S, Romano F, Doye J P K, Self-assembly of two-dimensional binary quasicrystals: a possible route to a DNA quasicrystal, *J. Phys.: Condens. Matter* **29** 014006 (12pp) 2017.

[735] Reisser R, Kronmüller H, High coercivity quasicrystals, *J. Magn. Magn. Mats.* **131** 90 (1994).

[736] Renner M, von Freymann G, Transverse mode localization in three-dimensional deterministic aperiodic structures, *Adv. Opt. Mater.* **2** 226–230 (2014).

[737] Renner M, von Freymann G, Spatial correlations and optical properties in three-dimensional deterministic aperiodic structures, *Sci. Rep.* **5** 13129 (2015). doi:10.1038/srep13129.

[738] Repetowicz P, Wolny J, Diffraction pattern calculations for a certain class of N-fold quasilattices, *J. Phys. A.* **31** 6873 (1998).

[739] Rephaeli E, Raman A, Fan S, Ultrabroadband photonic structures to achieve high-performance daytime radiative cooling. *Nano Lett.* **13** 1457—1461 (2013).

[740] Reynolds G A M, Golding B, Kortan A R, Parsey Jr. J M, Isotropic elasticity of the Al-Cu-Li quasicrsytal, *Phys. Rev. B* **41** 1194–1195 (1990).

[741] Ricciardi A, Gallina I, Campopiano S, Castaldi G, Pisco M, Galdi V, Cusano A, Guided resonances in photonic quasicrystals, *Opt. Express* **17** 6335 (2009).

[742] Rippa M, Capasso R, Mormile P, De Nicola S, Zanella M, Manna L, Nenna G, Petti L, Bragg extraction of light in 2D photonic Thue-Morse quasicrystals patterned in active CdSe/CdS nanorod-polymer nanocomposites, *Nanoscale* **5** 331–336 (2013).

[743] Rippa M, Capasso R, Pannico M, Musto P, Tkachencko V, Zhou J, Petti L, Engineeried plasmonic Thue-Morse nanostructures for LSPR detection of the pesticide Thiram, *Nanophotonics* **6** 1083–1092 (2017) ISSN (Online) 2192–2864.

[744] Rippa M, Castagna R, Marino A, Tkachenko V, Palermo G, Pane A, Umeton C, Thue-Morse nanostructures for tunable light extraction in the visible region, *Opt. Lasers Engineering* **104** 291–299 (2018).

[745] Rippa M, Castagna R, Zhou J, Paradiso R, Borriello G, Bobeico E, Petti L, Do-decagonal plasmonic quasicrystals for phage-based biosensing, *Nanotech.* **29** 405501 (7pp) 2018.

[746] Rivier N, Crystallography of spiral lattices, *Mod. Phys. Lett. B* **2** 953–960 (1988).

[747] Roche S, Mayou D, Conductivity of quasiperiodic systems: a numerical study, *Phys. Rev. Lett.* **79** 2518 (1997).

[748] Roche S, Fujiwara T, Fermi surfaces and anomalous transport in quasicrystals, *Phys. Rev. B* **58** 11338 (1998).

[749] Rodríguez A W, McCauley A P, Avniel Y, Johnson S G, Computation and visualization of photonic quasicrystal spectra via Bloch's theorem, *Phys. Rev. B* **77** 104201 (10pp) 2008.

[750] Roichman Y, Grier D G, Holographic assembly of quasicrystalline photonic heterostructures, *Optics Express* **13** 5434–5439 (2005).

[751] Rokhsar D S, Mermin N D, Wright D C, Rudimentary quasicrystallography: The icosahedral and decagonal reciprocal lattices, *Phys. Rev. B* **35** 5487–5496 (1987).

[752] Rokhsar D S, Wright D C, Mermin N D, Scale equivalence of quasicrystallographic space groups, *Phys. Rev. B* **37** 8145–8149 (1988).

[753] Romerio M V, Almost periodic functions and the theory of disordered systems, *J. Math. Phys.* **12** 552–562 (1971).

[754] Röntgen M, Morfonios C V, Wang R, Dal Negro L, Schmelcher P, Local symmetry theory of resonator structures for the real-space control of edge states in binary aperiodic chains, *Phys. Rev. B* **99** 214201 (24 pp) 2019.

[755] Rosenbaum R, Lin S T, Su T I, Hopping transport in an insulating quasicrystal bar of $Al_{70}Pd_{22.5}Re_{7.5}$ near the metal-insulator transition, *J. Phys. Condens. Matter* **15** 4169 (2003).

[756] Rosenbaum R, Mi S, Grushko B, Przepiórzyński B, Low temperature electronic transport in AlCuRu quasicrystalline alloys, *J. Low Temp. Phys.* **149** 314–329 (2007).

[757] Rotenberg E, Theis W, Horn K, Gille P, Quasicrystalline valence bands in decagonal AlNiCo, *Nature (London)* **406** 602 (2000).

[758] Roth C, Schwalbe G, Knöfler R, Zavaliche F, Madel O, Haberken R, Häussler P, A detailed comparison between the amorphous and the quasicrystalline state of Al-Cu-Fe, *J. Non-Cryst. Solids* **250–252** 869–873 (1999).

[759] Rouxel D, Pigeat P, Surface oxidation and thin film preparation of AlCuFe quasicrystals, *Prog. Surf. Sci.* **81** 488–514 (2006).

[760] Ruelle D, Do turbulent crystals exist?, *Physica* **113A** 619–623 (1982).

[761] Ryu C S, Oh G Y, Lee M H, Extended and critical wave functions in a Thue-Morse chain, *Phys. Rev. B* **46** 5162 (1992).

[762] Ryu C S, Kim I M, Oh G Y, Lee M H, Localized and extended states in a deterministically aperiodic chain, *Phys. Rev B* **49** 14991–14995 (1994).

[763] Sadoc J F, Rivier N, Boerdijk-Coxeter helix and biological helices as quasicrystals, *Mater. Sci. Eng.* **294–296** 397–400 (2000).

[764] Sadoc A, Majzoub E H, Huett V T, Kelton K F, Local structure in hydrogenated Ti-Zr-Ni quasicrystals and approximants, *J Alloys Compd.* **356–357** 96–99 (2003).

[765] Sahnoune A, Ström-Olsen J O, Zaluska A, Quantum corrections to the conductivity in icosahedral Al-Cu-Fe alloys, *Phys. Rev. B* **46** 10629 (1992).

[766] Sakai S, Takemori N, Koga A, Arita R, Superconductivity on a quasiperiodic lattice: extended-to-localized crossover of Cooper pairs, *Phys. Rev. B* **95** 024509 (2017).

[767] Salazar G, Naumis G, Electrical fields on quasiperiodic potentials, *J. Phys.: Condens. Matter* **22** 115501 (9pp) 2010.

[768] Saldana X I, López-Cruz E, Contreras-Solorio D A, Self-similar optical transmittance for a deterministic aperiodic multilayer structure, *J. Phys.: Condens. Matter* **21** 155403 (2009).

[769] Samson S, The crystal structure of the phase β Mg_2Al_3, *Acta Crystallogr.* **19** 401 (1965).

[770] Sánchez A, Maciá E, Domínguez-Adame F, Suppression of localization in Kronig-Penney models with correlated disorder, *Phys. Rev B* **49** 147 (1994).

[771] Sánchez A et al., in *Quasicrystals: Preparation, Properties and Applications,* J M Dubois, P A Thiel, A P Tsai et al. eds., (Materials Research Society, Warrendale, PA 1999).

[772] Sánchez F, Sánchez V, Wang C, Renormalization approach to the electronic localization and transport in macroscopic generalized Fibonacci lattices, *J. Non-Cryst. Solids* **450** 194 (2016).

[773] Sánchez F, Sánchez V, Wang C, Ballistic transport in aperiodic Labyrinth tiling proven through a new convolution theorem, *Eur. Phys. J. B* **91** 132 (12pp) 2018.

[774] Sánchez V, Pérez L A, Oviedo-Roa R, Wang C, Renormalization approach to the Kubo formula in Fibonacci systems, *Phys. Rev. B* **64** 174205 (11pp) 2001.

[775] Sánchez V, Wang C, Application of renormalization and convolution methods to the Kubo-Greenwood formula in multidimensional Fibonacci systems, *Phys. Rev. B* **70** 144207 (2004).

[776] Sánchez V, Wang C, Resonant AC conducting spectra in quasiperiodic systems, *Int. J. Comput. Mater. Sci. Eng.* **1** 1250003 (8pp) 2012.

[777] Sánchez V, Ramírez C, Sánchez F, Wang C, Non-perturbative study of impurity effects on the Kubo conductivity in macroscopic periodic and quasiperiodic lattices, *Physica B* **449** 121–128 (2014).

[778] Sánchez V, Wang C, Improving the ballistic AC conductivity through quantum resonance in branched nanowires, *Pjil. Mag.* **95** 326–333 (2015).

[779] Sánchez V, Wang C, Real space theory for electron and phonon transport in aperiodic lattices via renormalization, *Symmetry* **12** 430 (2020).

[780] Sánchez-Palencia L, Santos L, Bose-Einstein condensates in optical quasicrystal lattices, *Phys. Rev. A* **72** 053607 (2005).

[781] Sandvik Nanoflex strip steel datasheet updated 5/07/2018.

[782] Santos L, A quasicrystal for quantum simulations, *Physics* **12** 31 (2019). doi: 10.1103/Physics.12.31.

[783] Sato T J, Takakura H, Tsai A P, Shibata K, Ohoyama K, Andersen K H, Antiferromagnetic spin correlations in the Zn-Mg-Ho icosahedral quasicrystal, *Phys. Rev. B* **61** 476 (2000).

[784] Schellenberger P, Demangeat G, Lemaire O, Ritzenthaler C, Bergdoll M, Oliéric V, Sauter C, Lorber B, Strategies for the crystallization of viruses: using phase diagrams and gels to produce 3D crystals of Grapevine fanleaf virus, *J. Struct. Biol.* **174** 344–351 (2011).

[785] Schenk S, Förster S, Meinel K, Hammer R, Paleschke M, Pantzer J, Dresler C, Schumann F O, Widdra W, Observation of a dodecagonal oxide quasicrystal and its complex approximant in the $SrTiO_3$-Pt(111) system, *J. Phys.: Condens. Matter* **29** 134002 (2017).

[786] Schmiedeberg M, Achim C V, Hielscher J, Kapfer S C, Löwen H, Dislocation-free growth of quasicrystals from two seeds due to additional phasonic degrees of freedom, *Phys. Rev. E* **96** 012602 (7pp) 2017.

[787] Schokker A H, Koenderink A F, Lasing in quasi-periodic and aperiodic plasmon lattices, *Optica* **3** 686–693 (2016).

[788] Schönleber A, Organic molecular compounds with modulated crystal structures, *Z. Kristallogr.* **226** 499–517 (2011).

[789] Schrödinger E, *What is life? The Physical aspects of the Living Cell*, Cambridge University Press, New York 1945.

[790] Schumayer D, Hutchinson D A W, Colloquium: Physics of the Riemann hypothesis, *Rev. Mod. Phys.* **83** 307–330 (2011).

[791] Senabulya N, Xiao X, Han I, Shahani A J, On the kinetic and equilibrium shapes of icosahedral $Al_{71}Pd_{19}Mn_{10}$ quasicrystals, *Scripta Materalia* **146** 218–221 (2018).

[792] Senabulya N, Shahani A J, Growth interactions between icosahedral quasicrystals, *Phys. Rev. Mats.* **3** 093403 (8pp) 2019.

[793] Senechal M, *Quasicrystals and Geometry* (Cambridge University Press, Cambridge 1995).

[794] Senechal M, What is ... a quasicrystal? *Notices of the AMS* **53**(8) 886–887 (2006).

[795] Senechal M, Mapping the aperiodic landscape, 1982–2007, *Philos. Mag.* **88** 2003–2016 (2008).

[796] Senechal M, Sphere packings as stem cells, *Struct. Chem.* **28** 27–31 (2017).

[797] Severin M, Dulea M, Riklund R, Periodic and quasiperiodic wavefunctions in a class of one-dimensional quasicrystals: an analyticaltreatment, *J. Phys. Condens. Matter* **1** 8851 (1989).

[798] Sgrignoli F, Wang R, Pinheiro F A, Dal Negro L, Localization of scattering resonances in aperiodic Vogel spirals, *Phys. Rev. B* **99** 104202 (12pp) 2019.

[799] Sgrignoli F, Gorsky S, Britton W A, Zhang R, Riboli F, Dal Negro L, Multifractality of light in photonic arrays based on algebraic number theory, *Commun. Phys.* **3** 106 (2020).

[800] Schaub Th M, Bürgler D E, Güntherodt H J, Suck J B, Audier M, The surface structure of icosahedral $Al_{68}Pd_{23}Mn_9$ measured by STM and LEED, *Appl. Phys. A* **61** 491 (1995).

[801] Shechtman D, Blech I, Gratias D, Cahn J W, Metallic phase with long-range orientational order and no translation symmetry, *Phys. Rev. Lett.* **53** 1951 (1984).

[802] Sheng Y, Koynov K, Dou J, Ma B, Li J, Zhang D, Collinear second harmonic generations in a nonlinear photonic quasicrystal, *Appl. Phys. Lett.* **92** 201113 (2008).

[803] Shubin M A, Almost periodic functions and partial differential operators, *Russian Math. Surveys* **33** 1–52 (1978).

[804] Shukla A K, Biswas C, Dhaka R S, Das S C, Krüger P, Barman S R, Influence of sp-d hybridization on the electronic structure of Al-Mn alloys, *Phys. Rev. B* **77** 195103 (2008).

[805] Shu-yuan L, Xue-mei W, Dian-lin Z, He X L, Kuo K X, Anisotropic transport properties of a stable two-dimensional quasicrystal: $Al_{62}Si_3Cu_{20}Co_{15}$, *Phys. Rev. B* **41** 9625 (1990).

[806] Shuyuan L, Guohong L, Dian-lin Z, Thermopower of decagonal $Al_{73}Ni_{17}Co_{10}$ single quasicrystals: Evidence for a strongly enhanced electron-phonon coupling in the quasicrystalline plane, *Phys. Rev. Lett.* **77** 1998 (1996).

[807] Sil S, Karmakar S N, Moitra R K, Chakrabarti A, Extended states in one-dimensional lattices: application to the quasiperiodic copper-mean chain, *Phys. Rev. B* **48** 4192(R) (1993).

[808] Silva J R M, Vasconcelos M S, Anselmo D H A L, Mello V D, Phononic topological states in 1D quasicrystals, *J. Phys. : Condens. Matter* **31** 505405 (9pp) 2019.

[809] Silva B, Costa C H, Tuning band structures of photonic multilayers with positive and negative refractive index materials according to generalized Fibonacci and Thue-Morse sequences, *J. Phys.: Condens. Matter* **32** 135703 (17pp), 2020.

[810] Simkin M V, Mahan G D, Minimum thermal conductivity of superlattices, *Phys. Rev. Lett.* **84** 927 (2000).

[811] Sing B, Kolakoski sequences – an example of aperiodic order, *J. Non-Cryst. Solids* **334&335** 100–104 (2004).

[812] Singh B K, Pandey P C, Tunable temperature-dependent THz photonic bandgaps and localization mode engineering in 1D periodic and quasi-periodic structures with graded-index materials and InSb, *Appl. Opt.* **57** 8171–8181 (2018).

[813] Singh B K, Pandey P C, Influence of graded index materials on the photonic localization in one-dimensional quasiperiodic (Thue-Morse and Double-Periodic) photonic crystals, *Opt. Commun.* **333** 84—91 (2014).

[814] Singh K, Saha K, Parameswaran S A, Weld D M, Fibonacci optical lattices for tunable quantum quasicrystals, *Phys. Rev. A* **92** 063426 (8pp) 2015.

[815] Singh V K, Mihalkovic M, Krajcí M, Sarkar S, Sadhukhan P, Maniraj M, Rai A, Pussi K, Schlagel D L, Lograsso T A, Shukla A K , Barman S R, Quasiperiodic ordering in thick Sn layer on i-Al-Pd-Mn: A possible quasicrystalline clathrate, *Phys. Rev. Research* **2** 013023 (15pp) 2020.

[816] Sire C, Electronic spectrum of a 2D quasi-crystal related to the octagonal quasiperiodic tiling, *Europhys. Lett.* **10** 483–488 (1989).

[817] Sirindil A, Quiquandon M, Gratias D, Mackay clusters and beyond in icosahedral quasicrystals seen from 6F space, *Struct. Chem.* **28** 123–132 (2017).

[818] Smerdon J A, Parle J K, Wearing L H, Lograsso T A, Ross A R, McGrath R, Nucleation and growth of a quasicrystalline monolayer: Bi adsoption on the fivefold surface of $i-Al_{70}Pd_{21}Mn_9$, *Phys. Rev. B* **78** 075407 (2008).

[819] Smerdon J A, Young K M, Lowe M, Hars S S, Yadav T P, Hesp D, Dhanak V R, Tsai A P, Sharma H R, McGrath R, Templated quasicrystalline molecular ordering, *Nano Lett.* **14** 1184–1189 (2014).

[820] Smetana V, Lin Q, Pratt D K, Kreyssig A, Ramazanoglu M, Corbett J D, Goldman A I, Miller G J, A sodium-containing quasicrystal: using gold to enhance sodium's covalency in intermetallic compounds, *Angew. Chem Int. Ed.* **51** 12699–12702 (2012).

[821] Smith D R, Pendry J B, Wiltshire M C K, Metamaterials and negative refractive index, *Science* **305** 788 (2004).

[822] Smith A P, Ashcroft N W, Pseudopotentials and quasicrystals, *Phys. Rev. Lett.* **59** 1365 (1987).

[823] Smontara A, Smiljanić I, Bilušić A, Jagličić Z, Klanjsek M, Roitsch S, Dolinšek J, Feuerbacher M, Electrical, magnetic, thermal and thermoelectric properties of the "Bergman phase" $Mg_{32}(Al,Zn)_{49}$ complex metallic alloy, *J. Alloys Compd.* **430** 29 (2007).

[824] Socolar J E S, Weak matching rules for quasi-crystals, *Comments Math. Phys.* **129** 599–619 (1990).

[825] Socolar J E S, Taylor J M, An aperiodic hexagonal tile, *J. Comb. Theory A* **118** 2207 (2011).

[826] Solbrig H, Landauro C V, Systems with icosahedral clusters: direct links between atoms in cluster-recursion methods and DOS spectral fine structure, *Physica B* **292** 47 (2000).

[827] Solbrig H, Landauro C V, Electronic transport, spectral fine structures, and atom clusters in quasicrystals and approximants, *Adv. Solid State Phys.* **42** 151 (2002).

[828] Sordelet D J, Besser M F, Logsdon J L, Abrasive wear behavior of Al-Cu-Fe quasicrystalline composite coatings, *Mater. Sci. Eng. A* **255** 54–65 (1998).

[829] Sordelet D J, Kim J S, Besser M F, Dry sliding of polygrained quasicrystalline and crystalline Al-Cu-Fe alloys, *Mater. Res. Soc. Proc.* **553** 459 (1999).

[830] Stadnik Z M, Pordie D, Garnier M, Baer Y, Tsai A P, Inoue A, Edagawa K, Takeuchi S, Electronic structure of icosahedral alloys studied by ultrahigh energy resolution photoemission spectroscopy, *Phys. Rev. Lett.* **77** 1777 (1996); Stadnik Z M, Pordie D, Garnier M, Baer Y, Tsai A P, Inoue A, Edagawa K, Takeuchi S, Buschow K H J, Electronic structure of quasicrystals studied by ultrahigh-energy-resolution photoemission spectroscopy, *Phys. Rev. B* **55** 10938 (1997).

[831] Stadnik Z M ed. *Physical Properties of Quasicrystals* (Springer Series in Solid-State Sciences **126**; Springer Verlag, Berlin 1999).

[832] Stagno V, Bindi L, Shibazaki Y, Tange Y, Higo Y, Mao H K, Steinhardt P J, Fei Y, Icosahedral AlCuFe quasicrystal at high pressure and temperature and its implications for the stability of icosahedrite, *Science Rep.* **4** 5869 (9pp) 2014. doi:10.1038/srep05869.

[833] Stakhov A, *The Mathematics of Harmony*, (World Scientific, Singapore 2009).

[834] Stampfli P, A dodecagonal quasiperiodic lattice in two dimensions, *Helv. Phys. Acta* **159** 1260–1263 (1986).

[835] Steinhardt P J, Bindi K, In search of natural quasicrystals, *Rep. Prog. Phys.* **75** 092601 (11pp) 2012.

[836] Steinhardt P J, Quasicrystals: a brief history of the impossible, *Rend. Fis. Acc. Lincei* **24** S85–S91 (2013).

[837] Steurer W, Haibach T, Crystallography of quasicrystals, in *Physical Properties of Quasicrystals* , Z. M. Stadnik ed. (Springer Series in Solid-State Sciences **126**; Springer Verlag, Berlin 1999) pp. 51–90.

[838] Steurer W, Quasicrystal structure analysis, a never-ending story? *J. Non-cryst. Solids* **334&335** 137–142 (2004).

[839] Steurer W, Boron-based quasicrystals with sevenfold symmetry, *Philos. Mag.* **87** 2707–2712 (2007).

[840] Steurer W, Deloudi S, *Crystallography of Quasicrystals – Concepts, Methods and Structures* (Springer Series in Materials Science **126**; Springer Verlag, Berlin 2009).

[841] Steurer W, Why are quasicrystals quasiperiodic? *Chem. Soc. Rev.* **41** 6719–6729 (2012).

[842] Steurer W, Quasicrystal structure and growth models: discussion of status quo and the still open questions. *IOP Conf. Series: J. Phys. Conf. Series* **809** 012001 (6pp) 2017.

[843] Steurer W, Quasicrystals: What do we know? What do we want to know? What can be known? *Acta Cryst. A* **74** 1–11 (2018).

[844] Stewart I, *Why Beauty is Truth. A History of Symmetry* (Basic Books, New York 2007).

[845] Strogatz S, *Infinite Powers - How Calculus Reveals the Secrets of the Universe* (Eamon Dolan Book, New York 2019).

[846] Strzalka R, Buganski I, Wolny J, Statistical approach to diffraction of periodic and non-periodic crystals- Review, *Crystals* **6** 104 (2016). doi: 10.3390/cryst6090104.

[847] Suck J B, Shreiber M, Häussler P (Eds.) *Quasicrystals: An Introduction to Structure, Physical Properties and Applications* (Springer Series in Materials Science **55**; Springer, Berlin 2002).

[848] Süto A, Singular continuous spectrum on a Cantor set of zero Lebesgue measure for the Fibonacci hamiltonian, *J. Stat. Phys.* **56** 525 (1989).

[849] Sutter-Widmer D, Steurer W, Prediction of band gaps in phononic quasicrystals based on single-rod resonances, *Phys. Rev. B* **75** 134303 (4pp) 2007.

[850] Sutter-Widmer D, Deloudi S, Steurer W, Prediction of Bragg-scattering-induced band gaps in phononic quasicrystals, *Phys. Rev. B* **75** 094304 (11pp) 2007.

[851] Sutter-Widmer D, Neves P, Itten P, Sainidou R, Steurer W, Distinct band gaps and isotropy combined in icosahedral band gap materials, *Appl. Phys. Lett.* **92** 073308 (2008).

[852] Swenson C A, Lograsso T A, Ross A R, Anderson Jr. N E, Linear thermal expansivity (1–300 K), specific heat (1–108 K), and electrical resistivity of the icosahedral quasicrystal i-Al$_{61.4}$Cu$_{25.4}$Fe$_{13.2}$, *Phys. Rev B* **66** 184206-1-11 (2002).

[853] Symko O G, Klein T, Kleda D, Formation and application of AlCuFe thin films, US Patent 6.294,030 B1 (2001).

[854] Talapin D V, Shevchenko E V, Bodnarchuk M I, Ye X, Chen J, Murray C B, Quasicrystalline order in self-assembled binary nanoparticle superlattices, *Nature* **461** 964–967 (2009).

[855] Takagiwa Y, Kamimura T, Hosoi S, Okada J T, Kimura K, Thermoelectric properties of polygrained icosahedral Al$_{71--x}$Ga$_x$Pd$_{20}$Mn$_9$ ($x = 0, 2, 3, 4$) quasicrystals, *J. Appl. Phys.* **104** 073721 (5pp) 2008.

[856] Takagiwa Y, Kamimura T, Okada J T, Kimura K, Thermoelectric properties of Al-Ga-Pd-Re quasicrystals. *Materials Trans.* **55** 1226–1231 (2014).

[857] Takagiwa Y, Kimura K, Metallic-covalent bonding conversion and thermoelectric properties of Al-based icosahedral quasicrystals and approximants. *Sci. Technol. Adv. Mater.* **15** 044802 (12pp) 2014. doi: 10.1088/1468–6996/15/4/044802.

[858] Takahashi Y, Quantum and spectral properties of the Labyrinth model, *J. Math. Phys.* **57** 063506 (14pp) 2016.

[859] Takakura H, Pay-Gómez C, Yamamoto A, de Boissieu M, Tsai A P, Atomic structure of the binary icosahedral Yb-Cd quasicrystal, *Nature Materials* **6** 58 (2007). See also Steurer W, Deloudi S, *Crystallography of Quasicrystals – Concepts, Methods and Structures*, Springer Series in Materials Science **126** (Springer Verlag, Berlin 2009) pp. 300–305.

[860] Takasaki A, Han C H, Furuya Y, Kelton K F, Synthesis of amorphous and quasicrystal phases by mechanical alloying of Ti$_{45}$Zr$_{38}$Ni$_{17}$ powder mixtures and their hydrogenation, *Phil. Mag. Lett.* **82** 353–361 (2002).

[861] Takeuchi T, Mizutani U, Electronic structure, electron transport properties, and relative stability of icosahedral quasicrystals and their 1/1 and 2/1 approximants in the Al-Zn-Mg system, *Phys. Rev. B* **52** 9300–9309 (1995).

[862] Takeuchi T, Onogi T, Otagiri T, Mizutani U, Sato H, Kato K, Kamiyama T, Contribution of local arrangements and electronic structure to high electrical resistivity in the Al$_{82.6--x}$Re$_{17.4}$Si$_x$ ($7 \leq x \leq 12$) $1/1 - -1/1 - -1/1$ approximant, *Phys. Rev. B* **68** 184203 (10pp) 2003.

[863] Takeuchi T, Thermal conductivity of the Al-based quasicrystals and approximants. *Z. Kristallogr.* **224** 35–38 (2009).

[864] Takeuchi T, Very large thermal rectification in bulk composites consisting partly of icosahedral quasicrystals. *Sci. Tech. Adv. Mater.* **15** 064801 (8pp) 2014.

[865] Tamura R, Waseda A, Kimura K, Ino H, Semiconductorlike transport in highly ordered Al-Cu-Ru quasicrystals, *Phys. Rev. B* **50** 9640(R) (1994).

[866] Tamura R, Asao T, Takeuchi S, Roles of quasiperiodicity and local environment in the electronic transport of the icosahedral quasicrystals in Al-Pd-TM (TM = Fe, Ru, Os) systems. *Materials Trans.* **42** 928–932 (2001).

[867] Tamura R, Murao Y, Takeuchi S, Tokiwa K, Watanabe T, Sato T J, Tsai A P, Anomalous transport behavior of a binary Cd-Yb icosahedral quasicrystal, *Jpn. J. Appl. Phys.* **40** L912–L914 (2001).

[868] Tamura R, Asao T, Takeuchi S, Composition-dependent electrical resistivity in an Al-Re-Si 1/1–cubic approximant phase: An indication of electron confinement in clusters, *Phys. Rev. Lett.* **86** 3104–3107 (2001).

[869] Tamura R, Araki T, Takeuchi S, Is the negative temperature coefficient of the resistivity of the quasicrystals due to chemical disorder? *Phys. Rev. Lett.* **90** 226401 (4pp) 2003.

[870] Tamura R, Murao Y, Kishino S, Takeuchi S, Tokiwa K, Watanabe T, Electrical properties of the binary icosahedral quasicrystal and its approximant in the Cd-Yb system, *Mat. Sci. Engineering A* **375–377** 1002–1005 (2004).

[871] Tamura R, Takeuchi T, Aoki C, Takeuchi S, Kiss T, Yokoya T, Shin S, Experimental evidence for the p-d hybridization in the Cd-Ca quasicrystal: origin of the pseudogap, *Phys. Rev. Lett.* **92** 146402 (2004).

[872] Tamura R, Properties of Cd-based binary quasicrystals and their 1/1 approximants, *Isr. J. Chem.* **51** 1263–1274 (2011).

[873] Tanabe T, Kameoka S, Tsai A P, A novel catalyst fabricated from Al-Cu-Fe quasicrystal for steam reforming of methanol, *Catal. Today* **111** 153 (2006).

[874] Tanaka K, Tanaka Y, Ishimasa T, Nakayama M, Matsukawa S, Deguchi K, Sato N K, Tsai-Type quasicrystal and its approximant in Au-Al-Tm alloys, *Acta Phys. Pol. A* **126** 603–607 (2014).

[875] Tanake K, Mitari Y, Koiwa M, *Phil. Mag. A* **73** 1715 (1996).

[876] Tanese D, Gurevich E, Baboux F, Jacquim T, Lemaitre A, Galopin E, Sganes I, Amo A, Bloch J, Akkermans E, Fractal energy spectrum of a polariton gas in a Fibonacci quasiperiodic potential, *Phys. Rev. Lett.* **112** 146404 (2014).

[877] Tang X P, Hill E A, Wonnell S K, Poon S J, Wu Y, Sharp feature in the pseudogap of quasicrystals detected by NMR, *Phys. Rev. Lett.* **79** 1070 (1997).

[878] Tang Z, Lei D, Huang J, Jin G, Qiu F, Yan W, One-way electromagnetic waveguide using multiferroic Fibonacci superlattices, *Opt. Commun.* **356** 21–24 (2015).

[879] Tarumi R, Yoshimoto Y, Shiomi S, Ogi H, Hirao M, Tsai A P, Low temperature elastic and inelastic properties of an i-AlPdMn single quasicrystal, *Mat. Sci. Eng. A* **442** 39–42 (2006).

[880] Tasaki A, Kelton K F, Hydrogen storage in Ti-based quasicrystal powders produced by mechanical alloying, *Int. J. Hydrogen Energy* **31** 183 (2006).

[881] Taylor J R, *Classical Mechanics* (University Science Books, Sausalito, CA, USA 2005).

[882] Thiel P A, Concluding remarks to Quasicrystals 2001, *J. Alloys Compd.* **342** 477 (2002).

[883] Thiel P A, Quasicrystals, *Prog. Surf. Sci.* **75** 69 (2004).

[884] Thiem S, Schreiber M, Renormalization group approach for the wave packet dynamics in golden-mean and silver-mean labyrinth tilings, *Phys. Rev. B* **85** 224205 (2012).

[885] Thiem S, Schreiber M, Wavefunctions, quantum diffusion, and scaling exponents in golden-mean quasiperiodic tilings, *J. Phys. Condens. Matt.* **25** 075503 (2013).

[886] Todd J, Merlin R, Clarke R, Mohanty K M, Axe J D, Synchroton x-ray study of a Fibonacci superlattice, *Phys. Rev. Lett.* **57** 1157 (1986).

[887] Toft P A, Srensen J A, *The Radon Transform - Theory and Implementation.* (Kgs. Lyngby, Denmark: Technical University of Denmark (DTU) 1996).

[888] Toledano J-C, Berry R S, Brown P J, Glazer A M, Metselaar R, Pandey D, Perez-Mato J M, Roth R S, Abrahams S C, Nomenclature of magnetic, incommensurate, composition-changed morphotropic, polytype, transient-structural and quasicrystalline phases undergoing phase transitions. II. Report of an IUCr Working Group on Phase Transition Nomenclature, *Acta Cryst.* **A57** 614–626 (2001).

[889] Toledano-Solano M, Palomino-Ovando M A, Lozada-Morales R, Localization of surface modes along a periodic/quasiperiodic structure containing left-handed material, *Superlatt. Microstruc.* **88** 338–343 (2015).

[890] Torquato S, Zhang G, De Courcy-Ireland M, Hidden multiscale order in the primes, *J. Phys. A: Math. Theor.* **52** 135002 (32pp) 2019.

[891] Torquato S, Zhang G, De Courcy-Ireland M, Uncovering multiscale order in the prime numbers via scattering, *J. Stat. Mech.* 093401 (15pp) 2018.

[892] Torres M, Pastor G, Jiménez I, Montero de Espinosa F, Five-fold quasicrystal-like germinal pattern in the Faraday wave experiment, *Chaos Solitons Fractals* **5** 2089 (1995).

[893] Torres M, Montero de Espinosa F R, García-Pablos D, García N, Sonic band gaps in finite elastic media: surface states and localization phenomena in linear and point deffects, *Phys. Rev. Lett.* **82** 3054–3057 (1999).

[894] Torres M, Aragón J L, Domínguez P, Gil D, Regularity in irregular echinoids, *J. Math. Biol.* **44** 330 (2002).

[895] Torres M, Adrados J P, Aragón J L, Cobo P, Tehuacanero S, Quasiperiodic Bloch-like states in a surface-wave experiment, *Phys. Rev. Lett.* **90** 114501(4pp) 2003.

[896] Trabelsi Y, Benali N, Bouazzi Y, Kanzari M, Microwave transmission through one-dimensional hybrid quasi-regular (Fibonacci and Thue-Morse)/periodic structures, *Photonic Sensors* **3** 246–255 (2013).

[897] Trambly de Laissardière G, Mayou D, Nguyen Manh D, Electronic structure of transition atoms in quasicrystals and Hume-Rothery alloys, *Europhys. Lett.* **21** 25 (1993).

[898] Trambly de Laissardière G, Nguyen Manh D, Magaud L, Julien J P, Cyrot-Lackmann F, Mayou D, Electronic structure and hybridization effects in Hume-Rothery alloys containing transition elements, *Phys. Rev. B* **52** 7920 (1995).

[899] Trambly de Laissardiére G, Mayou D, Clusters and localization of electrons in quasicrystals, *Phys. Rev. B* **55** 2890 (1997); Trambly de Laissardiére G, Roche S, Mayou D, Electronic confinement by clusters in quasicrystals and approximants, *Mater. Sci. Eng. A* **226–228** 986 (1997).

[900] Trambly de Laissardière G and Mayou D, Magnetism in Al(Si)-Mn quasicrystals and related phases, *Phys. Rev. Lett.* **85** 3273 (2000).

[901] Trambly de Laissardière G, Nguyen-Manh D, Mayou D, Electronic structure of complex Hume-Rothery phases and quasicrystals in transition metal aluminides, *Prog. Materials Sci.* **50** 679–788 (2005).

[902] Trevino J, Cao H, Dal Negro L, Circularly symmetric light scattering from nanoplasmonic spirals, *Nano Lett.* **11** 2008–2016 (2011).

[903] Tritt T M, Pope A L, Chernikov M, Feuerbacher M, Legault S, Gagnon R, Strom-Olsen J, Potential of quasicrystals and quasicrystalline approximants for utilization in small scale thermoelectric refrigeration and power generation applications, *Mat. Res. Soc. Symp. Proc.* **553** 489 (1999).

[904] Tsai A P, Inoue A, Masumoto T, A stable quasicrystal in Al-Cu-Fe system, *Jpn. J. Appl. Phys.* **26** L1505–L1507 (1987).

[905] Tsai A P, Inoue A, Masumoto T, Stable decagonal Al-Co-Ni and Al-Co-Cu quasicrystals, *Mater. Trans. Jpn. Inst. Metal.* **30** 463 (1989).

[906] Tsai A P, Inoue A, Masumoto T, Stable icosahedral Al-Pd-Mn and Al-Pd-Re alloys, *Mater. Trans. Jpn. Inst. Metal.* **31** 98 (1990).

[907] Tsai A P, Niikura A, Inoue A, Masumoto T, Nishida Y, Tsuda K, Tanaka M, Highly ordered structure of icosahedral quasicrystals in Zn-Mg-RE (RE = rare earth metals) systems, *Philos. Mag. Lett.* **70** 169–175 (1994).

[908] Tsai A P, Guo J Q, Abe E, Takahura H, Sato T J, A stable binary quasicrystal, *Nature* **408** 537–538 (2000).

[909] Tsai A P, Icosahedral clusters, icosahedral order and stability of quasicrystals - a view of metallurgy, *Sci. Tech. Adv. Mater.* **9** 013008 (30pp) 2008.

[910] Tsai A P, Discovery of stable icosahedral quasicrystals: progress in understanding structure and properties, *Chem. Soc. Rev.* **42** 5352 (2013).

[911] Tsuei C C, Nonuniversality of the Mooij correlation - the temperature coefficient of electrical resistivity of disordered metals, *Phys. Rev. Lett.* **57** 1943 (1986).

[912] Turkiewicz A, Paley D W, Vesara T, Elbaz G, Pinkard A, Siegrist T, Roy X, Assembling hierarchical cluster solids with atomic precision, *J. Am. Chem. Soc.* **136** 15873–15876 (2014).

[913] Tuz V R, Optical properties of a quasi-periodic generalized Fibonacci structure of chiral and material layers, *J. Opt. Soc. Am. B* **26** 627–632 (2009); A peculiarity of localized mode transfiguration of a Cantor-like chiral multilayer, *J. Opt. A: Pure Appl. Opt.* **11** 125103 (2009).

[914] Tuz V R, Batrakov O D, Localization and polarization transformation of waves by a symmetric and asymmetric modified Fibonacci chiral multilayer, *J. Mod. Opt.* **57** 2114–2122 (2010).

[915] Twarock R, A tiling approach to virus capsid assembly explaining a structural puzzle in virology, *J. Theor. Bio.* **226** 477–482 (2004).

[916] Twarock R, Mathematical virology: a novel approach to the structure and assembly of viruses, *Phil. Trans. R. Soc.* **364** 3357–3373 (2006).

[917] Ueda K, Dotera T, Gemma T, Photonic band structure calculations of two-dimensional Archimedean tiling patterns, *Phys. Rev. B* **75** 195122 (2007).

[918] Uflyand I E, Drogan E G, Burlakova V E, Kydralieva K A, Shershneva I N, Dzhardimalieva G I, Testing the mechanical and tribological properties of new metal-polymer nanocomposite materials based on linear low-density polyethylene and $Al_{65}Cu_{22}Fe_{13}$ quasicrystals, *Polymer Testing* **74** 178–186 (2019).

[919] Unal B, Lograsso T A, Ross A, Jenks C J, Thiel P A, Terrace selection during equilibration at an icosahedral quasicrystal surface, *Phys. Rev. B* **71** 165411 (2005); Ledieu J, Cox E J, McGrath R, Richardson N V, Chen Q, Fournée V, Lograsso T A, Ross A R, Caspersen K J, Unal B, Evans J W, Thiel P A, Step structure on the fivefold Al-Pd-Mn quasicrystal surface, and on related surfaces, *Surf. Sci.* **583** 4 (2005).

[920] Ungar G, Percec V, Zeng X, Leowanawat P, Liquid quasicrystals, *Isr. J. Chem.* **51** 1206–1215 (2011).

[921] Van Tendeloo G, Amelinckx S, Electron Microscopy of Fullerenes and Related Materials in *Characterization of Nanophase Materials*, edited by Z. L. Wang (Wiley-VCH Verlag GmmH 2000).

[922] Varma V K, Pilati S, Kravtsov V E, Conduction in quasiperiodic and quasirandom lattices: Fibonacci, Riemann and Anderson models. *Phys. Rev. B* **94** 214204 (11 pp) 2016.

[923] Vasconcelos M S, Albuquerque E L, Mariz A M, Optical localization in quasi-periodic multilayers, *J. Phys. Condens. Matter* **10** 5839 (1998).

[924] Vasconcelos M S, Albuquerque E L, Transmission fingerprints in quasiperiodic dielectric multilayers, *Phys. Rev. B* **59** 11128–11131 (1999).

[925] Vasconcelos M S, Mauriz P W, de Medeiros F F, Albuquerque E L, Photonic band gaps in quasiperiodic photonic crystals with negative refractive index, *Phys. Rev. B* **76** 165117 (2007).

[926] Vasconcelos M S, Cottam M G, Plasmon-polariton fractal spectra in quasiperiodic photonic crystals with graphene, *Eur. Phys. Lett.* **128** 27003 p7 (2019).

[927] Velasco V R, García-Moliner F, Miglio L, Colombo L, Phonon calculations in super-periodic structures:the surface Green-function matching approach, *Phys. Rev. B* **38** 3172 (1988).

[928] Vekilov Y K, Isaev E I, Arslanov S F, Influence of phason flips, magnetic field, and chemical disorder on the localization of electronic states in an icosahedral quasicrystal, *Phys. Rev. B* **62**, 14040 (2000).

[929] Vekilov Y K, Isaev E I, Livanov D V, Electronic transport in quasicrystals, *J Exp. Theor. Phys.* **94** 172–177 (2002).

[930] Vekilov Y K, Isaev E I, Electron localization and conductivity in quasicrystals at low temperatures, *Phys. Lett. A* **300** 500 (2002).

[931] Vekilov Y K, Isaev E I, Godoniuk A V, Electronic spectrum of the three-dimensional Penrose lattice, *J Exp. Theor. Phys.* **95** 1005–1009 (2003).

[932] Venkatasubramanian R, Silvola E, Colpitts T, O'Quinn B, Thin-film thermoelectric devices with high room-temperature figures of merit, *Nature* **413** 597 (2001).

[933] Vidal J, Mosseri R, Generalized Rauzy tilings: construction and electronic properties, *Mat. Sci. Engineer.* **294–296** 572–575 (2000).

[934] Viebahn K, Sbroscia M, Carter E, Yu Jr-Ch, Scheneider U, Matter-wave diffraction from a quasicrystalline optical lattice, *Phys. Rev. Lett.* **122** 110404 (6pp) 2019.

[935] Vigneron B, Perrot A, Machizaud F, Dubois J M, Jendel G, Wyncke B, Hierarchical nature of the Brillouin zones in icosahedral $Al_{59}Cu_{25.5}Fe_{12.5}B_3$, *Phil. Mag. B* **77** 849–857 (1998).

[936] Vignolo P, Bellec M, Böhm J, Camara A,Gambaudo J M, Kuhl U, Energy landscape in a Penrose tiling, *Phys. Rev. B* **93** 075141 (7pp) 2016.

[937] http://viperdb.scripps.edu/

[938] Wagner J L, Biggs B D, Wong K M, Poon S J, Specific-heat and transport properties of alloys exhibiting quasicrystalline and crystalline order, *Phys. Rev. B* **38** 7436–7441 (1988).

[939] Wagner J L, Wong K M, Poon S J, Electronic properties of stable icosahedral alloys, *Phys. Rev. B* **39** 8091–8095 (1989).

[940] Wagner J L, Biggs B D, Poon S J, Band-structure effects on the electronic properties of icosahedral alloys, *Phys. Rev. Lett.* **65** 203–206 (1990).

[941] Wälti Ch, Felder E, Chernikov M A, Ott O R, de Boissieu M, Janot C, Lattice excitations in icosahedral Al-Mn–Pd and Al-Re-Pd, *Phys. Rev. B* **57** 10504–10511 (1998).

[942] Wan R, Fu X, Localization properties of electronic states of one-dimensional Galois sequences, *Solid State Commun.* **150** 919–922 (2010).

[943] Wang C, Ramírez C, Sánchez F, Sánchez V, Ballistic conduction in macroscopic non-periodic lattices, *Phys. Status Solidi B* **252** 1370–1381 (2015).

[944] Wang H, Lan X, Huang Y, Jiang X, Lithium storage property of graphite/AlCuFe quasicrystal composites, *Chin. Phys. Lett.* **36** 098201(5pp) 2019.

[945] Wang K, Garoche P, Phason-strain-field influences on low-temperature specific heat in icosahedral quasicrystals Al-Li-Cu and Al-Cu-Fe, *Phys. Rev. B* **55** 250–258 (1997).

[946] Wang K, David S, Chelnokov A, Lourtioz J M, Photonic band gaps in quasicrystal-related approximant structures, *J. Mod. Opt.* **50** 2095 (2003).

[947] Wang K, Light wave states in two-dimensional quasiperiodic media, *Phys. Rev. B* **73** 235122 (2006).

[948] Wang K, Structural effects on light wave behavior in quasiperiodic regular and decagonal Penrose-tiling dielectric media: a comparative study, *Phys. Rev. B* **76** 085107 (2007).

[949] Wang K, Light localization in photonic band gaps of quasiperiodic dielectric structures, *Phys. Rev. B* **82** 045119 (2010).

[950] Wang N, Chen H, Kuo K H, Two-dimensional quasicrystal with eightfold rotational symmetry, *Phys. Rev. Lett.* **59** 1010–1013 (1987).

[951] Wang R, Pinheiro F A, Dal Negro L, Spectral statistics and scattering resonances of complex primes arrays, *Phys. Rev. B* **97** 024202 (11pp) 2018.

[952] Wang X, Grimm U, Schreiber M, Trace and antitrace maps for aperiodic sequences: extensions and applications, *Phys. Rev. B* **62** 14020 (2000).

[953] Wang X, Xu J, Lee J C W, Pang Y K, Tam W Y, Chan C T, Sheng P, Realization of optical *periodic* quasicrystals using holographic lithography, *Appl. Phys. Lett.* **88** 051901 (2006).

[954] Wang Y, Liu J, Zhang B, Feng S, Li Z Y, Simulations of defect-free coupled-resonator optical waveguides constructed in 12-fold quasiperiodic photonic crystals, *Phys. Rev. B* **73** 155107 (2006).

[955] Wasio N A, Quardokus R C, Forrest R P, Lent C S, Corcelli S A, Christie J A, Henderson K H, Kandel S A, Self-assembly of hydrogen-bonded two-dimensional quasicrystals, *Nature* **507** 86–89 (2014).

[956] Wei Q P, Ma L, Ye J, Yu Z M, Growth mechanism of icosahedral and other five-fold symmetric diamond crystals, *Trans. Nonferrous Met. Soc .China* **25** 1587–1598 (2015).

[957] White M A, Meingast C, David W I F, Matsuo T, Anharmonic interactions in C_{60}, as determined by the Grüneisen parameter, *Solid State Comm.* **94** 481 (1995)

[958] Widmer R, Gröning O, Ruffieux P, Gröning P, Low-temperature scanning tunneling spectroscopy on the 5-fold surface of the icosahedral AlPdMn quasicrystal, *Phil. Mag.* **86** 781 (2006).

[959] Williams R C, Smith K M, The polyhedral form of the tipula iridescent virus, *Biochim. et Biophys. Acta* **28** 464 (1958).

[960] Wolny J, Buganski I, Strzalka R, Model refinement of quasicrystals. *Crystallography Rev.* **24** 22–64 (2018). doi: 10.1080/0889311X.2017.1340276.

[961] Wrigley N G, An electron microscope study of the structure of Serecesthis iridescent virus, *J. Gen. Virol.* **5** 123–134 (1969).

[962] Wu S, Brien V, Brunet P, Dong C, Dubois J M, Electron microscopy study of scratch-induced surface microstructures in an Al-Cu-Fe icosahedral quasicrystal, *Phil. Mag. A* **80** 1645–1655 (2000).

[963] Wu S, Sun Y, Tessellating tiny tetrahedrons: A tiling rule guides the formation of quasicrystalline superlattices of nanocrystals, *Science* **362** 1354–1355 (2018).

[964] Wu X, Homes C C, Burkov S E, Timusk T, Pierce F S, Poon S J, Cooper S L, Karlow M A, Optical conductivity of the icosahedral quasicrystal $Al_{75.5}Mn_{20.5}Si_4$ and its 1/1 crystalline approximant α-$Al_{72.5}Mn_{17.4}Si_{10.1}$, *J. Phys. Condens. Matter* **5** 5975–5990 (1993).

[965] Xavier J, Boguslawski M, Rose P, Joseph J, Denz C, Reconfigurable optically induced quasicrystallographic threedimensional complex nonlinear photonic lattice structures, *Adv. Mater.* **22** 356 (2010).

[966] Xia T, Chang S, Yan J, Tao S, Modified Thue-Morse zone plates with arbitrarily designed high-intensity twin main foci, *Laser Phys.* **27** 125001 (7pp) 2017.

[967] Xia T, Tao S, Cheng S, Twin equal-intensity foci with the same resolution generated by a modified precious mean zone plate, *J. Opt. Soc. Am. A* **37** 1067–1074 (2020).

[968] Xiao C, Fujita N, Miyasaka K, Sakamoto Y, Terasaki O, Dodecagonal tiling in mesoporous silica. *Nature* **487** 349—353 (2012). https://doi-org.bucm.idm.oclc.org/10.1038/nature11230.

[969] Xie X C, Das Sarma S, "Extended" electronic states in a Fibonacci superlattice, *Phys. Rev. Lett.* **60** 1585 (1988); Ananthakrishna G, Kumar V, Ananthakrishna and Kumar reply, *Phys. Rev. Lett.* **60** 1586 (1988).

[970] Xie Z X, Liu J Z, Yu X, Wang H B, Deng Y X, Li K M, Zhang Y, Tunability of acoustic phonon transmission and thermal conductance in three dimensional quasi-periodically stubbed waveguides, *J. Appl. Phys.* **117** 114308 (6pp) (2015).

[971] Xie Z X, Zhang Y, Chen X K, Zhou W X, Yi G J, Shi Y M, Zhang J X, Zhang L F, Phonon transport in periodically and quasi-periodically modulated cylindrical nanowires, *J. Phys.: Condens. Matter* **31** 505303 (7pp) 2019.

[972] Xu J, Ma R, Wang X, Tam W Y, Icosahedral quasicrystals for visible wavelengths by optical interference holography, *Opt. Express* **15** 4287 (2007); Tam W Y, Icosahedral quasicrystals by optical interference holography, *Appl. Phys. Lett.* **89** 251111 (2006).

[973] Xudong F, Bursill L A, Lin P J, Fourier transforms and structural analysis of spiral lattices, *Int. J. Mod. Phys. B* **2** 131 (1988).

[974] Yadav T P, Mukhopadhyay N K, Srivastava O N, Quasicrystals: bulk to nano, *Banaras Metall* **19** 53–64 (2014).

[975] Yamada T, Takakura H, Euchner H, Pay Gómez C, Bosak A, Fertey O, de Boissieu, Atomic structure and phason modes of the Sc-Zn icosahedral quasicrystal, *IUCrJ* **3** 247–258 (2016).

[976] Yang J K, Boriskina S V, Noh H, Rooks M J, Solomon G S, Dal Negro L, Cao H, Demonstration of laser action in a pseudorandom medium, *Appl. Phys. Lett.* **97** 223101 (2010).

[977] Yao M, Zebarjadi M, Opeil C P, Experimental determination of phonon thermal conductivity and Lorenz ratio of single crystal metals: Al, Cu, and Zn, *J. Appl. Phys.* **122** 135111 (7pp) 2017.

[978] Ye F, Sprengel W, Zhang X Y, Uhrig E, Assmus W, Schaefer H E, High temperature vacancy studies of icosahedral ZnMgEr quasicrystal, *J. Phys.: Condens. Matter* **16** 1531–1537 (2004).

[979] Yokoyama Y, Inoue A, Masumoto T, Mechanical properties, fracture mode and deformation behavior of $Al_{70}Pd_{20}Mn_{10}$ single-quasicrystal, *Mat. Trans. JIM* **34** 135 (1993).

[980] Yoshida K, Yamada T, Taniguchi Y, Long-period tetragonal lattice formation by solid-state alloying at the interfaces of Bi-Mn double-layer thin films, *Acta Crystallogr. B* **45** 40–45 (1989).

[981] Yoshimura M, Tsai A P, Quasicrystal application on catalyst, *J Alloys Compd.* **342** 451 (2002).

[982] You J, Nori F, The real-space renormalization group and generating function for Penrose lattices, *J. Phys. Condens. Matter* **5** 9431–9438 (1993).

[983] Yu S, Piao X, Hong J, Park N, Block-like waves in random-walk potentials based on supersymmetry, *Nature Commun.* **6** 8269 (2015).

[984] Yu S, Piao X, Hong J, Park N, Metadisorder for designer light in random systems, *Sci. Adv.* **2** e1501851 (2016).

[985] Zahoor A, Aziz T, Zulfiqar S, Sadiq A, Ali R, Shahid R N, Tariq N H, Shah A, Shehzad K, Ali F, Awais H B, Antimicrobial behavior of leached Al–Cu–Fe-based quasicrystals, *Applied Physics A* **126** 434 (9pp) 2020.

[986] Zárate J E, Fernández-Álvarez L, Velasco V R, Transverse elastic waves in Fibonacci superlattices, *Superlatt. & Microstruct.* **25** 519 (1999).

[987] Zeng L, Wang G, Modeling golden section in plants. *Prog. Nat. Sci* **19** 255–260 (2009).

[988] Zeng X, Liu Y, Ungar G, Percec V, Dulcey A E, Hobbs J K, Supramolecular dendritic liquid quasicrystals, *Nature* **428** 157–160 (2004).

[989] Zeng X, Ungar G, Inflation rules of square-triangle tilings: from approximants to dodecagonal liquid quasicrystals, *Phil. Mag.* **86** 1093–1103 (2006).

[990] Zhai T R, Liu D H, Zhang X D, Photonic crystals and microlasers fabricated with low refractive index material, *Front. Phys. China* **5** 266–276 (2010).

[991] Zhai J, Zhang Q, Chang Z, Ren J, Ke Y, Li B, Anomalous transparency induced by cooperative disorders in phonon transport, *Phys. Rev. B* **99** 195429 (2019).

[992] Zhai X, Li Z, Zhou X, Liu H, Sun J, Su Z, Liu W, Zhao J, Improved electrochemical hydrogen storage properties of $Ti_{49}Zr_{26}Ni_{25}$ quasicrystal alloy by doping with Pd and MWCNTs, *Int. J. Hydrogen Energy* **44** 29356–29364 (2019).

[993] Zhang D, Electronic properties of stable decagonal quasicrystals, *Phys. Status Solidi A* **207** 2666 (2010).

[994] Zhang G, Martelli F, Torquato S, The structure factor of primes, *J. Phys. A: Math. Theor.* **51** 115001 (2018).

[995] Zhang L, Fang K, Du G, Jiang H, Zhao J, Transmission properties of Fibonacci quasi-periodic one-dimensional photonic crystals containing indefinite metamaterials, *Opt. Commun.* **284** 703–706 (2011).

[996] Zhang X, Stroud R M, Libbert J L, Kelton K F, The icosahedral and related crystal approximant phases in Ti-Zr-Ni alloys, *Phil. Mag B* **70** 927 (1994).

[997] Zhang X D, Universal non-near-field focus of acoustic waves through high-symmetry quasicrystals, *Phys. Rev. B* **75** 024209 (4pp) 2007.

[998] Zhao D, Xu B, Guo H, Xu W, Zhong D, Ke S, Low threshold optical bistability in aperiodic PT-symmetric lattices composited with Fibonacci sequence dielectrics and graphene, *Appl. Sci.* **9** 5125 (2019) doi:10.3390/app9235125.

[999] Zhou Z, Harris K D M, Design of a molecular quasicrystal, *Chem.Phys.Chem.* **7** 1649–1653 (2006).

[1000] Zhou Z, Chen Q, Bermel P,, Prospects for high-performance thermophotovoltaic conversion efficiencies exceeding the Shockley-Queisser limit. *Energy Conver. Management* **97** 63–69 (2015).

[1001] Zhu S N, Zhu Y Y, Qin Y Q, Wang H F, Ge C Z, Ming N B, Experimental realization of second harmonic generation in a Fibonacci optical superlattice of $LiTaO_3$, *Phys. Rev. Lett.* **78** 2752–2755 (1997).

[1002] Zhu S N, Zhu Y Y, Ming N B, Quasi-phase-matched third-harmonic generation in a quasi-periodic optical superlattice, *Science* **278** 843 (1997).

[1003] Zhukovsky S V, Lavrinenko A V, Gaponenko S V, Optical filters based in fractal and aperiodic multilayers. In *Optics of Aperiodic Structures: Fundamentals and Device Applications*, Dal Negro L., Ed. (Pan Stanford, Boca Raton, FL, USA 2014) pp. 91—142.

[1004] Zia R K P, Dallas W J, A simple derivation of quasi-crystalline spectra, *J. Phys. A: Math. Gen.* **18** L341–345 (1985).

[1005] Zijlstra E S, Janssen T, Localization of electrons and influence of surface states on the density of states in a tight-binding model on the Penrose tiling, *Mater. Sci. Eng. A* **294–296** 886 (2000).

[1006] Zijlstra E S, Janssen T, Density of states and localization of electrons in a tight-binding model on the Penrose tiling, *Phys. Rev. B* **61** 3377 (2000).

[1007] Zijlstra E S, Janssen T, Non-spiky density of states of an icosahedral quasicrystal, *Europhys. Lett.* **52** 578 (2000).

[1008] Zijlstra E S, Bose S K, Detailed ab initio electronic structure study of two approximants to Al-Mn based icosahedral quasicrystals, *Phys. Rev. B* **67** 224204 (2003).

[1009] Ziman J M, *Electrons and Phonons*, (Clarendon, Oxford 1960).

[1010] Zito G, Piccirillo B, Santamato E, Marino A, Tkachenko V, Abbate G, Two-dimensional photonic quasicrystals by single beam computer-generated holography, *Opt. Express* **16** 5164 (2008).

[1011] Zito G, Rusciano G, Sasso A, De Nicola S, Symmetry-induced light confinement in a photonic quasicrystal-based mirrorless cavity. *Crystals* **6** 111 (9 pp) 2016. doi:10.3390/cryst6090111.

[1012] Zou Y, Kuczera P, Sologubenko A, Sumigawa T, Kitamura T, Steurer W, Spolenak R, Superior room-temperature ductility of typically brittle quasicrystals at small sizes, *Nature Commun.* **7** 12261 (7pp) 2016.

[1013] Zoorob M E, Charlton M D B, Parker G J, Baumberg J J, Netti M C, Complete photonic bandgaps in 12-fold symmetric quasicrystals, *Nature (London)* **404** 740 (2000).

[1014] Zurkirsh M, Atrei A, Erbudak M, Hochtstrasser M, Critical test for the icosahedral structure of the quasicrystal Al70Pd20Mn10, *Phil. Mag. Lett.* **73** 107 (1996).

Index

Note: The page references in *italics*, **bold** and with "n" represents figures, tables, and notes respectively.

Printed in the USA
CPSIA information can be obtained
at www.ICGtesting.com
LVHW082139190424
777912LV00008B/1185